Polymeric Membrane Synthesis, Modification, and Applications

Electro-Spun and Phase Inverted Membranes

Polymeric Membrane Synthesis, Modification, and Applications

Electro-Spun and Phase Inverted Membranes

Chandan Das
Kibrom Alebel Gebru

CRC Press
Taylor & Francis Group
Boca Raton London New York

CRC Press is an imprint of the
Taylor & Francis Group, an **informa** business

CRC Press
Taylor & Francis Group
6000 Broken Sound Parkway NW, Suite 300
Boca Raton, FL 33487-2742

Printed on acid-free paper

International Standard Book Number-13: 978-1-138-58579-9 (Hardback)

Library of Congress Cataloging-in-Publication Data

Names: Das, Chandan, author. | Gebru, Kibrom Alebel, author.
Title: Polymeric membrane synthesis, modification, and applications : electro-spun and phase inverted membranes / Chandan Das and Kibrom Alebel Gebru.
Description: First edition. | Boca Raton, FL : CRC Press/Taylor & Francis Group, 2019. | Includes bibliographical references and index.
Identifiers: LCCN 2018023694| ISBN 9781138585799 (hardback : acid-free paper) | ISBN 9780429505065 (ebook)
Subjects: LCSH: Polymeric membranes. | Membrane filters. | Electrospinning.
Classification: LCC TP159.M4 D373 2019 | DDC 668.9/2--dc23
LC record available at https://lccn.loc.gov/2018023694

Contents

Preface

Membranes and membrane separation techniques have grown from a simple laboratory tool to an industrial process with considerable technical and commercial impact. Today, membranes are used on a large scale to produce potable water from the sea by reverse osmosis; to clean industrial effluents and recover valuable constituents by electrodialysis; to fractionate macromolecular solutions in the food and drug industry by ultrafiltration; to remove urea and other toxins from the bloodstream by dialysis in an artificial kidney; and to release drugs at a predetermined rate in medical treatment. To maintain membrane separations as economical alternatives to conventional water and wastewater treatment technologies, we must produce a high-quality permeate at a fast rate, and be able to maintain that production for an extended period of time. However, the relationship between flux and selectivity along with fouling introduce a challenge that must be addressed for all membrane applications.

This book deals with both the fundamental concepts and practical applications of electrospun and phase inverted polymeric membranes. Moreover, this book covers the research works of the authors as an example already published/presented from the Membrane and Bioremediation Lab, Department of Chemical Engineering, IIT Guwahati during the last few years. A selection of these works is accumulated along with the fundamental concepts of the prepared membranes. Some of the knowledge from the present research is used as "practical examples" in the book. The book is divided into two broad parts.

Membranes with the capacity to remove heavy metals and other contaminants while less susceptible to membrane fouling were developed using electrospinning and phase inversion techniques are discussed in this book. This book is organized and divided into seven chapters. Chapter 1 is a platform to provide valuable information on the background knowledge of this research work for better understanding of the types of membranes, membrane preparation techniques, functionalization, characterization, and applications. The functionalization and grafting methods for the modification of polymer and/or membrane are discussed. Emphasis is given to the phase inversion and electrospinning techniques of membrane preparations. Moreover, the possible scopes for further enhancement on the preparation and modifications of the membranes are also identified. Chapter 2 provides detailed information on the membrane preparation procedures and physico-chemical and instrumental characterization techniques. An empirical exploration into the effects of time duration, voltage supply, concentration, and flow rate on the membrane average fiber diameter and surface pore size distribution using response surface methodology (RSM) based on central compact design (CCD) are presented in Chapter 3. Hybrid membranes from cellulose acetate (CA) and titanium oxide (TiO_2) nanoparticles (NPs) were fabricated using a novel electrospinning technique and evaluated as an adsorbent for the elimination of lead and copper metal ions. The impacts of various adsorption parameters, namely, pH, the amount of TiO_2 nanoparticles, contact time, temperature, and kinetics on metal uptake were investigated using batch adsorption experiments. The model isotherms, such as Dubinin–Radushkevich (D-R), Freundlich, and Langmuir, were used to analyze the adsorption equilibrium data. Pseudo first-order and pseudo-second-order were preferred for kinetic model study. Furthermore, preparation and characterization of electrospun polysulfone membranes; preparation and characterization of electrospun polyvinyl alcohol membranes; and preparation and characterization of electrospun polyvinylidene fluoride membranes are also presented in this chapter. Chapter 4 presents the investigation on the effects of two different hydrophilic additives and two solvents on the morphological structure, permeability property, and anti-fouling performances of CA ultrafiltration membranes. The experimental studies of fouling/rinsing cycles, rejection, and permeate fluxes were used to investigate the effect of PEG and PVP additives and effect of the two solvents on the fabricated membranes using bovine serum albumin (BSA) as a model protein. Effects of PEG additive and TiO_2 NPs on the preparation of phase inverted CA ultrafiltration membrane were investigated. The influences of PVP and TiO_2 on the preparation of phase inverted CA ultrafiltration membrane were also explored. Furthermore, preparation and characterization of polysufone membranes, preparation and characterization of polyvinylidene fluoride membranes, and preparation and characterization of polymeric membranes are also covered in this chapter. In Chapter 5, TiO_2 NPs were modified using different amine groups, namely, ethylenediamine (EDA), hexamethylenetetramine (HMTA), and tetra ethylene pentamine (TEPA), using an impregnation process. The prepared amine modified TiO_2 composites were explored as an additive to fabricate ultrafiltration membranes with enhanced capacity toward the removal of chromium ions. The graft copolymerization of CA and poly (methyl methacrylate) (PMMA) was synthesized through free radical polymerization with the presence of cerium sulfate (CS) as initiator under nitrogen atmosphere in an aqueous solution. During the grafting reactions the effect polymerization time and temperature on the grafting were investigated. Furthermore, functionalization of the synthesized product was done using amine group. The membranes prepared from the modified polymer were investigated for the ultrafiltration of humic acids. Additionally, topics such as functionalization and characterization of polysulfone membranes; functionalization and characterization of polyvinylidene fluoride membranes; and grafting copolymerization of monomers on polymers are also covered extensively in this chapter. Polymeric membranes for industrial effluent treatments applications and polymeric membranes for biomedical applications are covered in Chapters 6 and 7, respectively.

Though a number of books with excellent quality are available on this topic, this book is an extra effort to offer additional knowledge on fundamental concepts of preparation, characterization, and modification of conventional polymeric membranes, and presents their recent advancements for specific applications to the readers. This book is a guideline for students, scientists, and engineers, and it provides new ideas for creative thinkers. It is the coalition of views into the past, the present, and the future.

We wish to thank Central Instrumental Facility, IIT Guwahati for conducting all the experiments for characterizing the prepared membranes. We would like to thank the entire departmental non-teaching lab staffs for their enormous support. Moreover, the authors would like to thank their parents and family members and all the well-wishers for their constant support.

Chandan Das
Kibrom Alebel Gebru
Guwahati and Adigrat

Authors

Dr. Chandan Das is an Associate Professor in the Department of Chemical Engineering at Indian Institute of Technology Guwahati (IITG). He received his PhD in Chemical Engineering from Indian Institute of Technology, Kharagpur (IITKGP) after completing his BTech and MTech in Chemical Engineering from University of Calcutta. He has guided, so far, 6 scholars for their doctoral degree, 24 MTech, and is guiding 8 more doctorate scholars. Dr. Das is the recipient of Dr. A.V. Rama Rao Foundation's Best PhD Thesis and Research Award in Chemical Engineering/Technology for the year 2010 from Indian Institute of Chemical Engineers (IIChE). Dr. Das has authored about 100 technical publications in peer-reviewed journals and proceedings. He has authored two books titled *Treatment of Tannery Effluent by Membrane Separation Technology* in Nova Science Publishers (USA) and *Advanced Ceramic Membranes and Applications* in CRC Press (USA) and three book chapters. He has two patents in his credit. He has handled five sponsored projects and five consultancy projects so far. He has visited Denmark, Malaysia, Sri Lanka, Japan, and Greece for exchanging ideas, etc.

Being associated with various research works in the area of water and wastewater treatment, such as treatment of tannery wastewater using membrane separation technology, as well as removal of pollutants using micellar-enhanced ultrafiltration, Dr. Das has gained expertise in membrane separation technology for removing various pollutants from contaminated water and wastewater. His research activity encompasses both understanding of fundamental principles during filtration as well as the development of technology based on membrane separation. In particular, his research areas are modeling of microfiltration, ultrafiltration, nanofiltration, reverse osmosis, treatment of oily wastewater, and tannery effluent using membrane based processes. He has explored the detailed quantification of flux decline from fundamentals. As an offshoot of the major research, he has fabricated ceramic membranes using low-cost precursors as sawdust. Catalyst is coated on the ceramic support for manufacturing catalytic membrane reactor.

He is also working on decontamination of chromium laden aqueous effluent using Spirulina platensis. He is actively involved in the productions of high value added products, namely, total phenolics, flavonoids, tocopherol, etc. from black rice as well as of 6-gingerol, vitamin C content, and essential oil content from ginger of North East India.

Dr. Kibrom Alebel Gebru has received his Bachelor Degree in Chemical Engineering from Bahir Dar University, Ethiopia; Master of Technology in Chemical Engineering from Addis Ababa Institute of Technology, Ethiopia; and PhD in Chemical Engineering from Indian Institute of Technology

Guwahati, India. Dr. Kibrom is currently an Assistant Professor in the Department of Chemical Engineering and Acting Director of AdU Consultancy and Business Enterprises at Adigrat University, Adigrat. He has two years of industrial experience, three years of university teaching experience, and more than four years of research experience in the field of membrane science and technology and environmental engineering. He is an expert in the area of preparation and modification of electrospun and phase inverted membranes for various applications. He has prepared novel ultrafiltration membranes using electrospinning and phase inversion membranes for wastewater treatments applications. Dr. Kibrom's current research interests are in the fabrication of novel membranes using electrospinning and phase inversion techniques, modification of membranes and membrane based separations for wastewater treatment, biotechnology, and biomedical and environmental engineering applications. He has published about eight articles in reputed peer-reviewed international journals like *Chemosphere, Journal of Water Process Engineering, Chinese Journal of Chemical Engineering, Journal of Environmental Management, Journal of The Institution of Engineers, Journal of Membrane and Separation Technology*, etc.

Aim and Scope of the Book

Reader of this book will have a brief overview on both the fundamental concepts and practical applications of electrospun and phase inverted polymeric membranes. The book mostly covers the fundamentals and applications of the novel electrospun and advanced phase inverted membranes. In addition, readers will have an idea about novel and affinity membranes (such as charged membranes, adsorbent members, and anti-fouling membranes), which could provide additional knowledge toward the industrial application of the membranes.

1

Membrane Technology

1.1 Introduction

Water is one of the most vital resources on the planet. Currently, fresh and clean water supplies have been falling at an alarming rate. More than one billion people do not have access to fresh water, and more than two billion individuals are living in water shortage areas [1]. Since the world population is rising and environmental pollution is increasing rapidly, water purification has to be made more efficient and cost-effective. Access to clean water has been recognized as a serious challenge for the world's social and economic growth. The advancement of alternative water supplies is crucial and essential. One of the current challenges is to develop effective and less energy consuming process through treating and recovering of pure water from groundwater, industrial, and brackish water [2]. Removal of heavy metals and other contaminants from water and wastewater is a very important factor with respect to environmental pollution control and human health. Several methods have been used for the removal of heavy metal ions and other contaminants from aqueous solutions, including ion exchange, chemical precipitation, electrodialysis, and solvent extraction. However, these techniques are associated with problems such as excessive time requirements, high costs, and high energy consumption. Moreover, adsorption method can be considered as an effective and widely used process for removal of heavy metals and other contaminants from wastewater due to its simplicity, moderate operational conditions, and economic feasibility. The most important properties of any adsorbent are its surface area and structure [3]. However, particulate or powder adsorbents might re-pollute treated water because of the tremendous problems in recovery.

Among the different water purification technologies, the pressure-driven membrane filtration processes—microfiltration (MF), ultrafiltration (UF), nanofiltration (NF), and reverse osmosis (RO)—are the most energy efficient processes. Distillation process consumes more energy as compared to the membrane filtration processes, which are relatively fast, efficient, and practical [4]. Depending on the impact of various constituents of water on human health crops and industrial processes, certain water and stream

TABLE 1.1

List of Inorganic Chemicals and Their Maximum Contaminant Level (MCL) in Drinking Water

Contaminant	MCL (mg/L)	Potential Health Effects: Above the Limit	Public Health Goal (mg/L)
Arsenic	0.01	Skin damage or problems with circulatory systems may have increased risk of getting cancer	Zero
Cadmium	0.005	Kidney damage	0.005
Chromium (total)	0.1	Allergic dermatitis	0.1
Copper	1.3	Liver or kidney damage	1.3
Fluoride	1.5	Pain and tenderness of the bones Children may get mottled teeth	1.0
Iron	0.3	Leave the water with brown-red color aesthetic problems	<0.3
Lead	0.015	Infants and children: Delays in physical or mental development. Adults: Kidney problems; high blood pressure	Zero
Mercury	0.002	Kidney damage	0.002

Source: US EPA's Safe Drinking Water Web site: http://www.epa.gov/safewater; USEPA's Safe Drinking Water Hotline: (800) 426-4791; USEPA 816-F-09-004 May 2009.

water standards have been laid down by different standard institutions such as Bureau of Indian Standards (BIS), Indian Council of Medical Research (ICMR), World Health Organisation (WHO), and Food and Agriculture Organization (FAO) of United Nations for deciding the suitability of water for drinking and irrigation use. Table 1.1 [5] summarizes the drinking water standards.

1.2 General Background

1.2.1 Preparation of Synthetic Membranes

Currently, membranes and membrane separation techniques have grown from a simple laboratory tool to an industrial process with considerable technical and commercial impact. Today, membranes are used on a large scale to produce potable water from the sea by reverse osmosis; to clean industrial effluents and recover valuable constituents by electrodialysis; to fractionate macromolecular solutions in the food and drug industry by ultrafiltration; to remove urea and other toxins from the bloodstream by dialysis in an artificial kidney; and to release drugs at a predetermined rate in medical treatment. Although membrane processes may be different in their mode of operation, in the structures used as separating barriers, and in the driving forces used for the transport of the different chemical components, they have several features in common that make them attractive as a separation tool. In several cases, membrane processes are faster, more efficient, and more economical than conventional separation techniques [6]. During membrane process, the separation is usually performed at ambient temperature, thus allowing temperature-sensitive solutions to be treated without the constituents being damaged or chemically altered.

1.2.2 Definition of Membrane

A synthetic membrane can be defined as a barrier that separates two phases and restricts the passage of numerous chemical species in a rather specific way. A membrane can be homogeneous or heterogeneous, symmetric or asymmetric in structure, solid or liquid, may carry positive or negative charges, or may be bipolar. Its thickness may vary between less than 100 nm to more than a centimeter. The mass transport through a membrane may be caused by convection or by diffusion of individual molecules, induced by an electric field, concentration, pressure, or temperature gradient. The term *membrane*, therefore, includes a great variety of materials and structures, and a membrane can often be better described in terms of what it does rather than what it is. Schematic diagram of the basic membrane separation process is shown in Figure 1.1.

1.2.3 Types of Membranes

A synthetic membrane is a synthetically created membrane usually intended for separation purposes in a laboratory or industry. Synthetic membranes have been successfully used for small and large-scale industrial processes

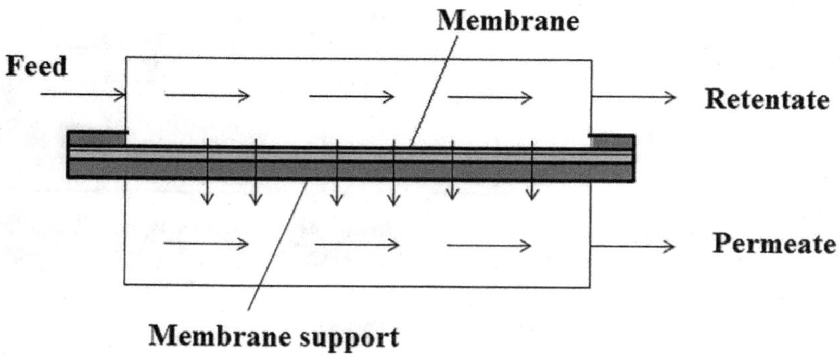

FIGURE 1.1
Schematic diagram of the basic membrane separation process.

since the middle of the 20th century. A wide variety of synthetic membranes are known. They can be produced from organic materials such as polymers and liquids as well as inorganic materials. Most commercially used synthetic membranes in separation industry are made of polymeric structures. A schematic diagram of the membrane types is presented in Figure 1.2.

Polymeric membranes lead the membrane separation industry market because they are competitive in performance and economics. A polymer has to have appropriate characteristics for the intended application. The polymer sometimes has to offer a low binding affinity for separated molecules (as in the case of biotechnology applications) and has to withstand harsh cleaning conditions. It has to be compatible with chosen membrane fabrication technology. The polymer has to be a suitable membrane former in terms of its chains rigidity, chain interactions, and polarity of its functional groups. The polymers can form amorphous and semi-crystalline structures, affecting the membrane performance characteristics. The polymer has to be obtainable and reasonably priced to comply with the low-cost criteria of membrane separation process. Many membrane polymers are grafted, custom-modified, or produced as copolymers to improve their properties. The most common polymers in membrane synthesis are cellulose acetate, nitrocellulose, cellulose esters (CA, CN, CE), polysulfone (PS), polyethersulfone (PES), polyacrylonitrile (PAN), polyimide (PI), polyethylene (PE), polypropylene (PP), polytetrafluoroethylene (PTFE), polyvinylidene fluoride (PVDF), and polyvinylchloride (PVC).

Ceramic membranes are made from inorganic materials (such as alumina, titania, zirconia oxides, recrystallized silicon carbide, or some glassy materials). By contrast with polymeric membranes, they can be used in separations where aggressive media (acids, strong solvents) are present. They also have an excellent thermal stability that makes them usable in high-temperature

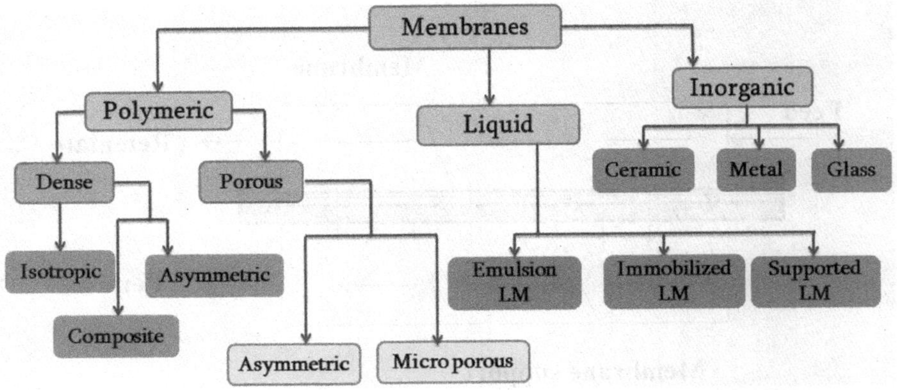

FIGURE 1.2
Schematic diagram for types of membranes.

membrane operations. Even though ceramic membranes have a high weight and substantial production costs, they are ecologically friendly and have long working life.

Liquid membranes refer to synthetic membranes made of non-rigid materials. Several types of liquid membranes can be encountered in industry: emulsion liquid membranes, immobilized (supported) liquid membranes, molten salts, and hollow fiber contained liquid membranes. Liquid membranes have been extensively studied but thus far have limited commercial applications. Maintaining adequate long-term stability is the problem, due to the tendency of membrane liquids to evaporate or dissolve in the phases in contact with them.

1.2.4 Preparation Techniques

The choice of a technique for polymeric membrane preparation depends on a selection of polymer and the required structure of the membrane [7]. There are many processing techniques to fabricate membranes, which include sintering, stretching, track etching, template leaching, interfacial polymerization, phase inversion (different ways include immersion precipitation, thermally induced phase separation, evaporation induced phase separation, vapor induced phase separation and precipitation by controlled evaporation), and electrospinning.

1.2.4.1 Sintering

The sintering process (Figure 1.3) is a unique and simple technique allowing porous material to be obtained from organic as well as from inorganic materials. The method involves compression of a powder consisting of particles of given size and sintering at elevated temperatures. The required temperature depends on the material used during sintering and the interfaces between the contacting particles. A wide range of different material can be used as powder of polymers (polyethylene, polytetrafluoroethylene, and polypropylene), metals (stainless steel and tungsten), ceramics (aluminium oxide and

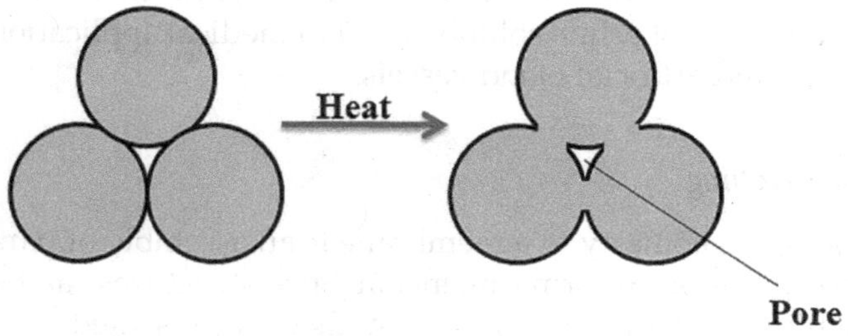

FIGURE 1.3
Sintering process.

zirconium oxide), graphite (carbon), and glass (silicates). The pore size of the resulting membrane is determined by the particle size and particle size destruction of the powder. Narrow particle size distribution gives a narrow pore size distribution in the resulting membrane. This technique allows pore sizes of about 0.1 to 10 μm to be obtained, the lower limit being determined by the minimum particle size.

Sintering is a suitable technique for preparing membranes from polytetrafluoroethylene because this chemically and thermally resistant polymer is not soluble. Only microfiltration membranes can be prepared via sintering. However, the porosity of porous polymeric membrane is low, normally in the range of 10 to 20% or sometimes a little higher.

1.2.4.2 Stretching

Another relatively simple procedure for preparing microporous membranes is the stretching of a homogeneous polymer film of partial crystallinity. This technique is mainly employed with films of polyethylene or polytetrafluoroethylene that have been extruded from a polymer powder and then stretched perpendicular to the direction of extrusion. This leads to a partial fracture of the film and relatively uniform pores with diameters of 1 to 20 μm. These membranes, which have a high porosity, up to 90%, and a fairly regular pore size are now widely used for microfiltration of acid and caustic solutions, organic solvents, and hot gases. They have to a large extent replaced the sintered materials used earlier in this application. Stretched membranes can be produced as flat sheets as well as tubes and capillaries. The stretched membrane made out of polytetrafluoroethylene is frequently used as a water-repellent textile for clothing, such as parkas, tents, sleeping bags, etc. This membrane has high porosity, high permeability for gases and vapors, but, because of the hydrophobic nature of the basic polymer, is up to a certain hydrostatic pressure completely impermeable to aqueous solutions. Thus, the membrane is repellent to rain water but permits the water vapor from the body to permeate. More recently, this membrane has also been used for a novel process, generally referred to as membrane distillation, that is, to remove ethanol from fermentation broths or wine and beer to produce low-alcohol products and for desalination of seawater. These membranes are also used for desalination of saline solutions and in medical applications, such as burn dressings and artificial blood vessels.

1.2.4.3 Track Etching

The simplest pore geometry in a membrane is an assembly of parallel cylindrical shaped pores of uniform dimension. Such structures can be obtained by track-etching, and the preparation method is presented in Figure 1.4.

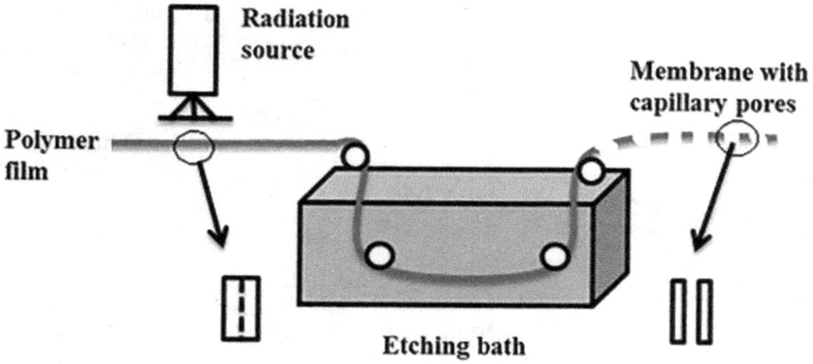

FIGURE 1.4
Preparation of porous membrane by track etching.

In this method, a film or foil (often a polycarbonate) is subjected to high energy particle radiation applied perpendicular to the film. The particles damage the polymer matrix and create tracks. The film is then immersed in an acid or alkaline bath, and the polymeric material is etched away along these tracks to form uniform cylindrical pores with a narrow pore size distribution. Pore sizes can range from 0.02 to 10 μm, but the surface porosity is low (about 10% at a maximum). The choice of the material depends mainly on the thickness of the film and the energy of the particles being applied (usually about 1 MeV). The maximum penetration thickness of particles with this energy is about 20 μm. When the energy of the particles is increased, the film thickness can also be increased, and even inorganic material (e.g., mica) can be used. The porosity is mainly determined by the radiation time whereas the pore diameter is determined by the etching time.

1.2.4.4 Template Leaching

Template leaching is another method of producing isotropic microporous membranes from insoluble polymers such as polyethylene, polypropylene, and poly (tetrafluoroethylene). In this process, a homogeneous melt is prepared from a mixture of the polymeric membrane matrix material and a leachable component. To finely disperse the leachable component in the polymer matrix, the mixture is often homogenized, extruded, and pelletized several times before final extrusion as a thin film. After formation of the film, the leachable component is removed with a suitable solvent, and a microporous membrane is formed. The leachable component can be a soluble, low-molecular-weight solid, a liquid such as liquid paraffin, or even a polymeric material such as polystyrene. A drawing of a template leaching membrane production machine is shown in Figure 1.5.

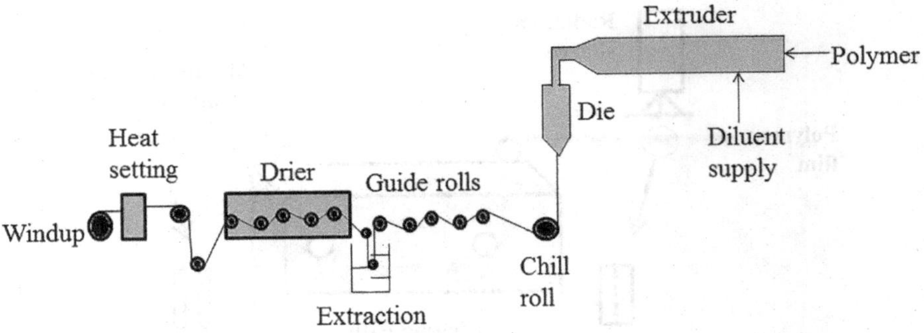

FIGURE 1.5
Template leaching process.

1.2.4.5 Phase Inversion

Phase inversion is a process whereby a polymer is transformed in a controlled manner from a liquid phase to a solid phase. The process of solidification is often initiated in the transition from one liquid phase into two liquid phase (liquid-liquid demixing). At a certain stage during demixing one of the liquid phases (the high polymer concentration phase) will solidify so that a solid matrix is formed. By controlling the initial stage of phase inversion, the membrane morphology can be controlled, that is, porous as well nonporous membrane can be prepared. The concept of phase inversion covers a range of different techniques, such as solvent evaporation, thermal precipitation, precipitation by controlled evaporation, precipitation from the vapor phase, and immersion precipitation. The majority of phase inverted membranes are prepared by immersion precipitation. In this process, a polymer solution is cast on a suitable support and then immersed in a coagulation bath containing a non-solvent, where an exchange of solvent and non-solvent takes place, and the membrane is formed. Schematic representation of phase inversion processes after polymer solution immersion in a non-solvent bath is shown in Figure 1.6. This type of membrane can be made from almost any

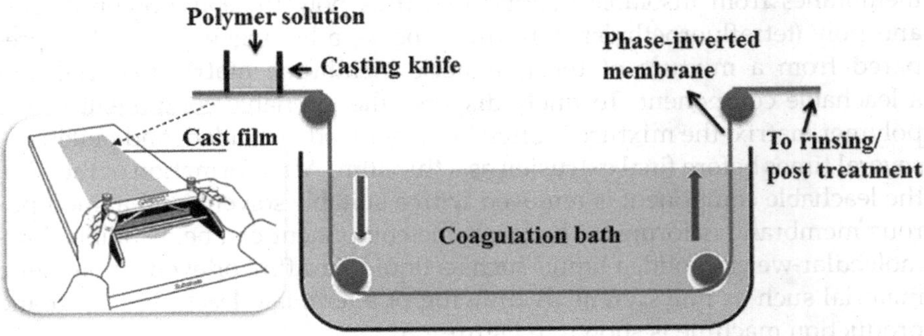

FIGURE 1.6
Membrane preparation byphase inversion methods.

polymer that is soluble in an appropriate solvent and can be precipitated in a non-solvent. By changing the type of polymer, the polymer concentration, the precipitation medium, and the precipitation temperature, microporous phase inverted membranes can be made with a large variety of pore sizes (from less than 0.1 to more than 20 μm) with varying chemical, thermal, and mechanical properties.

1.2.4.6 Electrospinning

Electrospinning is a process that fabricates fibers through an electrically charged jet of polymer solution or polymer melt. To understand electrospinning, one can look at the mechanism behind the production of polymer fibers. Conventional fibers of large diameter involve the drawing of molten polymer out through a die. The resultant stretched polymer melt will dry to form individual elements of fiber. Similarly, electrospinning also involves the drawing of fluid, either in the form of molten polymer or polymer solution. However, unlike conventional drawing methods where there is an external mechanical force that pushes the molten polymer through a die, electrospinning makes use of charges that are applied to the fluid to provide a stretching force to a collector where there is a potential gradient. When a sufficiently high voltage is applied, a jet of polymer solution will erupt from a polymer solution droplet. The polymer chain entanglements within the solution will prevent the electrospinning jet from breaking up. While molten polymer is used in both conventional fiber production method and electrospinning method, it cools and solidifies to yield fiber in the atmosphere. The electrospinning of polymer solution relies on the evaporation of the solvent for the polymer to solidify to form polymer fiber. Figure 1.7 shows the schematic diagram electrospinning process and fiber formation.

Since electrospinning is the drawing of a polymer fluid, there are many different types of polymers and precursors that can be electrospun to form fibers.

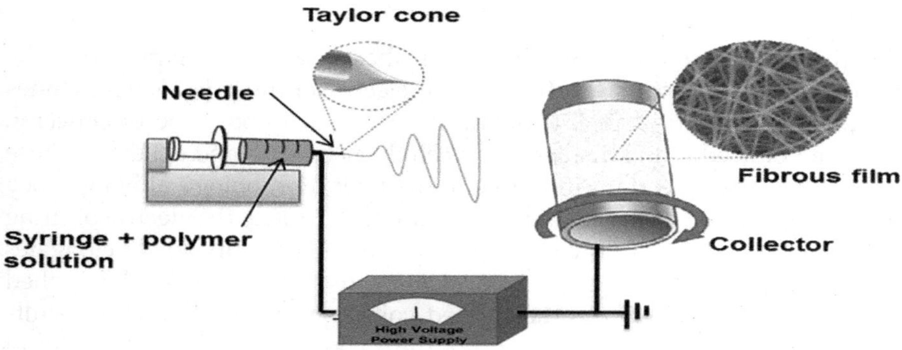

FIGURE 1.7
Basic schematic drawing of electrospinning process.

The materials to be electrospun will depend on the applications. Materials such as polymers and polymer nanofibercomposites can be directly produced by electrospinning technique. Other materials such as ceramics and carbon nanotubes require post processing of the electrospun fibers.

1.2.4.6.1 Polymer Solution Parameters

The properties of the polymer solution have the most significant influence in the electrospinning process and the resultant membrane morphology. The surface tension has a part to play in the formation of beads along the fiber length. The viscosity of the solution and its electrical properties will determine the extent of elongation of the solution [8]. One of the factors that affect the viscosity of the solution is the molecular weight of the polymer. When a polymer of higher molecular weight is dissolved in a solvent, its viscosity will be higher than solution of the same polymer but of a lower molecular weight. One of the conditions necessary for electro-spinning to occur where fibers are formed is that the solution must consist of polymer of sufficient molecular weight and the solution must be of sufficient viscosity. This will, in turn, have an effect on the diameter of the resultant electrospun fibers. Koski et al. [9] have studied the effects of polymer average molecular weight on the fiber structure of electrospun polyvinyl alcohol (PVA). They have reported an average fiber diameter between 250 nm and 2 μm and the fiber diameter increases with M_W and concentration. At lower M_W and/or concentrations, the fibers exhibit a circular cross-section. They have also observed flat fibers at high M_W and concentrations. The effect of viscosity on the membrane morphology has also studied by Mohammad et al. [10]. According to this study, viscosity of the spinning solutions has played an important role on the morphology of the mullite nanofibers. Continuous electrospun nanofibers with common cylindrical morphology were obtained when PVA content was 6 wt.%. Further increasing the amount of PVA in the pre-spinning solution led to excessively high viscosity level, making the shape of the resulting mullite nanofibers wide and flat ribbon.

1.2.4.6.2 Processing Conditions

Another important parameter that affects the electrospinning process is the various external factors exerting on the electrospinning jet. This includes supplied voltage, feed rate, temperature of the solution, type of collector, diameter of needle, and distance between the needle tip and collector. These parameters have a certain influence in the fiber morphology although they are less significant than the solution parameters. During the electrospinning process, the droplet of polymer solution at the needle tip slowly elongates from a semicircular shape to a conical shape or Taylor cone, as the applied voltage is increased [11]. As the applied voltage increases further, it results in the eruption of a jet from the tip of the Taylor cone and the diameter of fibers collected decreases. Barhate et al. [12] studied the effect of electrospinning process parameters on structural and transport properties of the

TABLE 1.2
Different Membrane Fabrication Techniques and Their Properties

Fabrication Technique	Pore Diameter (μm)	Porosity (%)	Properties
Sintering	0.1–10	10–20	Outstanding chemical, thermal, and mechanical stability Both organic and inorganic material can be used
Stretching	01–20	90	High permeability for gases and vapors Can be produced as flat sheets, tubes, or capillaries
Track-etching	0.02–10	Max 10	Low tendency to plug Good long-term flux stability
Template leaching	Min 0.05	Porous membrane	High surface area
Phase inversion	<0.21–>20	Both porous and non-porous	Properties varied according to polymer used
Electrospinning	Porous	Highly porous membrane (>80)	Relatively high flux and less tendency to fouling High surface area

electrospun membranes in detail. In the absence of an aerodynamically driven bending instability, electrical bending instability is the only mechanism that influences the way in which the jet is oriented before its deposition over the collector. The fiber crossing and pore size can be optimized to attain improved structural (pore size distribution, pore interconnectivity, and porosity) and transport (permeability) properties of the electrospun filtering media. Finally, they have suggested that optimization can be attained by coordination of the applied voltage and collection rates. Since the jet follows a bending, winding, spiraling, and looping path in three dimensions due to electrical bending instability before its deposition over the collector, the tip-to-collector distance is another parameter that may alter the morphology of the membrane. The area of the Taylor cone (resulting from the electrically driven bending instability) is an important parameter to consider when targeting uniform deposition of nanofibers. A summary of the various membrane fabrication techniques, range of pore size/porosity, and their resultant properties are presented in Table 1.2.

1.2.5 Functionalization of Membranes

To maintain membrane separations as economical alternatives to conventional water and wastewater treatment technologies, we must produce a high-quality permeate at a fast rate, and be able to maintain that production for an extended period. However, the relationship between flux and selectivity along with fouling introduce a challenge that must be addressed for all membrane applications. There are three main areas of interest when it comes to improving membrane performance:

Synthesis process: Synthesis process improvement involves using the techniques, methods, and materials of the manufacturing processes to produce a high-performance membrane.

Application process: The application process involves the specific operating parameters for a membrane system. These include selecting the raw water characteristics, operating pressure, and cleaning intervals to allow the system to operate at maximum efficiency.

Modification (functionalization): Post- or pre-synthesis modification involves functionalization of the polymer before synthesis and the modification of membrane after the initial preparation process is completed.

Functionalization or modification of membranes are mostly related to the introduction of functional groups into membranes such as carboxylic (–COOH), amine (–NH_2), hydroxyl (–OH), thiol (–SH), epoxide, aldehyde, etc. [13]. The functionalized membranes with proper activating groups can cover a great variety of applications such as an anti-fouling membrane (tunable permeation and separation), affinity membrane (toxic metal capture, nanoparticles immobilization for toxic organics degradation), membrane based biosensor, biomedical compatible biomaterials, and super hydrophobic or hydrophilic surfaces, etc. Many membranes are naturally hydrophobic due to the polymers that are used in the manufacturing process. A drawback of a membrane exhibiting a hydrophobic nature is that it increases the required transmembrane pressure (TMP) that must be used for operation since the solvent (i.e., water) is repelled by the surface of the membrane.

However, if a membrane can be rendered hydrophilic, the solvent is attracted to and transported through the membrane at a much faster rate, which reduces the required TMP for operation. In addition, many solutes found in feed waters are hydrophobic, so using a hydrophilic membrane could increase the membrane's selectivity and fouling resistance [14]. Moreover, membranes with higher surface roughness have different peaks and valleys and are more susceptible to fouling, and with a reduction of roughness, foulants are less likely to adhere to the membrane. Post-synthesis modifications may ultimately lead to a higher flux and selectivity and improved fouling resistance.

1.2.5.1 Functionalization Techniques

Membrane modification or functionalization can be achieved through either covalent or non-covalent attachment. The membrane modification can be completed using different techniques such as physical coating or blending, graft polymerization, plasma treatment, chemical treatment, etc.

1.2.5.1.1 Physical Coating or Blending

This technique is the easiest and the most straightforward way to modify polymer surface by blending functional groups into the bulk polymer or just coating it on the polymer surface. The biggest drawback of this technique is the instability of the bulk polymer surface composition caused by the loss of the functional materials from the polymer. But it is still a good choice if this loss is too slow to be considered or not fast enough to affect the application of the materials. Physical coating or blending of functional molecules into/onto the polymers was significantly developed by some new methods, such as self-migration (a), self-assembly (b), and layer by layer (c) (Figure 1.8). In a self-migration method, the functional group with special interest is mixed into the bulk polymer. By making it reach the lowest free energy state, the functional materials or molecules can automatically migrate toward the polymer surface and finally accumulate onto the polymer surface so that it can change the properties of the bulk polymer surface considerably [15]. In self-assembly method, the surfactant molecules coated onto substrate surface self-assembled into a thin film, driven by hydrophobic interaction, hydrogen bond, electrostatic interaction, or chemical reaction [16]. Furthermore, the layer-by-layer method is a special case of the self-assembly method and negatively and positively charged macromolecules are alternatively introduced onto the polymer surface through the strong electrostatic interaction [17].

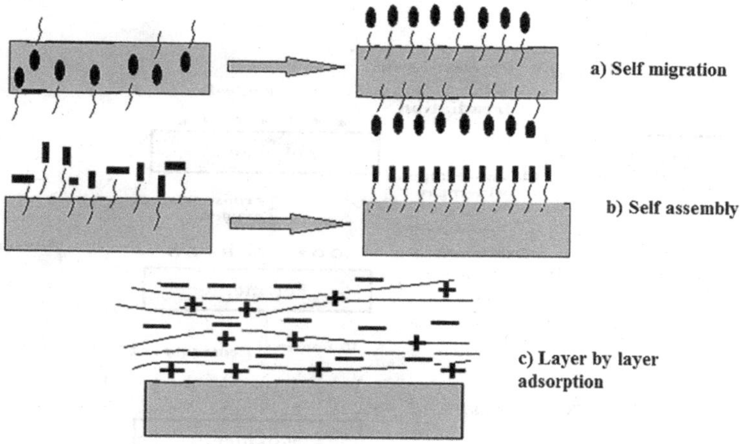

FIGURE 1.8
Schematic representation of (a) self-migration, (b) self-assembly, and (c) layer-by-layer adsorption techniques.

1.2.5.1.2 Graft Copolymerization

Grafting is one of the chemical modification methods developed to date and has emerged as a simple, useful, and versatile approach to improving surface properties of polymers for a wide variety of applications. Grafting has numerous advantages:

- The ability to modify the polymer surface to have distinct properties through the choice of different monomers
- The ease and a controllable introduction of graft chains with a high density and exact localization of graft chains to the surface with the bulk properties unchanged.
- Covalent attachments of graft chains onto a polymer surface assuring the long-term chemical stability of introduced chains, in contrast to physically coated polymer chains

For the graft copolymerization to be occurring, that which can produce radicals must be introduced onto the polymer surface first to activate the polymer surface. Most of the chemically inert polymers can be activated via irradiation (UV, γ-ray, electron beams, etc.), plasma treatment, ozone or H_2O_2 oxidization, or Ce^{4+} oxidization (Figure 1.9).

1.2.5.1.3 Plasma Treatment

For plasma induced graft polymerization, an inert gas is often used to produce radicals on the polymer surface. While reactive gas such as O_2, SO_2, NH_3, and CO_2 are used, chemical functional groups such as hydroxyl, carboxyl, carbonyl, and sulfonate can be yielded on the polymer surface directly. However, the functional groups introduced on the polymer surface are not

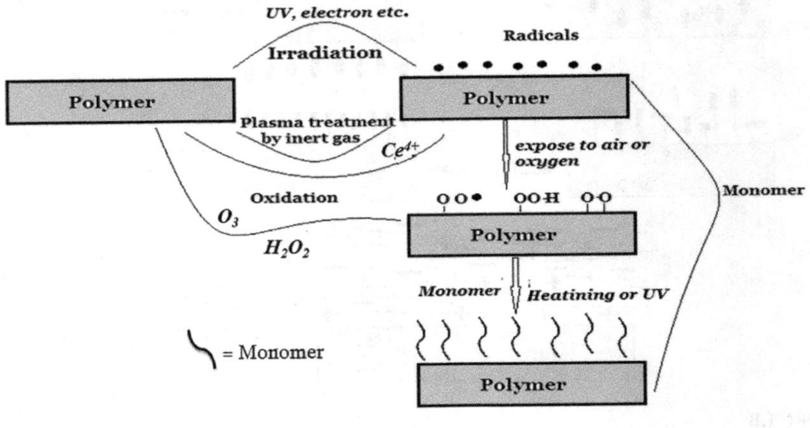

FIGURE 1.9
The general scheme of how the surface of polymer can be grafted.

distinct. As mentioned above, plasma consists of a unique mixture of different positively charged or negatively charged ions, electrons, free radicals, atoms, and molecules. The variety of the particles in a plasma chamber also causes a variety of the functional groups yielded on the polymer surface. Plasma treatment is usually not expected to introduce a specific functional group, but just to increase the polarity, hydrophilicity, and charge of the polymer surface to benefit applications like adhesion, dye removal ability, blood compatibility, etc.

1.2.5.1.4 Chemical Treatment

Polymer molecules that possess functional groups such as hydroxyl, carboxyl, amino and ester, etc., can be directly modified by chemical treatment. The chemical modification involves the introduction of one or more chemical species to a given surface so as to produce a surface that has enhanced chemical and physical properties. Chemical reactions can be carried out at sites that are vulnerable to electrophilic or nucleophilic attack. Structures such as benzene rings, hydroxyl groups, double bonds, halogen, ester groups, etc. qualify for such attacks. Wet chemical oxidation treatments are also commonly employed to introduce oxygen-containing functional groups (such as carbonyl, hydroxyl, and carboxylic groups) at the surface of the polymer. This can be conducted using gaseous reagents or with solutions of vigorous oxidants. The oxygen-containing functional groups increase the polarity and the ability to hydrogen bond, thus in turn resulting in the enhancement of wettability and adhesion.

1.3 Preparation and Applications of Electrospun Membranes

In the electrospinning process, a number of parameters can affect the morphology of the attained fibers. The main parameters are polymer solution (concentration, viscosity, surface tension, and conductivity of the polymer solution), process conditions (applied voltage, tip-to-collector distance, and feed rate), and ambient parameters (temperature, relative humidity, and velocity of the surrounding air in the spinning chamber) [18]. Electrospun membranes or fibers are normally collected in the form of nonwoven mats, which are interesting for a variety of applications such as semipermeable membranes, filters, composite reinforcement, and scaffolding used in tissue engineering, etc. [19]. Figure 1.10 represents the various applications of electro-spun membranes.

Deitzel et al. [19] have reported the controlled deposition of electrospun poly (ethylene oxide) fibers and the feasibility of dampening the instability of submicron polymer fibers less than 300 nm in diameter. Accordingly, poly(ethylene oxide) (PEO) with 10 wt.% in water was electrospun at high

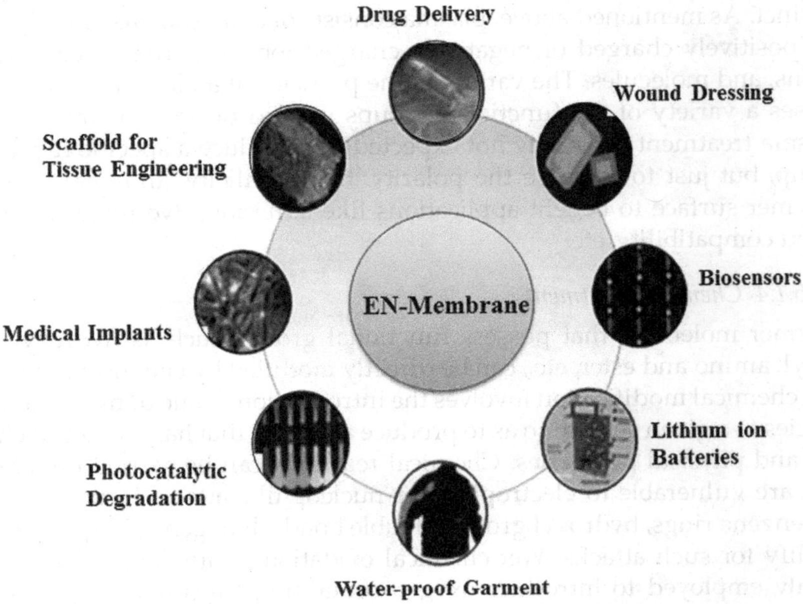

FIGURE 1.10
Various applications of electrospun membranes.

voltage supply from 5 to 15 kV to the vertically oriented syringe tip and collected to the plate situated at the bottom of the apparatus 17 cm apart. In their conclusion, they have shown the possibility to control the deposition of electrospun fibers through the use of an electrostatic lens element and biased collection target. Moreover, the possible control or elimination of the bending instability occurring in electrospinning experiments by applying a secondary external field of the same polarity as the surface charge on the jet was suggested. Krishnappa et al. [20] have electrospun bisphenol-A polycarbonate using two solvents: chloroform and a 1:1 mixture of tetrahydrofuran (THF) and dimethylformamide (DMF). The polycarbonate concentrations in the solutions were 14% and 15% by weight; the tip to target distances was maintained between 80–120 mm; voltage was varied from 6 kV to 30 kV. The morphological features of the electrospun polycarbonate fibers have been studied as a function of the solvent and processing voltage and the diameters of the fibers range from 100 nm to 500 nm. Finally, they have concluded that the surface tension, electrostatics, and viscosity, which are a function of the concentration of the polymer in the solution, play a major role in determining the morphology of the spun material. Choi et al. [21] also studied the application of electrospun PVDF nanofiber webs as an electrolyte binder or a separator for a battery where PVDF of 25 wt.% was dissolved in dimethylacetamide (DMAc); the distance between the needle tip and the rotating collector was 15 cm, and the applied voltage was kept at 10 kV. The diameters of the electrospun PVDF nanofibers were found to be 100–800 nm. Following thermal treatment at 150–160 °C, the membrane's physical property, dimensional stability, tensile strength, and elongations at break as well as the tensile modulus were notably improved. Yang et al. [22] have spun polyvinyl alcohol (PVA) cross-linked with maleic anhydride (MA) at a positive voltage of 20 kV and a tip to collector distance of 20 cm. From water durability tests they have found that the average mass loss and standard deviation of electrospun 8% PVA/MA (20/1, mole/mole) membrane was the least after boiling in water for 1 h, and they have also observed that the average diameter of fibers in PVA/MA membrane was larger than that in PVA membrane. Moreover, their experimental results indicated that rapid evaporation of water and high electric field during electrospinning process could promote esterification reaction. Tsioptsias et al. [23] have prepared cellulose acetate–Fe_2O_3 composite nanofibrous materials by electrospinning process where 20% (w/v) CA solutions with dispersed Fe_2O_3 nanoparticles varying between 1.4% and 4.5% (w/v), and N,N–dimethylacetamide/acetone was used as solvent. During the electrospinning process, they have kept the solution flow rate as 0.38 mL/h; needle to target distance was 5 cm; high voltages (13, 16.5, and 20 kV); and the collecting target was a rotating drum covered with aluminium foil. Finally, they have recommended that these composite materials with enhanced thermal stability compared to pure cellulose acetate and fairly uniform composite fibrous structures for potential use in separation processes or biomedical applications. Tsai et al. [24] have prepared electrospun chitosan–gelatine–polyvinyl alcohol hybrid nanofibrous mats using only a mild concentration (20 wt.%) of acetic acid (solvent) by electrospinning. During the preparation, different concentrations were used, and the solution was loaded into a 3 mL syringe with a flow rate of (0.15–0.3 mL/h) that was controlled by a syringe pump. High electric voltages (15–30 kV) were applied, and the fiber mats were collected on a vertical collector where the distance was kept to 15 cm. The optimum weight ratio of chitosan:gelatine:PVA to be 2:2:4, with resultant fiber diameters of around 150 nm and uniform nanofibers, were obtained at an applied voltage of 20 kV. The stability and tensile strength of the nanofibrous mats were both improved by glutaraldehyde vapor-crosslinking for 12 h. From cyto compatibility and 3-dimethylthiazol-2,5-diphenyltetrazolium bromide (MTT) assay (which measured cell activity) tests, they have found good results as compared to sponges and chitosan–gelatine–PVA nanofibrous mats were suggested as suitable for attachment and proliferation of KP-hMSCs, probably because the morphology and structure of nonwoven electrospun nanofibers are similar to the extracellular matrix (ECM). Kimura et al. [25] have prepared poly (vinylidene fluoride-co-hexafluoropropylene) (PVDF-HFP-6/1/3) nanofiber membranes with different amounts of polyethylene glycol (PEG) and polyethylene glycol dimethacrylate (PEGDMA) oligomer via electrospinning. The experiments were performed at room temperature, and the collecting roller distance was 12 cm from the tip, and a voltage of 11 kV was applied. The prepared nanofiber membrane had a higher porosity, electrolyte uptake, and bulk resistance value than existing separator, and this membrane was suggested to be a promising candidate as a separator for lithium-ion batteries.

1.4 Electrospun Membranes for Water and Wastewater Treatment Applications

Perm-selectivity and flux are important specificity in membrane separation. Until now it has been a dream for membranologists to simultaneously improve both perm-selectivity and flux, which generally show a trade-off relationship. An advancement in membrane separation would be realized by adopting membranes with higher surface area and higher porosity [26]. Polymer nanofibers are a main class of nanomaterials, which can be easily tuned to exhibit novel and significantly improved physical and chemical properties. In the last 10 years, they have received increasing attention because of their high surface-to-mass ratio and special characteristics that are attractive for advanced applications. Specifically, electrospun nanofiber membranes have high porosity, interconnected open pore structure, and tailorable membrane thickness [27]. Though nanofiber membranes have been commercially used for air filtration applications, their potential in water treatment is still largely unemployed [28]. Gopal et al. [29] have prepared electrospun membranes of polyvinylidene fluoride nanofibers by using electrospinning process and characterized to relate its structural properties to membrane separation properties and performance. They have confirmed that the membranes have similar properties to that of conventional microfiltration membranes with fiber diameter 106–308 nm; thickness 300 μm; trans membrane pressure 0.5 bar; pore size 10.6–4.0 μm; and flux of 200 kg/m^2h. The electrospun membranes were used to separate 1, 5, and 10 μm polystyrene particles and they were successful in rejecting more than 90% of the microparticles from the solution. The authors suggested that these membranes could be employed in the separation technology as a potential membrane for pretreatment of water prior to reverse osmosis or as prefilters to minimize fouling and contamination prior to ultra- or nano-filtration. Wang et al. [30], on the other hand, have developed high flux ultrafiltration (UF) membrane based on poly (vinyl alcohol) (PVA) electrospun nanofibrous scaffold support and PVA hydrogel coating. They have found that the electrospun scaffold fabricated by 96% hydrolysed PVA with relatively high molecular weight showed good overall mechanical performance before and after crosslinking. The PVA hydrogel coating was also fabricated with different crosslinking conditions and used as the anti-fouling layer for high-throughput water ultrafiltration. Their ultrafiltration test specified that the hydrophilic nanofibrous composite membranes showed a flux rate

(>130 L/m^2 h) significantly greater than commercial UF membranes but with similar filtration efficiency (rejection rate >99.5%). Yoon et al. [31] have reported about conventional ultrafiltration (UF)/nanofiltration (NF) filters for water treatments based on porous membranes, typically manufactured by the phase immersion method, where their torturous porosity usually results in a comparatively low flux rate. They have fabricated a new type of high flux UF/NF medium based on an electrospun nanofibrous scaffold (polyacrylonitrile, PAN) attached with a thin top layer of hydrophilic, water-resistant, and water-permeable coating (chitosan). The average diameter ranged from 124 to 720 nm and porosity of about 70% together with a chitosan top layer having a thickness of about 1 μm. The flux and rejection studies also evaluated in this report in which the flux rate was an order of magnitude higher than commercial NF membranes in 24 h of operation, while maintaining the same rejection efficiency (>99.9%) for oily wastewater filtration. Kim et al. [32] have investigated electrospun polycarbonate (PC) with quaternary ammonium salt (benzyl triethylammonium chloride, BTEAC) as an additive to develop antimicrobial nanofibrous membranes for ultrafiltration. They have found that the conductivity of the PC solution was a key parameter affecting the morphology and diameter of the electrospun PC fibers. Finally their report showed that the PC nanofibrous membrane with BTEAC had an excellent antimicrobial activity and filtration performance. Zhang et al. [33] also have assessed an adsorptive membrane made from electrospun cellulose acetate nanofibers as an ion-exchange medium for protein separations. They have found that the electrospun cellulose acetate nanofibers had diameters ranging from tens of nanometers to microns, and the pore sizes within the nanofiber felts were ranging from submicrons to microns. The results have indicated that the diethylaminoethyl (DEAE) surface functionalized nanofiber felt showed the highest static binding capacity of 40.0 mg/g for BSA, compared to 33.5 mg/g, 14.5 mg/g, and 15.5 mg/g for the functionalized commercial membrane, cellulose microfiber medium, and cotton balls, respectively. The porous nature of the nanofiber felts caused in a higher permeability compared to the commercial membranes. Aussawasathien et al. [34] have prepared electrospun nylon-6 nanofibrous webs with fiber diameters in the range of 30–110 nm and was employed as a membrane material for water filtration due to its excellent chemical and thermal resistance as well as high wettability. They have reported that this membrane was able to separate all particles with sizes from 10 μm down to 1 μm whereas roughly 90% separation was obtained for 0.5 μm particles, where low content of particles in the filtrate, indicated by the high separation factor. Finally, they have recommended that the prepared membrane could be used as prefilters prior to ultrafiltration or nanofiltration to increase the filtration efficiency and prolong the life of downstream membranes. Yoon et al. [35] have demonstrated a high flux thin-film nanofibrous composite (TFNC) membrane system based on polyacrylonitrile (PAN) electrospun scaffold (porosity of about 85%) attached with a thin barrier layer of cross-linked polyvinyl alcohol (PVA) on which its thickness was about 0.5 μm. According to the authors, the TFNC membrane system will be useful for ultrafiltration (UF) applications, exhibiting a high flux (12 times higher than that of conventional PAN UF membranes) and excellent rejection ratio, >99.5% for separation of oil/water mixture (1500 ppm in water). The membranes were tested up to 190 h in a practical pressure range of up to 130 psig. In addition, the MWCO characteristics of PVA–PAN TFNC membrane (between 9–11 k and 35–45 k) was studied and a slightly lower than that of PAN10 UF membrane was attained. Xu et al. [36] have developed hierarchically nanofibrous membrane of thermal plastic elastomer ester (TPEE) and iron alkoxide with high crystallinity as well as hierarchical structure. The synthesized membranes have indicated high efficiency for the removal of Cr (VI) from water and were mainly attributed to the adsorption and reduction of Cr (VI) to Cr (III), because of the hierarchical structure and the exposed iron oxide active sites. First, the adsorption of Cr (VI) on the surface of iron oxides by the electrostatic adsorption was achieved; second, the reduction of Cr (VI) to Cr (III), which is much less toxic, and almost all of Cr (VI) removal has been reported. Wang et al. [4] have prepared electrospun nanofibrous membranes for high flux microfiltration of polyacrylonitrile (PAN) on nonwoven polyethylene terephthalate (PET) support, with a total thickness of 200 ± 10 μm; average fiber diameter of 100 ± 20 nm; high effective porosity (up to 80%); interconnected pores; and optimized small pore size (maximum pore size of 0.62 ± 0.03 μm and mean flow pore size of 0.22 ± 0.01 μm). They have evaluated the membrane performance for microfiltration and showed significantly better in flux (1.5 L/m^2h) (i.e., 2–3 times) over the Millipore GSWP 0.22 μm membranes, while maintaining a high rejection level in the microparticle retention test, and achieved complete bacteria removal. Mahapatra et al. [37] have prepared iron oxide–alumina mixed nanocomposite fiber by electrospinning with fibers diameters range of 200–500 nm for heavy metal ion adsorbent. The electrospun nanofiber membranes were sintered at 1000 °C to convert them to pure oxide. The best isotherm fit for copper and lead was obtained with Langmuir adsorption isotherm while Freundlich was the best option for nickel and mercury, and complete removal of mixed oxide nanocomposite fibers was reported as copper (21%), lead (52%), nickel (67%), and Hg (89%). Li et al. [38] have fabricated novel high-performance hydrophilic poly (vinyl alcohol-coethylene) (PVA-co-PE) nanofiber membrane for filtration media. The investigation of the rejection rate of PVA-co-PE nanofibrous membrane to TiO_2 suspension was above 99%, and such unique hydrophilic nanofibrous membrane showed a higher flux rate than the commercial microfiltration membrane. According to their conclusion, the fabricated membrane indicated good performances of flux and low fouling, and was suggested to be used for potential applications in the microfiltration fields. You et al. [39] have developed a new class of high-performance thin-film nanocomposite (TFNC) ultrafiltration membrane using electrospinning technique combined with solution treatment method to separate an oil/water emulsion based on a polyacrylonitrile (PAN) nanofibrous substrate coupled with a thin hydrophilic nanocomposite barrier layer. The PVA–MWNT/PAN TFNC (10 wt.% MWNT) membrane showed a high water flux (270.1 L/m^{2}h) and high rejection rate (99.5%) at a low feeding pressure (0.1 MPa). Aliabadi et al. [3] have prepared poly ethylene oxide (PEO)/chitosan nanofiber membrane by electrospinning technique with the average diameter and surface area of 98 nm and 312.2 m^2 g^{-1}, respectively. The prepared membrane was evaluated for potential adsorption of nickel (Ni), cadmium (Cd), lead (Pb), and copper (Cu) from aqueous solution. It was found that the adsorption of metal ions onto the PEO/Chitosan nanofiber membrane was feasible, spontaneous, and endothermic within the deceasing order of Pb (II), Cd (II), Cu (II), and Ni (II) and the adsorption percentages of these metal ions were obtained as 68%, 72%, 82%, and 89%, respectively. Liu et al. [40] have presented a novel class of high-flux microfiltration filters consisting of an electrospun nanofibrous membrane and a conventional nonwoven microfibrous support with an average fiber diameter of 100 ± 19 nm; mean pore size 0.30 μm to 0.21 μm; and thickness varying from 10 μm to 100 μm. The water permeability was found to be 3 to 7 times than the Millipore GSWP 0.22 μm membrane due to the high porosity of the microfiltration filters based on these electrospun membranes. The nanofibrous PVA membranes with an average thickness of 20 μm could successfully reject more than 98% of the polycarboxylate microsphere particles with a diameter of 0.209 ± 0.011 μm, and still maintain 1.5 to 6 times higher permeate flux than that of the Millipore GSWP 0.22 μm membrane.

1.5 Preparation and Application of Phase-Inverted Membranes

Most commonly available commercial membranes are prepared by phase-inversion technique [41]. This technique is suitable for preparing membranes with all types of morphological structures. In this method, a casting solution comprising of polymer and solvent is immersed into a nonsolvent coagulation bath, and a polymer is transformed from liquid phase to a solid phase in a controlled condition. The solidification process is frequently started by the change from one liquid phase to two liquids (liquid-liquid de-mixing). At certain period during de-mixing, the polymer rich phase solidifies so that a solid membrane matrix is formed [41,42]. Many researchers have tried to describe the membrane's fabrication mechanisms in the phase inversion technique for several years. Membrane morphologies, particularly the pore size distributions, can be controlled by selecting the different solvent, nonsolvent, polymer, pore former, and fabrication parameters depending on the particular application [43–46]. Many researchers have prepared different membranes for various applications. Cellulose acetate is broadly applicable for the synthesis of membranes because of having tough, biocompatible, hydrophilic characteristics, good desalting nature, high flux, and it is moderately less expensive to be employed for reverse osmosis, ultrafiltration, microfiltration, and gas separation applications [47–50]. For this reason, cellulose acetate polymer based studies have been conducting for different applications. Cellulose acetate blended with polyvinylpyrrolidone (PVP) asymmetric membranes were prepared and evaluated for their performance. Membranes with different polyvinylpyrrolidone concentration and coagulation-bath-temperature were prepared by a group of researchers [51]. They have found that the addition of 0 to 3 wt.% of PVP to the polymer solution increased the macro-voids development and consequently the pure water flux was high. Nevertheless, due to further addition of PVP to 6 wt.% the macro-void formation was suppressed, and pure water flux (PWF) was reduced. Increasing the coagulation-bath-temperature up to 25 °C resulted in increasing macro-voids formation. Another study was conducted to evaluate the effect of molecular weight of polyethylene glycol on membrane morphology and transport properties. In this study, the authors [52] prepared flat sheet polysulfone membranes from casting solutions by using NMP and DMAc solvents. Polyethylene glycol (PEG) of average molecular weights of 400 Da, 6000 Da, and 20000 Da were used as an additive. They have confirmed that PEG 6000 can be an appropriate additive for preparing asymmetric membranes with dense skin layer and a relatively macro-void free sponge-type support layer. The PWF and hydraulic permeability (P_m) were enhanced significantly as the PEG molecular weight was increased. A substantial increase in rejection of bovine serum albumin (BSA) solution was also reported due to increasing the molecular weight of PEG from 400 Da to 6000 Da in the membrane casting solutions. Kim et al. [53] have investigated the effect of PEG additive as a pore-former on polysulfone/N-methyl-2-pyrrolidone (NMP) membrane formation using phase inversion process. Characterizations like light transmittance, coagulation value, and viscosity were used to study the kinetic and thermodynamic characteristics of membrane preparation method. The coagulation value was decreased as the PEG additive molecular weight (M_w) was increased. This result was explained due to the occurrence of the thermodynamically less stable casting solution. Furthermore, the relationship between the observed precipitation rate and thermodynamic data was analyzed using the viscosity of the casting solution. It was also observed that the membrane surface pore size becomes larger and the top layer appears more porous, and the distance from the top surface to the starting point of macro-void development becomes greater. In the case of using 15 wt.% of the polysufone casting solution, symmetric cross-sections and macro void free structures were found. It was reported that with increasing the PEG additive M_w, water flux increased and solute rejections were observed to decrease. It is well known that additives are important to improve membrane properties such as pore structure, mechanical stability, thermal stability, hydrophilicity, flux, and rejection capabilities. According to the literature, the most common additives are polyethylene glycol (PEG) [52–55] and polyvinylpyrrolidone (PVP) [51,56–58].

Using organic-inorganic membranes, the separation properties of polymeric membranes can be enhanced and may possess properties such as selectivity, good permeability, thermal, and chemical stability and mechanical strength [59]. Another challenge in the field of membrane process is fouling. Since membrane fouling is causing a severe decline of the solvent flux, it becomes essential to fabricate membranes less susceptible to fouling by making some modification during preparation. In ultrafiltration processes, several attempts have been accomplished to decrease fouling, which in general include feed solution pretreatment, membrane surface enhancements, and process modifications [60,61]. In recent research, it is confirmed that the introduction of nanoparticles in a membrane matrix develops the membrane hydrophilicity, anti-fouling property, and permeability. Accordingly, several inorganic oxide nanoparticles such as Al_2O_3, ZnO, TiO_2, and SiO_2 have been added with in polymer casting solution [62–64]. Current researchers have paid attention to TiO_2 due to its stable nature, ease of availability, and the potential for different applications. Moreover, TiO_2 can enhance the hydrophilicity of different polymers to improve flux and decrease the fouling problem, which are important parameters in water and wastewater treatment [61,65]. In membrane filtration process factors such as thermal, fouling, and flux Z are important properties, and our current study is focused on improving these properties simultaneously.

1.6 Modification of Membranes for Specific Application

Ma et al. [66] have fabricated nonwoven polyethylene terephthalate nanofiber mats (PET NFM) by electrospinning technique and that were surface modified to mimic the fibrous proteins in native extracellular matrix toward constructing a biocompatible surface for endothelial cells (ECs). Therefore, first, the electrospun PET/NFM was treated in formaldehyde solution to yield hydroxyl groups on its surface, followed by grafting polymerization of methacrylic acid (MAA) initiated by Ce(IV). Finally, the PMAA-grafted PET/NFM was again grafted with gelatin using water-soluble carbodiimide as coupling agent and ECs were cultured on the original and gelatin-modified PET/NFM, and the cell morphology, proliferation, and viability were studied. It was concluded that the gelatin grafting method can obviously improve the spreading and proliferation of the ECs on the PET/NFM, and moreover, can preserve the EC's phenotype. Zhang et al. [33] have described a facile approach for the surface modification of polypropylene nonwoven fabric (NWF) by polyvinyl alcohol (PVA) to determine its filterability where the NWF surface modification involved the physical adsorption of PVA to immobilize PVA on the NWF surface. The protein fouling property of the modified NWF with PVA was studied and results showed that after PVA modification, the polar groups such as C–O and C–O–C were introduced to the NWF surface; hydrophilicity was improved; and water static contact angles were decreased from 86 ± 1° to 43 ± 3°; the amount of bovine serum albumin (BSA) static adsorption on modified NWF was decreased by 83.4%. The filterability nature of modified NWF was also tested by using membrane bioreactor for the treatment of a pharmaceutical wastewater and results revealed that flux declination of modified NWF was only 12% in comparison with the original NWF of 40%. The authors concluded that the anti-fouling property for the modified NWF was enhanced greatly. Saeed et al. [67] have prepared electrospun polyacrylonitrile (PAN) nanofibers for metal adsorption on which the nitrile group in the PAN had chemically modified with amidoxime groups (hydroxylamine hydrochloride). It was confirmed that adsorption of the amidoxime-modified PAN (PAN-oxime) (25% conversion) nanofibers followed Langmuir isotherm and adsorption capacities for Cu(II) and Pb(II) was found to be 52.70 and 263.45 mg/g, respectively, indicating that the monolayer adsorption occurred on the nanofiber mats. Moreover, over 90% of metals were recovered from the metal-loaded PAN-oxime nanofibers in a 1 mol/L HNO_3 solution after 1 h showing that the potential use of the PAN-oxime nanofibers as a filter for recycling metals from wastewater. Zhang et al. [68] have prepared electrospun polyacrylonitrile (PAN) nanofiber membranes and surface modified. This membrane was studied as a novel affinity membrane, using bromelain as a research model. A series of modification were completed where initially, chitosan (CS) was tethered onto the electrospun membrane surface to form a dual-layer biomimetic membrane through the use of glutaraldehyde (GA). Second, the dye cibacronblue F3GA (CB) as a ligand was then covalently immobilized on the CS-coated membranes where the content of CB attached onto the membrane was 370 μmol/g. The adsorption isotherm fitted to the Freundlich model well and the CB-attached PAN nanofibers membrane showed a capturing capacity of 161.6 mg/g toward bromelain. Finally, it was demonstrated that the modified electrospun PAN nanofiber membrane has potential for affinity membrane applications in wastewater treatment system. Shi et al. [69] have developed a facile graft polymerization of methacrylic acid (MAA) onto polyethersulfone (PES) flakes using benzoyl peroxide (BPO) as chemical initiator. The polymerization was carried out in a heterogeneous polymer–monomer reaction system using water as reaction medium where the crucial parameters affecting the graft yield, such as monomer concentration, reaction time, and temperature, were studied. It was confirmed that the amount of grafted PMAA on PES was increased with increasing monomer concentration and reaction time. The optimum temperature for the grafting was around 75 °C. The synthesized PMAA-g-PES was cast into ultrafiltration membranes via phase inversion process. Finally, the results showed that the grafting of PMAA onto PES could not only increase the hydrophilicity where it occurred during the coagulation step of PES membranes but also give the membranes distinct pH sensitivity. Due to the pH-induced conformational change of the PMAA chains, the permeability of the membranes displayed pronounced pH dependence, which was large under low pH values and small under high pH values. Teng et al. [70] have reported the preparation of mesoporous polyvinyl alcohol (PVA)/SiO_2 composite nanofiber membrane functionalized with cyclodextrin groups by sol–gel/electro-spinning process. The mesoporous PVA/SiO_2 composite nanofiber membranes showed good performance in adsorption of indigo carmine dye in which the maximum adsorption capacity reached 495 mgg^{-1} where the adsorption equilibrium was obtained in less than 40 min. Furthermore, it was confirmed that the membranes exhibited good recycling properties for practical usage. Taha et al. [71] have successfully prepared a novel NH_2-functionalized cellulose acetate (CA)/silica composite nanofibrous membranes by sole-gel combined with electrospinning technology. They have employed tetraethoxysilane (TEOS) as a silica source, CA as precursor, and 3-ureidopropyltriethoxysilane as a coupling agent in membrane preparation. The results have confirmed that the composite nanofibrous membranes showed high specific surface area and porosity that were used for Cr (VI) ion removal from aqueous solution through static and dynamic adsorption experiments. The adsorption behavior of Cr (VI) was well described by the Langmuir adsorption model and the maximum adsorption capacity for Cr (VI) was estimated to be 19.46 mg/g where for CA and CA/SiO_2 was found to be 1 mg/g and 3 mg/g, respectively, under a pH of 1.0, and membrane can be conveniently regenerated by alkalization. Abbasizadeh et al. [72] have synthesized a novel polyvinyl alcohol (PVA)/titanium oxide (TiO_2) nanofiber adsorbent modified with mercapto groups by electrospinning. They have studied the influence of several variables such as TiO_2 and mercapto contents, adsorbent dose, pH, contact time, initial concentration of U (VI) and Th (IV) ions, and temperature in batch experiments. From the adsorption experiments they have presented that the sorption capacities of both metal ions for the modified PVA/TiO_2 nanofibers were remarkably greater than those of the unmodified nanofibers, and maximum sorption capacities of U (VI) and Th(IV) by Langmuir isotherm were estimated to be 196.1 and 238.1 (mg/g) at 45 °C with pH of 4.5 and 5.0, respectively. Yu et al. [73] have reported a multi-amino cellulose acetate adsorbent for arsenic adsorption. First cotton cellulose was treated with 100 mL of aqueous NaOH solution (20%, w/v) at 25 °C for 6 h to obtain alkali cellulose, and glycidyl methacrylate (GMA) was grafted onto the surface of cotton cellulose using ceric ammonium nitrate (CAN) as the chemical initiator. Then, the introduced epoxy groups reacted with tetraethylenepentamine (TEPA) to obtain a multi-amino adsorbent. Then, the adsorbent was investigated for adsorption of arsenic, and the results showed that Langmuir model could fit the experimental data perfectly and the adsorption capacities were 5.71 mg/g for As (III) and 75.13 mg/g for As (V), respectively. The adsorbent could be effectively regenerated for four cycles with 0.1 mol/L NaOH solution. Cellulose acetate is mostly abundant organic material and broadly applicable for the synthesis of different products because of having tough, biocompatible, hydrophilicity characteristics, and moderately less expensive [47–49]. The drawback of cellulose acetate membranes is that it is susceptible to thermal and mechanical stabilities depending on the environments and conditions of application [74]. Therefore, modification of cellulose acetate using graft copolymerization process gives a substantial way to modify the chemical and physical properties [75]. Recently, the modifying of polymers has received great attention, and grafting is one of the promising approaches. Grafting co-polymerization is an attractive technique to introduce different functional groups to the backbone of a polymer [76]. Moreover, grafting is one of the chemical modification methods advanced to date and has appeared as a simple and versatile method to develop surface properties of polymers for a various applications. Grafting method has numerous benefits such as the capacity of modifying the polymer surfaces to have a distinctive properties by choosing of various monomers; the ease and well-regulated introduction of grafting chains with a high density and an exact localization on the substrate surface without changing the bulk properties; and graft chains onto a substrate surface with covalent attachments promising a long-term chemical stability of introduced chains, in contrast to physically coated polymer chains. For the graft copolymerization to be occurring, groups that can generate radicals should be introduced onto the backbone of the polymer first to activate the polymer surface. Most of the polymers having chemically inert properties can be activated via UV irradiation [77], plasma treatment [78], ion and electron beam [79–82], benzoic per oxide oxidization [83–88], and Ce^{4+} oxidization [89–92]. In most of the grafting processes, the free radicals are initiated on the backbone of a polymer by numerous irradiations and free radical polymerizations of vinyl monomers or chemical initiators.

1.7 Economic Analysis of Membranes

While membrane technologies are attractive, these may not be appropriate for all separation processes. Further, they need to be economical. There are plenty of examples where membrane-based separation techniques are economical. The economic benefits of ion transport membrane (ITM) Oxygen technology for the Integrated Gasification Combined Cycle (IGCC) application are a 2.9% improvement in thermal efficiency and a 6.5% decrease in the cost of generated electric power. The efficiency increase also produces a simultaneous reduction in carbon dioxide and sulfur emissions. Therefore, integration of ITM Oxygen technology with IGCC offers the benefits of further improving system efficiency, resulting in better environmental performance and lower costs [93]. Lee et al. have shown the techno-economic feasibility of one-stage membrane-based propylene/propane separation, which produces 99.6% purity of propylene with 97% recovery. They have claimed that with high stage cut, membrane separation has higher economic potential than the distillation [94]. The membrane processes used to control the quality of products for several biological and chemical industries are successful as these are effectively and economically implemented at the large stage. Integrated membrane separation technology also shows simplicity of the units and possibility of advanced levels of autoimmunization. Membrane hybrid technologies work best when developed as one concept. Among the impending applications of cross-flow microfiltration, the clarification of rough beer represents a huge potential market (approximately 200,000 m^2 membrane areas). The wastewaters generated in fish meal production comprise a large quantity of potentially valuable proteins. A suitable treatment for fish meal effluents consisting of a microfiltration pretreatment, followed by ultrafiltration, produced a permeate flux of 7.78×10^{-6} $m^3/m^2.s$ and 62% proteins rejection with a volume reduction factor of 2.3 [95].

1.8 Summary

This chapter presents a platform to provide valuable information on the background knowledge of this work for better understanding of the types of membranes, membrane preparation techniques, functionalization, characterization, and applications. Membrane technology presents an attractive character comprised of easy manufacturing, modification, operation, and compactness. Several membrane preparation techniques are assessed and compared in this chapter in detail. Moreover, the functionalization using physical blending and grafting methods for the modification of polymer and/or membrane are discussed. Finally, the possible applications of the membrane processes have been extensively discussed for water and wastewater treatments. A detailed literature review about the preparation, modification, and application of the polymeric membranes was included. A particular emphasis was given to the phase inversion and electrospinning techniques of membrane preparations. The functionalization and/or grafting effects on the resultant membrane structures were assessed in terms of their specific applications. Moreover, the possible scopes for further enhancement on the preparation and modifications of the membranes are also identified. Furthermore, the specific objectives on the preparation, characterization, modification and application of the polymeric membranes have been discussed extensively.

References

1. Service, R.F., Desalination Freshens Up, in *Science*. 2006. pp. 1088–1090.
2. Huang, W.L., Welch, E.W., and Corley, E.A., Public sector voluntary initiatives: The adoption of the environmental management system by public waste water treatment facilities in the United States. *Journal of Environmental Planning and Management*, 2013. 57(10): pp. 1531–1551.
3. Aliabadi, M., Irani, M., Ismaeili, J., Piri, H., and Parnian, M.J., Electrospun nanofiber membrane of PEO/Chitosan for the adsorption of nickel, cadmium, lead and copper ions from aqueous solution. *Chemical Engineering Journal*, 2013. 220: pp. 237–243.
4. Wang, R., Liu, Y., Li, B., Hsiao, B.S., and Chu, B., Electrospun nanofibrous membranes for high flux microfiltration. *Journal of Membrane Science*, 2012. 392–393: pp. 167–174.
5. US EPA's Safe Drinking Water Web site: http://www.epa.gov/safewater; USEPA's Safe Drinking Water Hotline: (800) 426-4791; USEPA 816-F-09-004 May 2009.
6. Porter, M.C., ed. *Handbook of Industrial Membrane Technology*. Noyes Publications: California, 1990
7. Lalia, B.S., Kochkodan, V., Hashaikeh, R. and Hilal, N., A review on membrane fabrication: Structure, properties and performance relationship. *Desalination*, 2013. 326: pp. 77–95.
8. Ramakrishna, S., Teo, K.F.W.E., Lim, T.C., and Ma, Z., *An Introduction to Electrospinning and Nanofibers*. World Scientific Publishing Co. Pte. Ltd., 2005.
9. Koski, A., Yim, K., and Shivkumar, S., Effect of molecular weight on fibrous PVA produced by electrospinning. *Materials Letters*, 2004. 58(3–4): pp. 493–497.
10. Zadeh, M.M.A., Keyanpour-Rad, M., and Ebadzadeh, T., Effect of viscosity of polyvinyl alcohol solution on morphology of the electrospun mullite nanofibres. *Ceramics International*, 2014. 40(4): pp. 5461–5466.
11. Fennessey, S.F., and Farris, R.J., Fabrication of aligned and molecularly oriented electrospun polyacrylonitrile nanofibers and the mechanical behavior of their twisted yarns. *Polymer*, 2004. 45(12): pp. 4217–4225.
12. Barhate, R.S., Loong, C.K., and Ramakrishna, S., Preparation and characterization of nanofibrous filtering media. *Journal of Membrane Science*, 2006. 283(1–2): pp. 209–218.
13. Singh, G., Rana, D., Matsuura, T., Ramakrishna, S., Narbaitz, R.M., and Tabe, S., Removal of disinfection byproducts from water by carbonized electrospun nanofibrous membranes. *Separation and Purification Technology*, 2010. 74: pp. 202–212.
14. Bhattacharyya, D., Schäfer, T., Wickramasinghe, S.R., and Daunert, S., ed., *Responsive Membranes and Materials*. John Wiley & Sons, Ltd.: New Delhi, 2013
15. Wang, D.A., Ji, J., and Feng, L.X., Surface analysis of poly(ether urethane) blending stearyl poly(ethyleneoxide) coupling polymer. *Macromolecules*, 2000. 33(22): pp. 8472–8478.
16. Zhang, S., Yan, L., Altman, M., Lässle, M., Nugent, H., Frankel, F., Lauffenburger, D.A., Whitesides, G.M., and Rich, A., Biological surface engineering: A simple system for cell pattern formation. *Biomaterials*, 1999. 20(13): pp. 1213–1220.
17. Zhu, Y., Gao, C., He, T., Liu, X., and Shen, J., Layer-by-layer assembly to modify poly (L-lactic acid) surface toward improving its cytocompatibility to human endothelial cells, *Biomacromolecules*, 2003. 4(2): pp. 446–452.

18. Mit-uppatham, C., Nithitanakul, M., and Supaphol, P., Ultrafine electrospun polyamide-6 fibers: Effect of solution conditions on morphology and average fiber diameter. *Macromolecular Chemistry and Physics*, 2004. 205(17): pp. 2327–2338.
19. Deitzel, J.M., Kleinmeyer, J.D., Hirvonen, J.K., and Tan, N.B., Controlled deposition of electrospun poly(ethylene oxide) fibers. *Polymer*, 2001. 42: pp. 8163–8170.
20. Krishnappa, R.V.N., Desai, K., and Sung, C., Morphological study of electrospun polycarbonates as a function of the solvent and processing voltage. *Journal of Materials Science*, 2013. 38: pp. 2357–2365.
21. Choi, S.S., Lee, Y.S., Joo, C.W., Lee, S.G., Park, J.K., and Han, K.S., Electrospun PVDF nanofiber web as polymer electrolyte or separator. *Electrochimica Acta*, 2004. 50(2–3): pp. 339–343.
22. Yang, E., Qin, X., and Wang, S., Electrospun crosslinked polyvinyl alcohol membrane. *Materials Letters*, 2008. 62(20): pp. 3555–3557.
23. Tsioptsias, C., Sakellariou, K.G., Tsivintzelis, I., Papadopoulou, L., and Panayiotou, C., Preparation and characterization of cellulose acetate–Fe_2O_3 composite nanofibrous materials. *Carbohydrate Polymers*, 2010. 81(4): pp. 925–930.
24. Tsai, R.Y., Hung, S.C., Lai, J.Y., Wang, D.M., and Hsieh, H.J., Electrospun chitosan–gelatin–polyvinyl alcohol hybrid nanofibrous mats: Production and characterization. *Journal of the Taiwan Institute of Chemical Engineers*, 2014. 45(4): pp. 1975–1981.
25. Kimura, N., Sakumoto, T., Mori, Y., Wei, K., Kim, B.S., Song, K.H., and Kim, I.S., Fabrication and characterization of reinforced electrospun poly(vinylidene fluoride-co-hexafluoropropylene) nanofiber membranes. *Composites Science and Technology*, 2014. 92: pp. 120–125.
26. Sueyoshi, Y., Fukushima, C., and Yoshikawa, M., Molecularly imprinted nanofiber membranes from cellulose acetate aimed for chiral separation. *Journal of Membrane Science*, 2010. 357(1–2): pp. 90–97.
27. Feng, C., Khulbe, K.C., Matsuura, T., Tabe, S. and Ismail, A.F., Preparation and characterization of electro-spun nanofiber membranes and their possible applications in water treatment. *Separation and Purification Technology*, 2013. 102: pp. 118–135.
28. Qu, X., Alvarez, P.J., and Li, Q., Applications of nanotechnology in water and wastewater treatment. *Water Research*, 2013. 47(12): pp. 3931–3946.
29. Gopal, R., Kaur, S., Ma, Z., Chan, C., Ramakrishna, S., and Matsuura, T., Electrospun nanofibrous filtration membrane. *Journal of Membrane Science*, 2006. 281(1–2): pp. 581–586.
30. Wang, X., Fang, D., Yoon, K., Hsiao, B.S., and Chu, B., High performance ultrafiltration composite membranes based on poly(vinyl alcohol) hydrogel coating on crosslinked nanofibrous poly(vinyl alcohol) scaffold. *Journal of Membrane Science*, 2006. 278(1–2): pp. 261–268.
31. Yoon, K., Kim, K., Wang, X., Fang, D., Hsiao, B.S., and Chu, B., High flux ultrafiltration membranes based on electrospun nanofibrous PAN scaffolds and chitosan coating. *Polymer*, 2006. 47(7): pp. 2434–2441.
32. Kim, S.J., Nam, Y.S., Park, H.S., and Park, W.H., Preparation and characterization of antimicrobial polycarbonate nanofibrous membrane. *European Polymer Journal*, 2007. 43(8): pp. 3146–3152.

33. Zhang, C.H., Yang, F.L., Wang, W.J., and Chen, B., Preparation and characterization of hydrophilic modification of polypropylene non-woven fabric by dip-coating PVA (polyvinyl alcohol). *Separation and Purification Technology*, 2008. 61(3): pp. 276–286.
34. Aussawasathien, D., Teerawattananon, C., and Vongachariya, A., Separation of micron to sub-micron particles from water: Electrospun nylon-6 nanofibrous membranes as pre-filters. *Journal of Membrane Science*, 2008. 315(1–2): pp. 11–19.
35. Yoon, K., Hsiao, B.S., and Chu, B., High flux ultrafiltration nanofibrous membranes based on polyacrylonitrile electrospun scaffolds and crosslinked polyvinyl alcohol coating. *Journal of Membrane Science*, 2009. 338(1–2): pp. 145–152.
36. Xu, G.R., Wang, J.N., and Li, C.J., Preparation of hierarchically nanofibrous membrane and its high adaptability in hexavalent chromium removal from water. *Chemical Engineering Journal*, 2012. 198–199: pp. 310–317.
37. Mahapatra, A., Mishra, B.G., and Hota, G., Electrospun Fe_2O_3–Al_2O_3 nanocomposite fibers as efficient adsorbent for removal of heavy metal ions from aqueous solution. *Journal of Hazardous Materials*, 2013. 258–259: pp. 116–123.
38. Li, M., Xue, X., Wang, D., Lu, Y., Wu, Z., and Zou, H., High performance filtration nanofibrous membranes based on hydrophilic poly(vinyl alcohol-co-ethylene) copolymer. *Desalination*, 2013. 329: pp. 50–56.
39. You, H., Li, X., Yang, Y., Wang, B., Li, Z., Wang, X., Zhu, M., and Hsiao, B.S., High flux low pressure thin film nanocomposite ultrafiltration membranes based on nanofibrous substrates. *Separation and Purification Technology*, 2013. 108: pp. 143–151.
40. Liu, Y., Wang, R., Ma, H., Hsiao, B.S., and Chu, B., High-flux microfiltration filters based on electrospun polyvinylalcohol nanofibrous membranes. *Polymer*, 2013. 54(2): pp. 548–556.
41. Mulder, M., *Basic Principles of Membrane Technology*. Second Ed. London: Kluwer Acadamic Publishers, 1996
42. Young, T.H., and Chen, L.W., Pore formation mechanism of membranes from phase inversion process. *Desalination*, 1995. 103: pp. 233–247.
43. Mohammadi, T., and Saljoughi, E., Effect of production conditions on morphology and permeability of asymmetric cellulose acetate membranes. *Desalination*, 2009. 243: pp. 1–7.
44. Strathmann, H., and Kock, K., The formation mechanism of phase inversion membranes. *Desalination*, 1977. 21: pp. 241–255.
45. Chakrabarty, B., Ghoshal, A.K., and Purkait, M.K., Preparation, characterization and performance studies of polysulfone membranes using PVP as an additive. *Journal of Membrane Science*, 2008. 315(1–2): pp. 36–47.
46. Zhu, Z., Xiao, J., He, W., Wang, T., Wei, Z., and Dong, Y., A phase-inversion casting process for preparation of tubular porous alumina ceramic membranes. *Journal of the European Ceramic Society*, 2015. 35(11): pp. 3187–3194.
47. Algarra, M., Vázquez, M.I., Alonso, B., Casado, C.M., Casado, J., and Benavente, J., Characterization of an engineered cellulose based membrane by thiol dendrimer for heavy metals removal. *Chemical Engineering Journal*, 2014. 253: pp. 472–477.
48. Konwarh, R., Karak, N., and Misra, M., Electrospun cellulose acetate nanofibers: The present status and gamut of biotechnological applications. *Biotechnology Advances*, 2013. 31(4): pp. 421–437.

49. Abedini, R., Mousavi, S.M., and Aminzadeh, R., A novel cellulose acetate (CA) membrane using TiO_2 nanoparticles: Preparation, characterization and permeation study. *Desalination*, 2011. 277(1–3): pp. 40–45.
50. Das, C., and Gebru, K.A., Cellulose acetate modified titanium dioxide (TiO_2) nanoparticles electrospun composite membranes: Fabrication and characterization, *Journal of The Institution of Engineers (India)*: Series E, 2017. 98(2): pp. 91–101.
51. Saljoughi, E., and Mohammadi, T., Cellulose acetate (CA)/polyvinylpyrrolidone (PVP) blend asymmetric membranes: Preparation, morphology and performance. *Desalination*, 2009. 249(2): pp. 850–854.
52. Chakrabarty, B., Ghoshal, A.K., and Purkait, M.K., Effect of molecular weight of PEG on membrane morphology and transport properties. *Journal of Membrane Science*, 2008. 309(1–2): pp. 209–221.
53. Kim, J.H., and Lee, K.H., Effect of PEG additive on membrane formation by phase inversion. *Journal of Membrane Science*, 1998. 138: pp. 153–163.
54. Saljoughi, E., Amirilargani, M., and Mohammadi, T., Effect of PEG additive and coagulation bath temperature on the morphology, permeability and thermal/chemical stability of asymmetric CA membranes. *Desalination*, 2010. 262(1–3): pp. 72–78.
55. Wongchitphimon, S., Wang, R., Jiraratananon, R., Shi, L., and Loh, C.H., Effect of polyethylene glycol (PEG) as an additive on the fabrication of polyvinylidene fluoride-co-hexafluropropylene (PVDF-HFP) asymmetric microporous hollow fiber membranes. *Journal of Membrane Science*, 2011. 369(1–2): pp. 329–338.
56. Yang, M., Xie, S., Li, Q., Wang, Y., Chang, X., Shan, L., Sun, L., Huang, X., and Gao, C., Effects of polyvinylpyrrolidone both as a binder and pore-former on the release of sparingly water-soluble topiramate from ethylcellulose coated pellets. *International Journal of Pharmaceutics*, 2014. 465(1–2): pp. 187–196.
57. Yoo, S.H., Kim, J.H., Jho, J.Y., Won, J., and Kang, Y.S., Influence of the addition of PVP on the morphology of asymmetric polyimide phase inversion membranes: Effect of PVP molecular weight. *Journal of Membrane Science*, 2004. 236(1–2): pp. 203–207.
58. Zhao, S., Wang, Z., Wei, X., Tian, X., Wang, J., Yang, S., and Wang, S., Comparison study of the effect of PVP and PANI nanofibers additives on membrane formation mechanism, structure and performance. *Journal of Membrane Science*, 2011. 385–386: pp. 110–122.
59. Shi, F., Ma, Y., Ma, J., Wang, P., and Sun, W., Preparation and characterization of PVDF/TiO_2 hybrid membranes with different dosage of nano-TiO_2. *Journal of Membrane Science*, 2012. 389: pp. 522–531.
60. Kim, S.H., Kwak, S.Y., Sohn, B.H., and Park, T.H., Design of TiO_2 nanoparticle self-assembled aromatic polyamide thin-film-composite (TFC) membrane as an approach to solve biofouling problem. *Journal of Membrane Science*, 2003. 211: pp. 157–165.
61. Cao, X., Ma, J., Shi, X., and Ren, Z., Effect of TiO_2 nanoparticle size on the performance of PVDF membrane. *Applied Surface Science*, 2006. 253(4): pp. 2003–2010.
62. Maximous, N., Nakhla, G., Wan, W., and Wong, K., Preparation, characterization and performance of Al_2O_3/PES membrane for wastewater filtration. *Journal of Membrane Science*, 2009. 341(1–2): pp. 67–75.
63. Yan, L., Li, Y.S., Xiang, C.B., and Xianda, S., Effect of nano-sized Al_2O_3-particle addition on PVDF ultrafiltration membrane performance. *Journal of Membrane Science*, 2006. 276(1–2): pp. 162–167.

64. Bottino, A., Capannelli, G., and Comite, A., Preparation and characterization of novel porous PVDF-ZrO_2 composite membranes. *Desalination*, 2002. 146: pp. 35–40.
65. Saffaj, N., Younssi, S.A., Albizane, A., Messouadi, A., Bouhria, M., Persin, M., Cretin, M., and Larbot, A., Preparation and characterization of ultrafiltration membranes for toxic removal from wastewater. *Desalination*, 2004. 168: pp. 259–263.
66. Ma, Z., Kotaki, M., Yong, T., He, W., and Ramakrishna, S., Surface engineering of electrospun polyethylene terephthalate (PET) nanofibers towards development of a new material for blood vessel engineering. *Biomaterials*, 2005. 26(15): pp. 2527–2536.
67. Saeed, K., Haider, S., Oh, T.J., and Park, S.Y., Preparation of amidoxime-modified polyacrylonitrile (PAN-oxime) nanofibers and their applications to metal ions adsorption. *Journal of Membrane Science*, 2008. 322(2): pp. 400–405.
68. Zhang, H., Nie, H., Yu, D., Wu, C., Zhang, Y., White, C.J.B., and Zhu, L., Surface modification of electrospun polyacrylonitrile nanofiber towards developing an affinity membrane for bromelain adsorption. *Desalination*, 2010. 256(1–3): pp. 141–147.
69. Shi, Q., Su, Y., Ning, X., Chen, W., Peng, J., and Jiang, Z., Graft polymerization of methacrylic acid onto polyethersulfone for potential pH-responsive membrane materials. *Journal of Membrane Science*, 2010. 347(1–2): pp. 62–68.
70. Teng, M., Li, F., Zhang, B., and Taha, A.A., Electrospun cyclodextrin-functionalized mesoporous polyvinyl alcohol/SiO2 nanofiber membranes as a highly efficient adsorbent for indigo carmine dye. *Colloids and Surfaces A: Physicochemical and Engineering Aspects*, 2011. 385(1–3): pp. 229–234.
71. Taha, A.A., Wu, Y.N., Wang, H., and Li, F., Preparation and application of functionalized cellulose acetate/silica composite nanofibrous membrane via electrospinning for Cr(VI) ion removal from aqueous solution. *Journal of Environmental Management*, 2012. 112: pp. 10–16.
72. Abbasizadeh, S., Keshtkar, A.R., and Mousavian, M.A., Preparation of a novel electrospun polyvinyl alcohol/titanium oxide nanofiber adsorbent modified with mercapto groups for uranium(VI) and thorium(IV) removal from aqueous solution. *Chemical Engineering Journal*, 2013. 220: pp. 161–171.
73. Yu, X., Tong, S., Ge, M., Wu, L., Zuo, J., Cao, C. and Song, W., Synthesis and characterization of multi-amino-functionalized cellulose for arsenic adsorption. Carbohydrate Polymers, 2013. 92(1): pp. 380–387.
74. Arthanareeswaran, G., Thanikaivelan, P., Srinivasn, K., Mohan, D., and Rajendran, M., Synthesis, characterization and thermal studies on cellulose acetate membranes with additive. *European Polymer Journal*, 2004. 40(9): pp. 2153–2159.
75. Shen, D., and Huang, Y., The synthesis of CDA-g-PMMA copolymers through atom transfer radical polymerization. *Polymer*, 2004. 45(21): pp. 7091–7097.
76. Bhattacharya, A., and Misra, B.N., Grafting: A versatile means to modify polymers Techniques, factors and applications. *Progress in Polymer Science*, 2004. 29(8): pp. 767–814.
77. Khayet, M., Seman, M.A., and Hilal, N., Response surface modeling and optimization of composite nanofiltration modified membranes. *Journal of Membrane Science*, 2010. 349(1–2): pp. 113–122.

78. Kim, M.M., Lin, N.H., Lewis, G.T., and Cohen, Y., Surface nano-structuring of reverse osmosis membranes via atmospheric pressure plasma-induced graft polymerization for reduction of mineral scaling propensity. *Journal of Membrane Science*, 2010. 354(1–2): pp. 142–149.
79. Linggawati, A., Mohammad, A.W., and Ghazali, Z., Effect of electron beam irradiation on morphology and sieving characteristics of nylon-66 membranes. *European Polymer Journal*, 2009. 45(10): pp. 2797–2804.
80. Linggawati, A., Mohammad, A.W., and Leo, C.P., Effects of APTEOS content and electron beam irradiation on physical and separation properties of hybrid nylon-66 membranes. *Materials Chemistry and Physics*, 2012. 133(1): pp. 110–117.
81. Mukherjee, P., Jones, K.L., and Abitoye, J.O., Surface modification of nanofiltration membranes by ion implantation. *Journal of Membrane Science*, 2005. 254(1–2): pp. 303–310.
82. Wanichapichart, P., Bootluck, W., Thopan, P., and Yu, L.D., Influence of nitrogen ion implantation on filtration of fluoride and cadmium using polysulfone/chitosan blend membranes. *Nuclear Instruments and Methods in Physics Research Section B: Beam Interactions with Materials and Atoms*, 2014. 326: pp. 195–199.
83. Sacak, M., and Oflaz, F., Benzoyl-peroxide-initiated graft copolymerization of poly (ethylene terephthalate) fibers with acrylic acid. *Journal of Applied Polymer Science*, 1993. 50: pp. 1909–1916.
84. Sacak, M., and Pulat, E., Benzoyl-peroxide-initiated graft copolymerization of poly(ethyleneTerephthalate) fibers with acrylamide. *Journal of Applied Polymer Science*, 1989. 38: pp. 539–546.
85. Pulat, M., and Babayiğit, D., Graft copolymerization of PU membranes with acrylic acid and crotonic acid using benzoyl peroxide initiator. *Journal of Applied Polymer Science*, 2001. 80: pp. 2690–2695.
86. Sacak, M., Sertkaya, F., and Talu, M., Grafting of poly (ethylene terephthalate) fibers with methacrylic acid using benzoyl peroxide. *Journal of Applied Polymer Science*, 1992. 44: pp. 1737–1742.
87. Pulat, M., and Isakoca, C., Chemically induced graft copolymerization of vinyl monomers onto cotton fibers. *Journal of Applied Polymer Science*, 2006. 100(3): pp. 2343–2347.
88. Sanli, O., and Ünal, H.I., Graft copolymerization of 2-hydroxy ethyl methacrylate on dimethyl sulfoxide pretreated poly(ethylene terephthalate) films using benzoyl peroxide. *Journal of Macromolecular Science, Part A*, 2002. 39(5): pp. 447–465.
89. Kulkarni, A.Y., and Mehta, P.C., Ceric ion induced redox polymerization of acrylonitrile in presence of cellulose. *Journal of Applied Polymer Science*, 1965. 9(7): p. 2633.
90. Kulkarni, A.Y., and Mehta, P.C., Ceric ion-induced redox polymerization of acrylonitrile on cellulose. *Journal of Applied Polymer Science*, 1968. 12(6): pp. 1321–1342.
91. Hebeish, A., and Mehta, P.C., Cerium-initiated grafting of acrylonitrile onto cellulosic materials. *Journal of Applied Polymer Science*, 1968. 12: pp. 1625–1647.
92. Joshi, J.M., and Sinha, V.K., Ceric ammonium nitrate induced grafting of polyacrylamide onto carboxymethyl chitosan. *Carbohydrate Polymers*, 2007. 67(3): pp. 427–435.

93. Dyer, P.N., Richards, R.E., Russek, S.L., and Taylor, D.M., Ion transport membrane technology for oxygen separation and syngas production. *Solid State Ionics*, 2000. 134: pp. 21–33.
94. Lee, U., Kim, J., Chae, I.S., and Han, C., Techno-economic feasibility study of membrane based propane/propylene separation process. *Chemical Engineering and Processing: Process Intensification*, 2017. 119: pp. 62–72.
95. Saxena, A., Tripathi, B.P., Kumar, M., and Shahi, V.K., Membrane-based techniques for the separation and purification of proteins: An overview. *Advances in Colloid and Interface Science*, 2009. 145: pp. 1–22.

2

Materials and Characterization Methods

2.1 Materials

2.1.1 Chemicals

Polyvinyl alcohol, MW 80,000, 98–99 mole% hydrolyzed, was purchased from M/s. Loba Chemie Private Limited, India, acetone (99.9% purity), glutaraldehyde (GA) (25% aqueous solution) and hydrochloric acid (35% aqueous solution) were purchased from M/s. Merck Specialities Private Limited, Mumbai, India. First, the PVA solutions were prepared by dissolving the polymer in distilled water at 70 °C with continuous stirring overnight and subsequently cooled to room temperature (28 ± 2 °C) prior to electrospinning process. Cellulose acetate polymer (acetyl: 29 to 45% content, MW = 50,000 g/mol) was purchased from M/s. Loba Chemie Private Limited, India. The degree of substitution for acetyl groups, DS (Ac), is usually defined as the number of acetyl groups per glycosidic ring and according to the ^{1}H NMR analysis, it is found to be 2.47. The two solvents, namely, N, N-dimethyl acetamide (DMAc) and acetone (both with an analytical purity of 99%), deuterated dimethyl sulfoxide (DMSO-d6, deuteration degree min. 99.8% for NMR spectroscopy) were obtained from M/s. Sigma-Aldrich Co., USA. Deionized water (DI) was used throughout this experiment. Titanium oxide (TiO_2) nanoparticles with 99.5% purity were obtained from M/s. Sigma-Aldrich Co., USA. Copper chloride ($CuCl_2.2H_2O$) and lead nitrate (Pb $(NO_3)_2$) were used as the source for copper (II) and lead (II) stock solution, respectively. All the required solutions were prepared with analytical reagents and deionized water; 2.7 g of 99% $CuCl_2.2H_2O$ and 1.6 g of 99% Pb $(NO_3)_2$ (M/s. Merck Specialities Private Limited, Mumbai, India) was dissolved in deionized water of 1.0 L volumetric flask up to the mark to obtain 1000 mg/L of Cu (II) and Pb (II) stock solution. Synthetic samples of 50 mg/L of Cu (II) and Pb (II) were prepared from this stock solution by appropriate dilutions. The pH values of the prepared solutions were adjusted using 1 (N) of HCl and NaOH. Deionized water (DI) was used throughout this experimentation to avoid side reactions and contamination of the solution. Polyethylene glycol 4000 and ethylenediamine (EDA) (Mw. 60 g/mol and with purity of 99%) were obtained from M/s. Merck Specialities Private Limited, Mumbai, India; tetra ethylene pentamine (TEPA) (Mw. 189.3 g/mol and with purity of 99%) and hexamethylenetetramine (HMTA) (Mw. 140.2 g/mol and with purity of 99%) were obtained from M/s. Sigma-Aldrich Co., USA. Polyvinyl pyrrolidone (PVP, M_w of 40,000 Da) polymers were purchased from M/s. Loba Chemie Private Limited, India. Deionized water (DI) was used as the non-solvent in the coagulation bath throughout this experiment, which was purified using Millipore system. Bovine serum albumin (BSA) protein having a molecular weight of 66 kDa was purchased from M/s. Sisco Research Laboratories Private Limited, India. Potassium chromate (K_2CrO_4) is used as the source for Cr (VI) (M/s. Merck Specialities Private Limited, Mumbai, India). Methyl methacrylate (MMA, 99% purity and MW = 86.09) was purchased from M/s. Chemika Biochemika Reagents, Mumbai, India. Ceric (IV) sulfate (CS), hydroquinone (MW = 110.11 g/mol with 99.5% purity), and sulfuric acid (98% purity) were obtained from M/s. Merck, Germany. Methanol (with 99.8% purity HPLC grade) was purchased from M/s. Sisco Research Laboratories Private Limited, India. Ceric (IV) sulfate and sulfuric acid were analytical reagent grade and used without further purification. Humic acid (60–70%, dry basis) was purchased from M/s. Loba Chemie Private Limited, India.

2.2 Characterization Methods

2.2.1 Physicochemical Characterization

2.2.1.1 Pure Water Flux and Hydraulic Permeability

In this study, a 400 mL stirred batch cell experimental setup was used for the permeability experiments, which is presented schematically in Figure 2.1. This experimental setup included (1) a pressure source to supply the pressure required for the filtration experiment; (2) feed tank where the feed is collected; (3) filtration cell; (4) membrane piece; and (5) permeate collector after membrane filtration. Membranes with the circular shape of 7×10^{-2} m diameter and with an effective filtration area of 3.848×10^{-3} m^2 were employed for this experiment. Compaction studies of each of the prepared membranes were done using deionized water for 2 h using a fixed pressure of 300 kPa, and PWFs were recorded with 10 min interval. The membrane compaction factors (*CF*) were found by calculating the ratio of initial PWF (J_{wi}) to steady state PWF (J_{wf}). PWF results were calculated using the following equation:

$$J_W = \left[\frac{Q}{(A \times \Delta t)} \right] \tag{2.1}$$

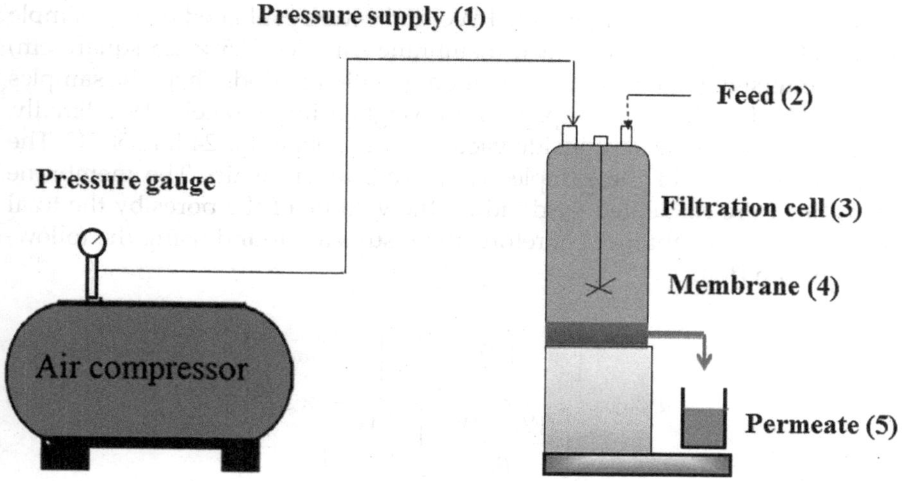

FIGURE 2.1
Schematic diagram of the batch cell ultrafiltration setup.

where J_w is PWF (L/m^2 h), Q is the volume of water permeated (L), A is the effective membrane area (m^2), and Δt is permeation time (h). Thus, PWFs were evaluated by passing deionized (DI) water through the membranes. PWF values at different operating pressures (ΔP) (ranging from 100 to 300 kPa) were recorded. The hydraulic permeability (P_m) (L/m^{2}h kPa) of the membranes was calculated from the inverse slope of J_w versus ΔP plot. The calculations for the hydraulic permeability were completed as follows:

$$P_m = \left(\frac{J_W}{\Delta P} \right) \tag{2.2}$$

The resistances (R_m) of the prepared membranes were calculated according to Darcy's law (Equation 2.3):

$$R_m = \left[\frac{\Delta P}{(\mu \times J_W)} \right] \tag{2.3}$$

where μ is the water viscosity (8.9×10^{-4}) (Pa. s).

2.2.1.2 Membrane Porosity Measurements

To study the membranes porosity results, two parameters were applied, namely, equilibrium water content (EWC) and membrane porosity (ε). The ε plays a significant role in the permeability and rejection nature of the membrane. On the other hand, the EWC and ε were evaluated using a simple gravimetric method, where each membrane samples (2.5 × 2.5 square cm) were immersed in DI water beakers for a specified period. Then, the samples were dabbed with dry filter paper and weighed immediately (W_w). Finally, the membranes were kept inside vacuum atmosphere for 24 h at 50 °C. The final dry weights of the samples (W_D) were taken again. The membrane porosities were calculated by dividing the volume of the pores by the total volume of the membrane. Therefore, the results are found using the following equation [1,2],

$$\varepsilon(\%) = \left\{ \frac{\left[\dfrac{W_W - W_D}{\rho_W}\right]}{\left[\dfrac{(W_W - W_D)}{\rho_W}\right] + \left(\dfrac{W_D}{\rho_P}\right)} \times 100 \right\} \tag{2.4}$$

where W_W is the wet membrane weight (g), W_D is the dry membrane weight (g), ρ_W is the pure water density at working condition (g cm^{-3}), and ρ_p is the polymer density (g cm^{-3}).

EWC results were calculated using the subsequent equation [3,4]:

$$EWC(\%) = \left\{ \left[\frac{(W_W - W_D)}{W_W}\right] \times 100 \right\} \tag{2.5}$$

The average values of EWC and membrane porosity were taken after measuring five different samples to reduce the error of the balancing measurement.

2.2.1.3 Membrane Fouling and Rejection Experiments

Membrane fouling experimentations were done using a stirred batch cell specified before to investigate the effect of PEG and PVP additives and effect of the two solvents on BSA rejection and permeate fluxes of the fabricated membranes. The BSA solution (1 g L^{-1}) was prepared using deionized water as a solvent. The pH of the BSA solutions was kept at 7.9. The flux values were calculated using Equation 2.3 and the % rejection of BSA solutions was evaluated by the next equation:

$$R_{BSA}(\%) = \left\{ \left[1 - \left(\frac{C_p}{C_f}\right)\right] \times 100 \right\} \tag{2.6}$$

where C_f and C_p are the BSA concentrations (mg L^{-1}) on the feed side and the permeate side, respectively. The ultrafiltration experiments for BSA solution was achieved through all prepared membranes at an operating pressure of 150 kPa. After that, the permeates were collected using a measuring container at a constant interval of time (20 min) to calculate flux values, and the length of each filtration test was for 2 h. Finally, the permeate concentration of BSA solution was evaluated by using UV–vis spectrophotometer (Make: M/s. Thermo Fisher Scientific, India; Model: UV 2300) at a wavelength of 278 nm. After filtration of BSA solutions, fouled membranes were soaked in DI water and rinsed for 30 min, and the water flux of the membranes was measured. These experimentations were repeated three times, and the graphs including error bars are presented. To investigate the anti-fouling properties of the prepared membranes, the flux losses due to irreversible fouling (F_{ir}), reversible fouling (F_r), and total fouling (F_t) of all the experiments were calculated using the following equations, respectively [5,6].

$$F_{ir} = \left[\frac{(J_{Wi} - J_{Wf})}{J_{Wi}}\right] \tag{2.7}$$

$$F_r = \left[\frac{(J_{Wf} - J_B)}{J_{Wi}}\right] \tag{2.8}$$

$$F_t = \left[1 - \left(\frac{J_B}{J_{Wi}}\right)\right] \tag{2.9}$$

where the fouling resistant capacity of the prepared membranes was evaluated using normalized flux ratio (NFR) as shown in Equation 2.10.

$$NFR(\%) = \left[\left(\frac{J_{Wf}}{J_{Wi}}\right) \times 100\right] \tag{2.10}$$

where J_{Wf} is the flux of the membrane after fouling (2 h) and J_{Wi} is the flux of the membrane found at the start of the fouling stage. Normally, greater NFR value (next to 1) indicates better anti-fouling nature of the membranes.

2.2.1.4 Average Pore Radius Determination

Image J software and water filtration velocity method were employed to determine the average pore size of the membranes, and the results were calculated at constant operating pressure (250 kPa). Water filtration velocity method was employed to determine the average pore size of the membranes, and the results were calculated at constant operating pressure (300 kPa).

Membrane average pore radius (r_m) is considered as an approximation of true pore size, and it denotes the average pore radius throughout the membrane's thickness (ζ). Average pore radius can be estimated by using the Guerout–Elford–Ferry equation [2,7],

$$r_m = \sqrt{\left[\frac{(2.9 - 1.75 \times \varepsilon)(8 \times \mu \times \zeta \times Q_W)}{(\varepsilon \times A_m \times \Delta P)}\right]} \tag{2.11}$$

where μ is the water viscosity (8.9×10^{-4}) in Pa. s, Q_W is the water flow ($m^3 s^{-1}$), and ΔP is the operating pressure (0.3 MPa).

2.2.1.5 Chromium Removal Performances

Ultrafiltration experimentations were done in the stirred ultrafiltiration batch cell to investigate the performances of prepared membranes' on the rejection and permeate flux Cr (VI) ions. The flux values and the rejection of Cr (VI) ions were evaluated by the following equations, respectively.

$$J_{Cr} = \left[\frac{Q}{(A \times \Delta t)}\right] \tag{2.12}$$

$$R_{Cr}(\%) = \left[\left(1 - \frac{C_p}{C_f}\right) \times 100\right] \tag{2.13}$$

where C_p and C_f are the concentrations (ppm) in the permeate side and the feed side, respectively. Subsequently, all the required solutions were prepared with analytical reagents and deionized water; 2.9 g of 99% K_2CrO_4 was dissolved in deionized water of 1.0 L volumetric flask up to the mark to obtain 1000 ppm of Cr (VI) stock solution. Synthetic samples of Cr (VI) with different concentrations (i.e., 10, 30, 50, 70, and 100 ppm) were prepared from this stock solution by appropriate dilutions. The pH of the Cr (VI) ion solutions was kept at 3.5 and 7. The pH values of the prepared solutions were adjusted using 1 N of HCl and NaOH. Deionized water (DI) was used throughout this experimentation to avoid side reactions and contamination of the solution. The initial and final concentrations of Cr (VI) ions before and after ultrafiltration process were determined by using an atomic absorption spectrophotometer (Make: M/s. Varian, Netherland; Model: Spectra AA 220 FS). The analytical wavelength for Cr (VI) was fixed at 357.9 nm. These experiments were repeated three times. The washing experiments of the adsorbed/deposited Cr (VI) ions in the membranes were performed usinig a 2.5 g/L KCl solution for four times, and the membranes were rinsed to a neutral condition by using DI water.

2.2.1.6 HA Ultrafiltration and Regeneration Performance

Ultrafiltration experimentations were done in the stirred ultrafiltiration batch cell to investigate the performances of prepared membranes on the removal and permeate flux of HA solution. The flux values were calculated using Equation 2.11, and the removal efficiencies of HA molecules were evaluated by Equation 2.14.

$$J_{HA} = \left[\frac{Q}{(A \times \Delta t)} \right] \tag{2.14}$$

$$\mathrm{R}_{HA}(\%) = \left[1 - \left(\frac{C_p}{C_f} \right) \times 100 \right] \tag{2.15}$$

where C_p and C_f are the concentrations (mg L^{-1}) in the permeate side and the feed side, respectively. The ultrafiltration experiments for HA solution were achieved through all prepared membranes at an operating pressure of 150 kPa. After that, the permeates were collected in a measuring container at a constant interval of time in order to calculate flux values, and the length of each filtration test was 2 h. A 1000 ppm stock solution of HA was prepared by dissolving 1 g HA in to100 ml DI water following by the addition of 900 ml DI water. This solution was stirred for 6 h at 200 rpm and then filtered through a filter paper. The pH values of the prepared solutions were adjusted using 1 N of HCl and NaOH. Finally, the permeate concentration of HA solution was evaluated by using UV–vis spectrophotometer (Make: M/s. Thermo Fisher Scientific, India; Model: UV 2300) at a wavelength of 254 nm. Later filtration of HA solutions, fouled membranes were cleaned with 0.05 M NaOH solution followed by rinsing using DI water for 30 min, and the DI water fluxes of the tested membranes were measured.

2.2.2 Instrumental Characterization

2.2.2.1 Field Emission Scanning Electron Microscopy (FESEM)

Morphological studies of the electrospun and phase inverted membranes were done using a high-resolution field emission scanning electron microscopy (Make: M/s. Zeiss, Germany; Model: Sigma), which provides visual information of the topographic view besides the cross-sectional structure of the membrane. The elemental composition analysis was done by using high energy electrons of the energy dispersive X-ray spectroscopy (Make: M/s. Zeiss, Germany; Model: Sigma). For both the analyses, a gold coating for the samples was done using a high-resolution sputter coater, Quorum, to protect the membranes from charging during the image analysis. Finally, the pieces of the membranes were attached to a plate holder using double-sided adhesive carbon tape in a horizontal position.

2.2.2.2 Thermo Gravimetric Analysis (TGA)

The thermal degradation analysis was conducted by thermo-microbalance (Thermo gravimetric Analyzer) (Make: M/s. Netzsch, Germany; Model: TG 209 F1 Libra®). The TGA results were found from 30 to 800 °C using nitrogen gas where the flow rate was kept as 40 mL/min at a heating rate of 10 °C/min. All the membranes were loaded into a platinum sample holding pan. The graphs were plotted as weight loss (%) vs. temperatures (°C).

2.2.2.3 Nuclear Magnetic Resonance (NMR) Analysis

^{1}H NMR spectra were recorded on a 600 MHz Nuclear Magnetic Resonance (NMR) Spectrometer (Make: M/s. Bruker, Japan). All the samples (20 mg) were dissolved in DMSO-d6. Each solution was placed in a 5 mm NMR glass tube (Make: M/s. WILMAD-Lab Glass Co., USA).

2.2.2.4 ATR-FTIR Spectroscopy Analysis

The chemical structures of all the samples were characterized by using attenuated total reflectance Fourier transform infrared (ATR-FTIR) spectrophotometer (Make: M/s. Shimadzu, Japan; Model: IRAffinity-1). Each spectrum was acquired in transmittance mode by an accumulation of 40 scans through resolutions of 4 cm^{-1}, and the wave number was ranged from 500 to 4000 cm^{-1}.

2.2.2.5 Zeta Potential (ζ) Analysis

Zeta potential (ζ) results of the samples were measured by dynamic and electrophoretic light scattering (Make: M/s. Beckman Coulter, Switzerland; Model: Delsa Nano C). Next, 10 mg of sample was dissolved in 40 mL water and then placed in the ultrasonic bath for 20 min at pH ranging from 1 to 10.

2.2.2.6 Water Contact Angle (WCA) Measurements

The hydrophilicity natures of all the prepared membranes were evaluated from the contact angle measurements. The contact angle (CA) measurements of the prepared membranes were conducted on a contact angle measuring instrument (Make: M/s. Kruss, Germany; Model: Drop Shape Analyzer – DSA100).

2.2.2.7 X-Ray Diffractometer (XRD)

X-ray diffraction patterns of the electrospun membranes, membrane textural properties, were recorded using an X-ray diffract meter (Make: M/s. Bruker, USA; Model: D8 Advance) working at 40 KV/50 mA with radiation of Cu Kα.

2.2.2.8 Leica Microscope

The thickness of the electrospun membranes was measured by Leica microscope (Make: M/s. Leica Microsystems, Germany; Model: DM 2500M).

2.2.2.9 Atomic Force Microscopy (AFM)

The Atomic force microscopy (AFM) (Make: M/s. Agilent, USA; Model: 5500 series) was employed to examine the smoothness of the electrospun membrane surfaces and top surface morphology in the size of 5 μm × 5 μm.

2.2.2.10 Transmission Electron Microscopy (TEM)

The particle size and microstructural information of U-TiO_2 (commercial) and M-TiO_2 NPs were characterized by using transmission electron microscopy (Make: Jeol, USA; Model: Jem 2100F) at 210 kV. First, the TiO_2 nanoparticle sample was prepared by dispersing in distilled water (500 mg/L) and then poured on a carbon tape covered plate. Finally, the sample was dried at room temperature and ready for TEM analysis.

2.2.2.11 Brunauer-Emmet-Teller (BET) Isotherm

The specific surface areas of the electrospun membranes were analyzed using N_2 adsorption-desorption isotherm measurements in surface area and pore size analyzer and high pressure analyzer (Make: M/s. Quantachrome®, USA; Model: ASiQwin™) by using Brunauer-Emmet-Teller (BET) isotherm model.

2.2.2.12 Atomic Absorption Spectrophotometer (AAS)

The initial and final concentrations of Pb (II) and Cu (II) ions before and after equilibrium adsorption were determined by using an atomic absorption spectrophotometer (Make: M/s. Varian, Netherland; Model: Spectra AA 220 FS). The analytical wavelengths were fixed at 217.0 and 324.7 nm for Pb (II) and Cu (II) ions, respectively. The initial and final concentrations of Cr (VI) ions before and after ultrafiltration process were determined by using an atomic absorption spectrophotometer (AA240FS, USA). The analytical wavelength for Cr (VI) was fixed at 357.9 nm.

2.2.2.13 UV–vis Spectrophotometer (UV-2600)

To examine the interaction between the adsorbent and the heavy metal ions and to evaluate the Ti leaching tendency, their UV–vis absorbance was determined using a Shimadzu UV–vis Spectrophotometer (Make: M/s. Shimadzu, Singapore; Model: UV-2600) over a wavelength of 250–600 nm.

The permeate concentration of HA solution was evaluated by using UV–vis spectrophotometer (Make: M/s. Thermo Fisher Scientific, India; Model: UV 2300) at a wavelength of 254 nm. BSA solution was evaluated by using UV–vis spectrophotometer (Make: M/s. Thermo Fisher Scientific, India; Model: UV 2300) at a wavelength of 278 nm.

References

1. Garcia-Ivars, J., Alcaina-Miranda, M.I., Iborra-Clar, M.I., Mendoza-Roca, J.A., and Pastor-Alcañiz, L., Enhancement in hydrophilicity of different polymer phase-inversion ultrafiltration membranes by introducing PEG/A12O3 nanoparticles. *Separation and Purification Technology*, 2014. 128: pp. 45–57.
2. Yuliwati, E., Ismail, A.F., Matsuura, T., Kassim, M.A., and Abdullah, M.S., Effect of modified PVDF hollow fiber submerged ultrafiltration membrane for refinery wastewater treatment. *Desalination*, 2011. 283: pp. 214–220.
3. Chakrabarty, B., Ghoshal, A.K., and Purkait, M.K., Preparation, characterization and performance studies of polysulfone membranes using PVP as an additive. *Journal of Membrane Science*, 2008. 315(1–2): pp. 36–47.
4. Das, C., and Gebru, K.A., Cellulose acetate modified titanium dioxide (TiO_2) nanoparticles electrospun composite membranes: Fabrication and characterization. *Journal of The Institution of Engineers (India): Series E*, 2017. 98(2): pp. 91–101.
5. Zhao, Y.F., Zhu, L.P., Yi, Z., Zhu, B.K., and Xu, Y.Y., Improving the hydrophilicity and fouling-resistance of polysulfone ultrafiltration membranes via surface zwitterionicalization mediated by polysulfone-based triblock copolymer additive. *Journal of Membrane Science*, 2013. 440: pp. 40–47.
6. Wang, Y.Q., Wang, T., Su, Y.L., Peng, F.B., Wu, H., and Jiang, Z.Y., Remarkable reduction of irreversible fouling and improvement of the permeation properties of poly(ether sulfone) ultrafiltration membranes by blending with pluronic F127. *Langmuir*, 2005. 21: pp. 11856–11862.
7. Wu, G., Gan, S., Cui, L., and Xu, Y., Preparation and characterization of PES/TiO_2 composite membranes. *Applied Surface Science*, 2008. 254(21): pp. 7080–7086.

3

Electrospun Composite Membranes: Preparation and Application

3.1 Electrospinning Process Parameters

3.1.1 Introduction

During electrospinning process, a high voltage (i.e., approximately 10 to 50 kV) is supplied in a polymer solution to the extent that charges are produced within the polymer solution. Once a sufficient amount of charge is supplied, a solution jet escapes from the needle-tip droplet, causing the development of the so-called Taylor cone. After that, the electrospinning jet travels to the grounded collector (i.e., with lower potential). The morphological structure of the electrospun fibers can be influenced by processing parameters. Several polymer types can be electrospun into nanofiber, providing that the molecular weight of polymers is sufficient and the solvent can be vaporized in a timely fashion throughout the jet journey time over the given distance between the needle tip and ground collecting plate. It is mentioned in the literature that more than 100 polymers, including polyvinyl alcohol, have been effectively electrospun into nanofibers commonly from a polymer solution [1,2]. The nanofiber jet can be condensed into a mat that forms a highly porous membrane structure. It is evident that the electrospun polymeric membrane possesses specific properties, including high specific surface area, high porosity, and continuous interconnected fibers. They can be easily modified to have good physical and chemical properties by blending polymer-inorganic nano composites prior to electrospinning and can be used for water and air filtration, tissue engineering, sensors, and other applications [3–5]. Therefore, as observed from the literature, further investigations are required to advance the electrospinning process, specifically to fabricate membranes for the purpose of water treatments. Electrospun nanofibrous membranes, such as PVA/Chitosan, polyvinyl alcohol, and PVA/cyanobacterial extracellular polymeric materials composite membranes, were fabricated for prospective water filtration applications using a microfiltration poly vinylidene fluoride as support membrane [6]. Electrospun nanofibrous membranes with a conventional non-woven microfibrous support showing higher flux microfiltration were also reported. The polyvinyl alcohol nanofibrous nonwoven membranes were directly fabricated on top of the porous support by using electrospinning technique [7]. Another study presented a nanofiber thin film composite membrane crosslinked electrospun polyvinyl alcohol nanofiber and was uniquely found to be a very effective support layer, specifically for forward osmosis applications, due to its low tortuosity, high porosity, and remarkable hydrophilic property [8]. It is known that PVA among synthetic polymers is easily soluble in water, nontoxic, relatively low-cost, stable in chemical and thermal conditions, and less degradable in most physiological environments. It is apparent that the polymer solution properties have the major impact in the electrospinning process and the resulting membrane morphological structures. The surface tension, viscosity, and electrical properties of the polymer solution play important roles in the degree of elongation of the solution and in the development of beads along the fiber length [9]. Koski et al. [1] have studied the effects of polymer average molecular weight on the fiber structure of electrospun polyvinyl alcohol (PVA). They have reported an average fiber diameter between 250 nm and 2 μm and the fiber diameter increases with MW and concentration. From their experimental results, they have confirmed that at lower MW and/or concentrations, the fibers exhibit a circular cross-section. They have also observed flat fibers at high MW and concentrations. Furthermore, the effect of viscosity on the membrane morphology has also studied by Mohammad Ali Zadeh et al. [2]. According to their study, viscosity of the spinning solutions had played an important role on the morphology of the mullite nanofibers. Continuous electrospun nanofibers with common cylindrical morphology were obtained when PVA content was 6 wt.%. Further increasing the amount of PVA in the pre-spinning solution led to excessively high viscosity level, making the shape of the resulting mullite nanofibers wide and flat ribbon. Other important parameters that influence the electrospinning process are voltage supply, feed rate, solution temperature, collector type, needle diameter, and the collector-needle tip distance. Barhate et al. [10] have studied the effect of electrospinning process parameters on structural and transport properties of the electrospun membranes. They have mentioned that the pore size and fiber entanglements can be optimized to enhance the pore size distribution, porosity, pore interconnectivity, and permeability performances of the electrospun membranes. Finally, they have suggested that an optimized structure of the membrane can be achieved by coordinating the collection rates and applied voltage. The shapes and the sizes of nanofiber membranes are controlled through numerous parameters, such as conductivity, concentration, viscosity, and surface tension of the polymeric solutions. Though all the parameters are significant factors, polymeric solution concentration and applied voltage are considered the most significant on the final characteristics of electrospun fibers. The polymeric solution properties had the main substantial effect in the electrospinning processes. The viscosity of the solution, surface tension, and its conductivity also determined the degree of elongations of the polymeric solution, which possibly had an influence on the diameters of the resulting electrospun fibers [9]. Electrospun poly vinyl alcohol was prepared using electrospinning technique and two parameters (i.e., membrane average fiber diameter and membrane surface pore diameter) were optimized using response surface methodology (RSM). Polymer concentration, voltage supply, time duration, and solution flow rate were optimized based on central composite design (CCD). An effort was also given to further investigate the effect of electrospinning process time duration on the resulting membranes, which was introduced as a new field along with the above important electrospinning parameters. On the other hand, this study also aims to realize the interaction between process parameters with the experimental responses (i.e., membrane average fiber diameter and membrane surface pore diameter). The experimental and predicted values (which were found from the mathematical model) were compared to validate the model and were able to predict the optimum independent parameters for the preparation of micro/ultra-filtration electrospun PVA membranes.

3.1.2 Preparation Methods: Electrospinning Technique

3.1.2.1 Preparation of Electrospun PVA and Cellulose Acetate

Electrospun PVA membranes were prepared by using the electrospinning technique. During electrospinning process, a high voltage is supplied to a polymeric solution in the syringe pump such that charges are induced within the polymer solution (Figure 3.1). When charges inside the solution become sufficiently high and able to overcome the surface tension of the solution droplet,

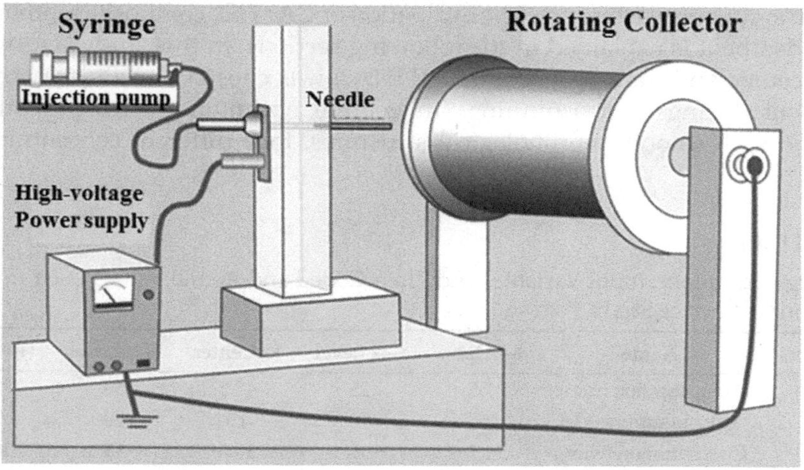

FIGURE 3.1
Schematic of the electrospinning equipment, syringe pump, needle, rotating collector, and voltage supply.

a solution jet erupts from the needle-tip drop resulting in the development of the so-called Taylor cone. Finally, the jet travels to the rotating collector with a state of lower potential. Electrospinning was performed using NABOND NEU (China) electrospinning unit at room temperature (28 ± 2 °C). The polymer solution was loaded into a 50 mL syringe with a needle (inner diameter of 0.4 mm). According to the design of experiments summery (Table 3.1), a polymer solution varied from 6 to 19.5 wt.%, an electric field of 5 to 25 kV was applied throughout the process, and the fluid feed rate was 1 to 3 mL/h and the time duration between 25 to 77.5 min. All the prepared fibers were deposited on a rotating collector, where the gap between the needle tip and the collecting plate was fixed at 100 mm. Using the optimized electrospinning parameters (concentration, voltage supply, and feed rate), the PVA membranes were prepared for the selected time durations (i.e., 25, 35, 45, and 60 min) to further investigate the impact of deposition time on the surface pore diameters, fiber diameter, surface areas, porosity, and morphological structures of the electrospun membranes. Surface tension, viscosity, and conductivities of the solutions were determined using a tensiometer (Make: M/s. Kwoya, Japan; Model: DY300), viscometer (Make: M/s. Anton Paar, Physica MCR 30, Austria), and a conductivity meter (Make: M/s. ELICO, India), respectively.

During the preparation of the CA/TiO_2 composite membrane, the electric field was 14 KV throughout the process, and the fluid flow rate was 2 mL/h. The fibers of cellulose acetate and CA_TiO_2 hybrid membranes were deposited on a rotating collector plate, where the distance from the collector plate to needle tip was kept at 100 mm. The fabrication and characterization of CA and CA/TiO_2 membrane are clearly mentioned below. Based on this study, the membrane was tested for adsorption of heavy metals (Pb (II) and Cu (II)) from synthetic wastewater. The evaluation of CA/TiO_2 composite membrane as adsorbent is explained in the following sections. In this study, a mixture of acetone and dimethylacetamide (DMAc) was chosen as cellulose acetate solvent during electrospinning. To get the optimum electrospun membrane with smooth morphological structures, four different concentrations

TABLE 3.1

Design Summary: Input Variables and Their Coded and Actual Values Used in the Response Surface Study

Factors	Name	(–)Alpha	(–)1 Level	(0)Center	(+1) Level	(+)Alpha
A	Time duration (min)	7.5	25	42.5	60	77.5
B	Voltage supply (kV)	5	10	15	20	25
C	Concentration (wt.%)	1.5	6	10.5	15	19.5
D	Flow rate (mL/h)	0	1	2	3	4

(25, 20, 15, and 13.5 wt.%) of cellulose acetate were prepared and 13.5 wt.% was selected for the next experiment. Cellulose acetate solutions were prepared by uniformly dissolving in 2:1 ratio of acetone/DMAc [11]. Subsequently, different amounts of TiO_2 nanoparticles were added to the cellulose acetate to enhance the thermal, specific surface area, and mechanical property of cellulose acetate. Therefore, different amount (0, 1.0, 2.5, 4.5, and 6.5 wt.%) of TiO_2 nanoparticles were dispersed in 13.5 wt.% solution of CA. The mixed solution was stirred for 4 h at 200 rpm speed and sonicated for 3 h to ensure the optimum distributions of the TiO_2 nanoparticles in the CA solution.

3.1.2.2 Response Surface Methodology Using Design of Expert

The design of experiment (DOE) has been known as an appropriate optimization tool to investigate and optimize the impact of electrospinning parameters [12,13]. The DOE technique is employed to reduce the number of experiments to be performed. Moreover, one of the effective optimization techniques to obtain the optimal conditions in a multivariable scheme is response surface methodology (RSM) [14,15]. Recently, the RSM optimization technique has been effectively employed in numerous processes to attain the optimal conditions. Furthermore, the central composite design (CCD) is a suitable experimental design technique among various approaches, which provides high-quality estimates in studying interaction, quadratic, and linear effects of parameters [16]. Therefore, the optimization of the above parameters for polyvinyl alcohol based solutions was the preliminary point of this work. The theory and definition of some terms regarding RSM optimization process have been explained in detailed by Bezerra et al. [14], Yordem et al. [13], and Ahmadipourroudposht et al. [17]. The design plan in this study includes four input factors—time duration (min), voltage supply (kV), polymer concentration (wt.%), and feed rate (mL/h)—and responses (R_1 and R_2; i.e., membrane fiber diameter and membrane surface pore diameter, respectively). Those four input factors have been selected for designing purpose, and their results are chosen based on preliminary investigation. The input factors are varied over five levels: high value (+1), the center point (0), low level (−1), and two outer points (−α and +α value); details are outlined in Table 3.1.

CCD is comprised of design points and axial points, consisting of a total of 30 experimental runs that are used to examine the experimental data. These data are finally used to optimize the electrospinning process parameters. The output variables (i.e., membrane fiber diameter and membrane surface pore diameter) are measured from the selected mathematical model with significant terms, and the model was not aliased. Table 3.2 presents the summary of the design of experiments and experimental responses for central composite design (CCD).

TABLE 3.2
Design of Experiments Summery and Experimental Responses for Central Composite Design (CCD)

Run	Factor 1, A: Time Duration (min)	Factor 2, B: Voltage Supply (kV)	Factor 3, C: Concentration (wt.%)	Factor 4, D: Flow Rate (mL/h)	Response 1, Fiber Diameter (nm)	Response 2, Surface Pore Diameter (nm)
1	42.5	15.0	10.5	4.0	179	155
2	25.0	20.0	6.0	3.0	95	90
3	42.5	15.0	10.5	2.0	105	109
4	25.0	10.0	15.0	1.0	117	180
5	25.0	10.0	6.0	1.0	106	144
6	7.5	15.0	10.5	2.0	129	267
7	60.0	10.0	15.0	1.0	125	207
8	60.0	20.0	6.0	1.0	76	132
9	42.5	15.0	10.5	2.0	100	107
10	42.5	15.0	10.5	2.0	103	111
11	60.0	10.0	15.0	3.0	78	192
12	25.0	20.0	15.0	1.0	50	193
13	42.5	15.0	1.5	2.0	46	90
14	60.0	10.0	6.0	1.0	47	149
15	60.0	20.0	15.0	3.0	39	270
16	42.0	15.0	10.5	2.0	107	108
17	42.5	15.0	10.5	0.0	135	200
18	25.0	20.0	15.0	3.0	25	290
19	60.0	20.0	6.0	3.0	55	90
20	60.0	10.0	6.0	3.0	150	50
21	42.5	15.0	10.5	2.0	104	110
22	60.0	20.0	15.0	1.0	133	230
23	42.5	15.0	19.5	2.0	20	310
24	42.5	5.0	10.5	2.0	160	60
25	25.0	20.0	6.0	1.0	80	98
26	77.5	15.0	10.5	2.0	78	169
27	42.5	25.0	10.5	2.0	56	100
28	42.5	15.0	10.5	2.0	103	109
29	25.0	10.0	6.0	3.0	256	80
30	25.0	10.0	15.0	3.0	138	200

3.1.2.3 Cross-Linking of Electrospun PVA

The effect of crosslinking of electrospun PVA nanofibers membranes was studied. The cross-linking procedure and optimization have been done following the procedure suggested by Wang et al., 2006. Accordingly, electrospun PVA samples were dipped in a solution containing 0.01N hydrochloric acid and 30 mM of glutaraldehyde/acetone for about 1 day. The cross-linked

PVA membranes were withdrawn and rinsed using distilled water for several times and then dried out.

3.1.3 Statistical Analysis (ANOVA) and Response Surface of Electrospinning Parameters

The statistical analyses of the electrospinning parameters were performed using design expert software. ANOVA test was employed to perform the statistical tests for significance of model, individual model terms, and lack of fit. It is obvious that a significant model is needed and "Prob. > F" values less than 0.05 shows significance of the model and the individual model terms. The ANOVA results for both fiber diameter (FD) distribution and surface pore diameter (SPD) distribution were analyzed. In both cases, the lack-of-fit test values were insignificant, which are desired as we need a model that fits.

3.1.3.1 Membrane Fiber Diameter

The relationship between electrospinning process factors (i.e., time duration, voltage supply, and concentration and flow rate) and the expected responses (i.e., membrane fiber diameter and membrane surface pore diameter) are plotted graphically after mathematical analysis of the experimental data. Figure 3.2 shows the three-dimensional response surfaces for membrane fiber diameter. As already observed from these figures, the factors involved in the membrane electrospinning process exhibited nonlinear effects on the membrane fiber diameter. The highest membrane average fiber diameter (i.e., 256 nm) was obtained at 25 min, 10 kV, 15 wt.% and 3 mL/h (run 29) of time duration, voltage supply, polymer concentration and flow rate, respectively. On the other hand, the lowest membrane average fiber diameter (i.e., 20 nm) was obtained at 42.5 min, 15 kV, 19.5 wt.% and 2 mL/h (run 23). As clearly showed in the three-dimensional plots (i.e., Figure 3.2a, b, c, d, and e), the membrane fiber diameter exhibited a decreasing trend from 140.6 nm to 72.9 nm as the voltage supply raised from 10 kV to 20 kV, where the polymer concentrations and flow rates were kept at 10.5 wt.% and 2 mL/h. As observed from the experimental results, continuous fibrous structures were obtained for 6 wt.% PVA aqueous solution concentrations, where the fibers contained many twists and branches and were highly interconnected. At higher concentrations, the cross-sections of the fibers were spherical, but once the concentration of the solution was increased above 19.5 wt.%, the fiber diameter and inter-fiber spacing was increased, and there was a slow change from circular to flat ribbon-like structure of the fibers [1,2]. When the PVA concentration was lowered below 6 wt.%, the formation of large beads and non-uniform entanglement of fibers was observed. This result may be accredited to an increasing in the surface tension and/or concentration of the polymer solution [6]. At relatively high voltage supply (15 kV), as the polymer

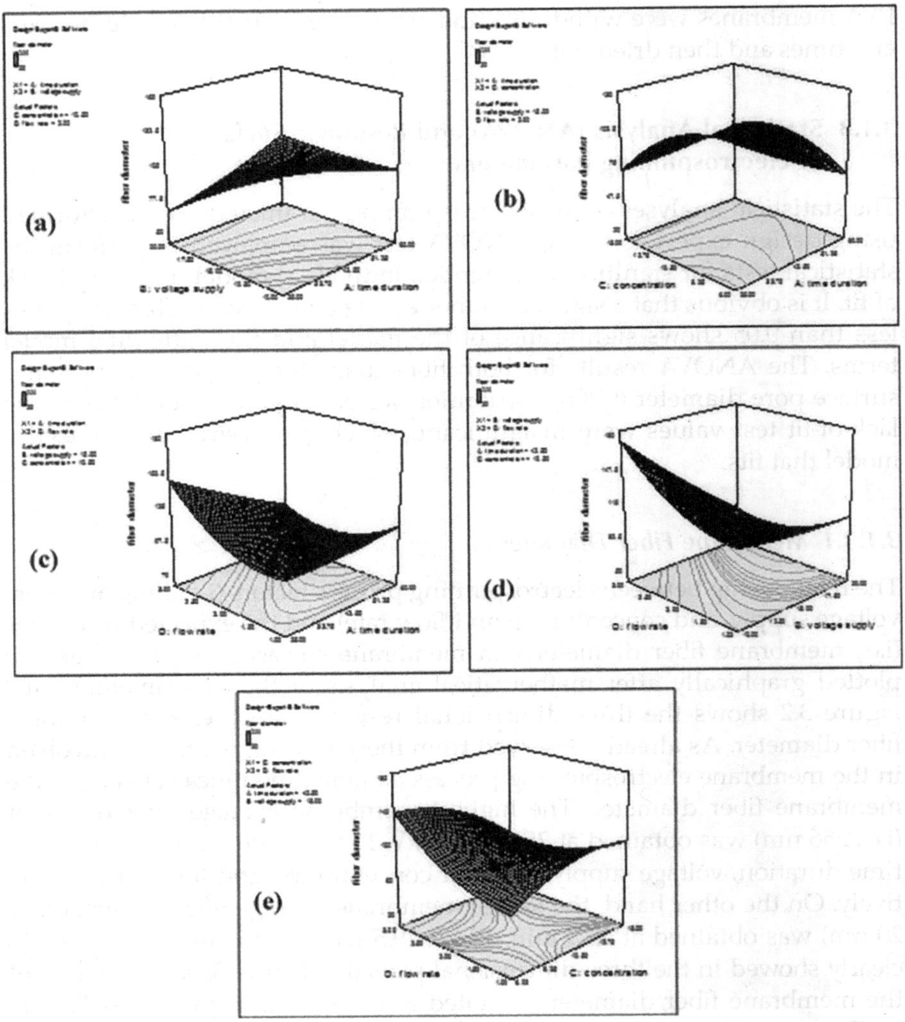

FIGURE 3.2
Three-dimensional response surface plot of (a) voltage supply and time duration, (b) concentration and time duration, (c) flow rate and time duration, (d) flow rate and voltage supply, and (e) flow rate and concentration for membrane fiber diameter.

concentration was raised from 6 wt.% to 8.25 wt.%, the average fiber diameter of the membrane was increased from 72.9 nm to 114 nm. Further increasing the concentration above 8.25 wt.%, the average fiber diameters seem to have a slight difference. Even though the average fiber diameter is expected to increase with increasing the polymer concentration, in this case, the high voltage supply seems to control the spinning of fibers with large diameters. The analysis of variance and regression model for the four model terms are presented in Table 3.3 where the model F-value of 22.63 implies the model

TABLE 3.3

Analysis of Variance (ANOVA) for Respective Response Surface Quadratic Models

	Fiber Diameter (nm)		Surface Pore Diameter (nm)	
Source	*p*-Value Prob > F	F Value	*p*-Value Prob > F	F Value
Model	<0.0001	22.63	<0.0001	27.15
A: time duration	0.0023	13.36	0.1265	2.62
B: voltage supply	<0.0001	85.26	0.0109	8.43
C: Concentration	0.0107	8.49	<0.0001	25.13
D: flow rate	0.0197	6.82	0.1051	2.98
AB	0.0004	20.64	0.4661	0.56
AC	0.0007	18.27	0.7281	0.13
AD	0.0021	13.71	0.0518	4.46
BC	0.4905	0.50	0.0122	8.11
BD	<0.0001	35.09	0.0058	10.34
CD	<0.0001	43.52	0.0003	21.70
A^2	0.5465	0.38	<0.0001	53.27
B^2	0.3269	1.03	0.0452	4.77
C^2	<0.0001	31.32	<0.0001	36.75
D^2	<0.0001	28.43	0.0004	20.39
Lack of fit	0.5012	1.07	0.0592	4.35

is significant. There is only a 0.01% chance that a "Model F-Value" this large could occur due to noise. As observed from the results, the model terms such as A, B, C, D, AB, AC, AD, BD, CD, C^2, D^2 are significant, where their Prob. > F values are below 0.05. Moreover, the voltage supply seems to be a highly significant factor when compared with other significant input factors (time duration, concentration, and flow rate). Values greater than 0.10 indicate that the model terms are not significant. The "Lack of Fit F-value" of 1.07 implies the lack of fit is not significant relative to the pure error. There is a 50.12% chance that a "Lack of Fit F-value" this large could occur due to noise. Nonsignificant lack of fit is good that we want the model to fit. The following equation is the final equation in terms of coded factors that is developed using central composite design to designate the curvatures around the optimal point.

$$FD = \begin{pmatrix} 95\text{-}11.08\ (A)\text{-}28\ (B)\text{-}8.83\ (C) + 7.92\ (D) + 16.88\ (A \times B) + 15.88\ (A \times C)\text{-}13.75\ (A \times D) \\ +2.63\ (B \times C)\text{-}22\ (B \times D)\text{-}24.5\ (C \times D)\ + 1.75\ (A^2) + 2.88\ (B^2)\text{-}15.88\ (C^2) + 15.12\ (D^2) \end{pmatrix} \quad (3.1)$$

The reliability of regression models for membrane fiber diameter was described on the basis of high values of R^2 (0.96), which shows that this model is well fitted to the experimental values. The "Pred R-Squared" of 0.80 is in reasonable agreement with the "Adj R-Squared" of 0.91. "Adeq Precision" measures the signal to noise ratio, and a ratio greater than 4 is desirable, where this model's ratio of 22.5 indicates an adequate signal.

3.1.3.2 Membrane Surface Pore Diameter

The three-dimensional response surface plots, which show the impact of input factors (i.e., time duration, concentration, and voltage supply and flow rate) on membrane surface pore diameter, are presented in Figure 3.3. The maximum average membrane surface pore diameter of 310 nm was attained at a time duration of 42.5 min, voltage supply of 15 kV, a concentration of 19.5 wt.%,

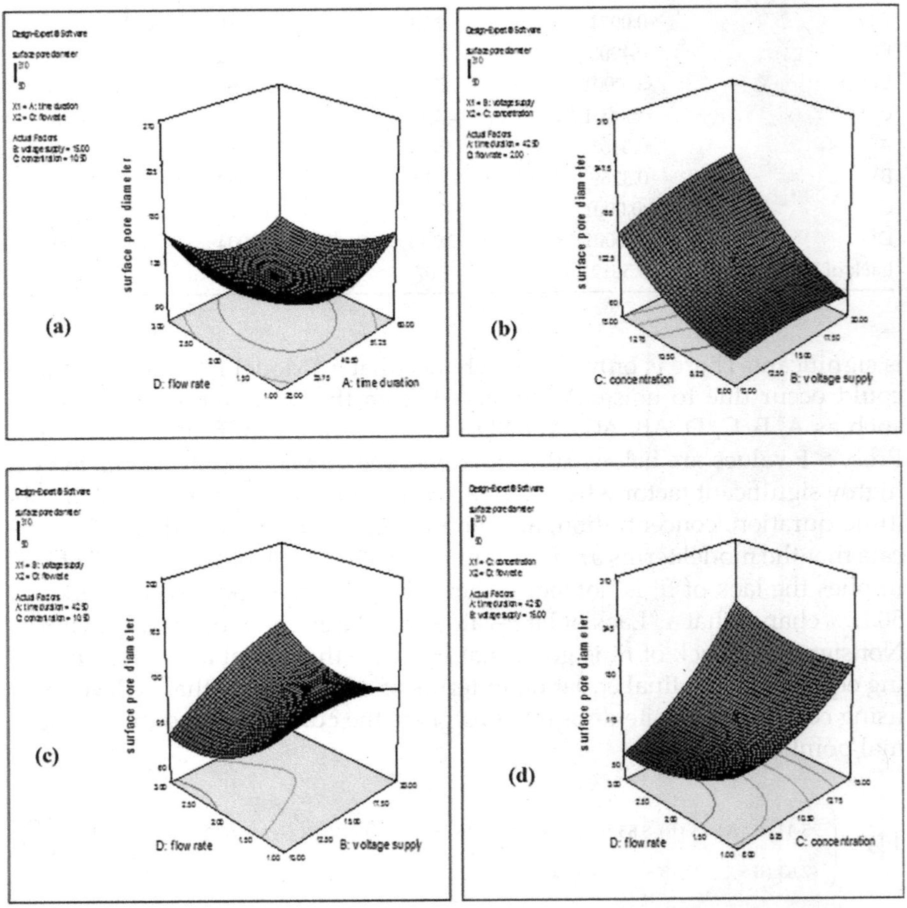

FIGURE 3.3
Three-dimensional response surface plot of (a) flow rate and time duration, (b) concentration and voltage supply, (c) flow rate and voltage supply, and (d) flow rate and concentration for membrane surface pore diameter.

and flow rate of 2 mL/h (run 23) whereas the minimum average membrane surface pore diameter of 50 nm were achieved at a time duration of 60 min, voltage supply of 10 kV, a concentration of 6 wt.%, and flow rate of 3 mL/h (run 20). These results indicate that the concentration seems to have a greater contribution to the variation in membrane surface pore diameter (Figure 3.3b and d). As shown from Figure 3.3a, the average surface pore diameter was decreased from 142 nm to 115 nm as the deposition time was increased from 25 to 51 min. These results can be explained due to the entanglement of fibers as the electrospinning time duration was increased. However, when the deposition time duration further increases beyond 51 min, the surface pore diameter showed a slightly increasing trend. At maximum polymer concentration of 19.5 wt.% (Run 23), it is observed that the surface pore diameter is high (310 nm), compared to the membranes prepared from less polymer concentration of (10.5) (Run 28), where the surface pore diameter was 100 nm. This incident is due to the increase in fiber diameter of the membrane, as the concentration of the solution was increased. The Model F-value of 27.15 implies the model is significant. There is only a 0.01% chance that a "Model F-Value" this large could occur due to noise. Values of "Prob > F" less than 0.05 indicate model terms are significant. In this case, B, C, BC, BD, CD, A^2, B^2, C^2, D^2 are significant model terms. Values greater than 0.10 indicate the model terms are not significant. The "Lack of Fit F-value" of 4.35 implies there is a 5.92% chance that a "Lack of Fit F-value" this large could occur due to noise. The following equation is the final equation in terms of coded factors that is developed using central composite design to designate the curvatures around the optimal point.

$$\mathrm{SPD} = \left(\begin{array}{l} 107.33 + 11.29\,(B) + 57.4\,(C) - 10.06\,(A \times D) + 13.56\,(B \times C) + 15.31\,(B \times D) \\ +22.19\,(C \times D) + 26.55\,(A^2) - 7.95\,(B^2) + 22.05\,(C^2) + 16.43\,(D^2) \end{array} \right) \tag{3.2}$$

The reliability of regression models for membrane surface pore diameter was described on the basis of high values of R^2 (0.96), which shows that this model is well fitted to the experimental values. On the other hand, the "Pred R-Squared" of 0.80 is in reasonable agreement with the "Adj R-Squared" of 0.93. "Adeq Precision" measures the signal to noise ratio. A ratio greater than 4 is desirable, and the ratio of 19.7 in this model indicates an adequate signal.

3.1.3.3 Model Verification on the Basis of Statistical Analysis

The effects of the interactions among the input electrospinning parameters were examined by using RSM optimization technique. This optimization tool is useful in investigating the effect of a binary combination of two input factors. The plots of the interaction between the input parameters are presented in Figures 3.4 and 3.5 for fiber diameters and surface pore diameter, respectively. As clearly observed from Figure 3.4, all the interaction plots exhibited nonparallel curvatures. From these results, it is suggested a strong interaction between the variables (i.e., AB, AC, AD, BD, and CD) for membrane fiber diameters. As presented in Table 3.3, BD and CD seemed to be highly significant model terms. Therefore, from this study, it can be suggested that the effect of voltage supply and concentration

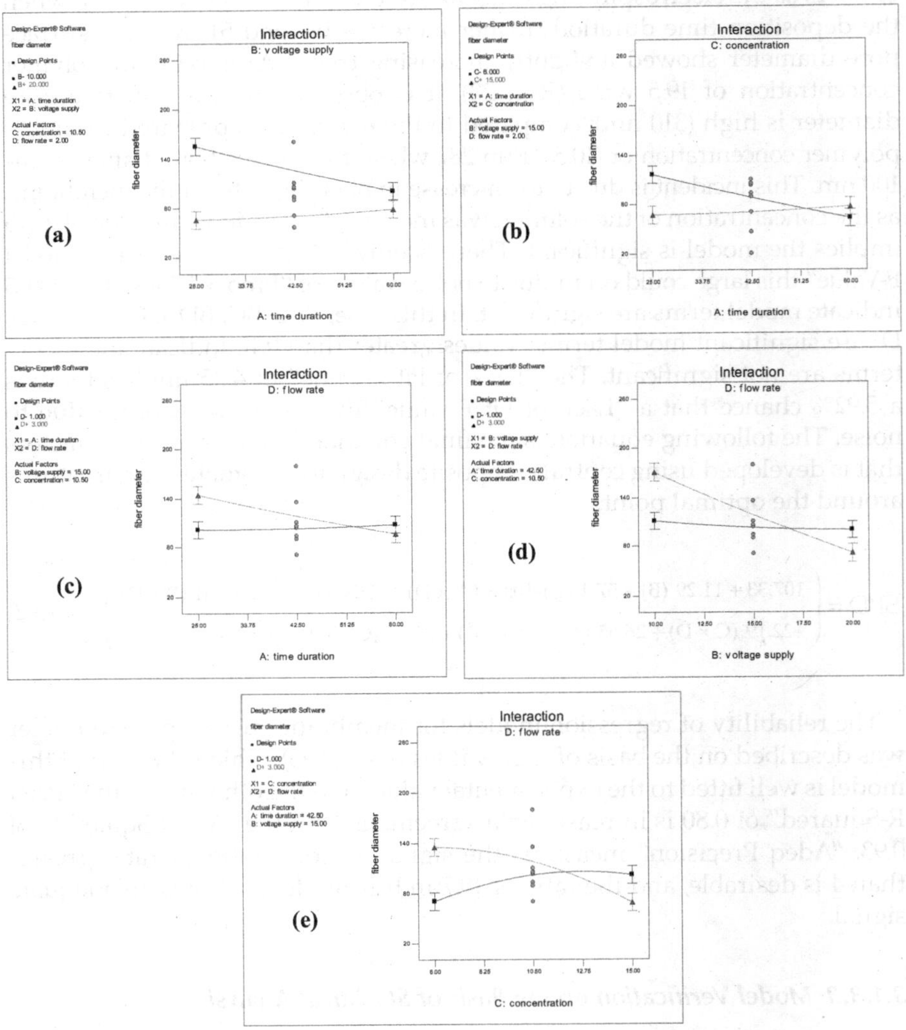

FIGURE 3.4
The interaction plots of (a) time duration and voltage supply, (b) time duration and concentration, (c) time duration and flow rate, (d) flow rate and voltage supply, and (e) flow rate and concentration for membrane fiber diameter.

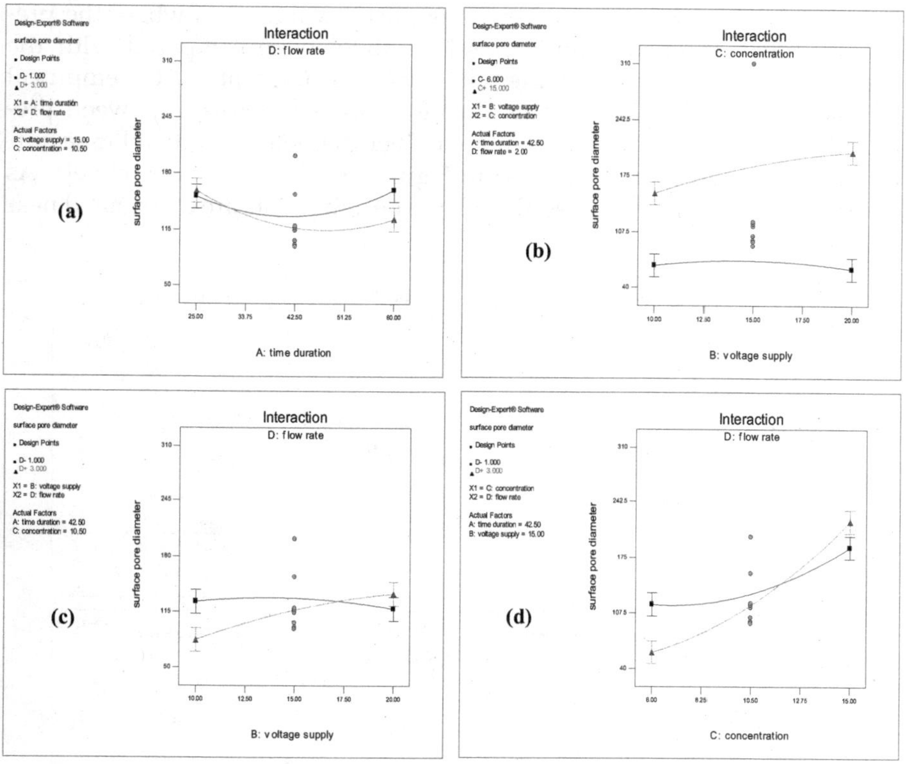

FIGURE 3.5
The interaction plots of (a) time duration and flow rate, (b) voltage supply and concentration, (c) voltage supply and flow rate, and (d) flow rate and concentration for membrane surface pore diameter.

are highly significant input parameters for average fiber diameter of the membrane.

It is clearly seen from Figure 3.5 that the interactions of all the input factors (i.e., time duration, voltage supply, concentration, and flow rate) have a significant effect on the surface pore diameter distribution. The interaction plots showed nonparallel curvatures, and it is suggested that there is strong interaction between the variables (i.e., AD, BC, BD and CD) for membrane surface pore diameters. As shown in Table 3.3, BD and CD seemed to be highly significant model terms when compared with the other model terms. In this case, the most significant factors are voltage supply and concentration. The effects of the interactions of time duration and flow rate with the other parameters on the surface pore diameter of the e-PVA membranes are investigated. This model suggested that the flow rate and time duration are more significant when they are in combination with other parameters than alone.

Figure 3.6 shows the actual versus predicted value plots where the predicted values attained from regression model were compared with the experimental values to check the reliability and suitability of the empirical model for individual responses. Therefore, the comparison between predicted and actual values for membrane fiber diameter and membrane surface pore diameters are presented in Figure 3.6a and 3.6d, respectively. As clearly shown from the figures, that all the design points are distributed near

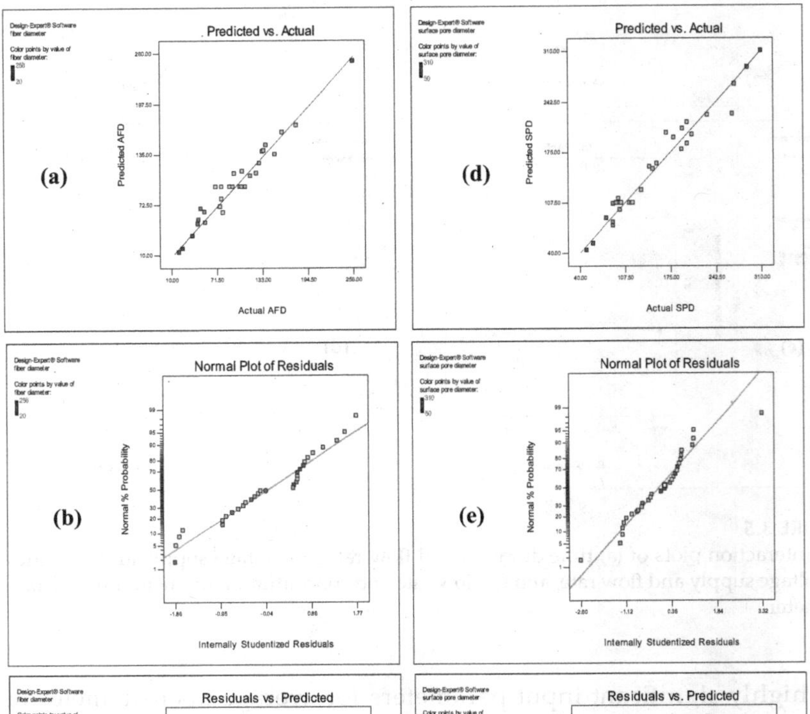

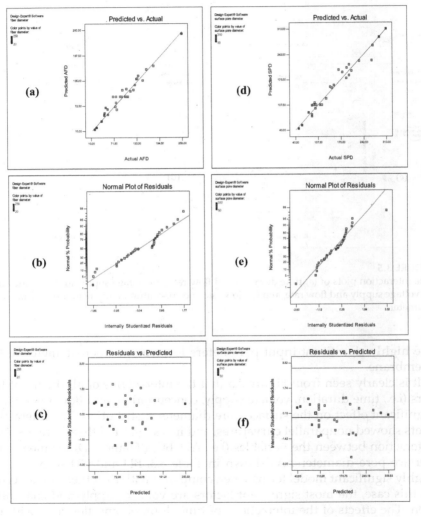

FIGURE 3.6
Plots of predicted versus actual values of membrane (a) fiber diameter ($R^2 = 0.96$) and (d) surface pore diameter ($R^2 = 0.96$); normal probability plot of residual (b) membrane fiber diameter and (e) membrane surface pore diameter; plot of residual vs. predicted of (c) membrane fiber diameter and (f) membrane surface pore diameter.

the straight line, where the points above the straight line are overestimated and below the straight line are underestimated. Based on these results we can conclude that the estimated models are acceptable and there are no violations of constant variance assumptions. Furthermore, according to the data, the empirical model attained from CCD can be used as a predictor for the optimization of the four input parameters to achieve a required average fiber diameter and surface pore diameter depending on the application interest.

Figure 3.6b and 3.6e show the normal probability plot of residuals for fiber diameter and surface pore size, respectively. The plots in Figure 3.6b and 3.6e ensured that no abnormality signal of the experimental results was observed. The falling of the residual points on a straight line suggests that the errors are normally distributed. Figure 3.6c and 3.6f show the plots of the residuals versus predicted responses for the membrane fiber diameter and membrane surface pore diameter, respectively. The random scattering of all experimental data points across the horizontal line of residuals suggests that the projected models are suitable.

3.1.3.4 Optimization Study

The investigation of the optimized input parameters (electrospinning parameters) was done through a desirability function (D) for two responses using Equation 3.3 [18]. The optimum time duration, concentration, voltage supply, and flow rate for preparation of the electrospun membrane predicted from all responses with a high or low limit of inputs can be satisfied with the desirability function (D).

$$D = \left[\prod_{i=1}^{N} d_i^{r_i} \right]^{1/\sum r_i} \tag{3.3}$$

where D is the desirability function, N is the number of responses, r_i refers to the significance of a particular response, and d_i indicates the partial desirability function for specific responses. The desirability plot presented in Figure 3.7 confirms that the desirable time duration, concentration, voltage supply, and flow rate are 43.48 min, 10.67 kV, 10.89 wt.%, and 1.01 mL/h, respectively, which gave optimized average fiber diameter of 110 nm and average surface pore diameter of 120 nm. The prediction of desirable input variables is also confirmed with the optimized input variables calculated from central composite design (listed in Table 3.4).

This study draws the conclusions of an empirical exploration into the effects of time duration, voltage supply, concentration, and flow rate (while the distance and collector rotating speed were kept constant) on fiber diameter and surface pore size distribution during the preparation of e-PVA membranes. Therefore, 10 wt.% aqueous PVA solutions were chosen as the

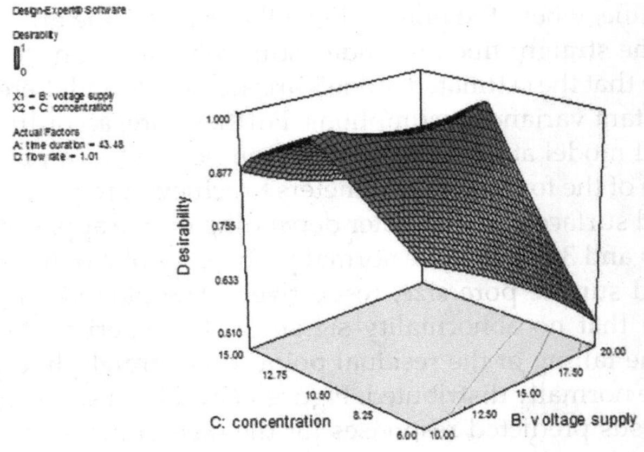

FIGURE 3.7
Response surface plot of desirability operating region: voltage supply and concentration of 10.67 kV and 10.89 wt.%, respectively.

optimized solution concentration; an applied voltage of 12 KV was selected throughout the process, and the feed rate was 1.0 mL/h by means of a 50 mL syringe using a needle of 0.4 mm internal diameter. For further investigation on the effect of time duration, membranes were prepared at 25, 35, 45, and 60 min. The conductivity, surface tension, and viscosity values of the PVA solution were 0.079 mS cm^{-1}, 73.7 mN/m, 0.299 Pa.s, respectively.

3.1.3.5 Morphological Study (FESEM)

The FESEM images of the electrospun PVA fibers obtained with 10 wt.% for different electrospinning durations are shown in Figure 3.8. Fabricated PVA nanofiber membranes showed a smooth morphological structure, without developing beads. The fiber diameter distributions and surface pore diameters of the membranes were measured using Image J software from FESEM images. We have measured more than 100 fibers of each membrane sample using Image J software to get the average fiber diameter. As shown from Figure 3.8 (a_2, b_2, c_2, d_2), the fiber diameter results indicated that diameters between 78 and 276 nm for 25 min (a_2); between 81 and 190 nm for 35 min (b_2); between 59 and 160 for 45 min (c_2); and between 37 and 199 nm for 60 min (d_2). The average fiber diameters were varied as 124, 117, 100, and 88 nm for samples designated as M_x (x = 25, 35, 45, 60 min), respectively. It can be seen from FESEM images presented in Figure 3.8 (a_1, b_1, c_1, d_1) that the slight decrease in average fiber diameter can be related to increasing in ambient temperature during the electrospinning duration. If the electrospinning duration of the surrounding electric field delays, the surrounding temperature may increase due to increase in an electron temperature because of collision between the charges. Therefore, increasing temperature has the effect of decreasing the

TABLE 3.4
Optimized Input Variables Calculated from CCD

No.	Factor 1: A, Time Duration (min)	Factor 2: B, Voltage Supply (kV)	Factor 3: C, Concentration (wt.%)	Factor 4: D, Flow Rate, (mL/h)	Response 1: Fiber Diameter (nm)	Response 1: Surface Pore Diameter (nm)	Desirability	Remarks
1	34.00	12.13	10.26	1.10	110	130	1.00	Selected
2	40.82	11.32	11.41	1.12	110	130	1.00	
3	38.68	11.57	11.48	1.17	110	130	1.00	
4	44.11	10.59	11.03	1.02	110	130	1.00	
5	46.33	10.14	11.41	1.07	110	130	1.00	
6	43.48	10.67	10.89	1.01	110	130	1.00	

Note: This optimization study shows that the desirability value of the model is D = 1.00.

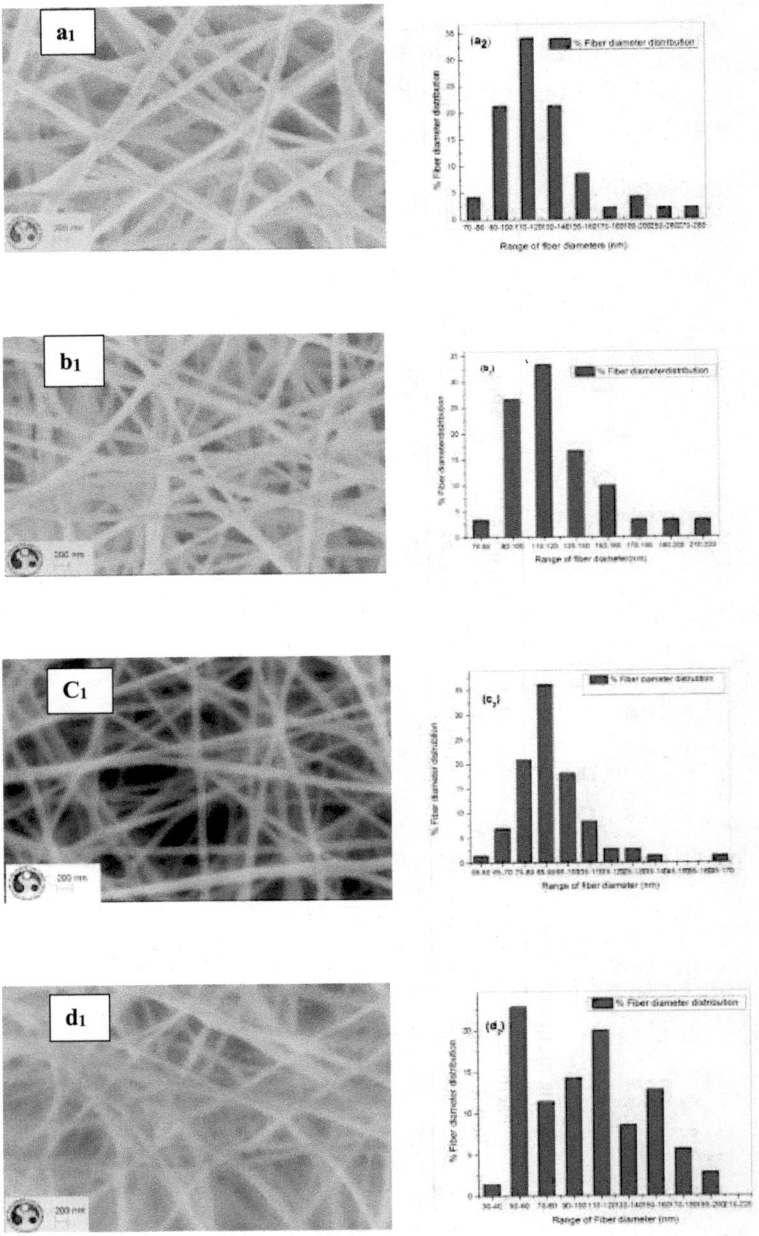

FIGURE 3.8
FESEM images of the ePVA fibers and their corresponding fiber diameter distribution at different process time, (a_1, a_2) 25 min, (b_1, b_2) 35 min, (c_1, c_2) 45 min, and (d_1, d_2) 60 min.

viscosities of the polymeric solutions, increasing solvent evaporation rate, and also may cause a high degree of polymer solubility within the solvent. As a result of these effects, the columbic forces would be able to affect the surface tension of the solution greatly and apply a larger stretching force within the solution, causing the fabrication of thinner fibers. Therefore the effect of the electrospinning duration alone on the fiber diameter distribution is not highly significant when compared with other parameters such as concentration and voltage supply. Due to this reason, the distribution of fiber diameters of the four different samples deposited at 25, 35, 45, and 60 min indicated similar distribution. These results show a similar agreement with statistical analysis of this study. The electrospinning method is an effective technique to produce nano meter range fibers and nanofibrous membranes with high porosity within nano to micrometer range pores [19]. The membrane surface pore size distribution was examined using Image J software from the FESEM images.

The pore size distributions are presented in Figure 3.9. The surface pore size of the electrospun PVA membranes increased as the electrospinning duration time decreased. This result can be explained due to the fact that increasing the

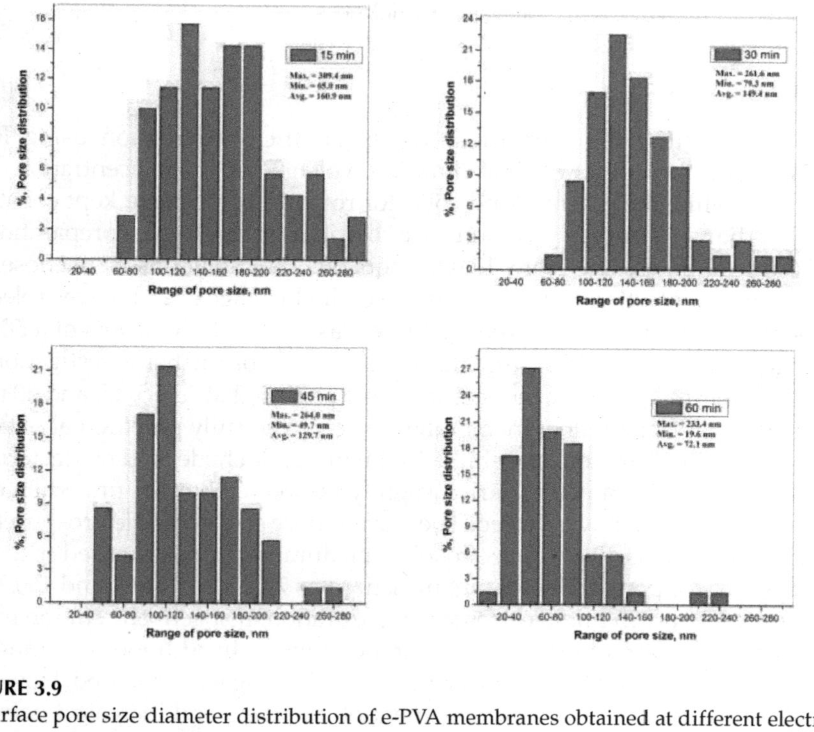

FIGURE 3.9
% Surface pore size diameter distribution of e-PVA membranes obtained at different electrospinning duration, 25 min, 35 min, 45 min, and 60 min.

electrospinning duration means that we are allowing more fibers to be collected on the collector plate and the entanglements of the fibers were practically increased. In other words, more fiber layers were formed by increasing electrospinning duration, where the surface pore was reduced due to the entanglements of the fibers layer-on-layer. Thus, the average surface pore diameters are measured as 72.1, 129.7, 149.4, and 160.9 nm for 60, 45, 35, and 25 min electrospinning duration, respectively. All the membranes fall into ultrafiltration/microfiltration ranges. The decreasing of the average surface pore size is due to an increasing in the number of entangled fibers over and over as the electrospinning duration increases from 25 to 60 min. Therefore, due to the increase in entanglement of these fibers, there is a possibility of formation of a narrower net-like structure that leads to the decrease of surface pore diameters of the membranes. The thickness of the membranes has increased as the electrospinning duration increases. The thickness measurement of the nanofiber membranes was performed using Lieca microscope as shown in Figure S 3.8 in the appendix. The images revealed that the thickness of the membranes increased from 20 to 38 μm for increasing electrospinning duration from 25 to 60 min, respectively, in which the surface pore size is inversely proportional to membrane thickness [20]. But the depth/flowing channel of the membrane pore is expected to increase with increasing thickness.

3.1.4 Summary

This study draws the conclusions of an empirical exploration using RSM method into the effects of time duration, voltage supply, concentration, and flow rate (while the distance and collector rotating speed were kept constant) on fiber diameter and surface pore size distribution during the preparation of e-PVA membranes. Therefore, 10 wt.% aqueous PVA solutions were chosen as the optimized solution concentration; an applied voltage of 12 KV was selected throughout the process, and the feed rate was 1.0 mL/h by means of a 50 mL syringe using needle of 0.4 mm internal diameter. For further investigation on the effect of time duration, membranes were prepared at 25, 35, 45, and 60 min. Electrospun PVA nanofiber membranes were successfully prepared at selected electrospinning duration, crosslinked with glutaraldehyde, and characterized. The average fiber diameters varied slightly between 88 to 124 nm when electrospinning duration was varied. The surface pore size of the electrospun PVA membranes increased as the electrospinning duration time decreased. Thus the average surface pore diameters are measured as 72.1, 129.7, 149.4, and 160.9 nm for 60, 45, 35, and 25 min electrospinning duration, respectively. All the membranes fall into ultrafiltration/microfiltration ranges. In addition, no change in surface pore size and fiber diameter after cross-linking was observed. However, crosslinking led to uniform arrangement of the fibers and increased network rigidity. The FESEM results agreed with the FTIR and TGA results in that the cross linker glutaraldehyde has reacted properly and confirmed the formation of an acetal-bridge. This study showed that membrane properties could be controlled by varying the electrospinning duration along with other process and solution parameters. However, future studies will be needed to fully investigate the performance characteristics of the electrospun PVA membranes for its flux, permeability, and fouling performances studies.

3.2 Preparation of CA_TiO_2 Electrospun Composite Membranes

3.2.1 Introduction

Cellulose acetate is broadly applicable for the synthesis of membranes because of tough, biocompatible, hydrophilicity characteristics, good desalting nature, high flux, and it is moderately less expensive [21–23]. Hybrid membranes of organic–inorganic are formed by the blending of inorganic oxide with the polymeric solution. Current researchers have paid attention to TiO_2 because of its stable nature, it is easily available, and it has potential for different applications [24,25]. Several research groups have developed different composite membranes of polymer–TiO_2 by using phase inversion method. Polyethersulfone/TiO_2 composite ultrafiltration membranes were prepared, where water flux and antifouling property were significantly enhanced [26]. A polyurethane/TiO_2/fly ash composite membrane was fabricated by using electrospinning technique for water purification [27]. Polyacrylonitrile–TiO_2 [28], TiO_2 and polyvinyl alcohol coated polyester [29], Polyethylene glycol/TiO_2/PVDF composite membranes [30] and enhancement of membrane properties of TiO_2 nano fibers by strengthening with polysulfone [31] were fabricated for different applications. Electrospun cellulose acetate_TiO_2 hybrid membranes with improved morphological structures were prepared by using electrospinning technique. The effects of TiO_2 on prepared membrane morphological structure were studied. The authors strongly believe that this work will have substantial contribution to the current state-of-the-art on the preparation and enhancement of the properties of conventional cellulose acetate membranes.

3.2.1.1 Solution Composition

The main aim of this study was to fabricate cellulose acetate_TiO_2 hybrid membranes using electrospinning technique. Organic–inorganic hybrid membranes are prepared using mixing inorganic oxide with the polymeric solution. Present researchers have paid attention to TiO_2 NPs because of its stable mechanical strength, it is less costly, and it has high potential for various applications. Furthermore, TiO_2 NPs can improve the hydrophilic nature of various polymers to enhance flux and decline fouling problems that are significant factors in water and wastewater treatments [24]. In this study, a mixture of acetone and dimethylacetamide (DMAc) was chosen as cellulose acetate solvent during electrospinning. To get the optimum electrospun membrane

TABLE 3.5
Solution Compositions of CA_TiO_2 Membranes

Membrane	CA (wt.%)	TiO_2 (wt.%)	Acetone/ DMAc (wt.%)
M_1	13.5	0	86.5
M_2	13.5	1.0	85.5
M_3	13.5	2.5	84.0
M_4	13.5	4.5	82.0
M_5	13.5	6.5	80.0

with good morphological structures, four different concentrations (25, 20, 15, 13.5 wt.%) of cellulose acetate were prepared and 13.5 wt.% was selected for the next experiment. Cellulose acetate solutions were prepared by uniformly dissolving in 2:1 ratio of acetone/DMAc [11]. As seen from Table 3.5, different amounts of TiO_2 NPs were added to the cellulose acetate to enhance the physical and mechanical property of cellulose acetate. Therefore, different amounts (0, 1.0, 2.5, 4.5, and 6.5 wt.%) of TiO_2 NPs were dispersed in 13.5 wt.% solution of CA. The mixed solution was stirred for 4 h at 200 rpm speed and sonicated for 3 h to ensure the optimum distributions of the TiO_2 NPs in the CA solution.

3.2.1.2 CA, CA_TiO_2 Membrane Fabrication

During the electrospinning of 20 and 25 wt.% solutions of CA, difficulty of getting uniform bending instability was observed, and not enough fibers were deposited due to the interruption of the spinning process due to the development of beads at the tip of the needle. Uniform bending instability and the continued deposition of fibers was observed as the solution concentration was reduced to 13.5 and 15 wt.% of CA throughout the electrospinning process. Finally, 13.5 wt.% solution of CA was selected as optimized solution, and the membrane from this solution was fabricated at process parameters of an electric field of 14 KV, solution feed rate of 2 mL/h, and the needle tip to collector plate distance was kept as 100 mm. As seen from Figure 3.10a, the CA fibrous membrane has shown sticky-cotton or sponge-like structure, shown by the rectangular shape, which was explained as a poor physical property where the fibers do not adhere to each other [32]. For this reason, we have chosen TiO_2 as additive to the CA membranes to improve their morphological properties. Then, the prepared hybrid solution (CA_TiO_2) was put into the electrospinning setup for the fabrication of CA_TiO_2 hybrid membranes. It was clearly observed that due to the addition of TiO_2, the sticky-cotton or sponge-like structure of the CA membrane was turned to smooth and tightened up membrane structure as shown in Figure 3.10b.

3.2.1.3 Study of TiO_2 Nanoparticles

As clearly shown in Figure 3.11, the size of TiO_2 was determined by transmission electron microscopy (TEM). The TiO_2 appeared in the form of spots. To measure

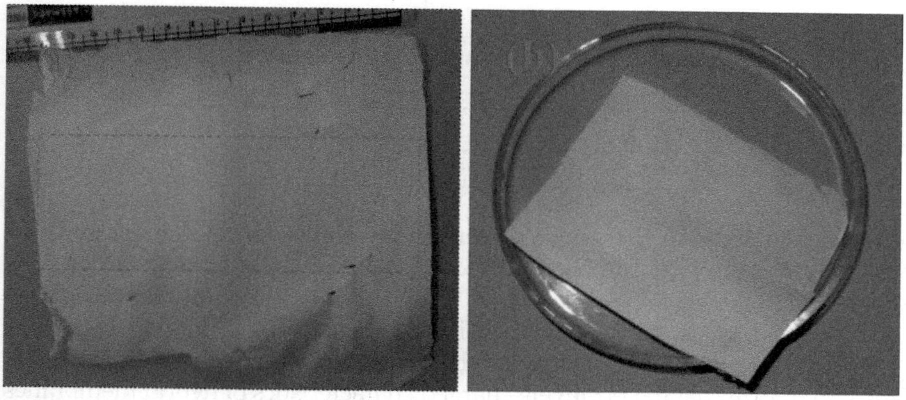

FIGURE 3.10
Images of (a) fabricated CA membrane and (b) CA_TiO_2 hybrid membrane.

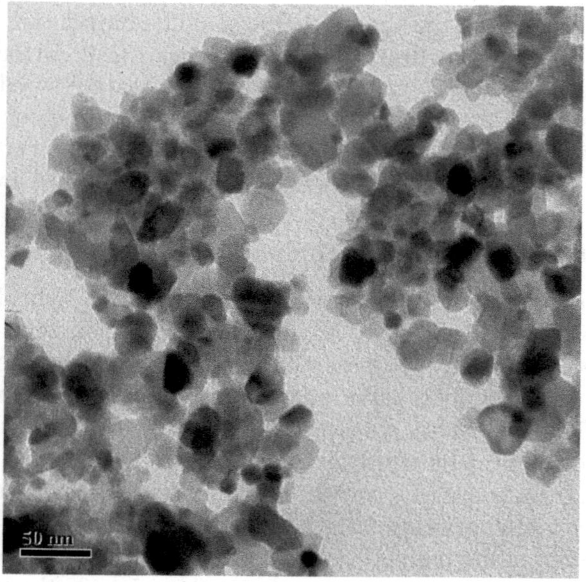

FIGURE 3.11
TEM image of commercial TiO_2 nanoparticles.

the size of each nanoparticle, Image J software was used, and their size ranged from 16 to 72 nm. The average particle size was approximately 29.8 nm.

3.2.1.4 Surface Charge Study

The outer surface zeta potential (ζ) values of the solutions of CA and CA_TiO_2 and TiO_2 at pH of 7 are presented in Table 3.6. The solution composition for CA membrane (M_1) had a negative outer surface ζ value (−5.10 mV) due to the

TABLE 3.6

Zeta Potential Values: Solution of CA and CA_TiO_2, and TiO_2

Samples	M_1	M_2	M_3	M_4	M_5	TiO_2
Zeta potential (mV)	−5.10	−4.13	−2.95	−1.74	+16.19	+33.4

dissociation of carboxylic groups present in the cellulose acetate chains and molecular structure [33]. The outer surface ζ values of the solution compositions for M_2, M_3, and M_4 were −4.13 mV, −2.95 mV, and −1.74 mV, respectively. However, further increasing the amount of TiO_2 blended with CA solution, a high positive ζ value was attained reaching a value of +16.19 mV at 6.5 wt.% of TiO_2. These results confirm that the positively charged characteristics of hybrid membranes were enhanced with increasing the amount of TiO_2 within the CA solution. On the other hand, the ζ value of the pure TiO_2 nanoparticles was positive (+33.4 mV) at pH = 7. The limitations, particularly the oxygen vacancies on the surface morphology and structure of the nanoparticles, could be introduced, and this may change the electronic structure of the nanoparticles [34]. Furthermore, water molecules can occupy the oxygen vacancies and produce adsorbed –OH groups, which indicated that the nanoparticles were positively charged. Nevertheless, the outer surface ζ values for the hybrid membranes containing TiO_2 were negative at the studied pH value (1.0 wt.%, 2.5 wt.%, and 4.5 wt.% of TiO_2).

3.2.1.5 Membrane Morphological Study

The FESEM images of the electrospun CA and CA_TiO_2 hybrid membranes, which are designated as M_1, M_2, M_3, M_4, and M_5, are shown in Figure 3.12 (a_1, b_1, c_1, d_1, e_1). The electrospun membranes with good morphological structures that have smooth fibers and uniform fiber entanglements were fabricated successfully. The FESEM images were taken to evaluate the impact of TiO_2 on the electrospun hybrid membranes morphological structures. According to these images, increasing TiO_2 content on cellulose acetate caused the formation of largely interconnected fiber networks that may have good effect on the enhancement of the membrane strength and membrane pore structures. When the addition of TiO_2 was increased from 0 to 6.5 wt.%, the entanglements of the fibers and the network between the fibers were increased. However for higher amount of TiO_2 (4.5 and 6.5 wt.%), large amount of agglomerations were observed on the membrane's top layer, and some of the fibers turned from a round shape to flat shape (Figure 3.12d_1, e_1). Therefore M_2 and M_3 were chosen as best hybrid membranes with their optimized morphological structures. The elemental results of the electrospun membranes are shown from the EDS images in Figure 3.12 (a_2, b_2, c_2, d_2, e_1). The EDS image clearly confirmed the existence of the TiO_2 NPs on the fibers within the membrane matrix. When the TiO_2 percentage increases from 0 to 6.5 wt.%, the composition of titanium atom increased from 0 to 10 wt.%. As shown in Figure 3.12a_2, only oxygen

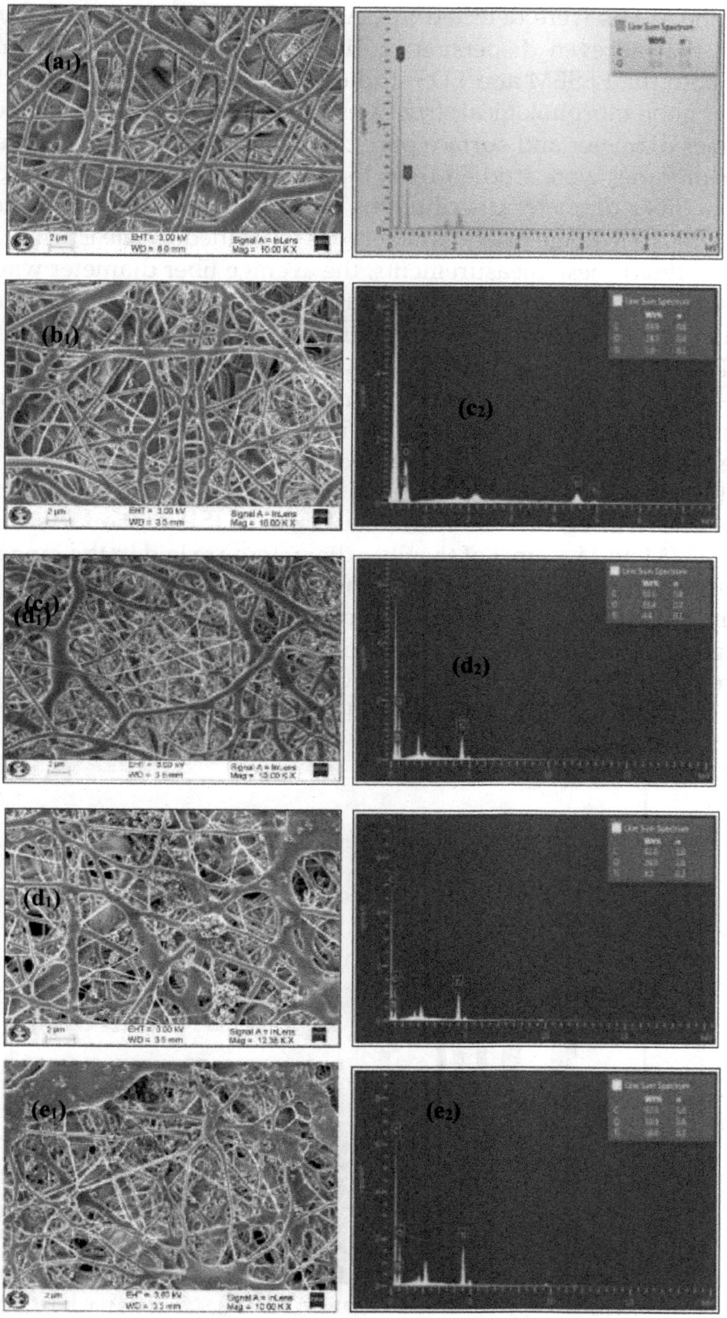

FIGURE 3.12
FESEM and EDS images of CA, CA_TiO_2 electrospun fibers M_1 (a_1, a_2), M_2 (b_1, b_2), M_3 (c_1, c_2), M_4 (d_1, d_2), and M_5 (e_1, e_2).

and carbon atoms were detected in CA membrane. The EDS mapping also revealed that an even dispersion of TiO_2 was observed in the membrane matrix. Both the FESEM and EDS studies revealed that the fabricated membrane has good morphological structures with a uniform distribution of TiO_2.

The fiber diameter and surface pore diameter distributions of the electrospun membranes were studied from the FESEM images using Image J software. The fiber diameter distributions of all the membranes are shown in Figure 3.13. An average fiber diameter was examined by measuring around 600 fibers. From these measurements, the average fiber diameter was found to be 0.43 μm. From the whole fiber diameter measurements, the minimum and maximum fiber diameters were found to be 0.06 μm and 1.2 μm, respectively. The effect of the increasing TiO_2 on the fiber diameters of the membrane was also investigated. A uniform decline in average fiber diameter was confirmed as the content of TiO_2 increases. This uniform decrease in average fiber diameter can be explained as some CA chains may be broken due to the ionization of the solution as the distribution of TiO_2 was further increased, as will be discussed in detail later in this section. As shown in Figure 3.13 (M_1 to M_5), most of the fiber diameters are laid in the range of 0.35 to 0.8 μm; large number of the fibers for M_3 were found in the range of 0.15 to 0.8 μm; the largest number of fibers for both M_4 and M_5 were found within the range of 0.15 to 0.3 μm; and the average fiber diameters for M_1, M_2, and M_3 were 0.6, 0.55, and 0.5 μm, respectively. The average fiber diameters for M_4 and M_5 were found to be 0.4 μm and 0.3 μm, respectively (defective fibers

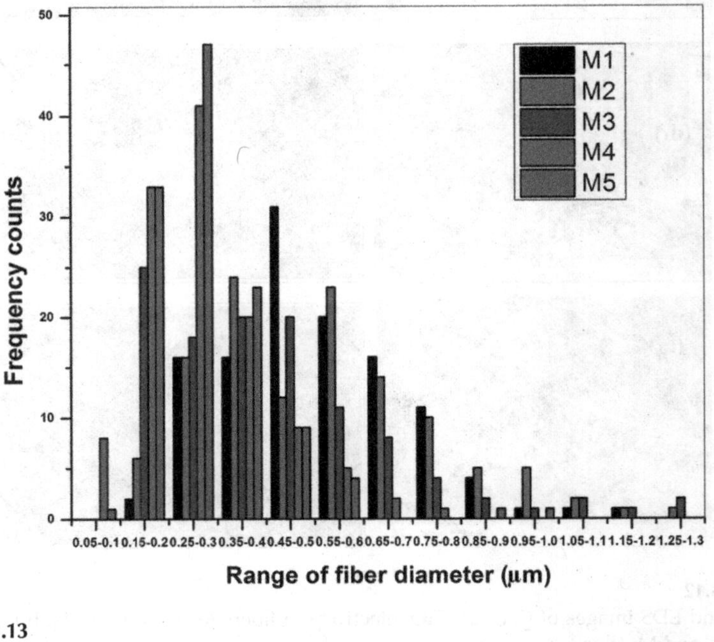

FIGURE 3.13
Fiber diameter distribution of CA (M_1) and CA_TiO_2 electrospun fibers (M_2, M_3, M_4, and M_5).

were excluded). Therefore, the impact of the TiO_2 on the fiber diameters of the hybrid membranes was evidently observed from this study.

Surface pore size distributions of the CA and CA_TiO_2 membranes were examined using Image J software from the FESEM images and are presented in Figure 3.14. One hundred surface pores of each membrane sample were measured. The results showed that the surface pore size of the electrospun

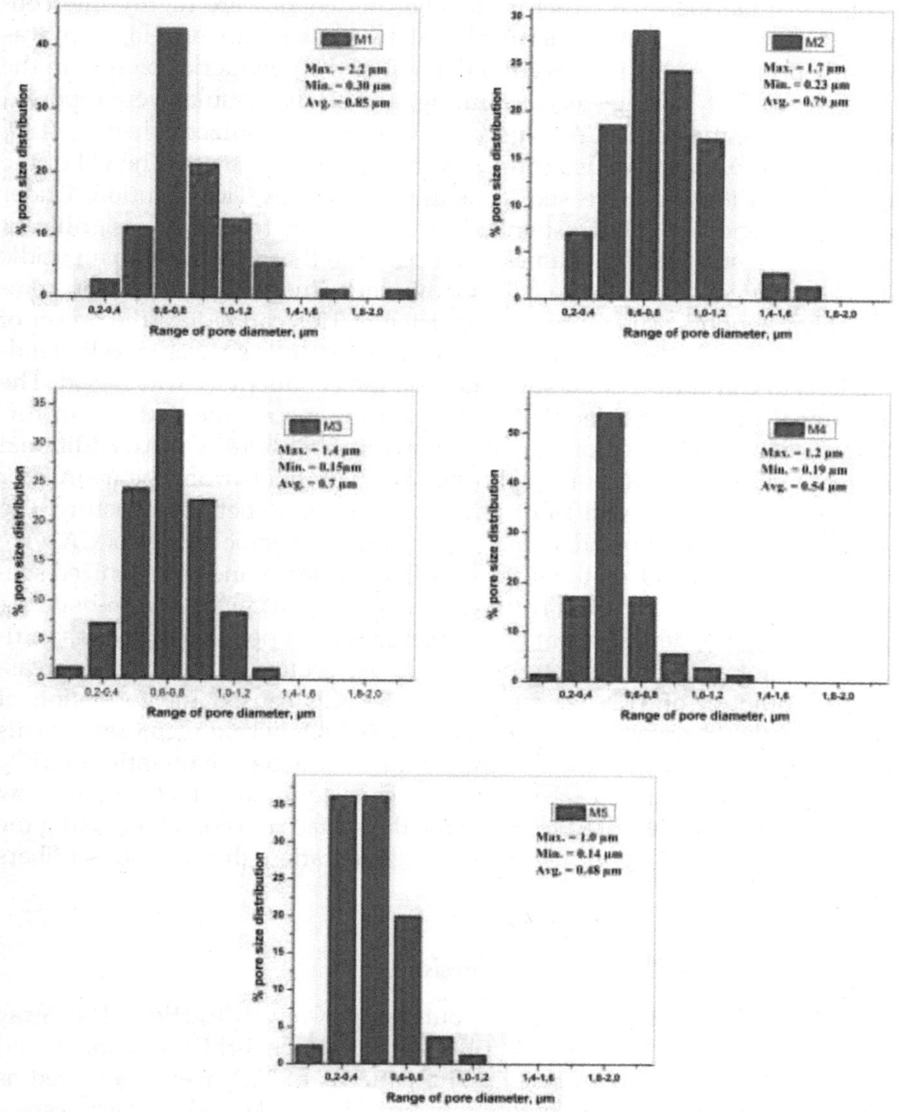

FIGURE 3.14
% Surface pore size diameter distribution of CA_TiO_2 membranes (M_1, M_2, M_3, M_4, and M_5).

hybrid membranes decreases as the content of the TiO_2 increase. For example, the average surface pore diameters are measured as 0.85, 0.79, 0.70, 0.54, and 0.48 μm for M_1, M_2, M_3, M_4, and M_5, respectively. Normally, all the membranes fall into microfiltration ranges. The decreasing of the average surface pore size is due to an increasing in the TiO_2 to the membrane matrix. From the FESEM image analysis, it was indicated that the cellulose acetate matrix comprises less spider-net like network, but, with increasing the TiO_2 content, the spider-net like structure or fiber networks were highly interconnected to each other. The spider-net like network structure development was mainly because of the improved ionization of the polymeric solution in the presence of TiO_2 during electrospinning, and similar results were reported by [35]. Thus, the acidic CA solution may be further ionized when the TiO_2 were added, which was also confirmed by pH measurement. The pH of the CA solution (6) was increased to neutral (7) for CA_TiO_2 solution. Under a neutral condition, electrospun CA_TiO_2 solution results in a significant reduction in surface pore diameter compared to those obtained in an acidic condition of cellulose acetate (CA) membranes, due to the increase in complex fiber network. The uniform dispersion of TiO_2 can reduce the power of the intermolecular hydrogen bonds among polymeric species, and by tending ions to travel freely and subsequently, the conductivity was raised. The additional uniform desperation of TiO_2 on the membrane matrix can further raise the number of ions in the electrospinning solutions. An additional increase of ions can cause the splitting-up of sub-fibers from the main fiber jets and solidification with TiO_2 in the form of a spider-net like structure. The surface hydroxyl group and acetate group part of the ionic species of CA were suggested to hold TiO_2 on the surfaces of fibrous membranes. The ionized species of the polymer and TiO_2 (Ti^{4+}) were linked either using hydrogen-bonding [36] or using the development of complexes within polymeric ligands initiated by the lone-pair of electrons in cellulose acetate. But, further increasing the amount of TiO_2 (4.5 and 6.5 wt.%) NPs caused the formation of thick and flat fibers on some parts of the CA_TiO_2 hybrid membranes. This can be explained due to the fact that, with increasing the quantity of TiO_2, the CA_TiO_2 was turned into a more viscous solution, which might grow the columbic force required to overcome the surface tension for erupting the electro-spinning jet, resulting in large agglomeration, thicker and flat fibers with rough surfaces.

3.2.1.6 Membrane Crystallinity Analysis

Crystallinity analysis was carried out using X-ray diffraction. The X-ray diffraction patterns of CA, CA_TiO_2 electrospun hybrid membranes and TiO_2 are presented in Figure 3.15. The patterns of TiO_2 were presented as $2\theta = 48.06°, 37.86°, 25.41°$, matching with brookite, rutile, and anatase, respectively, where these are characteristics of crystalline peaks [37]. According to these results, it was indicated that the TiO_2 are largely comprised of anatase.

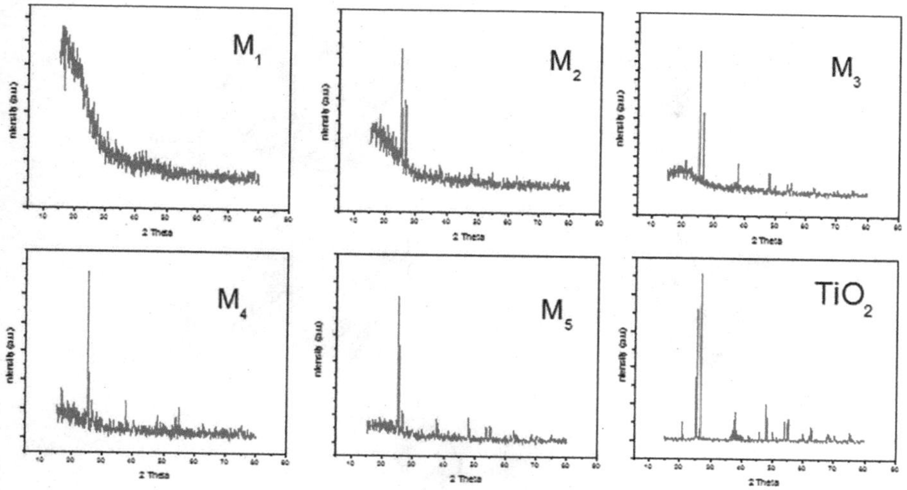

FIGURE 3.15
X-ray diffraction patterns of TiO_2 nanoparticle, CA and CA_TiO_2 electrospun membranes (M_1, M_2, M_3, M_4, and M_5).

Among the three types, anatase has an excellent stability, anti-fouling characters, and hydrophilic nature, which are the important characteristics of filtration membranes. In addition, this is suitable for membrane modification [23]. As seen in Figure 3.15, the wide peak observed below $2\theta = 20°$ for pure CA membrane (M_1), corresponds to the semi-crystalline arrangement of cellulose acetate membrane [38]. For the CA_TiO_2 hybrid membrane (M_2) only one crystalline characteristic peak is observed at 2θ of 24.15° and as the TiO_2 content increases, the CA_TiO_2 hybrid membranes (M_3, M_4, and M_5) showed characteristic of three crystalline peaks at $2\theta = 24.06°$, 37.85°, 48.15° that are similar with the TiO_2 characteristic peaks including the semi-crystalline peak of CA. However, 2θ was slightly shifted for the main peak of TiO_2 NPs as shown in Figure 3.15 (M_2–M_5). This slight shift of TiO_2 peaks in electrospun CA_TiO_2 hybrid membranes may be due to the interactions between cellulose acetate and TiO_2 [39]. The strong and sharp characteristic peaks in the hybrid membranes indicate the good crystallinity of the fabricated membrane.

3.2.1.7 Surface Roughness Study

The surface roughness study of the prepared membranes was conducted using atomic force microscopy (AFM). The three-dimensional surface AFM images of electrospun CA and CA_TiO_2 hybrid membrane are shown in Figure 3.16. The AFM analysis software was used to examine the membrane roughness parameters within a scanning area of 5 μm×5 μm. The mean value of the Z value (R_a), the root mean square of the Z value (Rms (R_q)) and the

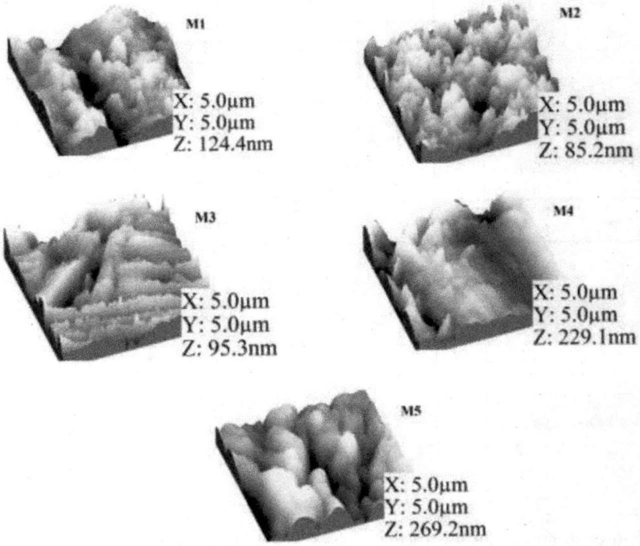

FIGURE 3.16
Three-dimensional AFM images of CA (M_1) and CA_TiO_2 (M_2, M_3, M_4, and M_5) electrospun membranes.

mean difference in the height between five highest peaks and five valleys (R_{max}) were evaluated. It was confirmed that all the roughness parameters of electrospun CA membrane (M_1) were higher than the electrospun CA_TiO_2 (M_2) membranes. As the content of the TiO_2 NPs on CA was raised from 0 to 2.5 wt.%, the membranes roughness also declined from 124.4 nm (M_1) to 95.3 nm (M_3). It might be considered that the uniform distribution of TiO_2 NPs within the matrix of the electrospun membranes made the membrane have denser skins, in which the rough and cotton like structure of CA membrane was changed into a smoother surface. However, as the TiO_2 NPs content increased to 4.5 and 6.5%, the roughness for M_4 and M_5 (Figure 3.16) was increased largely to 229.1 and 269.2 nm, respectively, which may be due to the high agglomeration of TiO_2 NPs on the membrane's matrix as already seen in Figure 3.12d_1 and 3.12e_1. Membranes with smoother surface have found to have better anti-fouling ability and good permeability properties. The membranes fouling tendency could rise, and the flux property would decline as the roughness of the membrane increases due to pollutants collecting on the valleys and peaks of the irregular membrane surface [24,30].

3.2.1.8 Thermal Stability Analysis

The thermo gravimetric analysis (TGA) was used to study the thermal analysis of (a) pure CA and (b–e) CA_TiO_2 hybrid membranes as shown in Figure 3.17. In addition to a slight change in decomposition temperatures,

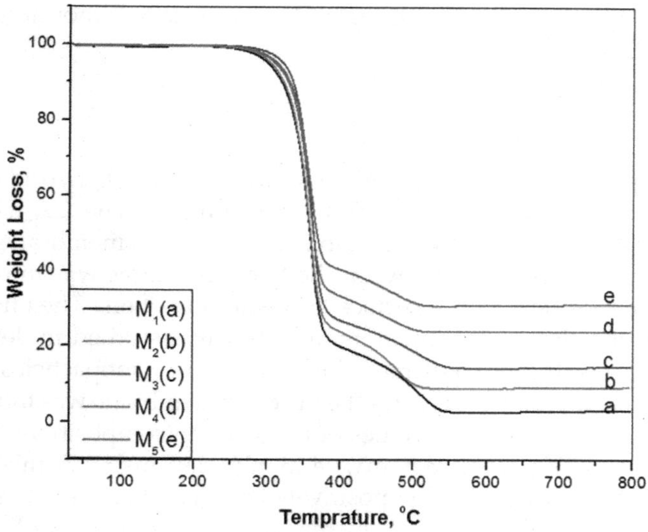

FIGURE 3.17
TGA analysis of (a) cellulose acetate and (b, c, d, and e) cellulose acetate_TiO_2 hybrid membranes.

the weight losses of the membrane samples were found to be 97, 91, 85, 76, and 69% for M_1, M_2, M_3, M_4, and M_5, respectively. The weight loss decreased uniformly with increasing TiO_2 amount in the hybrid membrane, and additional heat was transferred to the TiO_2 in the membrane throughout the analysis. The residual mass of CA_TiO_2 membrane is higher than that of TiO_2 free membranes where the degradation amount for the CA membrane was improved by the addition of TiO_2 nanoparticles. Thus, the delay in decomposition was due to the slight enhancement of CA_TiO_2 membrane's decomposition temperature (T_d). This result may be due to the increase in rigidity of the polymer chain due to the interaction between CA and TiO_2 NPs, which improves the energy of CA chain from simple breaking down. This interaction between CA chains and TiO_2 NPs was suggested, due to the hydrogen or covalent bonds. Similar results were reported by [23,36]. The results confirmed that the TiO_2 NPs were evenly distributed in the electrospun membrane fibers. This good compatibility between the CA and TiO_2 NPs was suggested due to the possible coordination bond of Ti^{4+} and acetate group and the possible hydrogen bond formation among the surface hydroxyl groups and acetate groups of CA [23]. Furthermore, the strong interaction between TiO_2 NPs and the CA chains could avoid TiO_2 NPs from being leached easily from membrane matrix. The results revealed that deposition of TiO_2 NPs into CA matrix had improved the rigidity of the polymer chain, which is significant property to increase the mechanical strength and thermal stability of the membranes. Also, EDS images of M_2–M_5 confirmed uniform distribution of TiO_2 NPs in the matrixes of the electrospun fibrous membranes, regardless of the large amount of TiO_2 NPs agglomeration for M_4 and M_5.

3.2.2 Summary

In this study, a cellulose acetate (CA) _titanium oxide (TiO_2) hybrid membrane was fabricated by using electrospinning technique. The CA_TiO_2 hybrid membranes showed a structure resembling the pure CA membrane. However, the spider-net like network between the fiber structures was improved due to the NPs deposition on the surface of membrane fibers. The effect of TiO_2 contents on the electrospun membrane matrix was studied in detailed. The characterization studies confirmed the existence of nanoparticles in the CA electrospun membrane matrix and the pure CA membrane was modified successfully. The outer surface ζ values of the solution compositions for M_1, M_2, M_3, M_4, M_5, and TiO_2 were –5.10 mV, –4.13 mV, –2.95 mV, –1.74 mV, +16.19 mV, and +33.4 mV, respectively. The positively charged characteristics of hybrid membranes were enhanced with increasing the amount of TiO_2 within the CA solution. The roughness surface observed for electrospun CA membrane (M_1) was enhanced when the content of the TiO_2 NPs on CA was raised from 0 to 1.0 wt.% and 2.5 wt.% (M_2 and M_3, respectively). The weight losses of the membrane samples were found to be 97, 91, 85, 76, and 69% for M_1, M_2, M_3, M_4, and M_5, respectively, and the degradation amount for the CA membrane was improved by the addition of TiO_2 nanoparticles. This modified CA_TiO_2 hybrid membrane will be used for water treatment applications. However, future studies will be needed to fully investigate the performance characteristics of CA-TiO_2 hybrid membranes for specific applications.

3.3 Removal of Heavy Metal Ions Using Composite Electrospun Cellulose Acetate/Titanium Oxide (TiO_2) Adsorbent

3.3.1 Introduction

The contamination of water resources in the presence of heavy metal ions causes major environmental health problems. Heavy metal ions, such as lead, copper, cadmium, chromium, mercury, cobalt, and nickel, are some examples of water contaminants. Therefore, the pollution of water due to the presence of heavy metal ions is a serious environmental concern because of their persistency, poisonousness, and bioaccumulation capability [40]. Among these heavy metals, excessive copper ions can cause lethargy, weakness, and anorexia [41]. Lead ion causes damages to the reproductive, nervous, blood circulation system, neurotoxicity, nephrotoxicity, hematological, and cardiovascular systems [42]. Manufacturing industries, such as leather processing, metal plating, mining, and glass processing wastes, are posing a risk to the community health due to their heavy metal contaminated effluents [43]. Since heavy metals are non-degradable and have a tendency to deposit in living animals or plants, they can cause numerous sicknesses and illnesses [44]. Therefore, the elimination of heavy metal ion pollutants is a significant stage in wastewater treatment processes. The adsorption method is one of the most common processes to eliminate those heavy metals, as a result of its ease, suitability, low cost, and highest removal capacity [45]. The adsorbent properties, such as specific surface area, porosity, and adsorption capacities, must be taken into consideration during the choice of the proper material for the adsorption process to be successful [19,46,47]. Several adsorbents have been employed for adsorption of heavy metals and particularly for the elimination of Pb (II) and Cu (II) ions including polymers [48], polymer/inorganic oxides [49], zeolites [40], and metal oxides [50]. However, reusing of powder adsorbent nanoparticles has remained challenging for researchers. Consequently, current studies have attracted by electrospun membrane materials because of having the satisfactory specific surface area, high porosity, interconnected pore structure, and tailorable thickness that are key factors for adsorption efficiencies and capacities [51]. The cellulosic adsorbent was prepared for effective arsenic elimination from the contaminated solutions using a two-step surface modification [52]. Cellulose adsorbents that showed a high removal potential for arsenic and adsorption capacities were reported as 5.71 mg/g and 75.13 mg/g for As (III) and As (V), for the optimal pH of 7 and 5, respectively. Polyvinyl alcohol and titanium oxide composite membranes were produced through electrospinning and investigated as adsorbents from contaminated solutions. The maximum adsorption capabilities of uranium and thorium were reported as 196.1 and 238.1 (mg/g) using optimum pH of 4.5 and 5.0 at 45 °C, respectively [53]. Polyacrylonitrile/TiO_2 adsorbent beads were prepared to eliminate Pb^{2+} ions from aqueous solutions. The synthesized adsorbents were found to have high porosity and highly stable in strong acids. The Pb^{2+} ion adsorption raised with increasing pH where total removal has attained at pH of 5.6 [28]. A blended solution of polyacrylonitrile, polysulfone with polydopamine membrane having a higher specific surface area was produced using the electrospinning technique. The composite membranes were examined as La (III) ion adsorbents. It was shown that the maximum equilibrium uptake capacity of the La (III) on the adsorbent membrane was 59.5 mg/g [54]. A poly(methacrylic acid) modified cellulose acetate membrane was fabricated by using electrospinning technique [41]. They have investigated the adsorption capacity of the prepared membrane for removal of heavy metals such as Cu^{2+}, Hg^{2+}, and Cd^{2+}. Their adsorption results indicated that higher initial pH values correspond to greater adsorption capability. Furthermore, this adsorbent showed highest adsorption selectivity for Hg^{2+}. On the other hand, blend hollow fiber membranes of chitosan and cellulose acetate were prepared and investigated for copper ion removal in a batch adsorption mode [55]. The adsorption experiments revealed that the blend hollow fiber membrane had good adsorption capability (up to 35.3–48.2 mg/g), short adsorption equilibrium time (20–70 min), and was effective at lower copper ion concentration (i.e., <6.5 mg/L).

Impacts of the amount of TiO_2 nanoparticles, contact time, pH, and temperature on the adsorption processes and adsorption kinetics were investigated. The prepared cellulose acetate/TiO_2 adsorbents with 2.5 wt.% of TiO_2 NPs exhibited smooth morphological structure with a relatively high surface area and porosity. For comparison, the heavy metal adsorption capacities of all the prepared adsorbents (with TiO_2 wt.% of 0, 1, 4.5, and 6.5) were also investigated. Increasing the TiO_2 amount beyond 2.5 wt.% declines the removal efficiency of heavy metals due to a decrease in specific area, porosity, and agglomeration of the NPs within the matrix of the adsorbents. The prepared cellulose acetate/TiO_2 adsorbents with 2.5 wt.% show higher adsorption capacity than other prepared adsorbents.

3.3.2 Batch Adsorption Experiments

The batch adsorption experiments were carried out in a 100 mL borosilicate beaker, and the experimental parameters are presented in Table 3.7. Adsorption experiment was done by adding 100 mg of adsorbents into 50 mL of Pb (II) and Cu (II) solutions separately using an incubated shaker at 150 rpm (solid/liquid ratio: 0.002 g/mL). The parameters, such as initial concentrations of Pb (II) and Cu (II) taken as 50 mg/L, contact time ranged from 0 to 300 min, and temperatures of the solution were varied from 25 °C to 55 °C. The effects of TiO_2 nanoparticles content on adsorption capability of the adsorbents for the elimination of heavy metals were investigated. The effects of pH on the elimination of Pb (II) and Cu (II) ions were studied within the range of 1 to 9 at 35 °C for 5 h agitation period. The effects of the contact time were also investigated by altering the time using a constant temperature of 35 °C.

The adsorbent phase concentrations after equilibrium, q (mg/g), were calculated using the well-known equation

$$q_e = \frac{(C_o - C_e)V}{M} \tag{3.4}$$

TABLE 3.7

Batch Adsorption Parameters

Initial conc. (mg/L)	Adsorbent Dose (mg)	Shaker Speed (rpm)	pH	Contact Time (min)	Temp. (°C)
50	100	150	1,2,3,4,5, 6,7,8,9	0, 30, 60, 90, 120, 150, 180, 210, 240, 270, 300	25, 35, 45, 55

The percentage of removal is calculated as Equation 3.5:

$$R(\%) = \frac{(C_o - C_e)}{C_o} \times 100 \quad (3.5)$$

where q_e is the adsorbent phase concentration after equilibrium in mg of adsorbate per g of adsorbent; V is the volume of the liquid in the solution (L); C_0 and C_e are the initial and the equilibrium concentrations of heavy metal ions in the liquid phase (mg/L), respectively; and M is the mass of adsorbent (g). Regeneration of adsorbent was done by rapidly washing the adsorbents with HNO_3 (1M) and then rinsed using deionized water for three times, followed by drying in a vacuum oven at 50 °C for 24 h [56].

3.3.3 Morphology and Diameter Distributions

In this work, electrospun adsorbents with smooth morphological structures (which have smooth fibers and uniform fiber entanglements) were fabricated using the electrospinning technique. As clearly shown from the FESEM images (Figure 3.18), increasing TiO_2 content on cellulose acetate caused the formation of largely interconnected fiber networks. When the addition of TiO_2 amount was increased from 0 to 2.5 wt.%, the entanglements of the fibers and the network between the fibers were increased. But for the higher amount of TiO_2 (i.e., 4.5 and 6.5 wt.%), some agglomerations were observed on the adsorbent top layer, and some of the fibers turned from a round shape to flat shape. Although the composition and solvents used in this study are different, similar results were reported by [38] on the effects of TiO_2 NPs during the preparation of CA/TiO_2 composite adsorbents.

As seen from Table 3.8 the addition of TiO_2 NPs improved the porosity of the plain adsorbent. An increase in porosity for CA/TiO_2 is because of the uniform network formation within the matrix of the adsorbent after the introduction of the TiO_2 NPs and the hydrophilic nature of the adsorbent, in which the TiO_2 NPs are known as hydrophilicity enhancer. And it is also believed that the occurrence of TiO_2 NPs creates the spaces within the adsorbent matrix by separating the CA chains that lead to the improvement of the adsorbent porosity. However, on the addition of higher TiO_2 NPs (i.e., 4.5 and 6.5 wt.%), the porosity value was observed to decrease. This result is due to the agglomeration of the NPs in the matrix of the adsorbent as clearly seen from the FESEM images (Figure 3.18). In this study, AM2 was chosen as best composite adsorbent due to its improved specific surface area, porosity, and overall best morphological structures.

3.3.4 Textural Properties of the Adsorbent

The textural properties (crystallinity) analysis was carried out using X-ray diffraction. The X-ray diffraction patterns of AM0, AM1, AM2, AM3, and

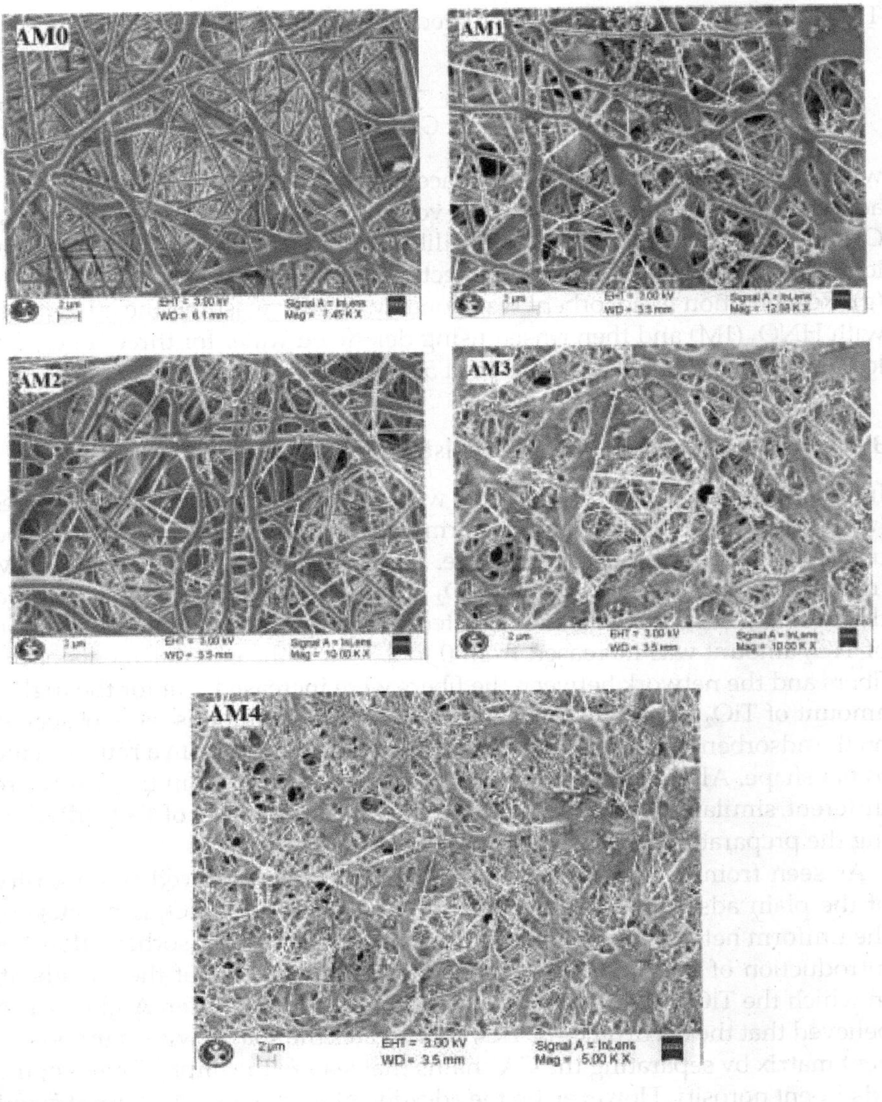

FIGURE 3.18
FESEM images of electrospun adsorbents (AM0, AM1, AM2, AM3, and AM4).

AM4 electrospun adsorbents and TiO_2 NPs are presented in Figure 3.19. The patterns of TiO_2 NPs were presented as $2\theta = 48.06°, 37.86°, 25.41°$, matching with brookite, rutile, and anatase, respectively, where these are characteristics of crystalline peaks [37]. According to these results, it was indicated that the TiO_2 NPs are largely comprised of anatase. As seen in Figure 3.19, the wide peak observed below $2\theta = 20°$ for AM0 corresponds to the semi-crystalline arrangement of cellulose acetate [38]. For AM1 only one crystalline characteristic peak is observed at 2θ of 24.15°, and as the TiO_2 NPs content increases,

TABLE 3.8

Effect of TiO_2 Content on Adsorbent Porosity and Specific Surface Area

Adsorbent	CA Content (wt.%)	TiO_2 (wt.%)	Solvent (wt.%)	Porosity, ε (%)	Specific Surface Area (m^2/g)
AM0	13.5	0	86.5	57.2	30.2
AM1	13.5	1.0	85.5	80.1	38.5
AM2	13.5	2.5	84.0	85.9	48.5
AM3	13.5	4.5	82.0	72.9	19.9
AM4	13.5	6.5	80.0	69.0	12.5

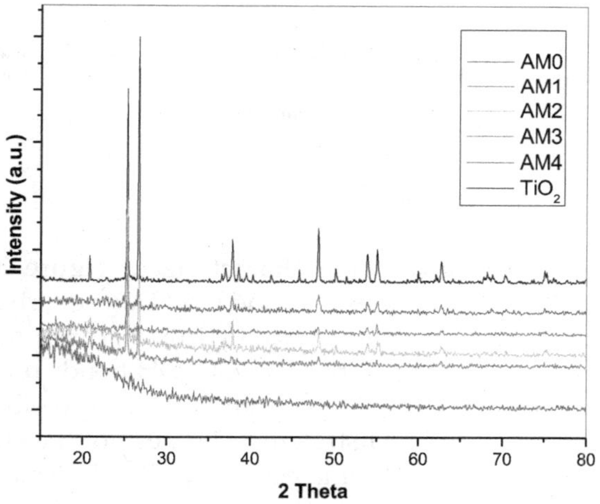

FIGURE 3.19
X-ray diffraction patterns of the TiO_2 nanoparticle, electrospun adsorbents (AM0, AM1, AM2, AM3, and AM4).

the AM2, AM3, and AM4 showed a characteristic of three crystalline peaks at $2\theta = 24.06°, 37.85°, 48.15°$ that are similar with the TiO_2 NPs characteristic peaks including the semi-crystalline peak of cellulose acetate. However, 2θ was slightly shifted for the main peak of TiO_2 NPs as shown in Figure 3.19 (AM1, AM2, AM3, and AM4). This slight shift of TiO_2 peaks may be due to the interactions between cellulose acetate and TiO_2 NPs [39]. The strong and sharp characteristic peaks in the adsorbents indicate the strong textural properties of the fabricated adsorbents that are an important property of the mechanical strength of adsorbents.

3.3.5 Infrared Spectroscopy Analysis

To investigate the chemical structure of the samples, the FTIR analysis was done. Therefore, the functional groups of TiO_2 nanoparticles, electrospun

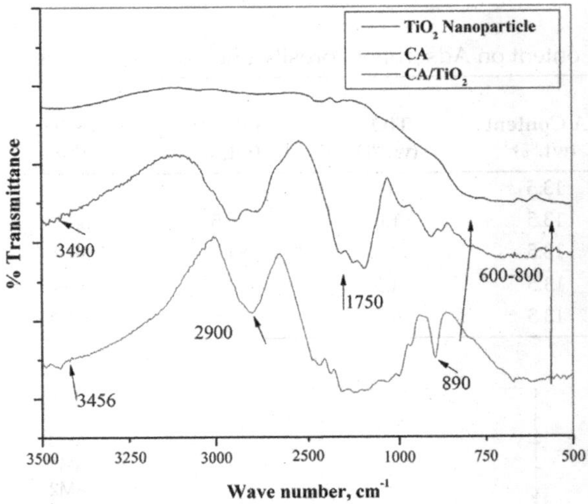

FIGURE 3.20
FTIR spectra of CA, TiO_2, and CA/TiO_2 adsorbents.

CA and CA/TiO_2 adsorbents were characterized. Figure 3.20 shows the FTIR bands for TiO_2 nanoparticles, CA and CA/TiO_2 adsorbents. The broad peak at 3490 cm^{-1} matches to the surface hydroxyl stretching and the peak at 1750 cm^{-1} corresponds to the vibrations of –OH bonds on the surfaces of TiO_2 nanoparticles [53]. In the spectra of pure CA, the broadband detected at 3456 cm^{-1} corresponded to –OH stretching because of the strong hydrogen bond of intermolecular and intramolecular kinds. The characteristic band of 2900 cm^{-1} attributed to the C–H stretching [57]. The patterns of the C–H bending bands in the region 890 cm^{-1} are likewise characteristics of the aromatic replacement patterns in CA. For the TiO_2 nanoparticles and CA/TiO_2 adsorbents, a new broadband around 600–800 cm^{-1} appeared, which is attributed to the Ti–O–Ti band [58]. This result shows that TiO_2 nanoparticles are successfully introduced to the cellulose acetate adsorbent matrix. In addition to this, some small peaks were disappeared after the introduction of TiO_2 nanoparticles.

3.3.6 Specific Surface Area Analysis

The nitrogen adsorption-desorption analysis was done for TiO_2 nanoparticles alone and CA/TiO_2 adsorbent with various TiO_2 doses. The adsorbents designated as AM0, AM1, AM2, AM3, and AM4, the amount of TiO_2 nanoparticles in the cellulose acetate matrix were varied as 0, 1.0, 2.5, 4.5, and 6.5 wt.%, respectively. Adsorption isotherm graphs were plotted as the volume of molecules adsorbed on solid surfaces versus partial pressures (P/P_o) at a constant temperature. Relative pressure is defined as the ratio of actual gas pressure (P) to the saturated vapor pressure of adsorbate (P_0) at a specific temperature. As observed from Figure 3.21, the results indicate that all the isotherm graphs are of type II, which is a characteristic of mesoporous structure. In the case of the isotherm graphs for nanoporous materials, these desorption curves retrace the adsorption curves. But, for macroporous and mesoporous materials, these desorption curves do not repeat the adsorption curves causing in a wide loop (Figure 3.21). In adsorption study, the specific surface area of the adsorbent is the most important factor [59]. In the BET analysis, the effects of TiO_2 nanoparticles on the specific surface areas of the adsorbents were observed. As shown in Table 3.8, the specific surface area of TiO_2 nanoparticles powder (i.e., 59.9 m^2g^{-1}) is higher than those of the adsorbents. The specific surface area of the adsorbent increased from 30.2 m^2g^{-1} to 48.5 m^2g^{-1} as the amounts of TiO_2 nanoparticles in the adsorbent matrix were increased from 0 to 2.5 wt.%, respectively. An additional increase of TiO_2 nanoparticles indicated a decline in specific surface area of AM3 and AM4 (i.e., 19.9 m^2g^{-1} and 12.5 m^2g^{-1}, respectively). This result is suggested due to an increase in agglomeration of the nanoparticles on the top layer and inside the channel of the adsorbent matrix. As already depicted from the morphological structures of the CA/TiO_2 adsorbents in the FESEM analysis (Figure 3.18), further increasing the amount of TiO_2 nanoparticles caused the formation of thick and flat fibers in some parts of the CA/TiO_2 adsorbent that could decrease the porosity of the adsorbent. The specific surface area of TiO_2 nanoparticles is greater than CA/TiO_2 as some of the vacant spaces of the nanoparticles were occupied by CA chains during the production

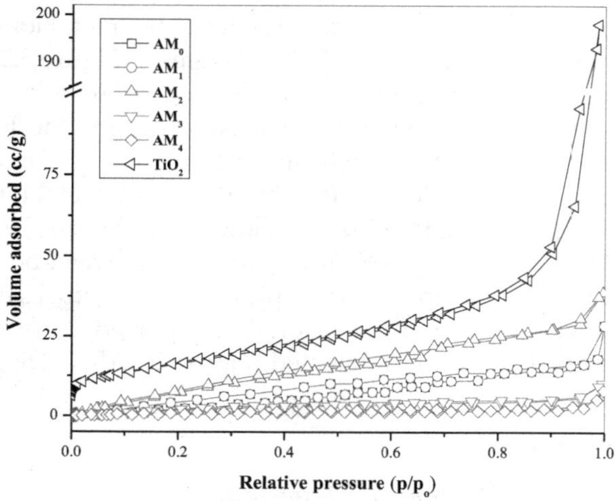

FIGURE 3.21
Adsorption-desorption isotherm graph of adsorbents (TiO_2 nanoparticles and AM0, AM1, AM2, AM3, AM4).

of CA/TiO_2 adsorbents. The main advantage of using films in the adsorption process is that it can be easily recovered without difficulty. Reusing of powder or particulate adsorbent nanoparticles has been challenging [51]. Throughout this study, AM2 was selected as the best adsorbent because of its enhanced specific surface area (Table 3.8), but for comparison, all the adsorbents were tested for their adsorption capacity.

3.3.7 Energy Dispersive X-Ray Analysis

The EDS characterization was used to study the elemental composition analysis of the adsorbents. The FESEM and EDS images of the AM2 (i.e., [a_1, a_2], [b_1, b_2], [c_3, c_2], [d_1, d_2], and [e_1, e_2] for pre-adsorption, after adsorption of Pb [II], after adsorption of Cu [II], after desorption of Pb [II], and after desorption of Cu [II], respectively) adsorbents before adsorption, after adsorption and after desorption of heavy metal ions are presented in Figure 3.22. The images were clearly able to evaluate the efficiency of the CA/TiO_2 adsorbents. As can be seen in Figure 3.22a_1, there is no deposition of metal ions on the surface of the adsorbent matrixes as this image was taken before metal adsorption analysis. During the EDS elemental analysis (Figure 3.22a_1), Pb and Cu elements were not detected at all. On the other hand, deposition/occurrence of lead and copper ions was clearly observed in Figure 3.22b_1 and 3.22c_1 after adsorption experiment. As discussed previously, the adsorbent had smooth morphological properties; it had a satisfactory surface area and porosity. Due to these favorable properties of the adsorbent, the metal ions were easily captured onto the surface of the fiber matrix within the adsorbents. The lead (Figure 3.22b_2) and copper (Figure 3.22c_2) elements were detected in the EDS analysis, which in turn confirmed the presence of the metal ions on the fibers of the matrix of the adsorbent. The FESEM images in Figure 3.22b_1 and 3.22c_1 depicted the development of adsorbent-metal complex ions on the surface of the adsorbent as clearly explains later. Accordingly, the FESEM results agreed with the experimental analysis, in which these adsorbents have shown high adsorption efficiencies for both the heavy metal ions. In the case of the desorption process, it is evidently shown from Figure 3.22d_1 and 3.22e_1 that the metal ions were desorbed successfully and a free fiber network was observed, unlike that of adsorbents morphological structures before desorption study. The elemental analysis strongly supported these results, where elements of Pb and Cu were not detected (Figure 3.22d_2, 3.22e_2).

3.3.8 Adsorption Study

3.3.8.1 Effect of Contact Time and Temperature

The study of the effects of contact period of the synthetic solution and solid phase adsorbent is one of the significant factors for effective use

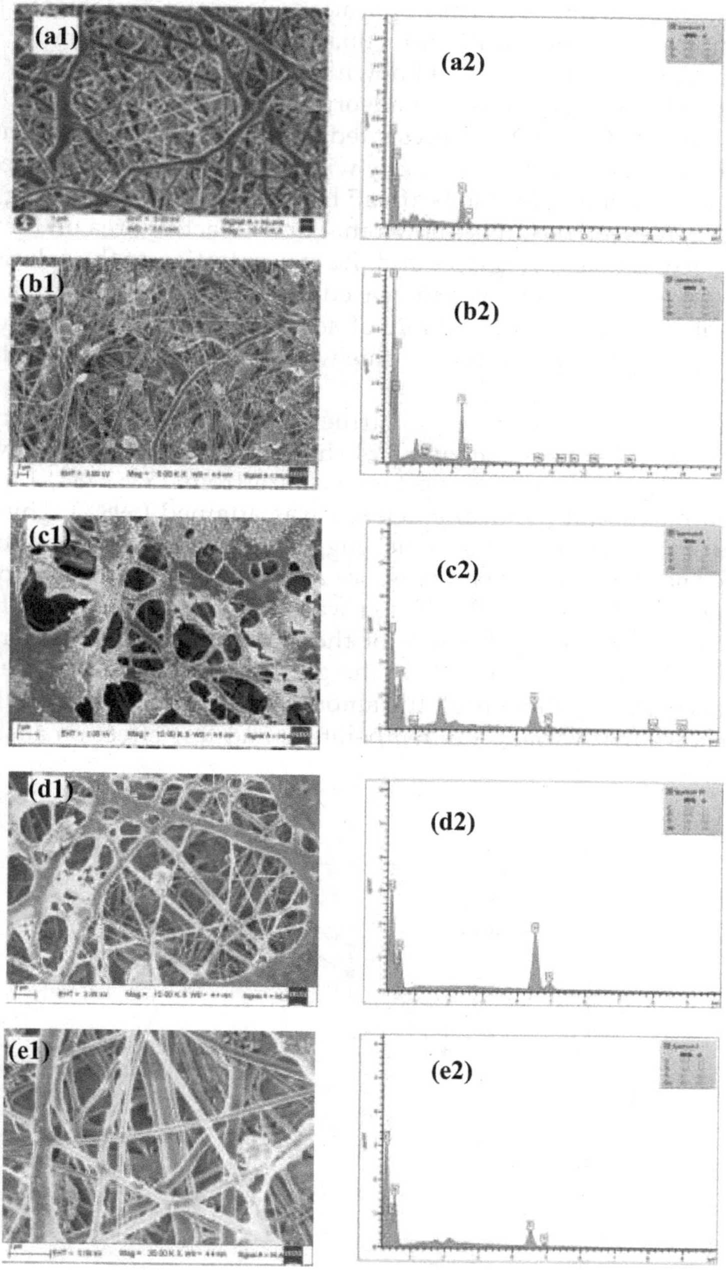

FIGURE 3.22
FESEM and EDS images of (a_1, a_2) AM2 adsorbent, (b_1, b_2) after adsorption of Pb (II), (c_1, c_2) after adsorption of Cu (II), (d_1, d_2) after desorption of Pb (II), (e_1, e_2) after desorption of Cu (II).

of the adsorbent for practical applications in water treatments [46,56]. The samples were shaken in an incubated shaker at 150 rpm for different times (30 to 300 min) with a known dosage of adsorbent at 35 °C. The results of kinetic studies for the adsorption of heavy metal ions Pb (II) and Cu (II) onto CA/TiO_2 are presented in Figure 3.23. The concentration of copper and lead decreases rapidly within 30 min and the final concentrations of copper and lead ions after 5 h contact time were 0.15 mg/L and 0.55 mg/L, respectively. It is known that first all active area on the adsorbent surfaces had empty spaces and the concentration of the solution was high. When contact time was increased, the available active area on the adsorbent was decreased, and rate of adsorption was found to decrease slowly. Consequently, 5 h contact time was chosen for all the equilibrium studies.

Adsorption experiments were furthermore studied at different temperatures (25 °C to 55 °C). Figure 3.24 shows that Cu (II) ion removal efficiency rises quickly with rising temperature from 25 °C (84%) to 35 °C and then maximum removal efficiency was attained (>98%). This rise is because of the stepping up of some sluggish adsorption stages or because of the formation of a new empty active area on the surface of adsorbents. The removal efficiency of Pb (II) ion was almost similar for all the temperatures (>99%), maybe because of the adsorption of Pb (II) ion using CA/TiO_2 adsorbent is favorable at the given temperatures. Increasing of temperature hardly influenced the amount adsorbed ions at equilibrium, which shows that variation in temperature had less effect on adsorption of Pb (II) ions.

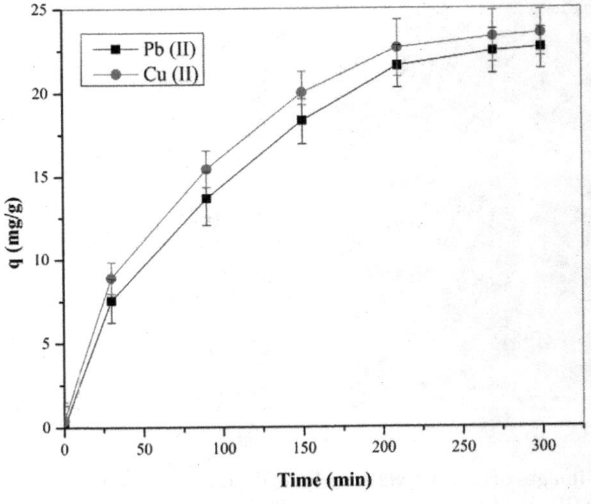

FIGURE 3.23
Effect of contact time on the Pb (II) and Cu (II) ions adsorption onto CA/TiO_2 adsorbent.

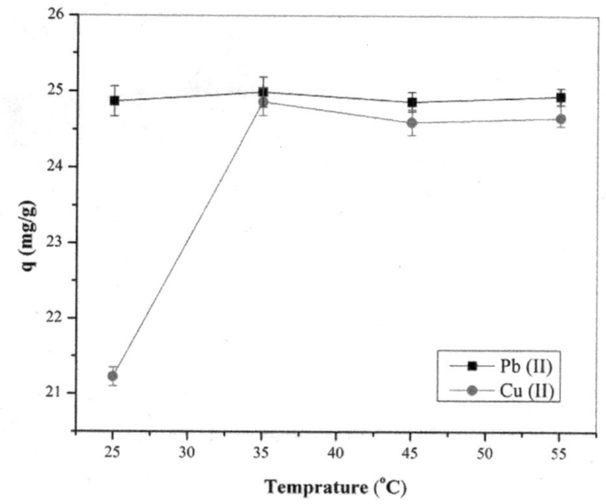

FIGURE 3.24
Effect of temperature on the Pb (II) and Cu (II) ions adsorption onto CA/TiO_2 adsorbent.

3.3.8.2 Effect of pH

To study the adsorption capacity of the adsorbents, the initial pH value of the solution is an important factor. Therefore, optimization of the pH value is required. The procedure of cation exchanges among adsorbent surface and the heavy metal cations comprising solutions are greatly dependent on the pH values of the intermediate. Therefore, the adsorptions of Pb (II) and Cu (II) were investigated at various pH values alternating from highly acidic to neutral (pH = 1.0 to 7.0). For comparison purpose, the adsorbent was tested at pH of 8 and 9. The influence of pH on the adsorption efficiency of metals is presented in Figure 3.25a. The adsorption occurred slightly at low pH values (up to pH = 1.0 to 3.0); beyond this pH level (between 3.0 and 5.0), the quantity of lead and copper ions adsorption on the CA/TiO_2 adsorbent was improved. With further increasing the pH level of the solutions, the adsorption develops significantly and maximizes at pH of 5.2 and 5.8 for both lead and copper ions, respectively. As the pH rises beyond 7, the TiO_2 surface –OH groups suggested making the adsorbent surface more negatively charged. These results were confirmed by analyzing the zeta potential (ζ) values of the solids, and the results are presented in Figure 3.25b. As clearly observed from the figure, at low pH (1–3) values, the surface of the adsorbent had positive charges. However, as the pH values were raised beyond 3, the surfaces of the solids were negative. High negative ζ values were attained with further increase in the pH values to 8 and 9. These results are accredited to an increase in the amount of –OH groups in the solid barrier due to alkaline nature of the solutions. Accordingly, the metal ions with positive charges easily interact with the adsorbent surface.

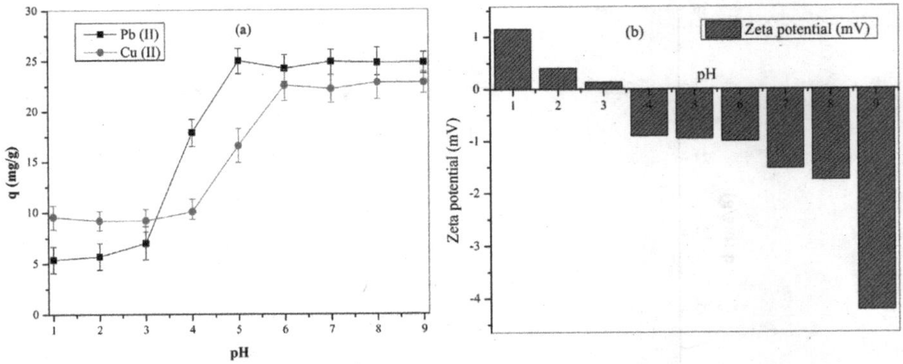

FIGURE 3.25
Effect of pH (a) on the Pb (II) and Cu (II) ions adsorption onto CA/TiO_2 adsorbent and (b) on zeta potential values of the adsorbent.

This condition helps to continue the adsorption process of the adsorbent since the competitions among H^+, and heavy metal ions are reduced [28,53]. Kim et al. [28] have suggested that the monodentate inner-sphere-type surface complexation is the possible mechanism of the elimination of Pb (II) ions by using TiO_2. The possible mechanisms for copper and lead in this study are suggested as

$$\text{Ti–OH} + Pb^{2+} \leftrightarrow Ti - O - Pb^{+} + H^{+} \tag{3.6}$$

$$\text{Ti–OH} + Cu^{2+} \leftrightarrow Ti - O - Cu^{+} + H^{+} \tag{3.7}$$

However, on further increase in the pH level of the solution, an acceptable removal of metal ions was observed. At high pH (alkaline solutions), the metal cations start to react with hydroxide ions to form metal hydroxide and are precipitated [60]. Since the adsorbent used for this study was highly porous, the heavy metal hydroxide precipitates can easily be adsorbed or captured on the surface as well as a depth channel of the adsorbent. Precipitation of each copper and lead ions in their solution as hydroxides was starting to rise at higher pH values (i.e., >7) [61].

3.3.8.3 Effect of TiO_2 Amount

The effect of TiO_2 content percentage on adsorption capacities or removal efficiencies of copper and lead ions onto CA/TiO_2 adsorbent was investigated. The adsorbents with various quantities of TiO_2 nanoparticles (0, 1.0, 2.5, 4.5, and 6.5 wt.%) are designated as AM0, AM1, AM2, AM3, and AM4, respectively. The removal efficiency of lead and copper ions by using CA/TiO_2 adsorbents are shown in Figure 3.26. As presented in Figure 3.26,

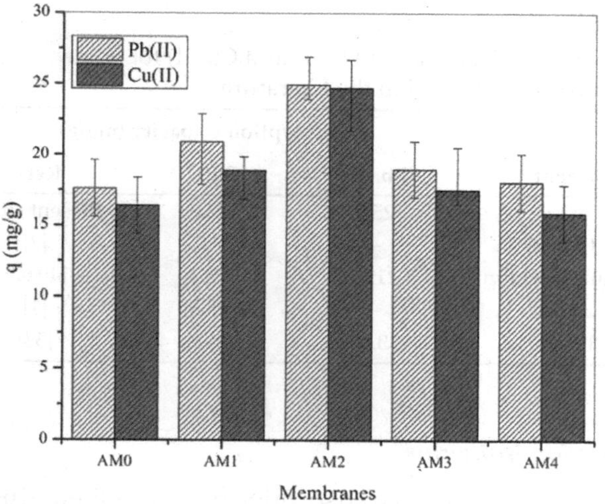

FIGURE 3.26
Effect of TiO_2 nanoparticles amount on removal efficiency of Pb (II) and Cu (II) ions.

the adsorption capacity of Pb (II) and Cu (II) ions rise with the increase of TiO_2 nanoparticles from 0 to 2.5 wt.%. This rise is because of the improved specific surface area of the adsorbent. An additional increase in TiO_2 amounts caused a reduction in the removal efficiency of the adsorbent. In the case of the Pb (II) ion, the removal efficiency of all the adsorbents showed a higher removal percentage, though the slight decrease was observed for TiO_2 contents of 4.5 and 6.5 wt.%. Because of the accumulation and thickening of TiO_2 nanoparticles, this decrease in removal efficiency for both metal ions is observed, which in turn diminishes the existing specific surface area and porosity of the adsorbent. Moreover, the agglomeration of nanoparticles causes difficulty to the Pb (II) and Cu (II) ions to diffuse within the internal and surface pores of the adsorbent matrix. As shown in Figure 3.26, maximum removal of 99.7% and 98.9% were attained for lead and copper ions, respectively, by using AM2 (2.5 wt.% of TiO_2 nanoparticles). Whereas 86.3% removal efficiency for Pb (II) at pH ranging from 4.5–5.0, adsorbent dose 0.5 g/L, initial concentration 30 mg/L and equilibrium time 90 min was reported by [42] and less selectivity for Cu (II) ion at pH of 3.4, initial concentration 50 mg/L and adsorbent dose 2 g/L was reported by [41]. Adsorption capacities of Pb(II) and Cu(II) for some adsorbents reported in the literature is presented in Table 3.9.

Comparatively, for the CA adsorbents without TiO_2, the corresponding removal efficiencies are 65.7% and 70.5% for Cu (II) and Pb (II), respectively, thereby confirming the efficiency of TiO_2 nanoparticles to enhance heavy metal adsorption. Therefore, CA/TiO_2 (2.5 wt.% of TiO_2) adsorbents were chosen for the next step of the adsorption study in which this result agreed with the theoretical conclusion of BET surface area analysis.

TABLE 3.9

Adsorption Capacities of Pb (II) and Cu (II) for Some Adsorbents Reported in the Literature

	Adsorption Capacity (mg/g)		
Adsorbent	**Pb (II)**	**Cu(II)**	**Refs.**
CA/TiO_2	25.0	23.0	Present study
CA/PMMA	–	3.0	[2]
Cellulose/chitin	<19.0	19.2	[40,62]
CA/zeolite	–	28.6	[1]
Sawdust	3.8	–	[33]

3.3.8.4 Adsorption Isotherms

For this work, three well-known adsorption models—Freundlich, Langmuir [50,63], and Dubinin–Radushkevich (D–R) [50]—were investigated to analyze the equilibrium data of Pb (II) and Cu (II) metal ions. In Langmuir model, the adsorption mechanism is based on the physical phenomenon in which the highest adsorption capacity involves of a mono-layer adsorption, where the energy of adsorption is constant; the surface is energetically uniform, with no interaction between neighboring adsorbed molecules. Therefore, the heat of adsorption is constant through the fractional superficial coverage. All adsorptions occur by the same mechanism and result in the same adsorbed structure. The general equation of the Langmuir isotherm is [64]:

$$q_e = \left[\frac{q_{\max} b C_e}{(1 + bC_e)}\right] \tag{3.8}$$

The linear form Equation 3.8 can be written as

$$\frac{C_e}{q_e} = \left(\frac{1}{q_{\max} b} + \frac{C_e}{q_{\max}}\right) \tag{3.9}$$

where C_e (mg/L) is the equilibrium metal ion concentration; q_e (mg/g) is the amount of heavy metal ions adsorbed per unit mass of adsorbent (mg/g) at equilibrium; q_{max} (mg/g) is the highest quantity of the metal ions per unit mass of adsorbents to develop a complete mono-layer, and b (L mg^{-1}) is the Langmuir constant related to the affinity of binding sites.

The Freundlich equation has been commonly used for isothermal adsorption and is dependent on a heterogeneous surface adsorption. It was derived empirically in 1912, and it is defined by the following equation [63]:

$$q_e(x/m) = \left[K_f(C_e)^{\frac{1}{n}} \right] \tag{3.10}$$

The linear form can be written as

$$log q_e = \left[\log K_f + \frac{1}{n}(\log C_e) \right] \tag{3.11}$$

where *x/m* is the weight of contaminants adsorbed per unit weight of adsorbents (mg/g), *n* and K_f (mg/g) $(mg/L)^n$ are the intensity parameter and Freundlich capacity factor of the adsorbent, respectively. The adsorption isotherms data of Pb (II) and Cu (II) were generated at pH of 5.2 and 5.8, respectively, for 300 min using 100 mg of CA/TiO_2 (AM2) adsorbent at 35 °C. The results of adsorption model from linear plots are given in Figure 3.27. All the parameters were computed by using the three isotherm model equations and the R^2 values (Table 3.10) for Freundlich (a), Langmuir (b), and D-R (c) isotherms were compared. The results from Freundlich isotherm revealed that the *1/n* values for both metals were not laid between 0 and 1. This result clearly indicates that the adsorptions of various metal ions using CA/TiO_2 adsorbents were not favorable for Freundlich. Therefore, by looking at the R^2 values (Table 3.10) of the Freundlich ($R^2 > 0.85, 0.91$), Langmuir ($R^2 > 0.98, 0.99$), and D-R ($R^2 > 0.44, 0.52$) models, the Langmuir isotherms model was well-fitted with the experimental data of both metal ions (Cu (II) and Pb (II), respectively) as compared to D–R and Freundlich isotherm models. The Langmuir isotherm model applicability for CA/TiO_2–metal ion adsorption

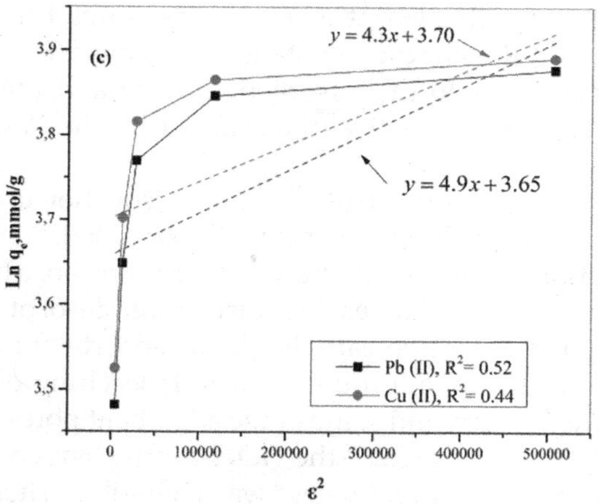

FIGURE 3.27
Adsorption model: (a) Freundlich, (b) Langmuir, (c) D-R plots for copper and lead adsorption onto the adsorbent.

TABLE 3.10

The Three Isotherms Model Parameters for Metal Adsorption onto CA/TiO_2 Adsorbent

	Freundlich Isotherm		Langmuir Isotherm		D-R Isotherm	
Metal Ion	K_f, (mg/g) (mg/L)^n	R^2	q_{max}, mg/g	R^2	q_{DR}, mmol/g	R^2
Pb (II)	6.12	0.91	31.9	0.99	38.55	0.52
Cu (II)	5.90	0.85	31.4	0.98	38.75	0.44

process suggests that the monolayer adsorption condition exists in the experimental condition.

The D–R adsorption model is more general than the Langmuir model and is used to show the adsorption method through a Gaussian energy distribution on a heterogeneous surface [50]. The common form of the D–R isotherm model equation is stated as:

$$q_e = \left[q_{DR}\exp(-K_{DR}\varepsilon^2)\right] \tag{3.12}$$

The linear form can be rewritten as

$$\ln q_e = \left[\ln q_{DR} + (-K_{DR}\varepsilon^2)\right] \tag{3.13}$$

where q_e is the quantity of metals adsorbed on the adsorbent surface per unit mass of adsorbents (mmol g^{-1}); q_{DR} is the highest adsorption capacity (mmol g^{-1}); ε is the polanyi potential ($\varepsilon = RT\ln(1 + 1/Ce)$); K_{DR} is the activity coefficient associated with adsorption free energy ($mol^2\ J^{-2}$); T is the absolute temperature (K), and *R* is the gas constant (8.314 J $mol^{-1}\ K^{-1}$). A fixed volume of adsorption space near to the adsorbent surface and the presence of an adsorption capacity of these sites are assumed by the Polanyi adsorption theory.

The results of four phases of adsorption and desorption of copper and lead metal ions onto the CA/TiO_2 adsorbents were examined. As shown in Figure 3.28a, the adsorption efficiencies of the adsorbents for those heavy metal ions were slightly declined with increasing adsorption desorption phases. This slight decreasing in adsorption capacity of the adsorbent might be because of losing a certain functional-groups and/or Ti leaching of the adsorbents through the acid cleavages, and some of the adsorbent pores may be blocked. However, as seen from Figure 3.28b the TiO_2 leaching tendency due to the acid cleavage during the desorption process was negligible. There were no obvious characteristic peaks for the AM2 adsorbent before adsorption took place. Nevertheless, after the adsorption of those heavy metals onto the adsorbent absorption peaks were observed at about 323 nm and 327 nm for Pb (II) and

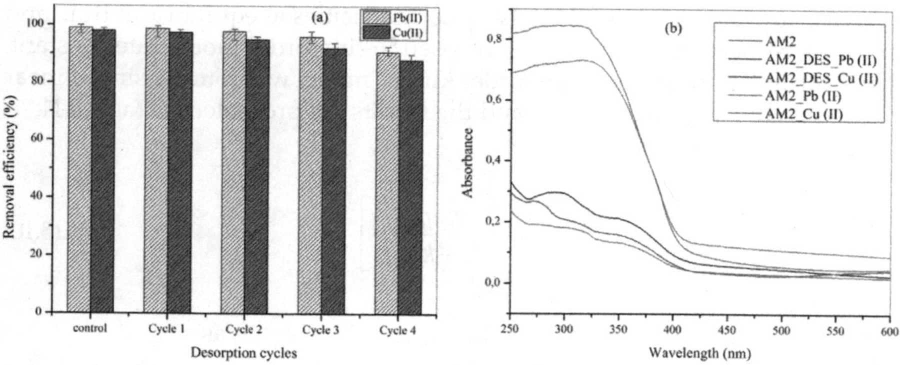

FIGURE 3.28
(a) Adsorption-desorption cycles, (b) UV–vis absorption spectra of before and after adsorption and after desorption of AM2 adsorbent.

Cu (II) ions, respectively. This confirms the formation of adsorbent metal ion complexion due to an efficient adsorption. The absorption peak for TiO_2 NPs was observed for all the samples at around 385 nm. On the other hand, no metal ion peaks were observed after the desorption process using HNO_3 acid and similar graphs with the AM2 adsorbent were detected. Therefore, the outcomes indicated that the CA/TiO_2 adsorbents could repeatedly be recycled without showing substantial damage in adsorption efficiency. The decreasing in the adsorption efficiencies was not significant when compared with the control. Moreover, the FESEM images in Figure 3.22d_1 and 3.22e_1 confirmed that the morphological structure of the adsorbent was not disturbed during the recycling procedure.

3.3.8.5 Adsorption Kinetics

Adsorption kinetics study is one of the most vital parameters of the adsorption study. Because the mechanisms of the adsorption processes can be modeled by using kinetic models [65]. Therefore, the experimental results were analyzed using the two kinetic models called pseudo-first-order and pseudo-second-order using the kinetic equations as follows [66–68].

Pseudo-first-order:

$$q_t = q_e(1 - \exp(-k_1 t)) \tag{3.14}$$

The linear form of first-order kinetic can be rewritten as

$$\log(q_e - q_t) = \left[\log q_e - \left(\frac{k_1}{2.303}\right)t\right] \tag{3.15}$$

where q_e and q_t (mg/g) are the adsorption capacities at equilibrium time and at time t, respectively; k_1 (min^{-1}) is the pseudo-first-order model rate constant. The constant of the pseudo-first-order kinetic model was found using a linear regression of $log\ (q_e - q_t)$ versus t, and the results are presented in Table 3.11.

Pseudo-second-order:

$$q_t = \left[\frac{k_2 q_e^2 q_t}{(1 + k_2 q_e t)}\right] \tag{3.16}$$

The linear form of second-order kinetic can be written as

$$\frac{t}{q_t} = \left(\frac{1}{k_2 q_e^2} + \frac{t}{q_e}\right) \tag{3.17}$$

where k_2 (g/mg min^{-1}) is the second-order adsorption rate constant. The plots of t/q_t versus t were employed to calculate the constants of the pseudo-second-order kinetic model. The graphs for both kinetic models are shown in Figure 3.29a and 3.29b and the pseudo-second-order kinetic model parameters are

TABLE 3.11

Kinetic Parameters of Metal Adsorption onto the CA/TiO_2 Adsorbent

		Pseudo-First-Order Model			Pseudo-Second-Order Model		
Metal Ion	q_{exp} **(mg/g)**	q_e **(mg/g)**	k_1 **(min^{-1})**	R^2	q_e **(mg/g)**	k_2 **(g/mg min)**	R^2
Pb (II)	20.5	5.40	0.0362	0.93	28.09	0.00057	0.99
Cu (II)	22	4.75	0.0193	0.88	33.11	0.00026	0.99

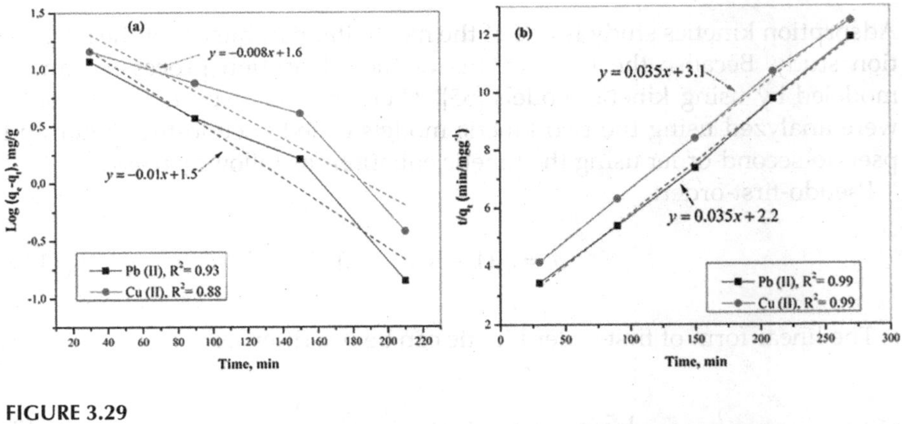

FIGURE 3.29

Adsorption kinetics: (a) Pseudo-first-order and (b) Pseudo-second-order plots for copper and lead adsorption onto the adsorbent.

given in Table 3.11. As indicated in Table 3.11, the pseudo-second-order equation fitted well with the experimental data of both copper ($R^2 = 0.99$) and lead ($R^2 = 0.99$). Since the CA/TiO_2 adsorbents employed in this experiment have relatively high equilibrium adsorption, the adsorption could be fast and thus the equilibrium time may be short. Due to this short equilibrium time together with high adsorption capacity, a high degree of affinity between the metal ions and the CA/TiO_2 adsorbents is indicated. On the other hand, the pseudo-first-order kinetic model fitted the experimental results poorly.

3.4 Summary

The characterization and adsorption study of the CA/TiO_2 adsorbent were completed successfully in this work. The specific surface areas of the adsorbent were increased from 30.2 to 48.47 m^2g^{-1} as the amounts of TiO_2 nanoparticles in the adsorbent matrix were increased from 0 to 2.5 wt.%. The highest removal efficiencies of lead and copper ions with CA/TiO_2 adsorbent were estimated to be 99.7% and 98.9% under the optimized conditions of TiO_2 amounts (2.5 wt.%), pH (5.2 for Pb (II) and 5.8 for Cu (II), respectively), agitation period (5 h), adsorbent dosage (2 g/L), and temperature (35 °C).

The Langmuir isotherm ($R^2 > 0.98$ for copper and > 0.99 for lead) model was well fitted to the equilibrium data of both metal ions than D–R and Freundlich isotherms.

The adsorption experimental data agreed perfectly with the pseudo-second-order kinetic equation where the regression coefficients were greater than 99% ($R^2 > 0.99$) for both copper and lead ions. Repeated adsorption–desorption experiments after four phases revealed that CA/TiO_2 adsorbent showed high elimination capacity for copper and lead ions. The CA/TiO_2 adsorbent provided in this study can be employed for the applied treatment of heavy metal contaminated water.

References

1. Koski, A., Yim, K., and Shivkumar, S., Effect of molecular weight on fibrous PVA produced by electrospinning. *Materials Letters*, 2004. 58(3–4): pp. 493–497.
2. Mohammad Ali Zadeh, M., Keyanpour-Rad, M., and Ebadzadeh T., Effect of viscosity of polyvinyl alcohol solution on morphology of the electrospun mullite nanofibres. *Ceramics International*, 2014. 40(4): pp. 5461–5466.
3. Wang, X., Fang, D., Yoon, K., Hsiao, B.S., and Chu, B., High performance ultrafiltration composite membranes based on poly(vinyl alcohol) hydrogel coating

on crosslinked nanofibrous poly(vinyl alcohol) scaffold. *Journal of Membrane Science*, 2006. 278(1–2): pp. 261–268.

4. Yoon, K., Hsiao, B.S., and Chu B., High flux ultrafiltration nanofibrous membranes based on polyacrylonitrile electrospun scaffolds and crosslinked polyvinyl alcohol coating. *Journal of Membrane Science*, 2009. 338(1–2): pp. 145–152.
5. Zhu, X., Loo, H.-E., and Bai, R., A novel membrane showing both hydrophilic and oleophobic surface properties and its non-fouling performances for potential water treatment applications. *Journal of Membrane Science*, 2013. 436: pp. 47–56.
6. Santos, C., Silva, C.J., Büttel, Z., Guimarães, R., Pereira, S.B., Tamagnini, P., and Zille, A., Preparation and characterization of polysaccharides/PVA blend nanofibrous membranes by electrospinning method. *Carbohydrated Polymers*, 2014. 99: pp. 584–592.
7. Yang Liu, R.W., Hongyang M., Benjamin S. H.*, Benjamin Chu, High-flux microfiltration filters based on electrospun polyvinylalcohol nanofibrous membranes. *Polymer*, 2013. 54(2): pp. 548–556.
8. Puguan, J.M.C., Kim, H.-S., Lee, K.-J., and Kim H., Low internal concentration polarization in forward osmosis membranes with hydrophilic crosslinked PVA nanofibers as porous support layer. *Desalination*, 2014. 336: pp. 24–31.
9. Ramakrishna, S., Fujihara, K., Teo, W.E., Lim, T.C., Ma, Z., An Introduction to Electrospinning and Nanofibers 2005, World Scientific: Singapore.
10. Barhate, R.S., Loong, C.K., and Ramakrishna, S., Preparation and characterization of nanofibrous filtering media. *Journal of Membrane Science*, 2006. 283(1–2): pp. 209–218.
11. Tungprapa, S., Puangparn, T., Weerasombut, M., Jangchud, I., Fakum, P., Semongkhol, S., Meechaisue, C., and Supapho, P., Electrospun cellulose acetate fibers: Effect of solvent system on morphology and fiber diameter. *Cellulose*, 2007. 14(6): pp. 563–575.
12. Baş, D., and İ.H. Boyacı, Modeling and optimization I: Usability of response surface methodology. *Journal of Food Engineering*, 2007. 78(3): pp. 836–845.
13. Yördem, O.S., Papila, M., and Menceloğlu, Y.Z., Effects of electrospinning parameters on polyacrylonitrile nanofiber diameter: An investigation by response surface methodology. *Materials and Design*, 2008. 29(1): pp. 34–44.
14. Bezerra, M.A., Santelli, R.E., Oliveira, E.P., Villar, L.S., and Escaleira, L.A., Response surface methodology (RSM) as a tool for optimization in analytical chemistry. *Talanta*, 2008. 76(5): pp. 965–977.
15. Ferreira, S.L.C., Bruns, R.E., Silva, E.G.P., Santos, W.N.L., Quintella, C.M., David, J.M., Andrade, J.B., Breitkreitz, M.C., Jardim I.C.S.F., and Neto, B.B., Statistical designs and response surface techniques for the optimization of chromatographic systems. *Journal of Chromatography A*, 2007. 1158(1–2): pp. 2–14.
16. Khanlou, H.M., Ang, B.C., Talebian, S., Barzani, M.M., Silakhori, M., and Fauzi, H., Multi-response analysis in the processing of poly (methyl methacrylate) nano-fibres membrane by electrospinning based on response surface methodology: Fibre diameter and bead formation. *Measurement*, 2015. 65: pp. 193–206.
17. Ahmadipourroudposht, M., Fallahiarezoudar, E., Yusof, N.M., and dris, A., Application of response surface methodology in optimization of electrospinning process to fabricate (ferrofluid/polyvinyl alcohol) magnetic nanofibers. *Materials Science and Engineering: C*, 2015. 50: pp. 234–241.

18. Bose, S., and Das, C., Role of binder and preparation pressure in tubular ceramic membrane processing: Design and optimization study using response surface methodology (RSM). *Industrial & Engineering Chemistry Research*, 2014. 53(31): pp. 12319–12329.
19. Feng, C., Khulbe, K.C., Matsuura, T., Tabe, S., and Ismail, A.F., Preparation and characterization of electro-spun nanofiber membranes and their possible applications in water treatment. *Separation and Purification Technology*, 2013. 102: pp. 118–135.
20. Ran Wang, Y.L., Brandon Li, Benjamin S. Hsiao, Benjamin Chu, Electrospun nanofibrous membranes for high flux microfiltration. *Journal of Membrane Science*, 2012. 392–393: pp. 167–174.
21. Algarra, M., Vázquez, M.I., Alonso, B., Casado, C.M., Casado, J., and Benavente, J., Characterization of an engineered cellulose based membrane by thiol dendrimer for heavy metals removal. *Chemical Engineering Journal*, 2014. 253: pp. 472–477.
22. Konwarh, R., Karak, N., and Misra, M., Electrospun cellulose acetate nanofibers: The present status and gamut of biotechnological applications. *Biotechnology Advances*, 2013. 31(4): pp. 421–437.
23. Abedini, R., Mousavi, S.M., and Aminzadeh, R., A novel cellulose acetate (CA) membrane using TiO2 nanoparticles: Preparation, characterization and permeation study. *Desalination*, 2011. 277(1–3): pp. 40–45.
24. Cao, X., Ma, J., Shi, X., and Ren, Z., Effect of TiO2 nanoparticle size on the performance of PVDF membrane. *Applied Surface Science*, 2006. 253(4): pp. 2003–2010.
25. Saffaj, N., Younssi, S.A., Albizane, A., Messouadi, A., Bouhria, M., Persin, M., Cretin, M., and Larbot, A., Preparation and characterization of ultrafiltration membranes for toxic removal from wastewater. *Desalination*, 2004. 168: pp. 259–263.
26. Rahimpour, A., Madaeni, S.S., Taheri, A.H., and Mansourpanah, Y., Coupling TiO2 nanoparticles with UV irradiation for modification of polyethersulfone ultrafiltration membranes. *Journal of Membrane Science*, 2008. 313(1–2): pp. 158–169.
27. Kim, H.J., Pant, H.R., Kim, J.H., Choi, N.J., and Kim, C.S., Fabrication of multifunctional TiO2–fly ash/polyurethane nanocomposite membrane via electrospinning. *Ceramics International*, 2014. 40(2): pp. 3023–3029.
28. Kim, H.T., Lee, C.H., Shul, Y.G., Moon, J.K., and Lee, E.H., Evaluation of PAN–TiO2 composite adsorbent for removal of Pb(II) ion in aqueous solution. *Separation Science and Technology*, 2003. 38(3): pp. 695–713.
29. Liu, L., Zhao, C., and Yang, F., TiO2 and polyvinyl alcohol (PVA) coated polyester filter in bioreactor for wastewater treatment. *Water Research*, 2012. 46(6): pp. 1969–1978.
30. Song, H., Shao, J., He, Y., Liu, B., and Zhong, X., Natural organic matter removal and flux decline with PEG–TiO2-doped PVDF membranes by integration of ultrafiltration with photocatalysis. *Journal of Membrane Science*, 2012. 405–406: pp. 48–56.
31. Kim, Y.B., Cho, D., and Park, W.H., Enhancement of mechanical properties of TiO2 nanofibers by reinforcement with polysulfone fibers. *Materials Letters*, 2010. 64(2): pp. 189–191.
32. Ma, Z., Kotaki, M., and Ramakrishna, S., Electrospun cellulose nanofiber as affinity membrane. *Journal of Membrane Science*, 2005. 265(1–2): pp. 115–123.

33. Nigmatullin, R., Lovitt, R., Wright, C., Linder, M., Nakari-Setälä, T., and Gama, M., Atomic force microscopy study of cellulose surface interaction controlled by cellulose binding domains. *Colloids and Surfaces B: Biointerfaces*, 2004. 35(2): pp. 125–135.
34. Liao, D.L., Wu, G.S., and Liao, B.Q., Zeta potential of shape-controlled TiO2 nanoparticles with surfactants. *Colloids and Surfaces A: Physicochemical and Engineering Aspects*, 2009. 348(1–3): pp. 270–275.
35. Pant, H.R., Bajgai, M.P., Nam, K.T., Seo, Y.A., Pandeya, D.R., Hong, S.T., and Kim, H.Y., Electrospun nylon-6 spider-net like nanofiber mat containing TiO2 nanoparticles. *Journal of Hazardous Materials*, 2011. 185: pp. 124–130.
36. Bae, T.H., Kim, I.C., and Tak, T.M., Preparation and characterization of fouling-resistant TiO2 self-assembled nanocomposite membranes. *Journal of Membrane Science*, 2006. 275(1–2): pp. 1–5.
37. Kim, S.H., Kwak, S.Y., Sohn, B.H., and Park, T.H., Design of TiO2 nanoparticle self-assembled aromatic polyamide thin-film-composite (TFC) membrane as an approach to solve biofouling problem. *Journal of Membrane Science*, 2003. 211: pp. 157–165.
38. Wang, S.D., Ma, Q., Liu, H., Wang, K., Ling, L.Z., and Zhang, K.Q., Robust electrospinning cellulose acetate@TiO2ultrafine fibers for dyeing water treatment by photocatalytic reactions. *RSC Advances*, 2015. 5(51): pp. 40521–40530.
39. Yang, Y., Zhang, H., Wang, P., Zheng, Q., and Li, J., The influence of nano-sized TiO2 fillers on the morphologies and properties of PSF UF membrane. *Journal of Membrane Science*, 2007. 288(1–2): pp. 231–238.
40. Ji, F., Li, C., Tang, B., Xu, J., Lu, G., and Liu, P., Preparation of cellulose acetate/zeolite composite fiber and its adsorption behavior for heavy metal ions in aqueous solution. *Chemical Engineering Journal*, 2012. 209: pp. 325–333.
41. Tian, Y., Wu, M., Liu, R., Li, Y., Wang, D., Tan, J., Wu, R., and Huang, Y., Electrospun membrane of cellulose acetate for heavy metal ion adsorption in water treatment. *Carbohydrate Polymers*, 2011. 83(2): pp. 743–748.
42. Abdelwahab, N.A., Ammar, N.S., and Ibrahim, H.S., Graft copolymerization of cellulose acetate for removal and recovery of lead ions from wastewater. *International Journal of Biological Macromolecules*, 2015. 79: pp. 913–922.
43. Li, S., Yue, X., Jing, Y., Bai, S., and Dai, Z., Fabrication of zonal thiol-functionalized silica nanofibers for removal of heavy metal ions from wastewater. *Colloids and Surfaces A: Physicochemical and Engineering Aspects*, 2011. 380(1–3): pp. 229–233.
44. Ahmaruzzaman, M., Industrial wastes as low-cost potential adsorbents for the treatment of wastewater laden with heavy metals. *Advances in Colloid and Interface Science*, 2011.
45. Afkhami, A., Saber-Tehrani, M., and Bagheri, H., Simultaneous removal of heavy-metal ions in wastewater samples using nano-alumina modified with 2,4-dinitrophenylhydrazine. *Journal of Hazardous Materials*, 2010. 181(1–3): pp. 836–844.
46. Teng, M., Li, F., Zhang, B., and Taha, A.A., Electrospun cyclodextrin-functionalized mesoporous polyvinyl alcohol/SiO2 nanofiber membranes as a highly efficient adsorbent for indigo carmine dye. *Colloids and Surfaces A: Physicochemical and Engineering Aspects*, 2011. 385(1–3): pp. 229–234.
47. Sang, Y., Li, F., Gu, Q., Liang, C., and Chen, J., Heavy metal-contaminated groundwater treatment by a novel nanofiber membrane. *Desalination*, 2008. 223(1–3): pp. 349–360.

48. Haider, S., and Park, S.Y., Preparation of the electrospun chitosan nanofibers and their applications to the adsorption of Cu(II) and Pb(II) ions from an aqueous solution. *Journal of Membrane Science*, 2009. 328(1–2): pp. 90–96.
49. Sang, Y., Gu, Q., Sun, T., Li, F., and Liang, C., Filtration by a novel nanofiber membrane and alumina adsorption to remove copper(II) from groundwater. *Journal of Hazardous Materials*, 2008. 153(1–2): pp. 860–866.
50. Günay, A., Arslankaya, E., and Tosun, I., Lead removal from aqueous solution by natural and pretreated clinoptilolite: Adsorption equilibrium and kinetics. *Journal of Hazardous Materials*, 2007. 146(1–2): pp. 362–371.
51. Saeed, K., Haider, S., Oh, T.J., and Park, S.Y., Preparation of amidoxime-modified polyacrylonitrile (PAN-oxime) nanofibers and their applications to metal ions adsorption. *Journal of Membrane Science*, 2008. 322(2): pp. 400–405.
52. Yu, X., Tong, S., Ge, M., Wu, L., Zuo, J., Cao, C., and Song, W., Synthesis and characterization of multi-amino-functionalized cellulose for arsenic adsorption. *Carbohydrate Polymers*, 2013. 92(1): pp. 380–387.
53. Abbasizadeh, S., Keshtkar, A.R., and Mousavian, M.A., Preparation of a novel electrospun polyvinyl alcohol/titanium oxide nanofiber adsorbent modified with mercapto groups for uranium(VI) and thorium(IV) removal from aqueous solution. *Chemical Engineering Journal*, 2013. 220: pp. 161–171.
54. Hong, G., Shen, L., Wang, M., Yang, Y., Wang, X., Zhu, M., and Hsiao, B.S., Nanofibrous polydopamine complex membranes for adsorption of Lanthanum (III) ions. *Chemical Engineering Journal*, 2014. 244: pp. 307–316.
55. Liu, C., and Bai, R., Adsorptive removal of copper ions with highly porous chitosan/cellulose acetate blend hollow fiber membranes. *Journal of Membrane Science*, 2006. 284(1–2): pp. 313–322.
56. Aliabadi, M., Irani, M., Ismaeili, J., Piri, H., and Parnian, M.J., Electrospun nanofiber membrane of PEO/Chitosan for the adsorption of nickel, cadmium, lead and copper ions from aqueous solution. *Chemical Engineering Journal*, 2013. 220: pp. 237–243.
57. Jin, X., Xu, J., Wang, X., Xie, Z., Liu, Z., Liang, B., Chen, D., and Shen, G., Flexible TiO2/cellulose acetate hybrid film as a recyclable photocatalyst. *RSC Advances*, 2014. 4(25): p. 12640.
58. Bagheri, S., Shameli, K., and Hamid, S.B.A., Synthesis and characterization of anatase titanium dioxide nanoparticles using egg white solution via sol-gel method. *Journal of Chemistry*, 2013. 2013: pp. 1–5.
59. Dąbrowski, A., Adsorption from theory to practice. *Advances in Colloid and Interface Science*, 2001. 93: pp. 135–224.
60. Kovačević, D., Pohlmeier, A., Özbaş, G., Narres, H.D., and Kallay, M.J.N., The adsorption of lead species on goethite. *Colloids and Surfaces A: Physicochemical and Engineering Aspects*, 2000. 166: pp. 225–233.
61. Sanchez, A.G., Ayuso, E.A., and Blas, O.J.D., Sorption of heavy metals from industrial waste water by low-cost mineral silicates. *Clay Minerals*, 1999. 34: pp. 469–477.
62. Zhou, D., Zhang, L., Zhou, J., and Guo, S., Cellulose/chitin beads for adsorption of heavy metals in aqueous solution. *Water Research*, 2004. 38(11): pp. 2643–2650.
63. Tchobanoglous, G., Burton, F.L., Stensel, H.D., Metcalf & Eddy, *Wastewater Engineering - Treatment and Reuse*. McGraw Hill Education: New York, 2004.

64. Yasemin B., Zeki, T., Removal of heavy metals from aqueous solution by sawdust adsorption. *Journal of Environmental Science*, 2007. 19: pp. 160–166.
65. Shahmohammadi-Kalalagh, S., Isotherm and kinetic studies on adsorption of Pb, Zn and Cu by kaolinite. *Caspian Journal of Environmental Sciences*, 2011. 9: pp. 243–255.
66. Lee, I.H., Kuan, Y.C., and Chern, J.M., Equilibrium and kinetics of heavy metal ion exchange. *Journal of the Chinese Institute of Chemical Engineers*, 2007. 38(1): pp. 71–84.
67. Chiou, M.S., and Li, H.Y., Equilibrium and kinetic modeling of adsorption of reactive dye on cross-linked chitosan beads. *Journal of Hazardous Materials*, 2002. 93: pp. 233–248.
68. Jiwalak, N., Rattanaphani, S., Bremner, J.B., and Rattanaphani, V., Equilibrium and kinetic modeling of the adsorption of indigo carmine onto silk. *Fibers and Polymers*, 2010. 11: pp. 572–579.

4

Phase Inverted Membranes: Preparation and Application

4.1 Preparation and Application of Phase Inverted Membranes

4.1.1 Introduction

Phase inversion is a suitable method for the preparing of polymeric membranes having all types of morphological structures. In this technique, a casting solution comprised of polymer and solvent is immersed into a non-solvent coagulation bath, and a polymer solution is phase separated into polymer rich and polymer lean phases in a controlled manner [1]. The solidification process is frequently started by the change from one liquid phase into two liquid phases (liquid-liquid demixing). At certain periods during de-mixing, the polymer rich phase solidifies so that a solid membrane matrix is formed [1,2]. Membrane morphologies, particularly the pore size distributions, can be controlled by selecting the different solvent, non-solvent, polymer, pore former, and fabrication parameters depending on the particular application [3–6]. Cellulose acetate is broadly applicable for the synthesis of membranes because of having tough, biocompatible, hydrophilic characteristics, good desalting nature, high flux, and relatively less expensive [7–9]. Cellulose acetate membranes with different polyvinylpyrrolidone (PVP) concentration and coagulation-bath-temperature were prepared by a group of researchers [10]. They have found that the addition of 0 wt.% to 3 wt.% of PVP to the polymer solution increased the macro-voids development and consequently the pure water flux was raised.

Nevertheless, due to further addition of PVP to 6 wt.% the macro-void formation was suppressed and pure water flux (PWF) was reduced. On the other hand, increasing in coagulation bath temperature up to 25 °C resulted in increasing of the macro-voids formation. Chakraborty et al. have prepared flat sheet polysulfone membranes from casting solutions by using NMP and DMAc solvents [11]. Polyethylene glycol (PEG) of average molecular weights of 400 Da, 6000 Da, and 20000 Da were used. The PWF and P_m were enhanced significantly as the PEG molecular weight was increased. A substantial increase in rejection of bovine serum albumin solution was also reported due to an increasing of the molecular weight of PEG from 400 Da to 6000 Da. This increasing tendency of the membrane with PEG 6000 may be due to the sponge-like and dense structure that can offer more resistance to the protein molecules. They have also reported that BSA rejection was observed to decrease when the PEG molecular weight was beyond 6000 Da (i.e., 20,000 Da). The high porosity and presence of finger-like cavities in the sublayer for membranes with PEG 20,000 Da have allowed more BSA transmission without much resistance and resulted in less BSA rejection. Kim et al. have also investigated the effect of PEG as a pore-former on polysulfone/N-methyl-2-pyrrolidone (NMP) membrane formation using phase inversion process [12]. They have found that the coagulation value was decreased as the molecular weight (M_w) of the PEG was increased. They have also observed that the membrane surface pore size becomes larger, and the top layer appears more porous and the distance from the top surface to the starting point of macro-void development becomes greater. They have also reported that with increasing of the PEG additive M_w, water flux was increased and PEG solute (PEG of 12,000 and 35,000 g/mole used as solutes) rejections were observed to decrease. Solute transmission increases if the electrostatic interaction is insignificant, where in turn solute rejection is expected to decrease with increasing in PEG molecular weight.

Generally, polymeric additives are essential to improve the membrane properties [11–14], mainly pore structure, mechanical stability, thermal stability, hydrophilicity, flux, and rejection capabilities [10,15–17]. An effort has been made to explore the effect of PEG and PVP additives, and the effect of solvents in the mixtures of CA/Ac: DMAc and CA/DMF solution discretely. CA was used as polymer and PVP and PEG were used as additives and pore forming agents to compare their effects on the morphological structure and permeability properties of the membrane. We put our effort to investigate the effect of the two solvents on the morphological structure and permeability properties of the membrane. All the experiments in this study were performed three times and all the graphs are presented with error bars. Moreover, all the experimental results were consistent with the membrane properties and agreed with each other.

4.1.2 Preparation Methods: Phase Inversion Process

4.1.2.1 Preparation of CA, CA-PEG, and CA-PVP Membranes

Phase inversion technique using immersion precipitation was used for the fabrication of asymmetric ultrafiltration membrane from cellulose acetate polymer. Initially 10.5 wt.% of CA powder was dissolved in acetone: DMAc (70/30 wt.%) and DMF separately with continuous magnetic stirring at room temperature (28 ± 2 °C). Particular attention was given to acetone and DMAc where the two solvent properties were then mixed to attain enhanced CA solubility due to a synergistic effect, and we have selected the mixture ratio according to the literature. DMF was selected due to its good solvent nature for CA where it avoids rapid evaporation of the solution [18,19]. After getting a uniform solution of the CA/solvent, a fixed amount of each additive (5 wt.%) was added and stirred continuously at least for 1 day until the solution was fully dissolved, and finally, homogeneous casting solution was obtained. The amount of polymer and additives were chosen based on the literature [13,16]. Membranes with different composition were designated as CA0, CA1, CA2, CA3, CA4, and CA5. Table 4.1 shows the solution casting compositions of polymeric additives with CA.

The membrane preparation process is shown in Figure 4.1. The casting solution was stirred overnight using a magnetic stirrer at room temperature. After that, a homogeneous solution of solvent, polymer, and the additives were attained and kept for 24 h at room temperature. Consequently, the casting

TABLE 4.1
Compositions and Viscosity of the Casting Solution of Polymeric Additives with CA

Membrane	CA wt.%	PEG wt.%	PVP wt.%	Ac: DMAc (70/30 wt.%)	DMF wt.%	Viscosity (mPa.s)
CA0	10.5	–	–	89.5	–	1680
CA1	10.5	5	–	84.5	–	1780
CA2	10.5	–	5	84.5	–	1720
CA3	10.5	–	–	–	89.5	1850
CA4	10.5	5	–	–	84.5	1920
CA5	10.5	–	5	–	84.5	1970

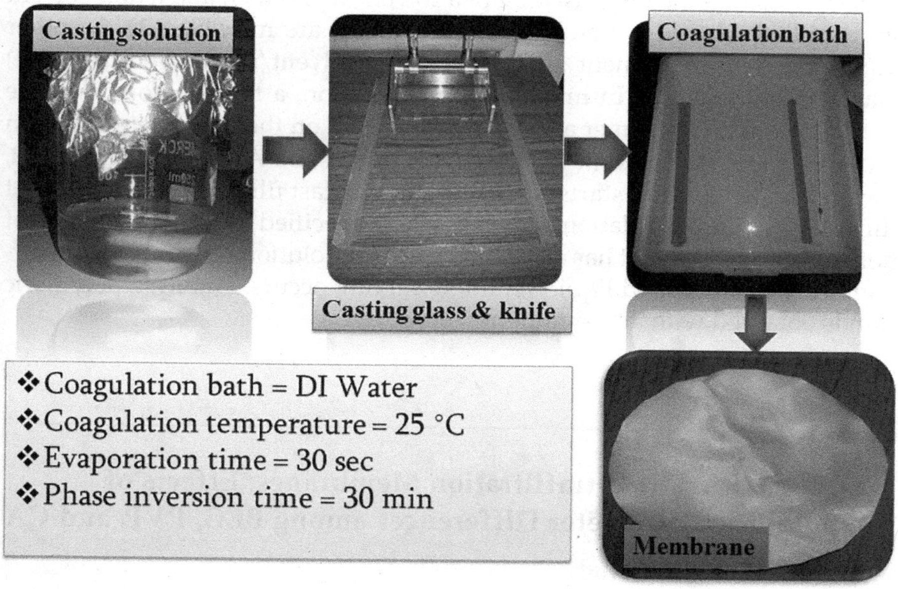

FIGURE 4.1
Membrane preparation process.

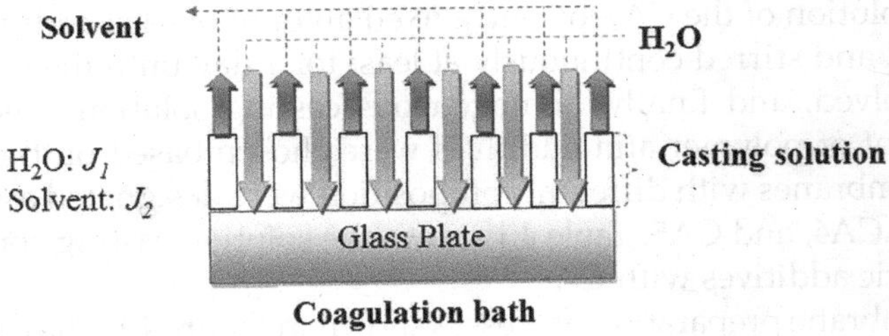

FIGURE 4.2
Schematic representation of casting film and coagulation bath interface: solvent (J_2) and non-solvent (J_1).

solutions were poured consistently on a glass sheet with the help of a casting knife keeping an allowance of roughly 0.25 mm between the knife and the glass plate. Then, the resultant films were exposed to air for about 30 s earlier to soaking in the coagulation bath containing DI water at room temperature.

In the coagulation bath, the cast solution turned from transparent to white color and the thin film detached from the glass plate. The membrane sheets were kept for 30 min in the coagulation bath. After that, the fabricated membranes were reserved in DI water beakers until use. Finally, the membrane sheet was cut into a circular disc to put in the filtration cell for pure water and BSA filtration experiments.

During the preparation of phase inverted membranes using immersion precipitation process, ternary and multicomponent systems were used. In ternary system, scheme consisting of a solvent, a polymer, and a non-solvent was applied where the non-solvent and solvent are miscible. On the other hand, the multicomponent system (polymer/solvent/non-solvent/additive) was also investigated. In immersion precipitation, a homogenous mixture of the solvent and polymer are cast to a thin film on the glass plate and then immersed in a non-solvent (water in this case) bath. As seen in Figure 4.2, the non-solvent (water) starts to diffuse into the cast film (J_1), and the solvent diffuses into the coagulation bath (J_2). After a specified time, the exchange of non-solvent and solvent has continued until the solution becomes thermodynamically unstable and liquid-liquid de-mixing occurs. Finally, a polymeric film is obtained with an asymmetric structure.

4.2 Preparation of Ultrafiltration Membranes: Effects of Solubility Parameter Differences among PEG, PVP, and CA

4.2.1 Morphological Study

The FESEM images of (a) top layer surface view and (b) the cross-sectional view of the prepared membly dissolving the polymer and additives in two different solvents (DMF and Ac: DMAc) are presented in Figure 4.3. As clearly shown from the figures, the prepared membranes are asymmetric in structures comprising a porous sublayer and a dense top layer for all types of the membranes (i.e., for CA/DMF and CA/Ac: DMAc). The sublayer portion of the membranes without additives (CA0 and CA3) seems to have finger-like voids as well as macro-void structures. According to the literature, there are so many parameters, such as evaporation step [20] and coagulation bath temperature (CBT) [13], that allow the development of macro-voids during membrane formation using immersion precipitation process. The development of

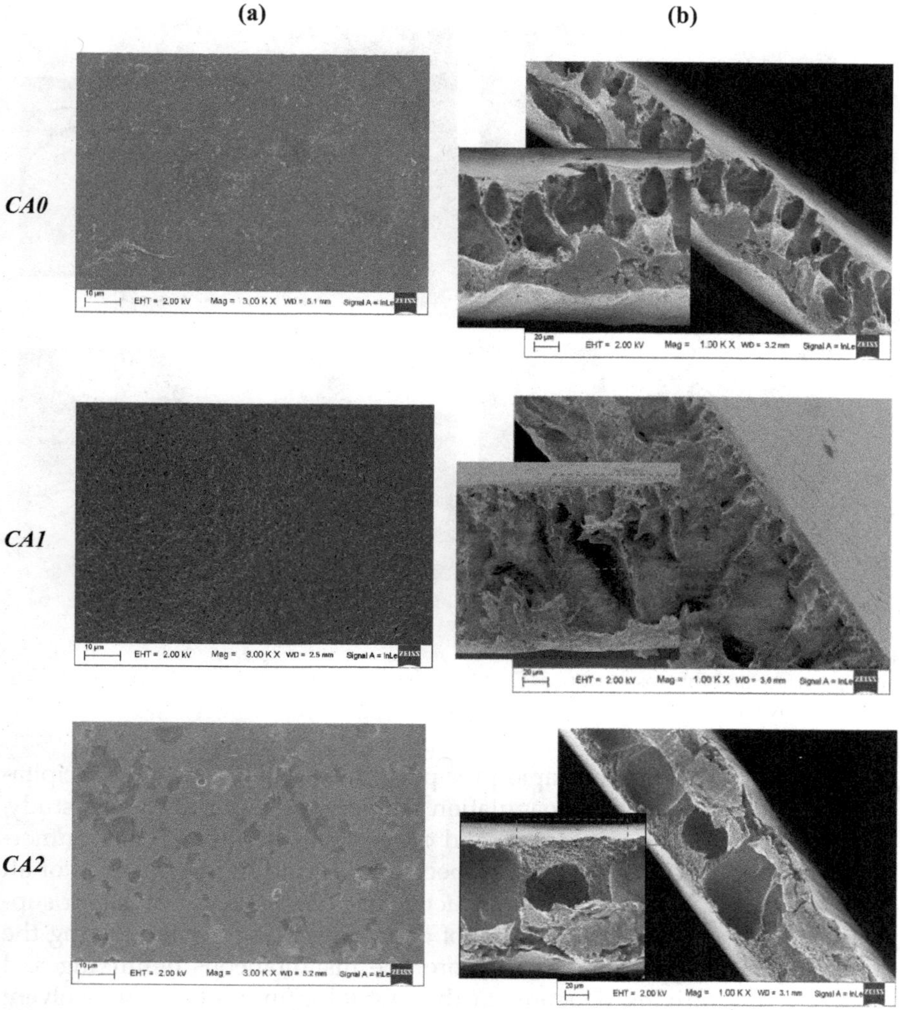

FIGURE 4.3
(a) Top surface and (b) cross-sectional FESEM images of CA membranes with different additives and solvents. *(Continued)*

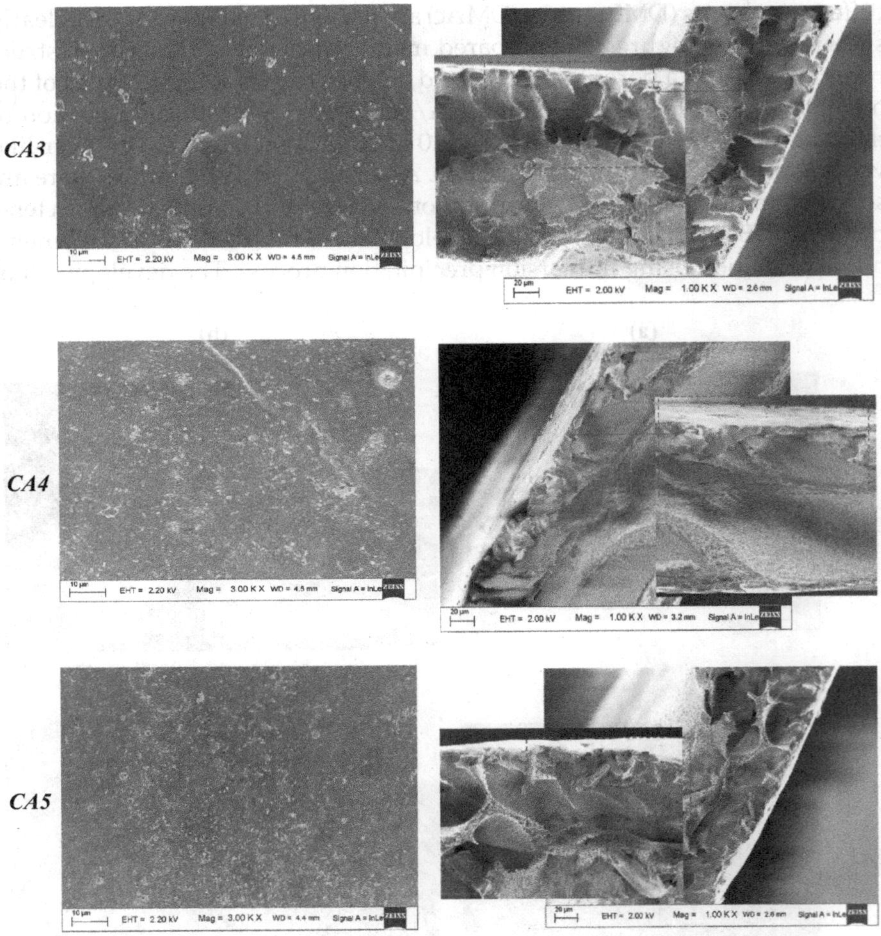

FIGURE 4.3 (CONTINUED)
(a) Top surface and (b) cross-sectional FESEM images of CA membranes with different additives and solvents.

macro-voids happen under rapid precipitation condition, and the precipitations are quicker at higher coagulation temperatures. In the current study, the coagulation bath temperature and evaporation step (i.e., before immersion) are fixed to 25 °C and 30 s, respectively, during the preparation of all the membranes. Another important factor that affects the formation or suppression of macro-voids is the type of de-mixing that occurred during the phase separation process. Furthermore, the instantaneous de-mixing and the delayed de-mixing depends on the mutual affinity of the non-solvent and solvent in the ternary system [1]. The solvents used in this study (DMF and Ac: DMAc) have a high mutual affinity with water. Therefore, apparently, the formation of the finger-like voids and macro-voids in the sublayer of the membranes without additives is due to the instantaneous de-mixing. On the other hand, when the polymeric additives are added to the ternary (polymer/solvent/non-solvent) system the formations of macro-voids are suppressed for membrane prepared using PEG (i.e., CA1 and CA4). However, for membranes prepared using PVP the formation of some macro-voids are still maintained (i.e., CA2 and CA5). However, the formation of microporous structures was observed for both the additives. In a quaternary system (polymer/solvent/non-solvent/additive), nevertheless, the exchange process of water and solvent between the coagulation bath and the casting solution taking place right upon immersing is considerably faster than the separation of the two polymers. However, after some instants of immersion process, polymer-polymer movement may occur. This is because of the separation of the two polymers (membrane forming and additive) that involves the movement of one polymer with respect to the other [21]. Therefore, a binary phase arises from the phase separation, where one involves the CA (i.e., membrane-forming polymer), Ac: DMAc/DMF (i.e., solvents), and non-solvent (i.e., water) and the second contains the polymer additives (i.e., PEG or PVP), solvents (i.e., Ac: DMAc or DMF), and non-solvent (i.e., water). The membrane forming polymer and additives have a driving force to separate completely into a CA-rich and a PEG-rich or PVP-rich phase. Consequently, the type of de-mixing process in this case can be influenced by the diffusion/movement of the two polymers on each other. Therefore, this phenomenon should be slow compared to the diffusion of solvent and non-solvent in the polymer solution. To maintain this, a polymeric additive with a certain minimum molecular weight in the system is required [22].

4.2.1.1 Effect of Additives

The introduction of hydrophilic additives that are soluble in the non-solvent (i.e., water) with high affinity to the solvents and low affinity to the membrane forming polymer tends to increase the thermodynamic instability of the casting film, which further leads to instantaneous de-mixing in the coagulation bath. This result encourages the development of macro-voids in the membrane structure. The presence of the polymeric additives increases the concentration/viscosity of the casting solution that may diminish the diffusional exchange rate of the solvent and non-solvent during the membrane formation process. Consequently, this may hinder the instantaneous liquid-liquid de-mixing process that suppresses the development of macro-voids [10,13,23]. It is clear from the figures that the finger-like voids in the sublayer are considerably reduced with the adding of polymeric additives (PEG and PVP) for both solvents. Particularly, for CA/Ac: DMAc/PEG scheme (CA1) and CA/DMF/PEG (CA4), the membranes are found to have almost macro-void free cross-sectional structures. The CA/Ac: DMAc/PVP system (CA2) and CA/DMF/PVP system (CA5) membranes showed the development of some finger-like and macro-voids. These results lead to the conclusion, although the instantaneous de-mixing is yet continued, the effect of those additives has considerable part on the suppression or development of macro-voids. The FESEM images of the top view of all the prepared membranes are presented in Figure 4.3a. Some of the open pore structures developed in the membranes are formed by nucleation and development of the polymer-lean phase in the metastable area between the bi-nodal and the spinodal curve [22,24]. Nevertheless, the top layer structure of the ultrafiltration membrane frequently doesn't reveal open pore structures and not a completely uniform gel-layer, which is stated as a nodular structure. The development of nodular structures can't be described by nucleation of the polymer lean part. It is also not common that nucleation of the polymer-rich part happens because this only occurs at initially low polymer concentrations, below the critical point. A possible explanation for the formation of a nodular structure on the top surface of the membranes could be due to the spinodal de-mixing because the diffusion process throughout the development of the top layer is faster for the homogenous system to develop greatly unstable and crosses the spinodal curvature [4,22]. The top surface with improved interconnected pores structures is observed more prominently in the case of PEG (i.e., CA1), which is attained because of the spinodal decomposition. Therefore, the interconnected pores may be regarded as a continuous CA-lean (PEG-rich) phase tangled by a constant CA-rich (PEG-lean) phase, which produces membrane with a uniform matrix. In the case of PVP (CA2), a nonuniform pore structure was observed. However, the remaining membrane surfaces (i.e., CA0, CA3, CA4, and CA5) depicted dense top surface structures.

4.2.1.2 Effect of Solvents

The effects of solvent:non-solvent and solvent:polymer interactions are also important parameters on the final morphological structures of the membranes. Therefore, the tendency of mixing of the solvent with non-solvent, the degree of swelling of the pure polymer in the non-solvent, and the solubility of the membrane forming polymer in the solvent should be considered during the investigation of the effect of different parameters on the final structure of the membranes. From the FESEM images, Figure 4.3b, it is clearly observed that the macro-void formation in the CA/DMF system is greater than Ac: DMAc system. This can be due to the transition from instantaneous to delayed onset of de-mixing attained by changing the type of solvents [5,21], which is also complemented by the diminishing of development of macro-voids. The miscibility of the solvent:non-solvent mixture, employed for the preparation of the membranes, has a significant effect on the final structure of the membrane [24]. According to Reuvers et al. [25], the delay time for the onset of liquid-liquid de-mixing increases with decreasing the tendency of mixing solvent with the non-solvent. The current observations are completely in agreement with these explanations that the absence or less macro-voids formation in the case of Ac: DMAc/CA system is due to the delayed onset of de-mixing. On the other hand, development of macro-voids or finger-like cavities is observed for DMF/CA systems. These results are because DMF has greater tendency of miscibility with water than that of Ac/DMAc and therefore the instantaneous de-mixing occurs that is responsible for the formation of macro-voids. The effect of these solvents can be explained from another point of view that the top layer thickness of the cross-sectional views depicted a visible difference. Therefore, the delayed and instantaneous de-mixing processes can affect the top layer surface of the finally prepared membrane. In this case, the membrane formed from DMF/CA system gives membranes with relatively lesser top layer thickness than the membranes prepared using AC: DMAc/CA system. However, the total thickness of the AC: DMAc/CA membranes are lesser than DMF-CA system membranes as seen from Table 4.1. It is clear that a thicker top layer is due to the delayed onset de-mixing during the phase separation. One point that should be clarified here is that the delayed de-mixing, in this case, is not extended delayed de-mixing. Normally, if the delayed de-mixing is extended, the polymer film thickness will decrease considerably, which may cause a decrease in the porosity of the sublayer membranes [25]. This result may further prevent the formation of interconnected pore structures that in turn can form membranes with homogenous and with dense layer structures. However, for the membrane prepared from Ac: DMAc/CA system (CA0, CA1, and CA2) the pores in the sublayer are interconnected due to the development of aggregates associated with the gelation of sublayer. Therefore, the resistance of the sublayer of these membranes is much lower than that of membranes and could prepare from extended delayed de-mixing (dense membranes).

4.2.2 Pure Water Flux and Hydraulic Permeability

4.2.2.1 Effect of Additives

The study of compaction factor (CF) is important to understand the pore structures of the membranes, specifically the membrane sublayer structure. Membranes with high CF implies that they are highly compacted, which may, in turn, indicate the availability of a large number of macro-voids or finger-like arrangements in the membrane's sublayer structure.

Therefore, the impacts of compaction time on PWF of the prepared membranes are depicted in Figure 4.4. All the membranes prepared using both solvents (Ac: DMAc and DMF), and both additives (PEG and PVP) are presented in Figure 4.4a and 4.4c, respectively. The PWF of all the membranes was observed to decline slowly with time because of the compaction, and finally, the steady state fluxes were reached after 80 min. This gradual decrease in PWF results can be explained due to the compaction of pore walls (attain a uniform and denser structures) causing the pore size and

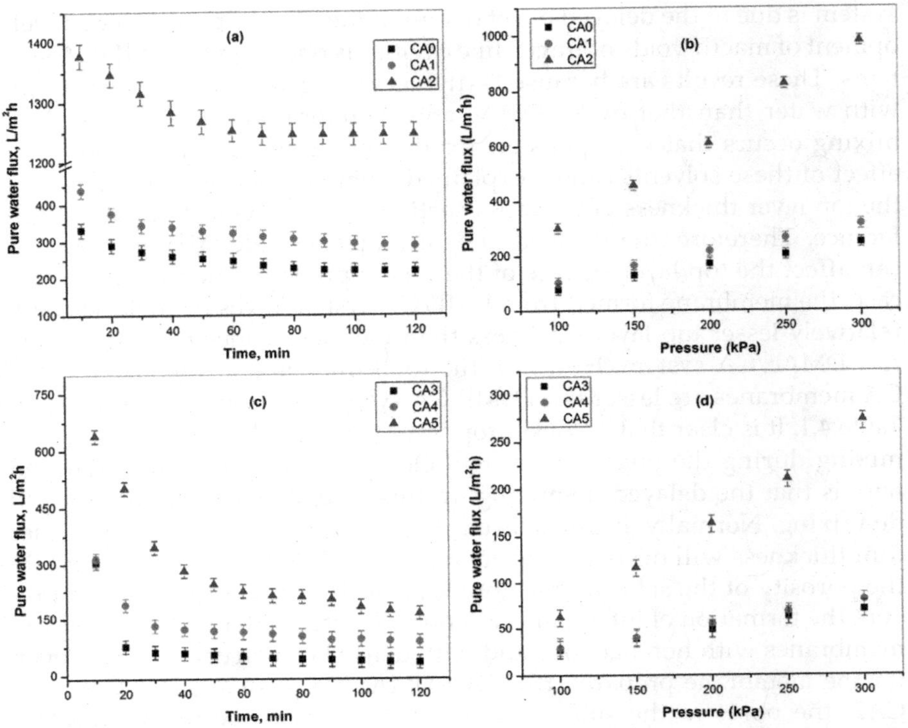

FIGURE 4.4
(a, c) PWF profile during compaction study and (b, d) effect of operating pressure on PWF.

the flux to decrease [1]. Therefore it is important to investigate the effect of additives on the compaction process, where the effects of the additives were significantly noticed from this study. Another point observed from this study was that the membranes without additives showed lesser PWF than membranes with additives for both the solvents. In this result it is apparent that the introduction of hydrophilic additives to the membrane casting solution makes the membrane porous and more hydrophilic [26]. The result evidently shows that the introduction of PEG and PVP impact the development of porous structure in the membrane, where the permeability properties of the ultrafiltration membranes were significantly affected [27,28]. Moreover, the whole removal of polymeric additive from the membrane-forming polymeric medium may not be achieved during the phase separation process in the coagulation bath and even after rinsing with DI water. Consequently, the residual additives (i.e., PEG and PVP) are forced to remain within the membrane matrix permanently, thus forming more hydrophilic membrane [28–31]. In the current study, due to the introduction of the hydrophilic additives, a small quantity of PVP and PEG may permanently stay entangled in the membrane's matrix. These results can improve the hydrophilic nature of the CA membranes (Figure 4.4).

Therefore, the polymeric additives can play an important role in the formation of porous membrane with improving its hydrophilic nature, which is again directly related to its water permeability performance. Comparing both the additives, the membranes with PVP gained higher flux than membranes with PEG for both solvents. It is clearly explained from the FESEM images (Figure 4.3) that the membranes formed using PVP additive (i.e., CA2 and CA5) showed the development of greater macro-voids when compared with membranes formed using PEG additive (CA1 and CA4). The decrease in water flux for membranes prepared from casting solutions containing PEG additive is due to the development of a denser top layer as the phase separation is delayed. These results are also supported by the porosity and pore size measurements reported in Table 4.3, where it clearly shows that a relatively lower affinity of PEG to the membrane forming polymer interferes with the de-mixing process and prevents the formation of large pores. Therefore, regardless of the effect of solvents, the membranes with PVP additive are more porous than the membranes with PEG additive, which influences their water flux performance.

4.2.2.2 Effect of Solvents

To develop the membrane skin layer with sharp polymeric concentration difference (lower concentration at the lowest surface and a higher concentration at the upper surface) should be attained. The polymeric concentration difference will be made because of the solvent evaporation before the immersion and due to the immersion of the casting solution into the non-solvent. Therefore, the investigation of the effects of organic solvents on preparation of membranes by phase-separation is an important matter [32]. The membrane characteristics, which are evidently interrelated to the membrane morphological structures, can be strongly impacted due to the level of interaction between the membrane forming polymers and solvents during the preparation of solution casting. An additional interesting point is that the residual solvents in the membrane after completing the phase separation may also alter the final membrane properties [33]. Normally, the morphological structure variation of the membranes can be associated with the solvent and polymer interaction. In a solvent and polymer interaction scheme, there are three kinds of interactions: polymer, polymer:solvent, and solvent:non-solvent. When we use good polymer solvents, the degree of polymer stretching reaches to its highest level, and more favorable polymer/solvent interactions can occur [34,35]. Furthermore, the affinity between the solvent and polymer can be predicted by presenting the "solubility parameter," δ, which is described as the square root of the cohesive energy density and expresses the strengths of attractive forces among the molecules. Solvent:polymer interaction in a polymer solution has been estimated based on the solubility parameter difference (Δ) between polymer and solvent as already clearly mentioned by Hansen [36]. The Hansen solubility parameters of the solvents, additives, and polymer used in this study

TABLE 4.2

Hansen Solubility Parameters for Selected Solvents, Non-Solvent, Additives, and Cellulose Acetate

Polymer/Solvent	δ_d (MPa$^{1/2}$)	δ_p (MPa$^{1/2}$)	δ_h (MPa$^{1/2}$)	δ (MPa$^{1/2}$)	Δ	Boiling Point (°C)
DMF	17.4	13.7	11.3	24.9	1.6	153
DMAc	16.8	11.5	10.2	22.8	2.3	165
Acetone	15.5	10.4	7.0	19.9	5.6	56
Cellulose acetate	18.6	12.7	11.0	25	–	–
PVP[a]	16.06	12.13	8.75	23.86	–	–
PEG[b]	16.8	10.2	8.6	21.45	–	–
Water	15.6	16.0	42.3	47.8	–	100

Sources: Guan, R. et al., *Journal of Membrane Science*, 277(1–2): pp. 148–156, 2006. [a] Nair, R. et al., *International Journal of Pharmaceutics*, 225(1–2): pp. 83–96, 2001; [b] Özdemir, C., and Güner, A., *European Polymer Journal*, 43(7): pp. 3068–3093, 2007.

with their boiling points are presented in Table 4.2. According to Hansen, the permanent dipole–dipole interactions (δ_p), dispersive (δ_d), and hydrogen bonding forces (δ_h) should be taken into consideration.

Consequently, the solubility parameters (δ) can be calculated as follows [36]:

$$\delta^2 = \left(\delta_d^2 + \delta_p^2 + \delta_h^2\right) \tag{4.1}$$

The Hansen solubility parameter differences (Δ) among the solvents and the membrane forming polymers were computed using the following equation:

$$\Delta = \sqrt{[(\delta_{P,d} - \delta_{S,d})^2 + (\delta_{P,p} - \delta_{S,p})^2 + (\delta_{P,h} - \delta_{S,h})^2]} \tag{4.2}$$

where S and P denote the solvent and polymer, respectively; and p, d, and h represent polar, dispersive, and hydrogen bonding elements of Hansen solubility parameter. The polymer-solvent interaction parameter is inversely proportional with Hansen solubility parameters (Δ), that is, the smaller the difference between the solubility parameters of polymer and solvent, the stronger the polymer–solvent interaction. The results of the solubility parameter difference are listed in Table 4.2 and show that the Δs of the solvents are in decreasing order (i.e., DMF < DMAc < Acetone).

Therefore, the solvent:polymer interaction of the three solvents increases in the order of DMF > DMAc > Acetone. From these results, it can be suggested that a relatively strong hydrogen bonding interaction happened between remaining DMF and CA when compared with CA/Ac/DMAc. Therefore, the strong interaction between DMF and CA was suggested during the preparation of solution casting process for membrane forming.

The effect of solvents on the PWF was also observed clearly from this study. Membranes prepared using Ac: DMAc (CA0, CA1, and CA2) showed higher fluxes than membrane fabricated with DMF (CA3, CA4, and CA5) regardless of the effect of additives. Moreover, the initial sharp decrease in PWF for CA/DMF membranes was significantly improved for CA/Ac: DMAc membranes. This result is evidently related to the effect of solvent on the morphological/sublayer structures of the membranes. All the membranes prepared with CA/DMF system depicted large flux decline due to non-uniform sublayer pore structures. The percentage flux declines were found to be as 89%, 72%, and 75% for CA3, CA4, and CA5, respectively. Second, the formation of larger macro-voids due to an instantaneous de-mixing may have resulted in the formation of non-uniform pore distributions within the membrane matrix.

However, for the membranes prepared from CA/Ac: DMAc system, the percentage flux decline for CA0, CA1, and CA2 was 34%, 32%, and 33%, respectively. Unlike the CA/DMF membranes, the uniformly interconnected pore structures of the sublayer with uniform pore distribution of CA/Ac: DMAc membranes showed better permeability characteristics. The difference in flux between the CA-DMF and CA-Ac: DMAc membranes can also be associated with the effect of the solvents on the hydrophilicity nature of the membrane. As it is believed that there are some residual solvents within the final membrane matrix [33,40], they may affect the water flux performance of the membrane. Guan et al. [33] also reported that the membranes with DMAc solvent showed a higher hydrophilic nature than membranes with DMF. Similar results were found in our study. Regardless of the effects of the polymeric additives, the CA/AC: DMAc showed more hydrophilicity nature than the CA/DMF membranes. It is clearly shown from the membranes prepared without additives (i.e., CA0 and CA3). The effects of the residual solvents are not significant as compared to the effect of additives. However, it is believed to have some influences on the membrane performance.

Table 4.3 presents the compaction factor and hydraulic characteristics of all prepared membranes. It was observed that for both CA/Ac: DMAc and CA/DMF membranes, the CF decreases with introduction of the additives.

TABLE 4.3

Compaction Factor, Hydraulic Characteristics, EWC, and Porosity of All Prepared Membranes at 300 kPa

Membrane	*CF*	P_m (L/m^2hkPa)	R_m ($\times 10^{-10} m^{-1}$)	*Jw* (L/m^2h)	*EWC* (%)	ε (%)	r_m (nm)	Thickness μm
CA-0	1.5 ± 0.5	0.8	1.1	248	79	83	14.4	72.9 ± 15
CA-1	1.3 ± 0.6	1.1	0.9	316	84	87	23.6	119.9 ± 10
CA-2	1.4 ± 0.5	3.3	0.8	978	86	89	28.9	116.4 ± 11
CA-3	4.0 ± 0.2	0.2	3.5	70	75	80	12.7	186.2 ± 8
CA-4	2.5 ± 0.3	0.3	1.9	81	82	85	13.9	166.5 ± 9
CA-5	3.3 ± 0.4	0.9	1.6	273	83	86	22.4	172.0 ± 8

However, for membranes prepared using both solvents, the CF for CA/PEG was less than CA/PVP. This can be described that an introduction of additive into the membrane forming solutions may suppress or increase formation of macro-voids in the membrane sublayer based on the type of additive and solvent [21]. In the current study, for both the solvents less formation of macro-voids was observed for CA/PEG than CA/PVP. Due to these reasons, membranes with PVP additives have resulted in highly porous sublayer structure because of the occurrence of more numbers of macro-voids for both the solvents. These results are also depicted from the FESEM images (Figure 4.3).

The effects of polymer additives and solvents on PWF at different operating pressures are presented in Figure 4.4b and 4.4d. The results of PWF for all the membranes were increased almost uniformly with increasing in an operating pressure from 100 to 300 kPa. It is also shown that the PWF for all the CA/DMF membranes is lesser than that of CA/Ac: DMAc membranes. Nevertheless, of the effect of the solvent, the membranes with PVP additives showed higher PWF than that of membranes with PEG additives for both solvents. These results are in good agreement with the conclusions of the compaction study in Figure 4.4a, c. The hydraulic resistance (R_m) of CA/DMF membranes was greater than that of CA/AC: DMAc membranes. In addition, the R_m for the membranes without additives was higher than all the membranes with additives for both solvents. The R_m for CA/PEG was higher than that of CA/PVP for both the solvents. Therefore, an increase in hydraulic resistance and decrease in flux with varying additives and solvents may be accredited to the decline in pore size during compaction. In general, both the additives, namely, PVP and PEG, are more hydrophilic than the CA polymer. Therefore the CA membranes with these additives showed improved flux results.

Comparing the two additives, the CA/PVP showed higher flux and lower Rm than membrane of CA/PEG. These results can be explained because of higher solubility and diffusivity of PVP than PEG, where most of the PVP can be washed out more easily and quickly (as compared to PEG) with the non-solvent (i.e., water) during the membrane fabrication period [10]. This result may also be explained due to the formation of some macro-voids because of the rapid penetration of non-solvent at certain weak spots in the top layer of the membrane; the PVP rich phase might act as the weak spots during the precipitation process and result in the development of macro-voids [17]. As a result, the suppressed sublayers of the PEG membranes are possible to give additional resistances to water permeability than PVP membranes resulting in low flux. The effects of additives as a pore former as well as hydrophilicity enhancer were also investigated, and the porosity and EWC results were calculated using Equations 2.4 and 2.5 (see Chapter 2). The wettability of the membranes was studied by measuring the water contact angle and the drop age, defined as the duration of the water droplet on the surface of the membrane and spreading and/or permeating through the membrane cross-section [41]. The variation of water contact angle of all the prepared membranes is presented in Figure 4.5a. The images of the water droplets with a volume of about 2 μL at 0.16 mL/min on the membrane surface after 60 s are shown in Figure 4.5b. CA1, CA2, CA4, and CA5 membranes have taken about 15 s, 10 s, 25 s, and 20 s, respectively, and show best water wettability, where most parts of the membrane surface were almost completely wetted and smaller spread radius of the water drops on the top side of the membrane were observed after 60 s. On the other hand, CA0 and CA3 membranes have taken about 34 s and 36 s to initiate the surface wetting and big water drops spread radius on the top side of the membrane were observed after 60 s (Figure 4.5b). The smaller the water drop spread radius wetting area between top and bottom surface and on the top membrane side, the better the water permeability is. As clearly observed from the graph, the contact

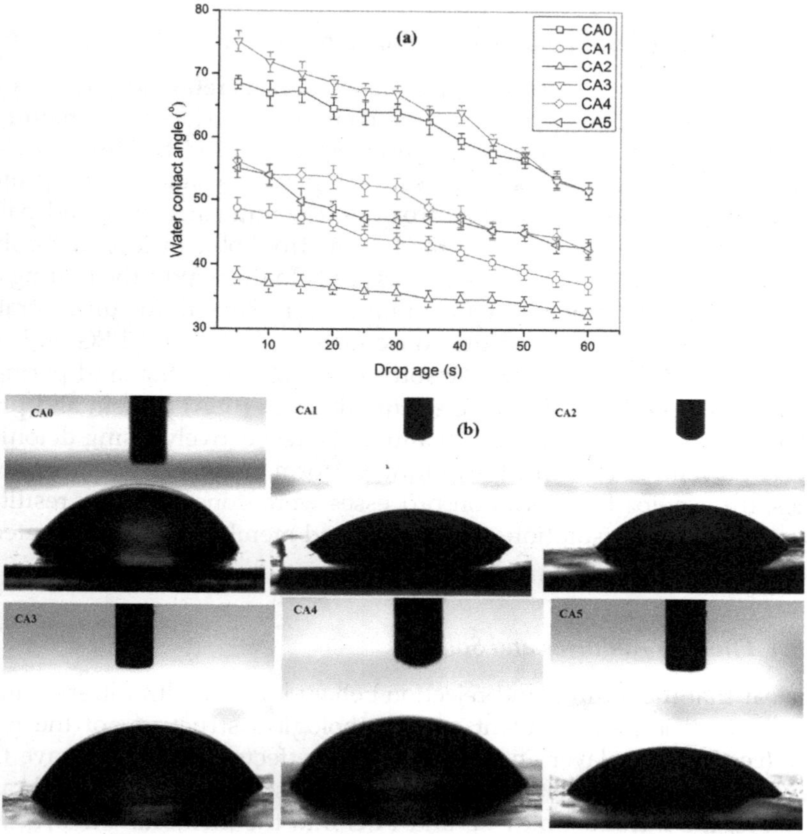

FIGURE 4.5
(a) Water contact angle values and (b) images of water droplets of the prepared membranes (water droplet volume is about 2 μL at 0.16 mL/min).

angle results of the pristine CA0 and CA3 membranes show water contact angle about 61.3 ± 1.6° and 64.1 ± 1.5°, respectively. In contrast, after the introduction of hydrophilic additives, all the CA1, CA2, CA4, and CA5 membranes showed greatly reduced water contact angle results (i.e., 43.3 ± 2.2°, 35.1 ± 3.1°, 49.6 ± 1.8°, and 47.6 ± 2.0°, respectively). Membranes having smaller contact angle results are considered as more hydrophilic membranes. Therefore, an increase in the hydrophilicity, EWC, and porosity of the membrane after adding of additives is due to the pore-forming and hydrophilicity properties of PEG and PVP. Besides, as already discussed above, using different solvents also influenced the hydrophilicity, EWC, and porosity of the membrane in some way. Thus, the CA/Ac: DMAc membranes showed better hydrophilicity nature than that of CA/DMF membranes irrespective of the effect of additives. Moreover, it is confirmed that all the membranes are in ultrafiltration range as seen from the pore radius (r_m) measurement results in Table 4.3.

4.2.3 Membrane Fouling and Rejection Experiments

Membrane fouling is defined as the deposition of retained particles colloids, macromolecules, salts, etc. at the membrane surface, pore mouth, or pore wall, causing flux decline. Fouling can be caused by three kinds of substances: organic (macromolecules, biological substances like protein, enzyme, etc.), inorganic (metal hydroxides, calcium salts, etc.), and particulates. Proteins are strongly adsorbed on hydrophobic or less hydrophilic surfaces but less on hydrophilic surfaces [1,42]. In this study, the fouling and rejection experiments of all the membranes were done in the ultrafiltration stirred batch cell to examine the consequence of PVP and PEG additives and effect of DMF and Ac: DMAc solvents on BSA rejection and permeate flux. The concentration of bovine serum albumin (BSA) protein and pH of the solutions were kept at (1 g L^{-1}) and at 7.9, respectively, using deionized water as solvent. To examine the fouling performance of the prepared membranes, three steps BSA filtration processes were done. The flux results of pure water and BSA solutions of the prepared membranes are presented in Figure 4.6.

4.2.3.1 Effect of Additives and Solvents

It is clear that the fouling and rejection behavior of the ultrafiltration membranes are strongly dependent on morphological structures of the membrane (i.e., both top layer and sublayer). The effects of additives have their potential role on the variation in morphological structure of the membranes. Therefore, the introduction PVP and PEG into the membrane matrix were observed to influence the flux and BSA rejection performance of the prepared membranes. The fluxes and rejections of protein in ultrafiltration membrane can be described using the theory of resultant pore narrowing and protein

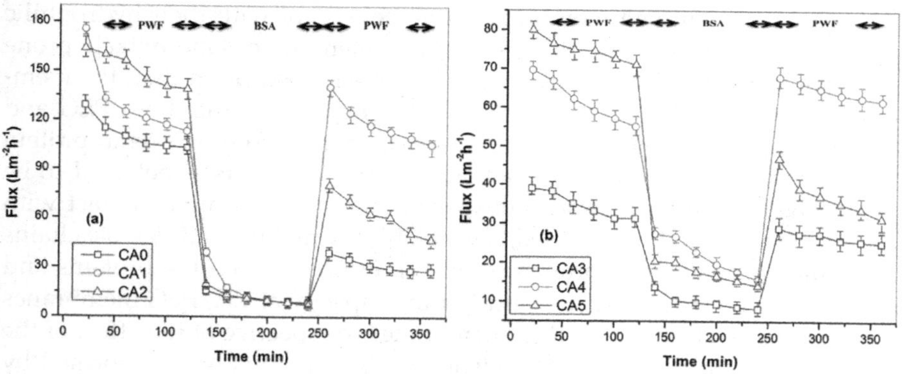

FIGURE 4.6
Permeate flux versus filtration time for (a) CA/Ac: DMAc and (b) CA/DMF membranes: effect of solvents and additives on fouling performance (25 °C, 150 kPa).

adsorption, due to electrostatic and hydrophobic interaction between BSA molecule and the membrane's surface [43]. Regardless of the effect of solvent, the effect of PEG and PVP were clearly observed from the fouling experiments. Therefore, the fouling experiment results of CA/Ac: DMAc membrane system and CA/DMF membrane system are presented in Figure 4.6a and 4.6b, respectively. The results showed that initially the CA/PVP membranes showed higher PWF fluxes than the CA/PEG membranes for both the solvents. However, relatively higher BSA fluxes results were observed for CA/PEG membranes. It is clear from the results that the membranes without additives (i.e., CA0 and CA3) showed less PWF as well as the BSA fluxes due to their less hydrophilic nature and less pore formation. The improvement of pure water and BSA fluxes for membranes with additives (i.e., CA1, CA2, CA4, and CA5) were attributed to the pore-forming effect and hydrophilic nature of PEG and PVP additives. On the other hand, the whole removal of additives may not be attained from the matrix of the membrane [44].

As clearly depicted from Figure 4.6, in the second BSA filtration operation (for 2 h), it is observed that with increasing time, the BSA fluxes gradually decreases for both CA/DMF and CA/Ac: DMAc membranes. The decrease in BSA fluxes with increasing time could be due to susceptible pore blocking of the membranes because of BSA protein adsorption and deposition on membrane surface, where the effect of concentration polarization was reduced by using a high molecular weight of BSA (66 kDa) molecules and rigorous stirring (200 rpm) on the surface of the membrane. Moreover, the rapid drop in the primary fluxes are realized, and the ending fluxes are slowly dropped, which is credited to the porosity loss of the membrane due to an interior adsorption of BSA protein that further leads to pore blocking. After membrane cleaning operation, the third step pure water filtration operation result showed that greater relative fluxes were recovered for both CA/PEG and CA/PVP membranes for both solvents. Normally, the orders of flux recovery ratios for the examined membranes were consistent with their hydrophilic and porosity nature. Thus, the additive-free membrane is more likely prone to pore-blockage and fouling because of protein adsorption than the membrane with additives. In the literature, it is mentioned that both PEG and PVP [45] have the potential to minimize membrane fouling due to protein adsorption. Moreover, the PEG among various polymers is believed to be an excellent material to resist nonspecific protein adsorption contact with the surfaces due to the hydrated, neutral, highly mobile, and flexible chains since they provide maximum entropic repulsion between the proteins and surfaces [46,47]. Therefore, the anti-fouling capacity of CA/PEG membranes was better than that of CA/PVP membranes irrespective of the effect of the solvents. The desorption of the adsorbed BSA proteins was performed by soaking the samples in DI water for 2 h and followed by rinsing using DI water. Moreover, the PWFs of CA-PEG-DMF and CA-PEG-Ac: DMAc were slightly increased after desorption of the BSA by washing the membranes for 2 h. Such a phenomenon is not common in the conventional trends for fouling ultrafiltration experiments. These results could be caused by the inherent interactions between foulant (BSA) and PEG additive entangled in the membrane matrixes and the membrane top layer [48–50]. It was also confirmed that PEG could efficiently avoid the irreversible adsorption of the protein on the surfaces. Therefore, due to the hydrophobic interaction between PEG and BSA, the proteins might be wrapped by PEG chains, forming a protective layer [47]. Membrane fouling consists of reversible and irreversible fouling. In the case of reversible fouling, deposited protein could be eliminated easily using hydraulic cleaning (i.e., back washing and cross flushing) [51].

Conversely, the irreversible fouling occurs due to an irreversible adsorption of proteins that can only be removed using chemical cleaning [52]. To further examine the anti-fouling properties of the prepared membranes and to study the effect of additives, the flux losses from total fouling (F_t), reversible fouling (F_r), and irreversible fouling (F_{ir}) were calculated, and the results are presented in Table 4.4. From these results, it can be shown that the F_{ir} for the CA/PEG membranes is less than CA/PVP membranes. However,

TABLE 4.4

Fouling Study Data of the Prepared Membranes

Membrane	J_{w1} (L/m² h)	J_{Bs} (L/m² h)	J_{w2} (L/m² h)	F_t $\left(1-\frac{J_B}{J_{Wi}}\right)$	F_r $\left(\frac{J_{Wi}-J_{Wf}}{J_{Wi}}\right)$	F_{ir} $\left(\frac{J_{Wf}-J_B}{J_{Wi}}\right)$	NFR $\left(\frac{J_{Wf}}{J_{Wi}}\right)$
CA0	111	15.9	76.7	0.86	0.55	0.31	0.69
CA1	130.7	21.5	118.4	0.84	0.74	0.09	0.91
CA2	150.7	17.2	81.8	0.89	0.43	0.46	0.54
CA3	34.7	10.0	26.6	0.71	0.48	0.23	0.76
CA4	64.6	11.5	61.4	0.82	0.77	0.05	0.95
CA5	75	11.1	37.1	0.85	0.35	0.51	0.49

the reversible fouling (F_r) for CA/PEG membranes are higher than CA/PVP membranes. This may be credited due to extra BSA protein buildup on membrane surfaces because of the better BSA rejection. In spite of the greater reversible fouling, total fouling (F_t) of CA/PEG membranes showed lesser results than CA/PVP membranes due to the decreasing of irreversible fouling. From this result, it can be suggested that the fouling resistances, particularly resistances due to irreversible fouling, of CA/PVP membranes are significantly higher than CA/PEG membranes irrespective of the solvent effects.

The effect of solvents on the membrane morphological structures and PWFs were already explained in detailed in previous discussions. Therefore, from these experiments, it was clearly observed that both the PWF and BSA flux of CA/Ac: DMAc membranes (Figure 4.6a) are higher than the CA/DMF membranes (Figure 4.6b). Similarly, the BSA protein rejection of CA/Ac: DMAc membranes is higher than of CA/DMF membranes irrespective of the effects of additives.

4.2.3.2 Rejection Performance

The rejection performances of the prepared membranes are showed in Figure 4.7. The maximum BSA rejection results for CA/Ac: DMAc and CA/DMF system (i.e., 94.4% and 82.9% rejections were attained for CA0 and CA3 membranes, respectively). These results can be explained due to the additive free membranes have less porous structures, where a better resistance to protein molecules was observed. On the other hand, the higher BSA rejection (91.9% and 69.9%) for CA/PEG (i.e., CA1 and CA4) membranes than CA/PVP

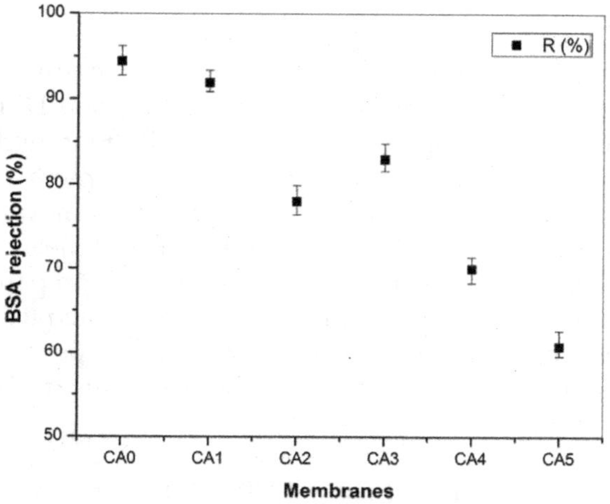

FIGURE 4.7
Rejection performance of different CA membranes: effect of solvents and additives (25 °C, 150 kPa).

(CA2 and CA5) membranes (77.9% and 60.7%), respectively, were achieved irrespective of the effect of the solvents. The characteristic of the BSA rejection can be described using the protein adsorption phenomenon as we have already explained above. The increase in BSA rejection for CA/PEG membranes is understood to be an indicator of substantial protein adsorption inside and on the membrane surfaces. Therefore, the surface depositions of BSA proteins provide an extra hindrance to solute transportation. Another reason for CA/PEG membranes to have higher rejections than CA/PVP membranes may be due to the adsorption of protein on pore-wall of CA/PVP membrane has lesser influence on pore-narrowing because of their highly porous structure and higher macro-voids. However, the rejection result for CA3 is higher than CA2 but less than CA0 and CA1 membranes. This result can be explained due to high resistance capacity of the membranes without additives having less porous properties. The high rejections and comparatively lower fluxes of CA/PEG membranes than CA/PVP for both the solvents can be assumed from the morphological study. The reasonably dense asymmetrical layer possibly describes the enhancement in the rejection rate whereas the sublayer with inhibited macro-voids gives resistance ensuing in moderately lower fluxes and higher BSA rejections. From the above discussions, it is clear that the type of additives and type of solvents showed significant effects on the membrane flux and BSA rejection performances. The results attained in this specific study seem to be reasonably fascinating that the membranes with PEG additives gained higher rejection results and moderate fluxes. On the other hand, the membranes prepared using Ac: DMAc solvent showed better characteristics (i.e., hydrophilicity, flux and rejection) than the membranes prepared using DMF as solvent.

4.2.4 Summary

Phase inverted CA membranes were prepared from casting solutions comprised of two different solvents, namely, Ac: DMAc and DMF individually using immersion precipitation process. Two different hydrophilic additives, PEG and PVP, were used as membrane pore formers and hydrophilic enhancers. The effects of additives and solvents on the morphological structures (i.e., top surface and a cross-sectional structure of the membranes) were studied in detail. The permeability properties of the prepared membranes using different additives and solvents were also estimated using CF, hydraulic permeability, hydraulic resistance, hydrophilicity, porosity, EWC, PWF, and BSA rejection performance. The outcomes from this study presented the following concluding points:

- The membranes prepared using PVP additive showed the development of greater macro-voids and finger-like structure when compared with membranes formed using PEG additive.

- The introduction of hydrophilic additives to the membrane casting solution makes the membrane highly porous and more hydrophilic where the permeability properties of the CA ultrafiltration membrane were significantly influenced.
- The difference in flux between the CA/DMF and CA/Ac: DMAc membranes are associated with the effect of the solvents on the hydrophilic nature of the membrane, as it is believed that there are some residual solvents within the final membrane matrix.

The fouling resistances, particularly resistances due to irreversible fouling, of CA/PVP membranes are significantly higher than CA/PEG membranes irrespective of the solvent effects. Higher BSA rejection and relatively moderate flux of CA/PEG membranes than CA/PVP for both the solvents were obtained, and the comparatively dense asymmetric layer of CA/PEG membrane possibly describes the enhancement in the rejection rate, whereas the sublayer with inhibited macro-voids gives resistance resulting in moderately lower fluxes and higher BSA rejections.

4.3 Preparation of CA–PEG–TiO_2 Membranes: Effect of PEG and TiO_2 on Morphology, Flux, and Fouling Performance

4.3.1 Introduction

Synthetic membranes in general and ultrafiltration (UF) membranes specifically have got more attentions in various applications, such as wastewater treatments, bio-separation, recovery of proteins, clarification of fruit juice and alcoholic beverages, separation of oil-water emulsions, and food and paper industry [53]. The principle of fabrication of asymmetric polymeric membranes is based on the phenomenon of phase inversion [17]. The common techniques of the precipitation of polymer are immersion precipitation, thermally induced phase inversion, vapor induced phase inversion, and dry casting of polymer solution [25]. Most commonly available commercial membranes are prepared by phase inversion [1]. Cellulose acetate is broadly applicable for the synthesis of membranes because of having tough, biocompatible, hydrophilic characteristics, good desalting nature, high flux, and moderately less expensive to be employed for reverse osmosis, microfiltration, ultrafiltration, and gas separation applications [7]. The drawback of cellulose acetate membranes is that they are susceptible to thermal and mechanical stabilities depending on the environments and conditions of application [44]. Generally, the presence of the polymeric additive increases the concentration/viscosity of the casting solution, which may diminish the diffusional exchange rate of the solvent and non-solvent during the membrane formation process [13]. Liu et al. [54] have studied the morphology controlled characterization of polyethersulfone hollow fiber membrane by the introduction of polyethylene glycol. In their study, they have shown that PEG can be used as an additive to improve the polymer (PES) dope viscosity and to develop the pore interconnectivity. Panda and De completed a detailed investigation on the effect of polyethylene glycol in polysulfone membranes. The influences of PEG molecular weight, solution concentration, the type of solvents, and thickness of casting solution were explored. As the solubility of solvent and non-solvent can play significant roles in the morphology of the resulting membrane, higher solubility of solvent: non-solvent leads to a rapid de-mixing. On the other hand, a delayed de-mixing may occur due to poor solubility and resulting in a membrane with denser top layer. Due to the higher solubility between NMP and water than that of the DMF and water, the exchange ratio between NMP/PES is higher than the DMF/PES ratio, which implicates more diffusion between solvent and polymer and the formation of a porous membrane. Thus a highly porous membrane was achieved using N-methyl-2-pyrrolidone (NMP) as a solvent when compared to N, N-dimethylformamide (DMF) [55]. Chakrabarty et al. [11], however, have prepared flat sheet asymmetric polymeric membranes from a homogeneous solution of polysulfone by phase inversion method. Their results have shown that the membranes with higher molecular weights of PEG resulted in high pure water flux because of higher porosity of the membranes. Saljoughi et al. [13] have prepared asymmetric CA membrane from blends of CA, PEG, and NMP using immersion precipitation phase-inversion process. From their observations, an increasing in PEG concentration in the casting solution alongside higher coagulation bath temperature resulted in an improvement of the pure water flux, membrane thickness, and human serum albumin transmission. Using organic-inorganic membranes, the separation properties of polymeric membranes can be enhanced and may possess properties such as selectivity, good permeability, thermal and chemical stability, and mechanical strength [56]. Another challenge in the field of membrane process is fouling. Since membrane fouling is causing a severe decline of the solvent flux, it becomes essential to fabricate membranes less susceptible to fouling by making modification during preparation. In ultrafiltration processes, there have been several attempts to decrease fouling, which in general include feed solution pretreatment, membrane surface enhancements, and process modifications [57]. In recent research, it is confirmed that the introduction of nanoparticles in a membrane matrix develops the membrane hydrophilicity, anti-fouling property, and permeability. Accordingly, several inorganic oxide nanoparticles such as Al_2O_3 [58], ZrO_2 [59], TiO_2 [57], and SiO_2 [60] have been added within polymer casting solution. Prince et al. have prepared functionally modified PES hollow fiber membrane using PEG 400 and silver (Ag) NPs through thermal grafting. The attachment of PEG additive and silver NPs on the surface of the PES hollow fiber membranes were completed by using poly (acrylonitrile-co-maleic acid) (PANCMA) as a chemical linker. The WCA results of the modified membranes were found to be decreased by about 75.5% from 62.6 ± 3.7° to 15.3 ± 1.2° and the PWF have improved by around 36% from 513 L/m^{2}h to 702 L/m^{2}h [61]. Garcia-Ivars et al. have prepared UV irradiation modified polyethersulfone (PES) UF membranes using two nano-sized hydrophilic compounds, PEG and Al_2O_3 NPs. The WCA and pore size results of the modified membranes have decreased, indicating an enhancement of the hydrophilicity nature of the resulting membranes. Furthermore, the PEG solute flux and rejection of the PES membranes were enhanced due to UV photo grafting. Moreover, the modified PES membranes (i.e., 2.0 wt.% PEG and 0.5 wt.% Al_2O_3) have exhibited superior anti-fouling performance than the other tested membranes [62]. Current researchers have paid attention to TiO_2 due to its stable nature, ease of availability, and the potential for different applications. Moreover, TiO_2 can enhance the hydrophilicity of different polymers to improve flux and decrease the fouling problem, important parameters in water and wastewater treatment [57]. In membrane filtration process factors such as thermal/chemical, fouling, and flux are important properties, and the current study is focused on improving these properties simultaneously. An effort was made to investigate the influences of PEG and TiO_2 on the morphological structure, permeability performance, and thermal stability property of the membrane in addition to the anti-fouling properties of CA-PEG-TiO_2 membranes.

4.3.2 Preparation of CA–PEG–TiO_2 Membrane

Phase inversion technique by immersion precipitation was used for fabrication of asymmetric UF membrane from cellulose acetate polymer. Initially, uniform solutions of CA in Acetone/DMAc (70/30; v/v) were prepared under continuous magnetic stirring at room temperature (28 ± 2 °C). The casting solution was stirred overnight using a magnetic stirrer at room temperature. After that, the solutions containing CA, solvents, and the additives (PEG and TiO_2 NPs) became homogeneous, and they were kept at room temperature for one day to avoid air bubbles. Subsequently, the casting solution was poured consistently on a glass sheet and carefully cast using a casting knife, keeping a gap of roughly 0.25 mm between the knife and glass plate. The resultant films were exposed to air for approximately 30 s earlier to immersing to the coagulation bath comprising of DI water at room temperature. In the coagulation bath, the cast solution turned from transparent to white color for membranes CA and CA-PEG. The milky color for CA-TiO_2 and CA-PEG-TiO_2 membranes was changed to white, and all the thin films were detached from the glass plate. The membrane sheets were kept for 30 min in the coagulation bath. After that, the prepared membranes were put in DI water-filled beakers until use. Last, the membrane films were cut into circular discs to put in the membrane cell for UF experiments. The membranes with different composition are designated as M_1, M_2, M_3, and M_4 (i.e., CA, CA-TiO_2, CA-PEG-TiO_2, and CA-PEG, respectively). Table 4.5 shows

TABLE 4.5
Solution Compositions and Viscosity of the Casting Solution: CA, PEG, and TiO_2 Nanoparticles

Membrane	CA (wt.%)	TiO_2 (wt.%)	PEG (wt.%)	AC: DMAc (70:30; v/v)	Viscosity (mPa.s)
M_1	10.5	–	–	89.5	1680
M_2	10.5	2	–	87.5	3180
M_3	10.5	2	4	83.5	1850
M_4	10.5	–	4	85.5	1781

the solution casting compositions of CA, PEG, and TiO_2 nanoparticles. The flow diagram for membrane preparation process is presented in Figure 4.8.

The amounts of CA, PEG, and TiO_2 NPs were selected based on the previous works [13,63].

4.3.2.1 Morphological Study

The FESEM analysis is an important method to study the membrane morphological structure and qualitative information about surface and cross-sectional morphology of the membranes to be achieved. The top layer surface view, the cross-sectional view, and elemental analysis results of the prepared membranes by dissolving the polymer and additives in AC: DMAc are presented in the FESEM and EDS images (i.e., Figure 4.9a, b, and c, respectively). As revealed by the figures, the synthesized membranes are

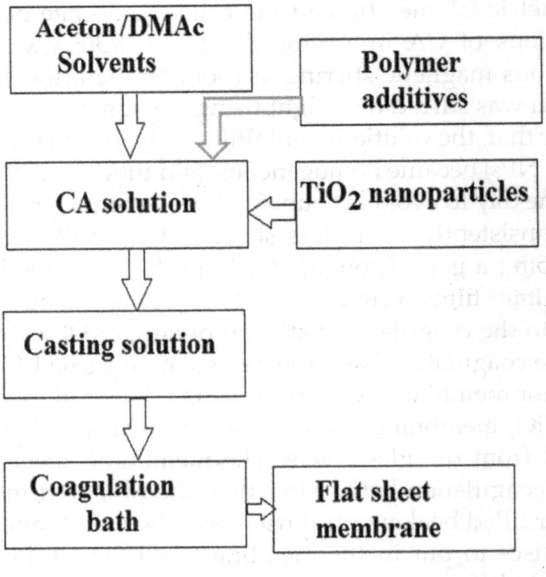

FIGURE 4.8
Flow diagram of the preparation method of flat sheet membranes.

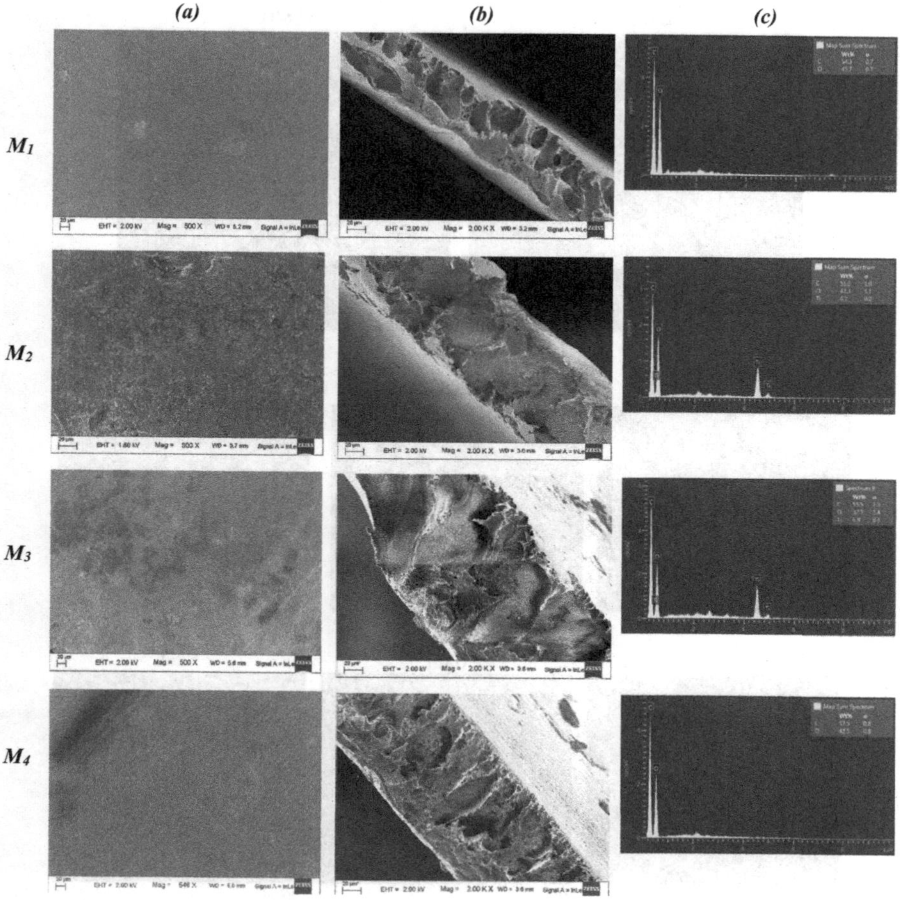

FIGURE 4.9
(a) Top surface and (b) cross-sectional FESEM images; (c) EDS results of M_1 (CA), M_2 (CA-TiO_2), M_3 (CA-PEG-TiO_2), and M_4 (CA-PEG) membranes. *(Continued)*

asymmetric in structure involving a dense top layer and a porous sublayer for all types of membranes. For additional information, the mapping or distributions of the elements within the matrix of the prepared membranes are presented in Figure 4.9d. PEG polymer and TiO_2 nanoparticle were added to the membrane forming polymer (i.e., CA) as an additive to examine their impacts on the morphological structures, thermal stability, and anti-fouling performance of the synthesized membranes. As revealed by the figures, the synthesized membranes are asymmetric in structure involving a dense top layer and a porous sublayer for all types of membranes. The sublayer portion of the membranes without additives (M_1) appears to have finger-like voids as well as macro-void structures.

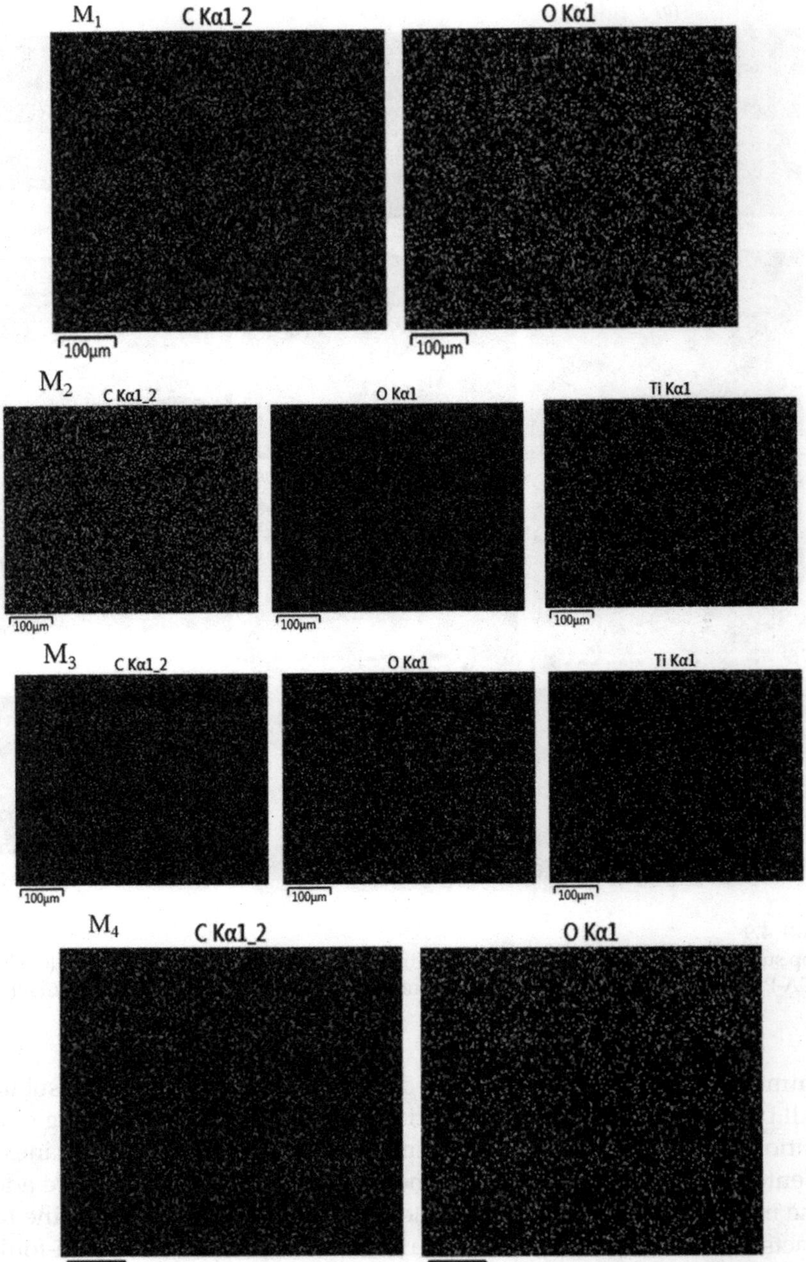

FIGURE 4.9 (CONTINUED)
(d) EDS mapping images of CA membranes with PEG and TiO_2 NPs.

4.3.2.1.1 Effect of PEG additive and TiO_2 NPs

The sublayer portion of the M_1 membrane appears to have finger-like voids as well as macro-void structures. Generally, the development of macro-voids happens under rapid precipitation condition, and the precipitations are quicker at high coagulation temperature [13]. In the present study, the coagulation bath temperature and evaporation time (i.e., before immersion) were fixed to 25 ± 2 °C and 30 s, respectively, during the preparation of all the membranes. An additional significant factor that can affect the formation or suppression of macro-voids is the type of de-mixing that occurred during the phase separation process, where the instantaneous de-mixing and the delayed de-mixing depend on the mutual affinity of the solvent and non-solvent in the ternary system [1]. In solvent and polymer interaction scheme, three kinds of interactions—polymer:polymer, polymer:solvent, and solvent:non-solvent—are applied. If good polymer solvents are used, the degree of polymer stretching reaches to its highest level, and more favorable polymer/solvent interactions can occur. The solvents used in this study (i.e., AC: DMAc) have high mutual affinity with water [33]. Thus, apparently, the development of the finger-like voids and macro-voids in the sublayer of the M_1 (additive free) membrane is due to the instantaneous de-mixing. In this case, the CA (additive free) stretches to its highest level where there is a maximum interaction between the CA and AC: DMAc, which tends to instantaneous de-mixing condition to happen. After adding the PEG and TiO_2 to the ternary (polymer/solvent/non-solvent) system, the developments of macro-voids are suppressed significantly (i.e., M_2, M_3, and M_4). The presence of the polymeric additives and TiO_2 NPs can increase the concentration/viscosity of the casting solution, which may diminish the diffusional exchange rate of the solvent and non-solvent during the membrane formation process. Accordingly, this may hamper the instantaneous liquid-liquid de-mixing process that suppresses the development of macro-voids [23]. However, for membranes prepared using PEG and TiO_2 nanoparticles (M_3) and without TiO_2 nanoparticles (M_4), the formation of micro-voids (microporous structure, which is important for the porosity of membrane) was observed. It is clear that the pore forming properties of PEG polymer have played the key role in the development of porous sublayer. A quaternary system (polymer/solvent/non-solvent/additive) of the phase separation involves de-mixing of the entangled polymers. In this quaternary system, two phases occur from the phase separation; one involves CA (i.e., membrane forming polymer), AC: DMAc (i.e., solvent), and non-solvent; the second contains the additive (i.e., PEG), AC: DMAc, and non-solvent. The membrane forming polymer and additive have a dynamic strength to be completely separated to CA-rich and PEG-rich phases. Consequently, the type of de-mixing process can also be decided by the diffusion of these polymers with respect to each other.

The addition of TiO_2 to the membrane formation system has its influence on the de-mixing process. However, almost macro-void free cross-sectional structures were observed for the M_4 system. From these results, it can be concluded that although the instantaneous de-mixing is still continued, the effect of the PEG has a substantial role in the suppression of the finger-like structure and macro-voids that occurred for the M_1 membrane. The presence of the polymeric additive increases the concentration/viscosity of the casting solution, which may diminish the diffusional exchange rate of the solvent and non-solvent during the membrane formation process. Consequently, this may hinder the instantaneous liquid-liquid de-mixing process that suppresses the development of macro-voids [13]. The comparatively low affinity of PEG to the solvents may take additional time to reach the top surface, allowing the polymeric molecule to get sufficient time to accumulate and rearrange and subsequently developing a relatively thicker and denser top layer. The open pore structures developed in the membranes are formed by nucleation and development of the polymer-lean phase in the metastable area between the bi-nodal and the spinodal curve [22,24]. A possible explanation for the formation of a nodular structure on the top surface of the membranes could be due to the spinodal de-mixing because the diffusion process throughout the development of the top layer is faster for the homogenous system to become highly unstable and crosses the spinodal curve [22]. In the present study, the top surface with some open pores structures is observed more prominently in the case of PEG (i.e., M_3 and M_4), which is attained because of the spinodal decomposition. Therefore the interconnected pores may be accounted as a constant CA-lean/PEG-rich phase entwined by a continuous CA-rich/PEG-lean phase that is responsible for developing the uniform matrix of the membrane [64]. On the other hand, in the case of M_1 and M_2 membranes, no open pore structure was detected, where a less porous top surface structure was observed instead. It is evident from these figures that the finger-like structures and macro-voids are suppressed after adding TiO_2 (M_2 and M_3) regardless of the effect of PEG for M_3 membrane. However, the M_2 membrane seems to have less microporous cross-sectional structures. These results are also confirmed by the porosity study (Table 4.6) where the M_2 membrane showed lower porosity than the other membranes. The introduction of both PEG and TiO_2 played an important role in the improvement of the membrane hydrophilicity and porosity. The porosity, EWC, and pore

TABLE 4.6
Compaction and Hydraulic Characteristics of Those Prepared Membranes at 250 kPa

Membrane	*CF*	R_m ($\times 10^{-10} m^{-1}$)	J_w (L/m^2h)	*EWC* (%)	ε (%)	r_m (nm)	Thickness (μm)
M1	1.50 ± 0.4	0.90	204.5	77.8 ± 1.5	81.9 ± 2.0	23.6 ± 5	83.5 ± 8.0
M_2	4.50 ± 0.2	4.30	21.1	76.7 ± 2.6	81.0 ± 2.5	15.6 ± 7	102.4 ± 4.2
M_3	1.66 ± 0.3	0.70	530.7	79.5 ± 1.2	83.4 ± 1.8	35.0 ± 3	84.2 ± 6.5
M_4	1.52 ± 0.4	0.67	265.3	79.0 ± 1.4	82.9 ± 1.9	31.4 ± 4	126.9 ± 3.0

radius results of the prepared membranes are calculated using Equations 2.4, 2.5, and 2.11 (see Chapter 2). The membrane average pore radius (r_m) is considered as an approximation of true pore size and the results are 23.6 nm, 15.6 nm, 35 nm, and 31.4 nm for M_1, M_2, M_3, and M_4, respectively.

Moreover, the lowest pore radius for M_2 membrane was attributed to aggregation of some of the TiO_2 on the surface of the membrane pores. It can also be seen from the top surface view of M_2 presented in Figure 4.9a. Adding of TiO_2 NPs to the polymeric solution can increase its viscosity. Therefore, the particles leaching problem is lesser, and subsequently, the pore forming effect of NPs can be declined in the case where the high viscosity of a solution hampers the development of pores and causes the porosity of the membrane to decrease [65]. To minimize the influence of thickness shrinkage of the polymer-phase due to the accumulation of NPs, in the present study, TiO_2 NPs having low concentration (i.e., 2 wt.%) were selected [66]. It may be worthy to mention here that the addition of NPs in the polymeric membrane may decrease or increase its viscosity depending on various parameters, such as concentration of additive and ligand stabilizer [67]. Due to the introduction of a relatively lower concentration of TiO_2, the finger-like structures and macro-voids present in M_1 were greatly suppressed and a relatively dense layer with small finger-like structures was observed for M_3 and M_4. The presence of small macro-voids with a little finger-like structure in the case of M_3 and M_4 is assumed to be related to the interference effect of NPs and PEG additive during the phase inversion process. Therefore, due to the interfacial stress between polymers and NPs, interfacial pores are formed as a result of shrinkage of polymer-phase during the de-mixing process [68]. However, the presence of TiO_2 in M_2 diminishes the presence of finger-like structures and a sublayer structure with almost macro-voids free is obtained. These results can be described in terms of NPs agglomeration on the membrane forming polymer matrix during phase inversion process. The occurrence of NPs agglomeration can be mainly caused due to the high surface energy of the NPs, which tend to aggregate for weakening their surface energy to reach a more stable state. Furthermore, NPs agglomeration leads to a non-uniform dispersion of the NPs within the polymer surface and structure. This phenomenon can negatively change the resulting membrane properties such as hydrophilicity and surface roughness [66]. However, in the case of M_3, the NPs agglomeration was significantly minimized due to the improved distribution of the NPs because of the introduction PEG additive [69].

4.3.2.2 Thermal Stability Studies

The thermal degradation analyses (i.e., TGA and DTG results) of all the prepared membranes are presented in Figure 4.10a and 4.10b, respectively. The graphs were plotted as weight loss (%) vs. temperatures (°C). It can be clearly seen from the TGA figure that the decomposition of M_1 shows three steps. The first degradation step of the M_1 membrane was detected between 30 and

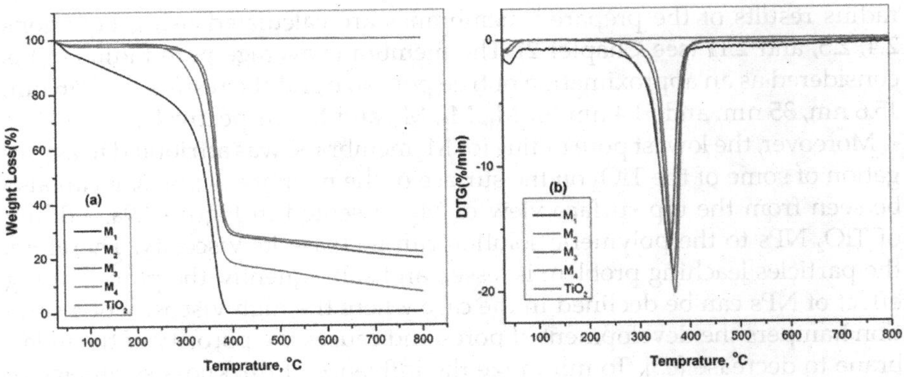

FIGURE 4.10
(a) TGA and (b) DTG analysis results of M_1, M_2, M_3, M_4 membranes and TiO_2 NPs.

60 °C, and the weight loss was about 4.5%. This degradation is due to the presence of some volatile material or because of the evaporation of absorbed moisture by the sample. During the second degradation step, a high weight loss of around 16.5% was observed between 60 to 260 °C, which is possibly due to the start of the main thermal breakdown of CA chains. A final degradation step started at 260 °C and ended at 380 °C with a weight loss of 79% due to the main degradation and possibly because of the carbonization of the decomposition of the residual materials to ash. The TGA results of the M_1 membrane clearly show that it could be highly unstable at high temperature, and similar results were reported by Zafar et al. and Chatterjee et al. [70]. The TGA results of the CA membrane with PEG additive and TiO_2 displayed two-step degradation procedures. Therefore, the start of the decomposition step for M_2, M_3, and M_4 were 278 °C, 271 °C, and 234 °C, respectively. The observed weight losses are due to the degradation of CA chains because of the pyrolysis of the backbone of the CA polymer and also followed by de-acetylation of CA [71]. During the last decomposition step, the degradation temperatures of M_2, M_3, and M_4 are 388 °C, 386 °C, and 382 °C, signifying the main thermal degradation of the CA chains, whereas the TiO_2 were totally stable until the end of the analysis (i.e., up to 800 °C). From these results, it is clearly noticed that the thermal degradation of M_1 membrane was significantly enhanced due to the addition of PEG and TiO_2. An interesting observation from this study is that the increase in thermal stability after addition of PEG to CA membrane (M_4) is due to the presence of a trace amount of the PEG in the membrane matrix. The improved thermal stability of the M_4 membranes was due to the strong interaction between the CA and PEG in the membrane matrix by creating hydrogen bonds [72]. Moreover, the residuals (i.e., 0.08%, 21.9%, 19.6%, and 7.3% for M_1, M_2, M_3, and M_4) are consistent with the thermal stability performances of the prepared membranes. Furthermore, the thermal stability of M_3 membrane exhibited higher degradation temperature than the M_4 membrane. This result is attained due

to the highly stable nature of TiO_2 at high temperatures and therefore the degradation temperature of the M_4 membrane was significantly improved after addition of TiO_2. The slight improvement in degradation temperature for the M_2 membrane when compared with the M_3 membrane is that the less porous sublayer structure of the membrane is responsible for having stronger resistant to heat flow than that of the porous membrane (M_3). In this study, the TGA and DTG plots clearly revealed that the thermal properties of CA membrane were significantly improved after incorporation of PEG and TiO_2 into the CA solution.

4.3.2.3 Pure Water Flux Performance

The membranes prepared using CA, CA-TiO_2, CA-PEG-TiO_2, and CA-PEG (M_1, M_2, M_3, and M_4, respectively), using AC: DMAc as solvent, were investigated to evaluate the influence of PEG additive and TiO_2 NPs on PWF properties of the membranes. The membranes were characterized in terms of compaction factor, PWF, and hydraulic resistance.

4.3.2.3.1 Effect of PEG and TiO_2 NPs

Studying of compaction factor (CF) of the prepared membranes is essential to recognize the morphological structures (i.e., pore arrangements) of the membranes, especially the membrane sublayer configuration. Membranes having high CF indicate that these are highly compacted and show the existence of some defective pores in the membrane sublayer structure. The compaction factor and hydraulic characteristics of all the prepared membranes are presented in Table 4.6. It was observed that the CF of M_2 was greater (i.e., 4.5 ± 0.2), which can also be explained due to the aggregation of NPs on the pore walls of the membrane could further block the pores after compaction. The CFs for the remaining membranes are almost similar except for the M_1 (1.5 ± 0.4), which is slightly lower than M_3 (1.66 ± 0.3) and M_4 (1.52 ± 0.4). This result is achieved due to the existence of more porous structure in case of M_3 and M_4 due to the introduction of PEG additive where some of the pores were compacted after the compaction process. The introduction of additives into the membrane casting solutions may either suppress or increase the formation of macro-voids in the membrane sublayer based on the type of additive [21]. The effects of PEG additive and TiO_2 NPs on PWF at different operating pressures are presented in Figure 4.11b. The results of PWF for all the membranes were increased almost uniformly with increase in operating pressure from 100 to 300 kPa. It is also shown that the PWF for membranes without PEG additive (M_1 and M_2) membranes are lesser than that of membranes with PEG additive (M_3 and M_4). Nevertheless, the effect of the PEG additive, the CA-PEG-TiO_2 membrane, shows higher pure water flux than that of CA-TiO_2 membrane. These results are in good agreement with the conclusions of the compaction study in Figure 4.11a. The hydraulic resistance (R_m), EWC, average pore radius, porosity, and thickness of all the

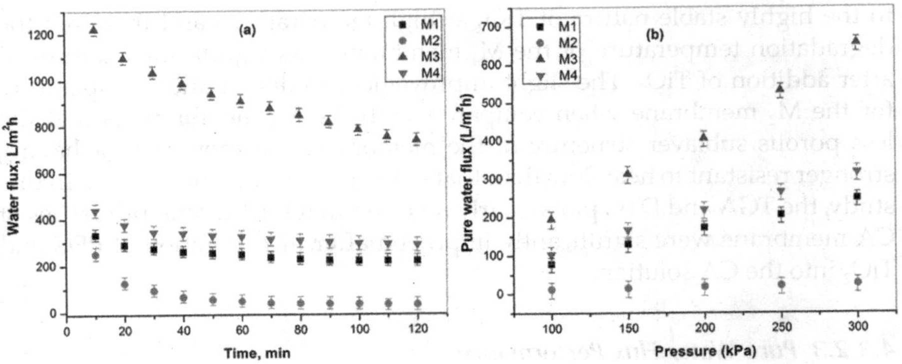

FIGURE 4.11
(a) PWF profile during compaction study (at 300 kPa), (b) Effect of transmembrane pressure on PWF.

prepared membranes are reported in Table 4.6. As seen from these results the hydraulic resistance of M_2 was greater than all the other membranes. This higher resistance of M_2 is because of its less porous nature than the other membranes, which can resist the water flux better than the other membranes. Additionally, this result is supported by the average pore radius calculation (Equation 2.11 in Chapter 2) where the result was 15.6 nm, which is again less compared with the other membranes. In addition, the R_m (m^{-1}) for the membranes without PEG additive (M_1) was higher than that of the membranes with PEG additive. Thus, the increase in hydraulic resistance and decrease in flux results of the membranes without PEG additive are obviously due to the decrease in average pore size and porosity as already explained in the previous sections. The PEG additive is more hydrophilic than the CA polymer. Therefore the membranes with PEG additive displayed improved flux results. As shown in Table 4.6, by comparing the two membranes with PEG additive (M_3 and M_4), the CA-PEG-TiO_2 membrane gained slightly higher resistance and higher flux than CA-PEG membrane. These results can be explained because the CA-PEG-TiO_2 membrane has higher and uniform porous and hydrophilic surface due to the introduction of TiO_2 NPs. On the other hand, as PEG additives were added and the whole additive wouldn't be washed away during the membrane development due to its low solubility and diffusivity, rather it could exist inside the pores and within the membrane matrices.

Therefore, the introduction of both PEG additive and TiO_2 NPs at the same time could play an important role in the improvement of the membrane hydrophilicity and porosity simultaneously. The porosity and EWC results of the prepared membranes results are calculated using Equations 2.4 and 2.5 (see Chapter 2) and presented in Table 4.6. From the porosity measurement results, it is shown that all the membranes have shown satisfactory results in the range of 81 to 83.4%, which is accredited to lower concentration of the membrane forming polymer (10.5 wt.%) as well as due to the addition of PEG additives and TiO_2 NPs together with the type of solvent used and type of de-mixing occurred. Therefore, the high porosity result for CA (additive free) membranes is due the instantaneous de-mixing happened during the phase inversion process as already explained. The effect of PEG additive as pore-former and TiO_2 NPs as hydrophilicity enhancer was also accredited to the highest pore sizes and porosity results of the membranes with PEG additive and TiO_2 NPs, though a relatively delayed de-mixing was observed. One point that should be clarified here is that the delayed de-mixing, in this case, is not extended delayed de-mixing. Normally, if the delayed de-mixing is extended, the polymer film thickness will decrease considerably, which may cause a decrease in the porosity of the sublayer membranes [25]. However, the CA-TiO_2 gained the lowest porosity due to the NPs agglomeration phenomenon as explained previously. It can also be seen from the results that an increase in the EWC of the membrane after addition of PEG additive is because the PEG is known for its hydrophilicity properties.

The influences of compaction time on pure water flux of the prepared membranes are shown in (Figure 4.11a). The PWFs of all the membranes were observed to decrease slowly with increasing time because of the pore compaction, and the steady state fluxes were reached approximately after 80 min filtration operation. The gradual decrease in PWF results can be described because the compaction of pore walls attain uniform and denser structures and causing the pore size and the flux to decrease [1]. Consequently, it is crucial to explore the effect of PEG additive on the membranes compaction process, where the effect of the PEG additive was significantly observed from this study. Another important point observed during the compaction study was that the membranes without PEG additive (i.e., M_1 and M_2) revealed lower PWFs when compared with the membranes having PEG additive (M_3 and M_4). The PWF values at 250 kPa, for M_3 and M_4, are 530.7 and 265 L/m^2 h, respectively; these values are 204.5 and 21.1 L/m^2 h, respectively, for M_1 and M_2 as presented in Table 4.6. It is clear from the results that the introduction of hydrophilic PEG additive to the membrane casting solution helps these membranes to be porous and more hydrophilic [26]. The lowest PWF for M_2 can be explained due to the blockage of pores due to aggregation of the TiO_2 NPs inside the membrane matrix. The membranes with PEG additive (M_3) have shown the highest PWF result. This result obviously indicates that the introduction of PEG additive influences the membrane in two ways: (1) development of pores in the membrane structure and (2) enhancement of the hydrophilic nature of the membrane [27]. In the present study, due to adding the hydrophilic additive, trace amount of PEG may permanently exist tangled in the membrane matrix. The presence of PEG additive can enhance the hydrophilic nature of the CA membranes. Hence, the PEG additive can play a significant role in the formation of porous membrane with improving its hydrophilic nature, which is again directly related to its water permeability performance. Comparing both the membranes with PEG additive (i.e., M_3 and M_4), the CA-PEG-TiO_2 membrane gained higher flux than CA-PEG membrane. These results are also confirmed by the porosity measurements reported in Table 4.6, where the presence of the TiO_2 NPs in the CA-PEG-TiO_2 membrane could interfere with the de-mixing process and help the formation of additional pores. The hydrophilic nature of TiO_2 NPs can also play an important role on increasing flux result of CA-PEG-TiO_2 membrane.

It is well known that the flux properties of the membrane can be influenced by several factors such as membrane pore-size, cross-sectional morphology, skin-layer thickness, and hydrophilic nature of the membrane. Thus, the PWF of the prepared membranes in this study could be influenced by TiO_2 NPs. The introduction of the TiO_2 NPs can affect the membrane in two ways: (1) due to its hydrophilic nature, which could improve the PWF, and (2) its effect on the membrane morphological structure would also influence the permeation properties negatively or positively. Figure 4.11a shows the PWF of all the prepared membranes with and without adding TiO_2 NPs. As TiO_2 is more hydrophilic than CA, the water has higher affinity for TiO_2, and therefore, PWF should increase in CA-TiO_2 membrane (M_2). But, it can be seen from the figure that the pure water flux value for M_2 shows the lowest result. Because of the addition of TiO_2 NPs to the membrane, forming polymer has caused pore blockage and pore failure in the membrane matrix due to the accumulation of the NPs [66]. Another important point that should be noted is that the hydrophilic nature of the CA membrane has its effect on the PWF of the membrane without additive (M_1) so that it has shown better flux result than the CA-TiO_2 (M_2) due to the aggregation of the TiO_2 NPs on to CA matrix. On the other hand, membranes prepared from CA-PEG-TiO_2 resulted in highest flux values. As already explained from the FESEM images, where the presence of PEG additive promotes the formation of porous structure. Thus, the TiO_2 NPs can play their significant role in the enhancement of the hydrophilicity of the membrane with less effect on the porosity of the prepared membrane. It is clear from the porosity (Table 4.6) data and the flux result (Figure 4.11) that the membranes with TiO_2 NPs (CA-PEG- TiO_2) gained higher porosity and water flux than the membranes without TiO_2 NPs (CA-PEG). Besides the enhancement of the hydrophilicity and porosity, morphological structure of the CA membranes with TiO_2 NPs may affect the permeability properties. As TiO_2 NPs have higher affinity to water than the membrane forming polymer, diffusion velocity of non-solvent (i.e., water) into the nascent membrane could be increased with TiO_2 NPs addition during the phase inversion process. Furthermore, the AC: DMAc (solvent) diffusion velocity from the membrane to non-solvent (water) could also be increased by TiO_2 NPs addition. Based on this fact the interaction between membrane forming polymer and the solvent molecules could be weakened by the hindrance of NPs so that the solvent molecules can be diffused simply from the polymer matrix to the coagulation bath [73]. Consequently, the porosity and pore size of TiO_2 entangled (i.e., CA-PEG-TiO_2) membrane were higher than those of the membranes without TiO_2 NPs (CA- PEG). The effect of PEG as a pore former as well as hydrophilicity enhancer was also accredited.

4.3.2.4 Membrane Hydrophilicity and TiO_2 NPs Stability

The hydrophilicity of the prepared membranes was studied by measuring the WCA and the drop age, defined as the duration of the water droplet on the surface of the membrane and spreading and/or permeating through the membrane cross-section [41]. The difference of WCA of all the prepared membranes is presented in Figure 4.12a. The images of the water droplets with a volume of about 2 µL at 0.16 mL/min on the membrane surface after 60 s are shown in Figure 4.12b. M_3 and M_4 membranes have taken about 15 s, and 25 s, respectively, and show best water wettability, where most parts of the membrane surfaces were almost fully wetted, and smaller spread radius of the water drops on the top side of the membrane were detected after 60 s. On the other hand, M_1 and M_2 membranes have taken about 36 s and 32 s to initiate the surface wetting and big water drops spread radius on the top side of the membrane were observed after 60 s (Figure 4.12b). The smaller the water drop spread radius wetting area between top and bottom

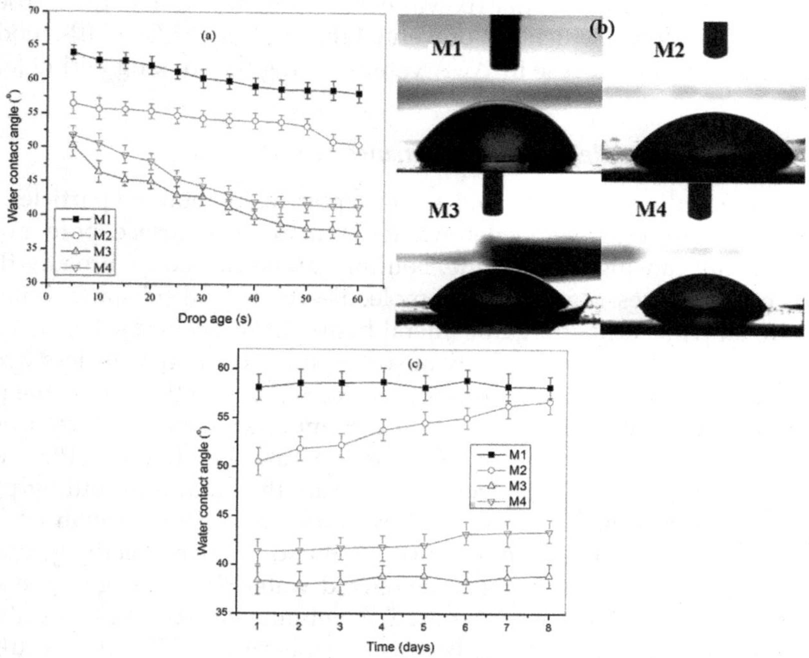

FIGURE 4.12
(a) Water contact angle values with different drop ages; (b) images of water droplets of the prepared membranes; and (c) TiO_2 NPs stability study (water droplet volume is about 2 µL at 0.16 mL/min).

surface and on the top membrane side, the better the water permeability is. As clearly observed from the graph, the WCA results of the pristine CA (M_1) membrane displays WCA about 60 ± 1.8°. Conversely, after the introduction of hydrophilic PEG additive and TiO_2 NPs, the M_2, M_3, and M_4 membranes displayed significantly reduced WCA results (i.e., 54.3 ± 2.3°, 42.1 ± 3.4°, and 45 ± 2.0°, respectively). From these results, it is clearly depicted that the contact angle results of the M_1 membrane were significantly reduced after the introduction of PEG and TiO_2 (M_2, M_3, and M_4). Membranes having smaller contact angle results are considered more hydrophilic membranes. The WCA results of the prepared membranes found after each soaking period are presented in Figure 4.12c. The WCA values for M_1 and M_3 and M_4 have remained almost constant (i.e., 58 ± 1.3°, 38 ± 3.3°, and 42 ± 2.6°, respectively) with increasing the soaking period in DI water. The WCA value of M_2 was observed to increase significantly from 51.5 ± 2.2° to 54.7 ± 2.0°, where this increase in WCA value can be accredited that the TiO_2 NPs could leach out from matrix of the CA-TiO_2 membrane with increasing the soaking period. Conversely, no significant change in the WCA values was observed in the case of M_3 (CA-PEG-TiO_2) with the soaking period. This interesting result is mainly accredited to the permanent existence of TiO_2 NPs within the membrane matrix, and the stability of the NPs was confirmed. The stability of the NPs within the membrane matrix was occurred mainly due to the introduction of PEG additive, which has prevented the leaching of TiO_2 NPs, tending to an insignificant increase in WCA values during the soaking period [62].

4.3.2.5 Fouling and Rejection Performance Study

Membrane fouling can be defined as the deposition of retained particles, colloids, macromolecules, and salts, etc. at the membrane surface, pore mouth, or pore wall, causing flux decline. Fouling can be caused by broadly three kinds of substances: organic (macromolecules, biological substances such as protein, enzyme, etc.), inorganic (metal hydroxides, calcium salts, etc.), and particulates. Proteins are strongly adsorbed on hydrophobic or less hydrophilic surfaces but less on hydrophilic surfaces [1,42]. In this study, the fouling and rejection experiments of all the membranes were conducted using the ultrafiltration stirred batch cell to investigate the effects of PEG additive and TiO_2 NPs on BSA rejection, permeate flux, and anti-fouling properties. The concentration of bovine serum albumin (BSA) protein and pH of the solutions were kept constant (1 g L^{-1}) and at 7.0, respectively, during the experiments by dissolving in deionized water. To investigate the anti-fouling performance of the prepared membranes, three cycle ultrafiltration processes were done using BSA as a model protein. The flux results of pure water and BSA solutions of the prepared membranes are presented in Figure 4.13a.

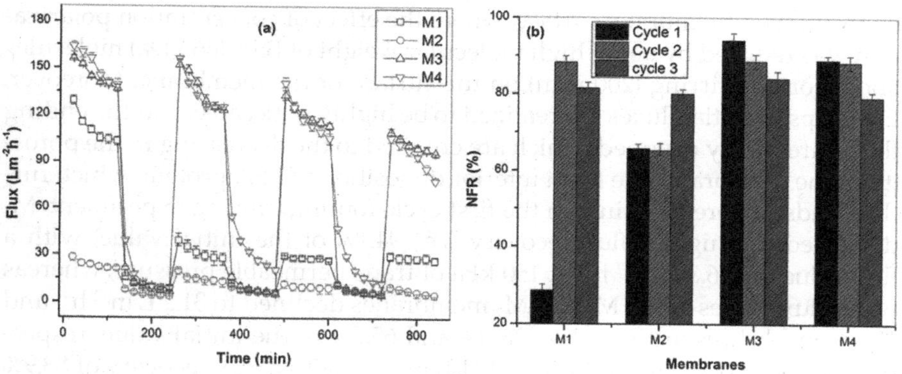

FIGURE 4.13
(a) Permeate flux versus filtration time for M_1 (CA), M_2 (CA-TiO_2), M_3 (CA-PEG-TiO_2), and M_4 (CA-PEG) membranes: effect of PEG and TiO_2 NPs on the anti-fouling performance of the membranes (25 ± 2 °C, 150 kPa), (b) NFR percentage results.

4.3.2.5.1 Effect of PEG Additive and TiO_2 NPs on Fouling

The fouling performances of all the prepared membranes are presented in Figure 4.13a. In this study, the effect of PEG and TiO_2 were evidently detected from the fouling experiments. It is clearly observed from the figure that the PWF and BSA flux results for M_1 and M_2 are less than that of M_3 and M_4. As already explained in previously, these results were attained due to the introduction of PEG and TiO_2 on the membrane matrix in M_3 and M_4, which modifies the hydrophilicity and porosity of the membranes. However, the lowest flux results for M_2 were observed because of the less porosity due to the aggregation of the TiO_2 on the membrane surfaces and pore channels of the membrane (Morphology study). It is clear from the results that the M_1 and M_2 membranes displayed less PWF as well as the BSA fluxes due to their less hydrophilic nature and less pore formation. The improvement of pure water and BSA fluxes for M_3 and M_4 membranes were attributed to the pore-forming effect and hydrophilic nature of PEG and TiO_2, respectively [69]. However, the hydrophilic effect of the TiO_2 in the case of M_2 membrane is dominated by the pore blockage (i.e., less porous membrane). On the other hand, apart from the formation of the porous membrane, a trace amount of PEG may entangle within the membrane matrix permanently, and due to this reason, the hydrophilic nature of the prepared membrane may be enhanced [44]. As clearly shown in Figure 4.13a, three BSA fouling/rinsing cycles are carried out for a total filtration time of 840 min. Each of the fouling experiments was performed with BSA solution with a concentration of 1 g L^{-1} for 2 h duration, and rinsing experiments were completed with deionized water for 30 min. The decrease in BSA fluxes with increasing time could be due to susceptible pore blocking of the membranes because of BSA protein deposition on membrane surface, where the effect of concentration polarization was reduced by using high molecular weight of BSA (66 kDa) molecules and rigorous stirring (200 rpm) on the surface of the membrane. Moreover, the drops in initial fluxes are realized to be highly noticeable, and the ending fluxes are slowly dropped, which are credited to the decreasing in the porosity of the membrane due to an interior deposition of BSA protein, which further leads to pore blocking. In the first cycle fouling/rinsing experiment, M_3 displayed the highest flux recovery (i.e., 94.1% of the initial value) with a flux value of 136.3 L m^{-2} h^{-1} at 150 kPa of trans permeable pressure, whereas water flux values of the M_1 and M_2 membranes declined to 31.2 L m^{-2} h^{-1} and 15.3 L m^{-2} h^{-1} respectively (i.e., 28.6% and 65.7% of the initial value, respectively). The flux result for M_4 was 134 L m^{-2} h^{-1} with a flux recovery of 88.9% of the initial flux value. These results are achieved due to the hydrophilic nature of PEG and TiO_2, which can minimize severe solute fouling of the membranes. The parameter normalized flux ratio (NFR) with filtration time (2 h) of the membranes operated for three cycles are presented in Figure 4.13b. From the figure, it is clearly seen that M_3 exhibited the highest NFR values 94.1%, 88.6%, and 84.1% for cycle 1, cycle 2, and cycle 3, respectively. Moreover, these results indicate that the lowest irreversible flux loss and less BSA deposition on the pore walls as well as on the surface of the membrane consistently during the three cycle operations. However, M_1 and M_2 show a significant decline in the permeate fluxes to about 28.6% and 65.7%, respectively, of the initial flux during the first cycle operation. Moreover, as seen from Table 4.7, the irreversible fouling (F_{ir}) results in the first cycle are high enough where their tendency to be fouled would be maximum. To further examine the anti-fouling properties of the prepared membranes and to study the effect of additives, the flux losses—total fouling (F_t), reversible fouling (F_r), and irreversible fouling (F_{ir})—were calculated, and the results are presented in Table 4.7. The increase in NFR results for M_1 and M_2 during the

TABLE 4.7

Results for Flux Losses Caused by Total Fouling (F_t), Reversible Fouling (F_r), and Irreversible Fouling (F_{ir}) of the Three Cycles

	First Cycle				Second Cycle				Third Cycle			
Membrane	F_t	F_r	F_{ir}	**NFR**	F_t	F_r	F_{ir}	**NFR**	F_t	F_r	F_{ir}	**NFR**
M_1	0.90	0.21	0.71	28.6	0.75	0.61	0.15	85.2	0.73	0.55	0.19	81.5
M_2	0.78	0.44	0.34	65.7	0.54	0.20	0.35	65.4	0.21	0.01	0.20	80
M_3	0.92	0.87	0.06	94.1	0.93	0.82	0.11	88.6	0.94	0.78	0.16	84.1
M_4	0.89	0.78	0.11	88.9	0.77	0.64	0.12	88.3	0.81	0.60	0.21	79.1

second and third cycle operations may be due to the development of a cake layer because of the high irreversible fouling where their flux was too low, and the flux value difference was insignificant. These results are achieved because the porosity and hydrophilicity effect could have played an important role on their anti-fouling and flux properties as already explained in the pure water flux study. Moreover, the NFR results of M_4 are 88.9%, 88.3% and 79.1% for first, second, and third cycle operations, respectively. However, these results are still less than that of M_3 membrane, where it is clear that the introduction of TiO_2 has played a crucial role in the enhancement of hydrophilicity and membrane anti-fouling property. Normally, the orders of flux recoveries for the examined membranes were consistent with their hydrophilicity and porosity nature. Therefore, the PEG free membranes are more likely to prone to pore-blockage and fouling because of protein deposition than those of membranes with PEG. It is mentioned in the literature that the PEG [45] has the potential to minimize membrane fouling because of protein deposition [46]. In this study, M_3 exhibited better anti-fouling properties in the dynamic fouling process than M_1, M_2, and M_4 membranes. Therefore, the combined effect of PEG and TiO_2 could have played a significant role in a higher resistance toward membrane fouling due to BSA deposition by reducing the hydrophobic interaction between BSA protein and membrane surface. Desorption of the deposited BSA proteins was performed by soaking the samples in water for 30 min. It was also confirmed that PEG could efficiently avoid the irreversible deposition of the protein on the surfaces. Therefore, due to the hydrophobic interaction between PEG and BSA, the proteins might be wrapped by PEG chains, forming a protective layer in addition to the anti-fouling properties of TiO_2.

Additional focus was given to the influences of filtration resistance due to concentration polarization for all membranes. Therefore the resistances due to concentration polarization were calculated using the resistance in series model Equation 2.10 (see Chapter 2).

The total fouling resistance of all the membranes thus was due to both internal membrane fouling (adsorption and deposition) and the formation of a cake/gel layer on the membrane surface. As clearly presented in Table 4.8, the results of the fouling resistance due to concentration polarization (R_p) for M_3 are low in the three cycles when compared to the other membranes (i.e., M_1, M_2, and M_4). Therefore, the effect of concentration polarization on the membrane surface was reduced due to the introduction of PEG and TiO_2 simultaneously. These results are consistent with the flux recovery results of the membrane, where the highest values were attained (Table 4.7), and it was accredited to the best anti-fouling property of the M_3 membrane as already discussed in the previous sections.

TABLE 4.8

Results for Resistances due to the Membrane (R_m), Fouled Layer (R_f), and Polarization (R_p) of the Three Cycles

	First Cycle				Second Cycle				Third Cycle			
Membrane	$\Delta p/\mu J_{bs}$	R_f	R_m	R_p	$\Delta p/\mu J_{bs}$	R_f	R_m	R_p	$\Delta p/\mu J_{bs}$	R_f	R_m	R_p
M_1	7.78	1.37	0.55	3.58	7.79	1.62	0.55	5.61	8.49	1.62	0.55	6.32
M_2	8.73	1.37	2.60	8.05	8.73	3.50	2.06	2.62	7.72	4.99	2.60	2.92
M_3	1.86	0.02	0.42	3.37	1.86	0.08	0.40	1.37	2.67	0.18	0.40	2.07
M_4	6.82	0.05	0.40	5.17	6.82	0.11	0.41	6.30	13.6	0.25	0.42	12.9

Note: All the units are in ($\times 10^{-10}\ m^{-1}$).

4.3.2.6 Rejection Performance

The BSA rejection performances of the prepared membranes are shown in Figure 4.14.

The maximum BSA rejection values, 98.4% and 91.6%, were attained for M_2 and M_1 membranes, respectively. The rejection results for M_3 and M_4 membranes are 88.9% and 85.9%, respectively. It is clearly explained in the morphology analysis section that the PEG free membranes have less porous structures, in which a better resistance to protein molecules was detected. However, the slight increase in BSA rejection for M_3 membrane could be due to the effect of TiO_2 addition to the membrane matrix. The characteristic BSA rejection can be described using the protein deposition/repulsion phenomenon as already explained above.

The prepared membranes could become more negatively charged after an introduction of TiO_2 NPs due to more negatively charged carboxylic groups along with –OH and Ti–OH groups present on the surfaces and within the matrices of the membranes [74]. Furthermore, as pH 7.0 is far from isoelectric point (IEP = 4.9), the BSA protein becomes more negatively charged, and a stronger electrostatic repulsion between BSA and the modified membranes was suggested [75]. However, some of the BSA removal characteristic can also be described using the protein deposition phenomenon onto the membrane surface in this pH range, which could be accredited to structural interaction. In addition to the electrostatic repulsion phenomena, the highest BSA removal for M_2 and M_3 membrane was also considered as an indicator of considerable protein removal due to its structural interaction. The BSA removal phenomenon for M_1 and M_4 membranes can be suggested due to a substantial protein deposition inside and on the membrane surfaces. Therefore, the

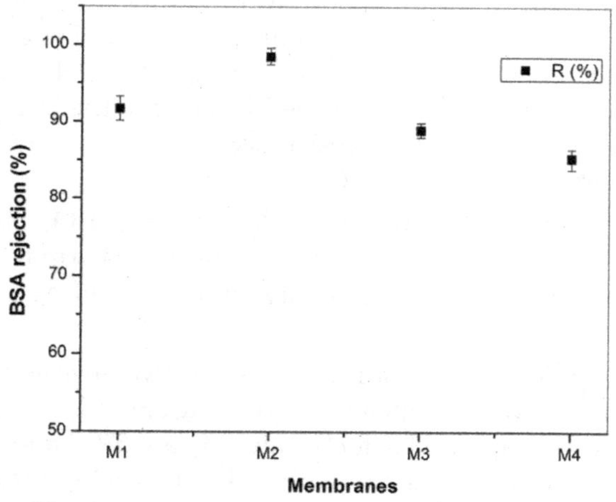

FIGURE 4.14
BSA rejection performances of M_1 (CA), M_2 (CA-TiO_2), M_3 (CA-PEG-TiO_2), and M_4 (CA-PEG) membranes: effect of PEG and TiO_2 NPs (25 ± 2 °C, 150 kPa).

surface depositions of BSA proteins can provide an extra hindrance to solute transportation. The high rejection and comparatively lower flux of M_2 and M_1 membranes than M_4 and M_3 can also be seen from the morphological study of their cross-section. Moreover, all the experimental results were consistent with the membrane properties and agreed with each other.

The results presented in this work clearly show that a detailed performance evaluation was done for the prepared ultrafiltration M_3 membrane (i.e., CA-PEG-TiO_2) in this study than the previous studies [9,13,44]. Therefore, the improved thermal stability and high anti-fouling properties of this membrane will help us to further investigate for specific ultrafiltration applications. Thus, the authors strongly believe that this work will have a substantial contribution to the current state-of-the-art on the modification and enhancement of the properties of conventional cellulose acetate membranes. Future studies will be needed to fully investigate the performance characteristics of CA-PEG-TiO_2 membranes for different ultrafiltration applications.

4.3.3 Summary

In the present work, the effects of PEG and TiO_2 on the preparation of phase inverted CA ultrafiltration membrane blended with TiO_2 (i.e., CA-PEG-TiO_2) were investigated, and the following conclusions were made:

- In the case of M_2 (CA-TiO_2) membrane, some of the TiO_2 observed to be aggregated on the surface of the membrane pores. The NPs agglomeration led to a non-uniform dispersion of the NPs within the polymer surface and structure.
- TiO_2 NPs are highly stable at high temperatures, and the degradation temperature of the M_1 and M_4 were significantly improved after addition of TiO_2. The thermal stability of M_3 exhibited higher degradation temperature than the M_1 and M_4. The slight improvement in degradation temperature for the M_2 is due to the less porous sub-layer structure and stronger resistance to heat flow than that of the porous membrane (M_3).
- The introduction of both PEG additive and TiO_2 NPs play an important role in the improvement of the thermal stability of membrane, hydrophilicity, porosity, and anti-fouling performance simultaneously.
- Membranes without PEG additive (i.e., M_1 and M_2) revealed lower PWFs when compared with the membranes having PEG additive (M_3 and M_4). The PWF values at 250 kPa, for M_3 and M_4, are 530.7 and 265 L/m2 h, respectively; these values are 204.5 and 21.1 L/m2 h, respectively, for M_1 and M_2. The highest PWF was attained for the M_3 (CA-PEG-TiO_2).

- The maximum BSA rejection values 98.4% and 91.6% were attained for M_2 and M_1 membranes, respectively. The rejection results for M_3 and M_4 membranes are 88.9% and 85.9%, respectively. CA-TiO_2 blended with PEG membrane (i.e., M_3) exhibited highest BSA flux permeates and flux recovery ratios for the three fouling/rinsing cycles.

4.4 Preparation of Fouling Resistant Ultrafiltration Membranes for Removal of Bovine Serum Albumin

4.4.1 Introduction

The fabrication of polymeric membrane based on the phenomena of phase-inversion is most commonly used process [17]. Ultrafiltration (UF) membrane has been getting greater attention for different applications such as bio separation recovery of proteins, wastewater treatments, clarification of fruit juice and alcoholic beverages, food and paper industry, and separation of oil water emulsions [53]. During a phase-inversion process, the precipitation of the polymer from a layer of the casting solution is occurred by suitably controlling the casting process. It is already mentioned in the literature that phase-inversion process can be occurred using several procedures based on reduction of the solubility of the membrane forming polymer in the solvent. Thus, separation of the casting solution into polymer-rich phase and polymer-lean (i.e., solvent-rich) occurs. The most commonly used precipitation methods for membrane forming are thermally induced phase inversion, immersion precipitation, vapor induced phase inversion, and dry casting of polymer solution [76]. Cellulose acetate polymer is widely applicable for the preparation of phase-inverted membranes because of its biocompatible, hydrophilic characteristics, high flux, good desalting nature, and it is moderately less expensive; and it can be employed for reverse osmosis, microfiltration, ultrafiltration, and gas separation applications [7]. Though cellulose acetate is commonly chosen for the preparation of phase-inverted membranes, it is susceptible to thermal and/or chemical stabilities. However, the limitations with cellulose acetate can be improved by introduction of appropriate additives based on the application requirements [44]. Saljoughi et al. [10] have studied the effects of PVP concentration and coagulation bath temperature on the morphology, permeability, and contact angle results of CA membrane. The initial addition of PVP to the cast film solution from 0 to 3 wt.% caused formation of macro-voids and increasing PWF. However, as the amount of PVP was increased from 3 to 6 wt.%, suppression of macro-voids and reduction of PWF was observed. An increase in coagulation temperature from 0 to 25 °C resulted in raising PWF. Further increasing in coagulation temperature from 25 to 50 °C, they have observed noticeable membrane hydrophilicity decrease and then resulted in PWF decline. Qin et al. [77] have synthesized a hydrophilic hollow fiber ultrafiltration membranes from a dope solution comprising of CA/PVP/N-methyl-2-pyrrolidone/water system using a dry-jet wet spinning process. The results showed that PVP would favor the suppression of macro-voids and would result in increased thickness of inner skin with increasing air gap. The influence of PVP on the morphological structures of poly (vinylidenefluoride) (PVDF)/thermoplastic polyurethane (TPU) blend membranes were studied by Yuan et al. [78]. The asymmetric blend hollow fiber membrane was prepared by using a casting dope, PVDF polymer, TPU, PVP additive, and DMAc solvent. The addition of lower concentration (i.e., <3 wt.%) of PVP increases the hydrophilicity of PVDF polymer. However, with further increase of PVP, the solution de-mixing was observed to be delayed and the kinetic hindrance because of the rise of viscosity was detected. On further increase of PVP (i.e., 5 wt.%), the suppression of macro-void rather than the expansion of structure in the membranes was obtained. In addition to using polymeric additives, the separation properties of polymeric membranes can be enhanced and may possess properties such as selectivity, good permeability, and thermal and chemical stability by preparing organic-inorganic membranes [56]. Polymer/inorganic composite membranes prepared by means of homogeneously dispersing the inorganic nanoparticles in a polymeric matrix have been receiving much more attention in the areas of ultrafiltration, gas-separation, and pervaporation membranes. Several inorganic oxide nanoparticles, such as Al_2O_3 [58], ZrO_2 [60,57], TiO_2 [57], and SiO_2 [79] have added within polymer casting solution for different applications. Membrane fouling has been considered a common problem in the field of membrane process. Especially in industrial processes, the solvent flux decline results from membrane fouling or concentration polarization. In ultrafiltration processes, several attempts have been accomplished to decrease fouling, which in general include feed solution pretreatment, membrane surface enhancements, and process modifications [57]. Moreover, present researchers have been paying attention to TiO_2 nanoparticles because of its stable nature, ease of availability, and potential for different applications. Ong et al. [63] have prepared PVDF hollow fiber ultrafiltration membrane using various concentrations of TiO_2 with the presence of PVP. All the composite membranes exhibited comparatively greater fluxes and removals results when compared to TiO_2 free PVDF membranes. It is stated that the PVDF membranes combined with 2 wt.% TiO_2 displayed the best separation performances. Su et al. [80], however, have evaluated the fouling mitigation performance of mixed cellulose ester and cellulose acetate by coating neutral TiO_2 on the surface of the membranes for membrane bioreactor application. Based on their results, they have concluded that the membrane fouling was reduced after coating of TiO_2 on the CA membrane surface. Therefore, introduction of TiO_2 has enhanced the fouling mitigation and the hydrophilic nature of the membrane surface. Furthermore, TiO_2 can enhance the hydrophilicity of different polymers to improve flux and decrease the fouling problem, which are important parameters in water and wastewater treatment [57,81]. An effort was made to investigate the influence of PVP and TiO_2 on the morphological structure, permeability performance, and thermal stability property of the membrane in addition to the anti-fouling properties of the prepared membranes.

4.4.2 Preparation of CA-PVP–TiO_2 Membrane

All the asymmetric ultrafiltration membranes from cellulose acetate polymer presented in this study were fabricated using phase-inversion technique by immersion precipitation. Initially of CA polymer was uniformly dissolved in Acetone/DMAc (70/30) to prepare homogeneous solution under continuous magnetic stirring at room temperature. After getting a uniform solution of the CA/solvent, a quantified amount of PVP additive (i.e., 4 wt.%) was added and stirred continuously at least for 1 day until the solution was fully dissolved, and finally, homogeneous casting solution was obtained.

Then 2 wt.% TiO_2 NPs was added and well dispersed for another 24 hours to form a homogeneous composite casting solution (i.e., CA-PVP-TiO_2) and also entrapped into the polymer matrix due to the high viscosity of the polymer solution. Finally, the resultant polymer solutions were kept for 24 hours to remove all of the bubbles before casting was done. Thus, the membranes with different composition were designated as CA, CAT, CATP, and CAP. Table 4.9 shows the solution casting compositions CA, PVP additive, and TiO_2 nanoparticles. The membrane preparation process is presented in Figure 4.15. The casting solution was poured consistently on a glass sheet with the help of a casting knife maintaining a clearance of approximately 0.25 mm between the knife and the glass plate. The resulting films were then exposed to air for about 30 s before immersing into the coagulation bath containing distilled water at room temperature. In the coagulation bath, the casted solution turned from transparent to white color for membranes CA and CA-PVP.

TABLE 4.9

Solution Compositions and Viscosity of the Casting Solution: CA, PVP Additive, and TiO_2 Nanoparticles

Membrane	CA wt.%	PVP wt.%	TiO_2 wt.%	Acetone/ DMAc (70/30 wt.%)	Viscosity (mPa.s)
CA	10.5	–	–	89.5	1020
CAT	10.5	–	2	87.5	2080
CATP	10.5	4	2	83.5	1790
CAP	10.5	4	–	85.5	1680

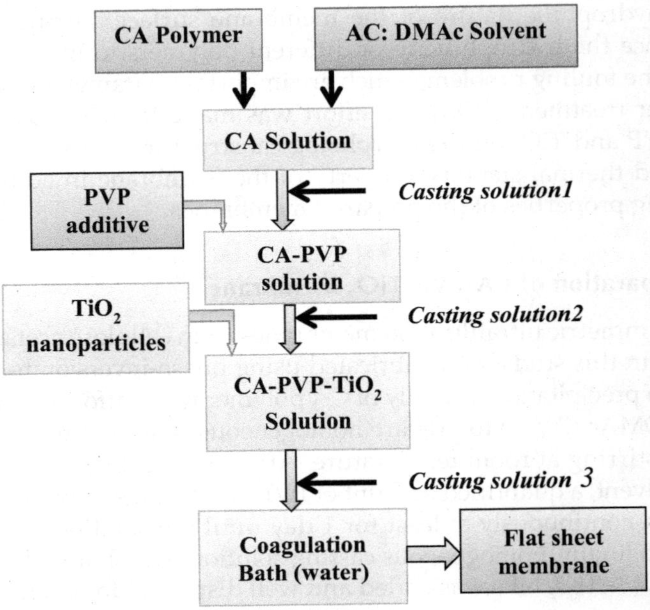

FIGURE 4.15
A flow diagram for the preparation process of flat sheet membranes.

The milky colors for CA-TiO_2 and CA-PVP-TiO_2 membranes were changed to white, and all the thin films were detached from the glass plate. The membrane sheets were kept for 30 min in the coagulation bath. After that, the prepared membrane sheets were kept in deionized water bottles until use.

4.4.2.1 Study of TiO_2 Nanoparticles

The size of commercial TiO_2 nanoparticles was determined using transmission electron microscopy (TEM) as presented in Figure 4.16. The TiO_2 nanoparticles appeared in the form of spots. To measure the size of each nanoparticle, Image J software was employed, and their size ranged from 15 to 71 nm. The average particles size was approximately 29 nm.

4.4.2.2 Morphological Study

The top layer surface view, cross-sectional view, EDS images, and maps of the prepared membranes are presented in Figure 4.16a, b, c, and d, respectively. The presence of the TiO_2 NPs within the matrixes of the membranes was confirmed from the EDS results. It can be seen from the figures that all the prepared membranes exhibited asymmetric structure involving of a dense top-layer and a porous sublayer. However, the sublayer structure of the CA membrane seems to have finger-like structures and macro-void.

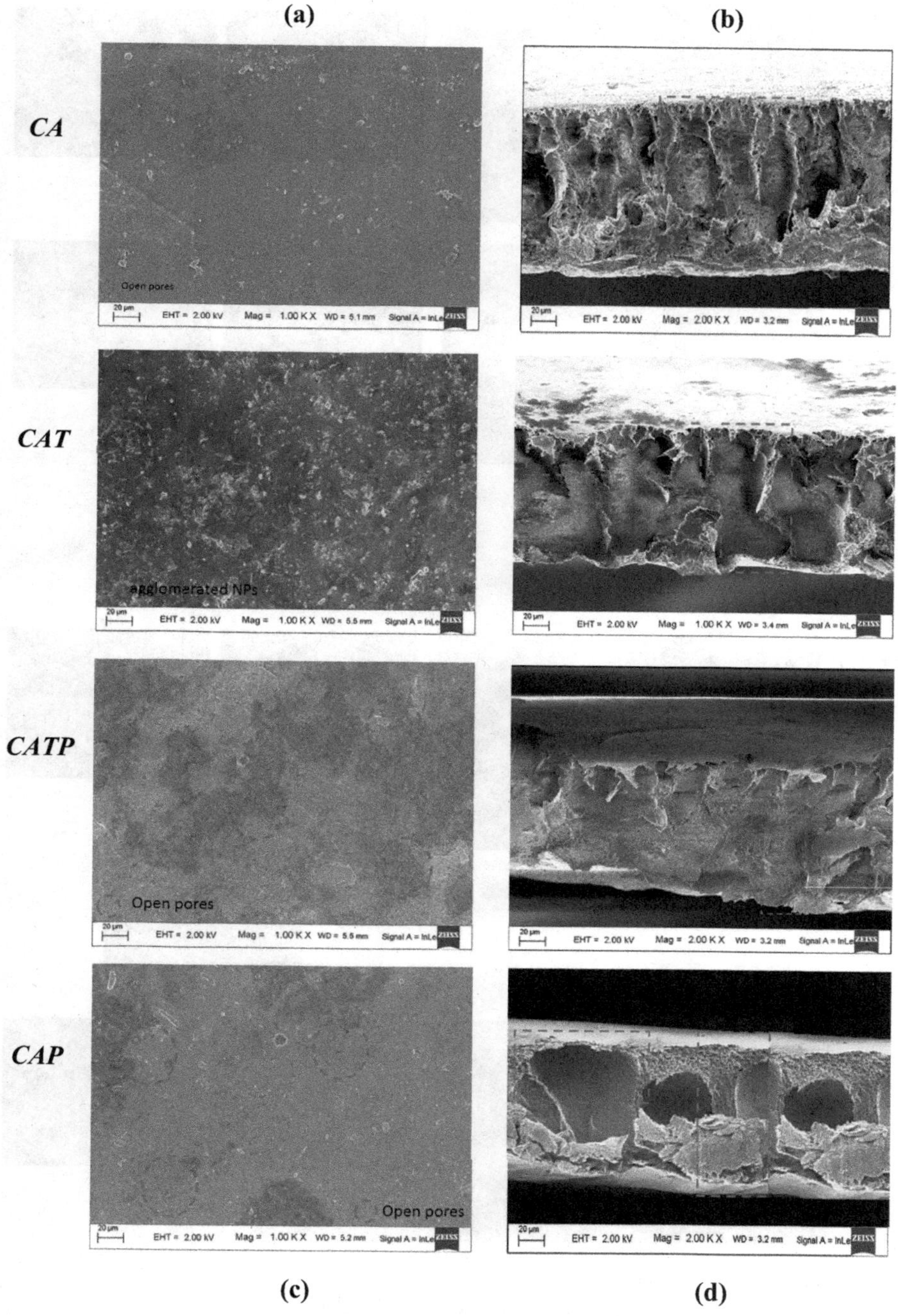

FIGURE 4.16
(a) The top surface and (b) cross-sectional.

(Continued)

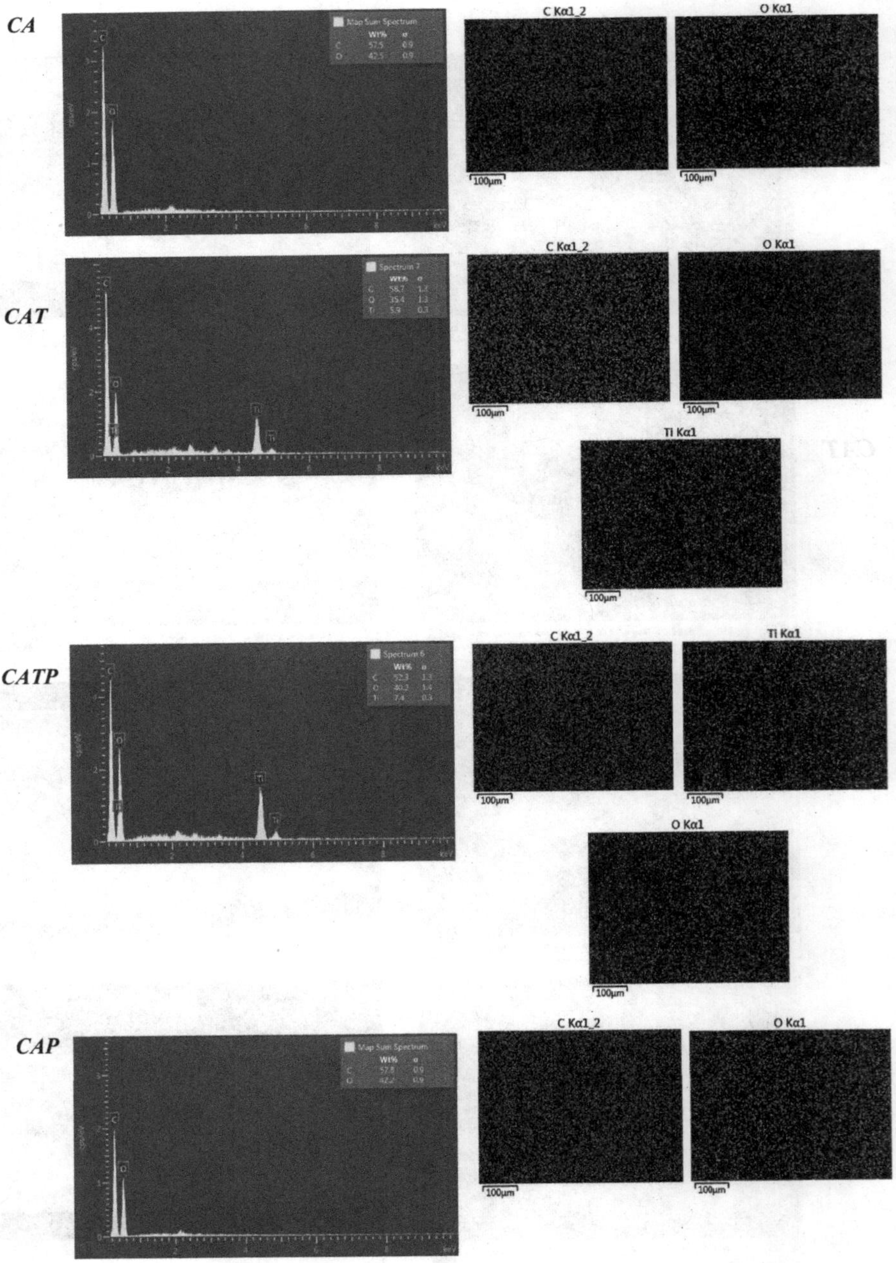

FIGURE 4.16 (CONTINUED)
(c) EDS, and (d) mapping images of CA, CA with PVP, and TiO_2 NPs membranes.

It is known that there are several factors that made the finger-like structures and macro-voids to occur during membrane preparation using immersion precipitation method. The effect of solubility parameter differences among organic additives and CA on the morphological structures of the resulting membranes were studied in our previous work [82]. The formation or suppression of macro-voids can significantly be influenced by the type of de-mixing that occurred during the phase separation process. The instantaneous de-mixing and the delayed de-mixing depend on the mutual affinity of the solvent and non-solvent in the ternary system [1]. The occurrence of macro-voids in membranes prepared from ternary systems: (1) membranes without macro-voids can be formed in the case of delayed de-mixing, except when the delay time is short; (2) membranes with macro-voids can be formed in the case of instantaneous de-mixing, except when the polymer concentration and/or the non-solvent concentration in the casting solution exceed a minimum value [24]. Generally, macro-voids formation occurs under rapid precipitation condition, and the precipitation is quicker at higher coagulation temperature [13]. In the current study, the coagulation bath temperatures and evaporation step prior to the immersion process were kept at 25 °C and 30 s, respectively, throughout the membrane preparation process [10].

4.4.2.2.1 *Effects of PVP and TiO_2 NPs*

In general, the addition of organic and/or inorganic additives to the membrane forming polymer can increase the concentration and then the viscosity of the casting solution. This increase in viscosity may, in turn, reduce the diffusional exchange rate of the non-solvent and solvent in the membrane preparation procedure (i.e., phase inversion method) [13]. Accordingly, this can delay the instantaneous liquid-liquid de-mixing process and hinders the formation of macro-voids. In solvent and polymer interaction scheme, three kinds of interactions—polymer:polymer, polymer:solvent, and solvent:non-solvent—are applied. If good polymer solvents are used, the degree of polymer stretching reaches to its highest level, and more favorable polymer/solvent interactions can occur. The solvents used in this study (i.e., AC: DMAc) have high mutual affinity with water [33]. Thus, evidently, the development of the finger-like voids and macro-voids in the sublayer of the CA membrane is due to the instantaneous de-mixing [83]. In this case, the CA stretches to its highest level where there is a maximum interaction between the CA and AC: DMAc, which tends to allow instantaneous de-mixing condition to happen. After adding the PVP and TiO_2 to the ternary (polymer/solvent/non-solvent) system, the developments of finger-like structures are suppressed significantly (i.e., CAT, CATP, and CAP). The presence of the polymeric additives and TiO_2 NPs can increase the concentration/viscosity of the casting solution, which may diminish the diffusional exchange rate of the AC: DMAc and water during the membrane formation process. Accordingly, this may hamper the instantaneous liquid-liquid de-mixing process that suppresses the development of finger-like and macro-voids [77]. However, the presence of large macro-voids along with micro porous structure in the case of CAP membrane can be explained because of its water soluble behavior, and PVP can simply assemble around the water molecule during immersion precipitation process, resulting in the development of PVP rich phase [4]. The introduction of water (i.e., non-solvent) soluble hydrophilic additive with high affinity to the solvents and low affinity to the membrane forming polymer tends to increase the thermodynamic instability of the casting solution, which further leads to instantaneous de-mixing to sustain in the coagulation bath. These results could make the macro-voids develop in the membrane sublayer structure [77]. In quaternary system two phases arise from the phase separation; one involves the CA (i.e., membrane forming polymer), AC:DMAc (i.e., solvent), and water (i.e., non-solvent), and the second contains the polymeric additive (i.e., PVP), solvent (i.e., AC:DMAc), and non-solvent (i.e., water). The membrane forming polymer and additive have a driving force to separate completely into a CA-rich and a PVP-rich phase. Accordingly, the type of de-mixing process, in this case, can be influenced by the diffusion/movement of the two polymers on each other. Therefore, this phenomenon should be slow compared to the diffusion of solvent and non-solvent in the polymer solution. To maintain this, a polymeric additive with a certain minimum molecular weight in the system is required [22]. The addition of TiO_2 to the membrane formation system has its effect on the de-mixing process. It was evidently observed from the FESEM images that the finger-like structures and macro-voids in the membrane sublayer were significantly reduced after the adding of TiO_2 NPs in CAT and CATP membranes. As shown from the FESEM images, the finger-like structures and macro-voids are reduced in the case of CAT and highly suppressed in the case of CATP after adding TiO_2 irrespective of the effect of PVP additive. Nevertheless, the CAT membrane seems to have less micro-porous cross-sectional structures with some defect structures. This result can be described due to some of the TiO_2 that could be accumulated on the surface and pore walls of the membrane. It is also observed from the top surface image of CAT (Figure 4.16a) that the accumulation of some of the TiO_2 NPs was clearly detected as indicated by the circles. The enhancement of membrane morphological structure occurs with the addition of a small amount of TiO_2 NPs [65]. In the current study, the concentration of the TiO_2 NPs was fixed at 2 wt.%. Therefore, the finger-like structures and macro-voids observed in the CA membrane were greatly suppressed, and a relatively dense layer with microporous structures was observed for CATP. However, minor macro-void and finger-like structures were still detected in the case of CAT membrane, and these results are suggested to be related to the interference effect of NPs during the phase inversion process. Thus, due to the interfacial stress between polymers and NPs, interfacial pores are formed as a result of shrinkage of polymer-phase during the de-mixing process [68]. The presence of TiO_2 in CAT diminishes the presence of finger-like structures and a sublayer structure with almost macro-voids free is obtained. These results can be described in terms of NPs agglomeration on the membrane forming polymer matrix during phase inversion process. The occurrence of NPs agglomeration can be mainly caused due to the high surface energy of the NPs, which tend to aggregate for weakening their surface energy to reach a more stable state. Furthermore, NPs agglomeration leads to a non-uniform dispersion of the NPs within the polymer surface and structure. This phenomenon can negatively change the resulting membrane properties such as hydrophilicity and surface roughness [66]. However, in the case of CATP, the NPs agglomeration was significantly minimized due to the improved distribution of the NPs because of the introduction organic additive [69]. From these results, it can be concluded that although the instantaneous de-mixing is still sustained, the PVP and TiO_2 NPs have a considerable role on the hindrance of the finger-like structure and macro-voids that occurred for CA membrane [77]. However, the presence of large macro-voids along with microporous structure in the case of CAP membrane is because of the water soluble behavior of PVP where it can simply assemble around the water molecule during immersion precipitation process and resulted in the development of PVP rich phase [4]. Therefore, the developments of macro-voids are as a result of rapid penetration of water at some weak spots in the top layer. Thus, the PVP rich phase may act as the weak spot in the precipitation process and resulted in the formation of macro-voids [17]. As indicated by the rectangular shapes for the CATP and CAP in Figure 4.16, the formation of microporous structures is related to the pore forming nature of PVP. However, the formations of big macro-void structures in CAP membrane are undesirable, so that they may cause lack of thermal and/or chemical stabilities in the resulted membrane.

4.4.2.3 Thermal Stability Analysis

It can be clearly seen from TGA and DTG figures (Figure 4.17a and 4.17b, respectively) that the decomposition of CA membrane showed three steps where the graphs were plotted as weight loss (%) vs. temperatures (°C). The first degradation step of CA membrane was detected between 30 and 260 °C, and the weight loss was about 21%. This degradation is due to the presence of some volatile material or because of the evaporation of absorbed moisture by the sample. During the second degradation step, a maximum weight loss of around 76% was detected between 260 to 380 °C, which is because of the main thermal breakdown of CA chain [84]. The last degradation step greater than 380 °C was due to the carbonization of the decomposed products to ash. The TGA results of the CA membrane clearly presented that it could not be stable at high temperature and similar results were reported by Zafar et al. [70]. The TGA results of the CA membranes with PVP and TiO_2 displayed two-step degradation procedures. Therefore, the first decomposition step for CAT, CATP, and CAP are 320 °C, 227 °C, and 208 °C, respectively. These observed weight losses are due to the decomposition of CA chains because of the pyrolysis of the skeleton backbone of the CA polymer, which is also

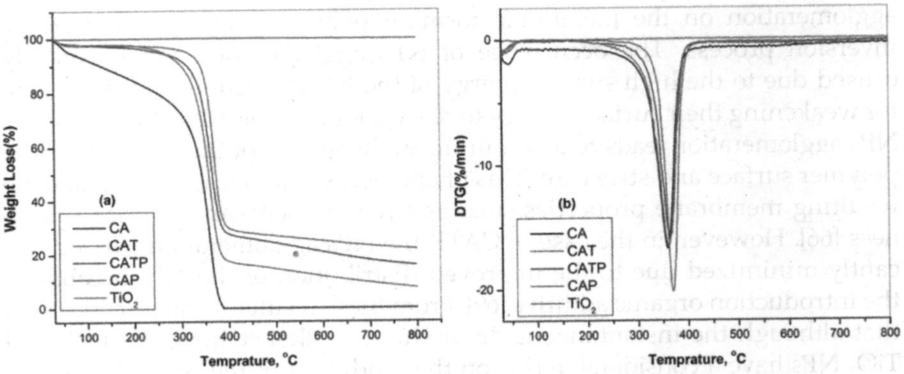

FIGURE 4.17
(a) TGA and (b) DTG analysis of all the prepared membranes and TiO_2 nanoparticles.

followed by de-acetylation of CA [71]. During the second decomposition step, the degradation temperatures of CAT, CATP, and CAP are 400 °C, 390 °C, and 387 °C, respectively, signifying the main thermal degradation of the CA chains, whereas the TiO_2 were highly stable up to 800 °C. From these results, it is clearly observed that the thermal degradation of CA membrane was significantly enhanced due to the addition of PVP and TiO_2. Another interesting observation from this study is that an increase in thermal stability after addition of PVP to CA membrane (i.e., CAP) is due to the presence of a trace amount of the PVP in the membrane matrix. The improved thermal stability of the CA membranes after addition of PVP was due to the strong interaction between the CA and PVP in the membrane matrix by creating hydrogen bonds [72]. Therefore, the improvement of the membrane hydrophilic nature due to the presence of PVP within the membrane matrix in addition to the enhancement of its thermal stability was credited. Furthermore, the thermal stability of CATP membrane exhibited higher degradation temperature than CAP membrane. This result is attained due to the fact that the TiO_2 nanoparticles are highly stable at high temperatures and therefore the degradation temperature of the CAP membrane was significantly improved after addition of TiO_2. The slight improvement in degradation temperature for the CAT membrane when compared with CATP membrane was that the less porous sublayer structure of the membrane is responsible for strong resistance to heat flow than that of the porous membrane (i.e., CATP). In this study, the TGA plots have clearly revealed that the thermal properties of CA membrane were significantly improved after incorporation of PVP and TiO_2 into the CA solution.

4.4.2.4 Pure Water Flux Study

Studying of membrane compaction factor (CF) is crucial to investigate the morphological structures, especially the pore arrangements of the

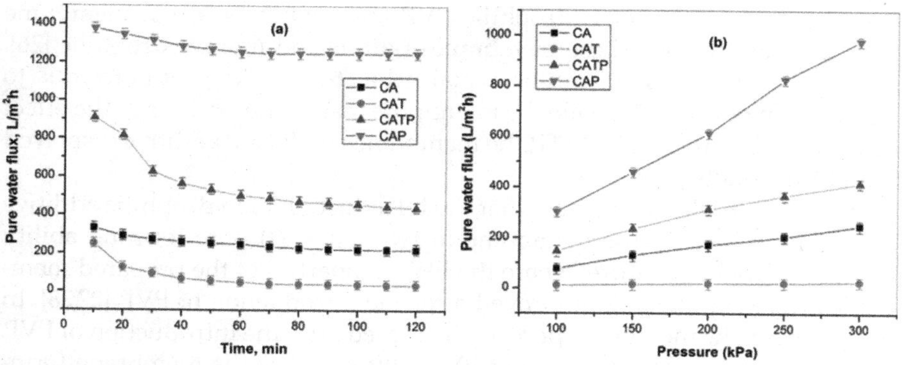

FIGURE 4.18
(a) PWF profile during compaction study, (b) effect of trans membrane pressure on PWF.

membranes and the membrane sublayer configuration. Membranes having high CF indicate that they are highly compacted, which further show the existence of some defect pores in the membrane sublayer structure [5]. Thus, the influences of compaction time on pure water flux of the prepared membranes are shown in Figure 4.18a. The figure shows that the PWF results of all the membranes were observed to decrease slowly with increasing time because of the pore compaction, and the steady state fluxes were reached approximately after 90 min filtration process. The gradual decrease in PWF results can be described due to the compaction of pore walls that attained uniform and denser structures and caused the pore size and the flux to decrease [1].

4.4.2.4.1 Effect of PVP and TiO_2 NPs

It is essential to discover the effect of PVP and TiO_2 NPs on the membranes compaction process, where the effects of PVP and TiO_2 were observed from this study. From the compaction study, it was observed that the CA and CAT membranes displayed lower PWF results when compared with the CATP and CAP membranes. As seen from Table 4.10, the PWF results at an operating pressure of 300 kPa for CA, CAT, CATP, and CAP are 248 L/m2 h, 26 L/m2 h, 416 L/m2 h, and 978 L/m2 h, respectively. These results show

TABLE 4.10
Compaction and Hydraulic Characteristics of Those Prepared Membranes at 300 kPa

Membrane	*CF*	R_m ($\times 10^{-10} m^{-1}$)	J_w (L/m^2h)	*EWC* (%)	ε (%)	r_m (nm)	Thickness μm
CA	1.5 ± 0.4	0.4	248	76.8 ± 1.6	82.0 ± 2.0	22.2 ± 4.0	71.1 ± 3.0
CAT	4.0 ± 0.2	4.6	26	66.7 ± 2.5	61.1 ± 2.5	14.9 ± 6.0	62.3 ± 7.0
CATP	1.1 ± 0.4	0.3	416	81.8 ± 1.7	85.3 ± 1.8	34.8 ± 3.1	68.7 ± 4.5
CAP	2.1 ± 0.3	0.1	978	83.7 ± 1.3	86.9 ± 1.9	37.9 ± 2.2	70.3 ± 3.2

that the introduction of hydrophilic PVP and TiO_2 NPs to the membrane casting solution made these membranes porous and more hydrophilic [26]. The lowest PWF for CAT can be explained as the blockage of pores due to aggregation of the TiO_2 inside the membrane matrix. Regardless of the effect of TiO_2 NPs in the case of CATP, the membranes with PVP additive displayed high PWF results.

These results clearly indicate that the introduction of hydrophilic additive (i.e., PVP) influences the membrane in two ways: (1) pore forming ability and (2) hydrophilic nature, where the PWF properties of the prepared membranes were significantly improved after the introduction of PVP [27,28]. In the first case, the membrane porosity increased after the introduction of PVP, and second, the whole removal of PVP additive from the membrane forming polymeric matrix may not be attained during the phase separation process in the coagulation bath and even after rinsing with deionized water [10]. Consequently, the trace amounts of PVP additive that are entrapped in the membrane matrix permanently, thus enhancing the membrane hydrophilic property.

Comparing CATP and CAP, the membrane without TiO_2 NPs (i.e., CAP) displayed higher flux than CATP membrane. The increase in water flux for membranes prepared from casting solutions CAP membrane is due to the presence of few macro-voids. On the other hand, the CATP showed moderate water flux and where the macro-voids and finger-like structures were significantly suppressed but still with the development of micro porous structures (important for the porosity of membrane). It is also mentioned in the literature that TiO_2 could considerably increase the porosity of the resulting membrane [85]. However, in this study, the influence of TiO_2 on membrane porosity was insignificant as all the prepared membranes exhibited a reasonably high porosity, that is, greater than 80% except for CAT membrane (61.1%).

Generally, adding of TiO_2 NPs alone to the polymeric solution can increase its viscosity. Therefore, the particles leaching problem is lesser, and subsequently, the pore forming effect of NPs can be declined in the case where the high viscosity of a solution hampers the development of pores and causes the porosity of the membrane to decrease [65]. As clearly seen from the results presented in Table 4.10, the introduction of TiO_2 into the CA membrane matrix (i.e., CAT) resulted in water flux decline as membrane average pore size was decreased from 22.2 ± 4.0 nm (i.e., CA) to 14.9 ± 6.0 nm (i.e., CAT). The decrease in average pore size of the membranes after the introduction of TiO_2 was mainly credited to the pore blocking caused by aggregation of TiO_2 on the membrane matrix. In addition to pore blockage, it is also explained that the aggregation of TiO_2 may reduce the contact area of hydroxyl groups carried by TiO_2 with water molecules, which could further influence the membrane water permeability properties [63]. The introduction of TiO_2 alongside with PVP has shown significant improvement on the average pore size and flux property of the resulted membranes. Thus, it is believed that the presence of hydrophilic PVP can hinder the pore blocking activity of TiO_2 due to its pore forming property as it is easily soluble in the non-solvent (i.e., water) regardless of the presence TiO_2 [69]. Though the flux of CAT membrane was expected to increase after the introduction of TiO_2 NPs due to improvement of membrane hydrophilicity [85], membrane pores plugging reduces the influence of hydrophilic nature on the PWF results. Another important point is that the hydrophilic nature and higher average pore size of the CA membrane have its effect on the PWF of the membrane, without additive, so that it has shown better flux result than the CAT.

The results of compaction factor of all the prepared membranes are presented in Table 4.10. The CF of CAT (i.e., 4.0 ± 0.2) is greater than all of the other membranes. This result can be explained due to the fact that the aggregated TiO_2 on the pore walls and surface of the membrane could further compact and may block the pores after compaction process. The CF results for CA, CATP, and CAP are 1.5 ± 0.4, 1.1 ± 0.4, and 2.1 ± 0.3, respectively. The relatively smaller CF value for CATP was suggested due to the uniformly interconnected pore structures that existed in the membrane matrix because of the simultaneous introduction of PVP and TiO_2 NPs [5]. The effects of PVP and TiO_2 NPs on PWF at different operating pressures are presented in Figure 4.18b. The results of PWF for all the membranes were increased almost uniformly with increasing the operating pressure from 100 to 300 kPa. It is also observed that the PWF results for CA and CAT membranes are lower than that of CATP and CAP. As seen from the results of the hydraulic resistance presented in Table 4.10, the CAT displayed higher resistance than all the other membranes. This higher resistance is due to the less porous structure of the CAT membrane as compared to the other membranes; in turn, it has a higher tendency to resist the water flux than the other membranes.

Moreover, this result is supported by the average pore radius calculation (Equation 2.11 in Chapter 2). Therefore, the membrane with less average pore radius is more likely to have higher resistance to water flux irrespective of the hydrophilic nature of the membranes. Thus, the increase in hydraulic resistance and decrease in flux with and without PVP is obviously due to the decrease in average pore size and porosity as already explained in the previous sections. The PVP is more hydrophilic than the CA polymer. Therefore, the membranes with PVP have displayed improved flux results. Comparing CATP and CAP membranes, the CAP membrane has shown less resistance and higher flux than CATP membrane. The EWC and porosity results of the prepared membranes results are also calculated using Equations 2.4 and 2.5 (see Chapter 2) and presented in Table 4.10. The results from the porosity measurement showed that all the membranes have a satisfactory porosity in the range of 61.1 to 86.9%, which can be accredited to low concentration of membrane forming polymer (10.5 wt.%) as well as due to the addition of PVP and TiO_2 together with the type of solvent used and type of de-mixing occurred [78]. The effect of PVP as a pore former as well as hydrophilicity enhancer was also accredited.

4.4.2.5 Membrane Hydrophilicity and TiO_2 NPs Stability

The hydrophilicity nature of the prepared membranes was studied by measuring the WCA and the drop age, defined as the duration of the water droplet on the surface of the membrane and spreading and/or permeating through the membrane cross-section [41]. The differences on WCA of all the prepared membranes are presented in Figure 4.19a. The images of the water droplets on the membrane surface after 60 s are shown in Figure 4.19b. As clearly observed from the graph, the contact angle results for CA, CAT, CATP, and CAP are 60.5 ± 1.3°, 54.1 ± 2.0°, 45.4 ± 3.2°, and 47.3 ± 2.8°, respectively. From these results, it is clearly depicted that the contact angle results of the CA membrane were significantly reduced after the introduction of PVP and TiO_2 (CAT, CATP, and CAP). CATP and CAP membranes have taken about 22 s and 30 s, respectively, and show best water wettability, where most parts of the membrane surfaces were almost fully wetted, and smaller spread radius of the water drops on the top side of the membrane was detected after 60 s. CA and CAT membranes have taken about 40 s and 32 s to initiate the surface wetting. Big water drops spread radius on the top side of the membrane was observed for CA after 60 s (Figure 4.19b). The smaller the water drop spread radius wetting area between

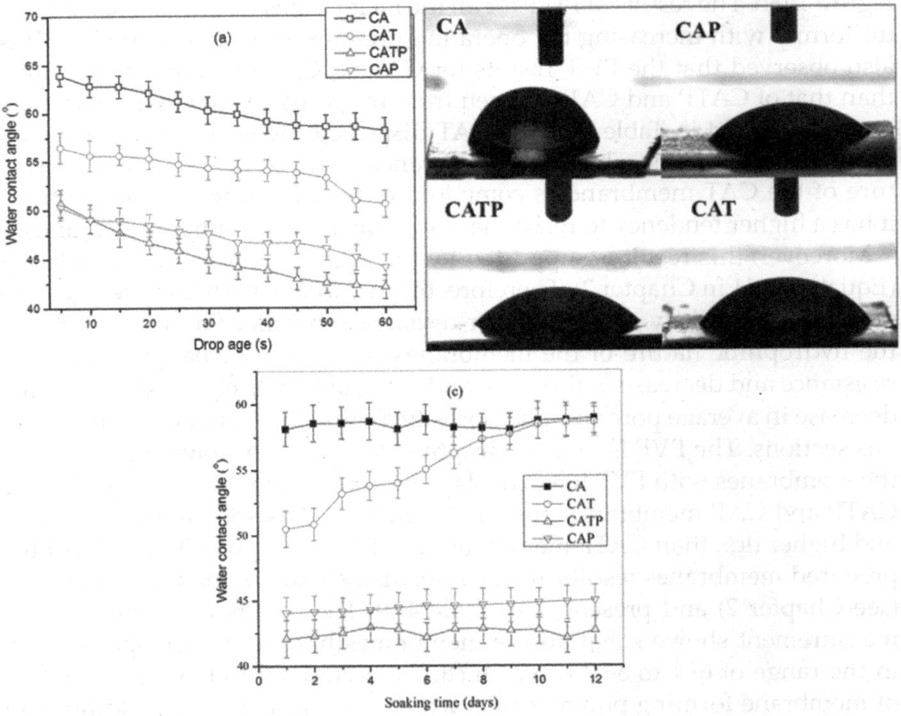

FIGURE 4.19
(a) Water contact angle values with different drop ages; (b) images of water droplets of the prepared membranes; (c) TiO_2 NPs stability study.

top and bottom surface and on the top membrane side, the better the water permeability [41]. From these results, it is clear that the contact angle results of the CA membrane were significantly reduced after the introduction of PVP and TiO_2 (CAT, CATP, and CAP). Membranes having smaller contact angle results are considered as more hydrophilic [66]. The stability of the TiO_2 NPs incorporated within the matrix of the modified membranes was studied qualitatively. All the prepared membranes were soaked in DI water for 12 days at room temperature (25 ± 2°C). The WCA values of all the membranes were completed, and the TiO_2 leaching tendency during each soaking period was evaluated [62]. The WCA results of the prepared membranes found after each soaking period are presented in Figure 4.19c. The WCA values for CA, CATP, and CAP have remained almost constant (i.e., 58.5 ± 1.5°, 42.6 ± 3.0°, and 44.5 ± 2.7°, respectively) with increasing the soaking period in DI water. The WCA value of CAT was observed to increase from 50.5 ± 2.9° to 58.6 ± 2.2°, where this increase in WCA value can be attributed that the TiO_2 NPs could leach out from matrix of the CA-TiO_2 membrane with increasing the soaking period. Conversely, no significant change in the WCA values was observed in the case of CATP with the soaking period. This interesting result is mainly accredited to the permanent existence of TiO_2 NPs within the membrane matrix, and the stability of the NPs was confirmed. The stability of the NPs within the membrane matrix occurred mainly due to the introduction of organic additive, which has prevented the leaching of TiO_2 NPs, tending to an insignificant variation in WCA values during the soaking period [62].

4.4.2.6 Membrane Fouling Experiments

Fouling can be caused by three kinds of substances: organic (macromolecules, biological substances like protein, enzyme, etc.), inorganic (metal hydroxides, calcium salts, etc.), and particulates. Proteins are strongly adsorbed on hydrophobic or less hydrophilic surfaces but less on hydrophilic surfaces [42]. Therefore, to investigate the BSA removal and anti-fouling performance of the prepared membranes, three cycle filtration processes were completed. The flux results of pure water and BSA solutions of the prepared membranes are presented in Figure 4.20, and the effects of PVP and TiO_2 are investigated.

4.4.2.7 Membrane Anti-Fouling Performance

The fouling of membranes due to BSA protein in ultrafiltration can be described using the theory of repulsion, adsorption of protein, and resultant pore-narrowing, due to electrostatic and hydrophobic interaction between the protein molecule and the membrane surface [43]. The fouling performances of all the prepared membranes are presented in the Figure 4.20. It is clear that the fouling and removal performance of the ultrafiltration membranes are strongly dependent on morphology of the membrane (i.e., both top layer and sublayer). As already discussed above, the effects of PVP and

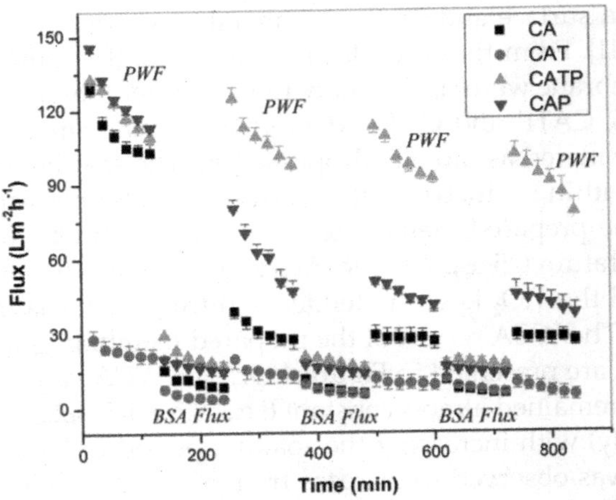

FIGURE 4.20
Permeate flux versus filtration time for CA, CAT, CATP, and CAP membranes: effect of PVP and TiO_2 NPs on BSA removal and fouling performance (25 °C, 150 kPa).

TiO_2 have a significant role in modifying the morphological structure of the membranes. Consequently, the introduction of both PVP and TiO_2 into the matrix of the CA membrane appeared to influence the flux and BSA removal performance of the prepared membranes. The effects of PVP and TiO_2 were clearly noticed from the fouling/rinsing experiments. The flux results in Figure 4.20 have displayed that the PWF (i.e., J_{W0}, J_{W1}, J_{W2}, and J_{W3}) and BSA flux (i.e., J_{B0}, J_{B1}, and J_{B2}) results for CATP and CAP are higher than that of CA and CAT. After completion of the PWF tests, the BSA filtration process was continued. To further investigate the fouling performance, the membranes were cleaned after BSA solution ultrafiltration, and the pure water flux of the cleaned membranes was measured. As clearly shown from the results, the membranes without PVP again have displayed low BSA flux results as well as PWF due to their less hydrophilic and pore formation nature. The enhancement of pure water and BSA fluxes for membranes for CATP and CAP were accredited to the pore-forming effect and hydrophilic nature of PVP and TiO_2 [69]. Nevertheless, the hydrophilic effect of the TiO_2 in case of CAT membrane is dominated by the pore blockage (i.e., less porous membrane). As clearly presented in Figure 4.20, we have carried out three BSA fouling/rinsing cycles for a total filtration time of 840 min. Each of the fouling experiments was performed with BSA solution with a concentration of 1 g L–1 for 2 h duration, and each rinsing experiment was completed with deionized water for 30 min. A decrease in BSA fluxes with increasing time could be due to susceptible pore blocking of the membranes because of BSA protein adsorption and deposition on membrane surface, where the effect of concentration polarization was reduced by using high molecular weight of BSA (66 kDa) molecules and rigorous stirring (200 rpm) on the surface of the membrane [44]. Furthermore, the initial flux drops are realized to be more noticeable and ending fluxes are slowly dropped, which is accredited to the loss of porosity of the membrane due to an interior adsorption of BSA protein, which further leads to pore blocking. In the first cycle of fouling/rinsing experiment, CATP has displayed the highest flux recovery (i.e., 90% of the initial value) with a flux value of 108.7 L m^{-2} h^{-1} at 150 kPa, whereas the flux values of the CA and CAT membranes have declined to 31.7 L m^{-2} h^{-1} and 16.4 L m^{-2} h^{-1}, respectively (i.e., 28.6% and 65.7% of the initial value, respectively). The flux result for CAP was 61.9 L m^{-2} h^{-1} with a flux recovery of 47.4% of the initial flux value. The highest flux and recovery ratio for CATP membrane is attributed to the hydrophilic nature of both PVP and TiO_2 NPs and anti-fouling property of TiO_2 loaded within the membrane, which could minimize severe solute fouling of the membranes [57].

In general, membrane fouling is comprised of irreversible and reversible fouling. In the case of reversible fouling, deposited protein could be removed easily using hydraulic cleaning (i.e., back washing and cross flushing) [51]. Conversely, the irreversible fouling is caused by an irreversible adsorption of proteins that can only be removed using chemical cleaning [52]. To further examine the anti-fouling properties of the prepared membranes and to study the effect of additives, the flux losses regarding total fouling (F_t), reversible fouling (F_r), and irreversible fouling (F_{ir}) were calculated, and the results are presented in Table 4.11. An increase in NFR results for CA, CAT, and CAP during the second and third cycle fouling/rinsing operations could be due to the development of the cake layer because of high irreversible fouling where their flux was too low. These results are because the porosity and hydrophilicity of the membrane have played an important role on the anti-fouling and flux properties as already explained in the pure water flux study. Additionally, relatively consistent NFR results for CATP for first, second, and third cycle operations are attributed to the introduction of TiO_2, and it has played a crucial role in the enhancement of hydrophilicity and membrane anti-fouling property. The surface pore size could also have a complex influence on membrane anti-fouling property [57]. It is considered that membranes having larger pore size are possibly more prone to pore-blocking or pore adsorption than those having smaller pore size [86]. Hence, membrane with smaller surface pore size and the more hydrophilic surface shows better anti-fouling property. In this study, CATP membrane exhibited better anti-fouling properties in the fouling process than other membranes. Thus, the combined effect of PVP and TiO_2 have played a significant role in a higher resistance toward membrane fouling due to BSA adsorption by reducing the hydrophobic interaction between BSA protein and membrane surface.

4.4.2.8 BSA Removal Performance

The BSA removal performances of the prepared membranes are shown in Figure 4.21a. The maximum BSA removal values 96.8%, and 94.3% were attained for membranes CAT and CATP, respectively. The removal efficiency

TABLE 4.11

Results for Flux Losses Caused by Total Fouling (F_t), Reversible Fouling (F_r), and Irreversible Fouling (F_{ir}) of the Three Cycles

	First Cycle				Second Cycle				Third Cycle			
Membrane	F_t	F_r	F_{ir}	NFR	F_t	F_r	F_{ir}	NFR	F_t	F_r	F_{ir}	NFR
CA	0.9	0.19	0.71	28.6	0.75	0.64	0.12	88.3	0.73	0.51	0.22	77.8
CAT	0.79	0.44	0.34	65.7	0.54	0.20	0.35	65.4	0.21	0.01	0.20	80.0
CATP	0.94	0.84	0.09	90.0	0.95	0.88	0.07	93.0	0.95	0.86	0.09	90.2
CAP	0.91	0.39	053	47.4	0.85	0.59	0.26	74.0	0.80	0.65	0.15	85.2

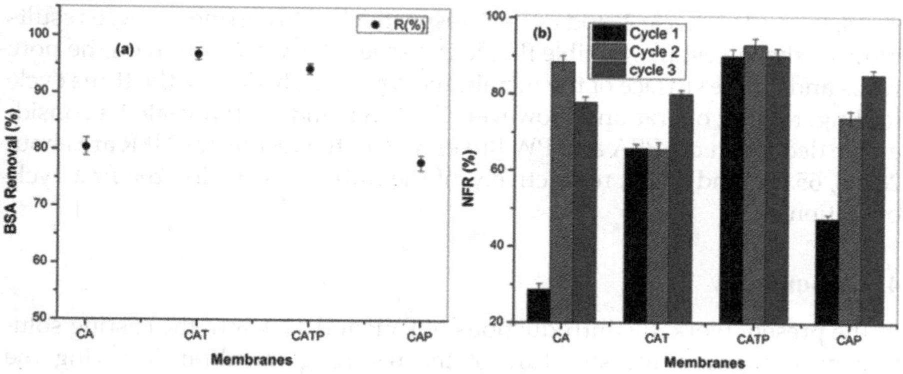

FIGURE 4.21
(a) BSA removal performances and (b) NFR percentage results of the prepared membranes: effect of PVP additive and TiO_2 NPs (25 °C, 150 kPa).

results for CA and CAP membranes are 80.4% and 77.9%, respectively. It is clearly explained in the morphology analysis section that the PVP-free membranes have less porous structures, in which a high resistance to protein molecules was detected. However, an increase in BSA removal for CATP and CAT membranes could be explained due to the effect of TiO_2 addition to the membrane matrixes. The prepared membranes could become more negatively charged after an introduction of TiO_2 nanoparticles due to more negatively charged carboxylic groups along with –OH and Ti– OH groups present on the surfaces and matrixes of the membranes [74]. As pH 7 is far from isoelectric point (IEP = 4.9), the BSA protein becomes more negatively charged, and a stronger electrostatic repulsion between BSA and the modified membranes was suggested [75]. However, some of the BSA removal characteristic can also be described using the protein adsorption phenomenon onto the membrane surface in this pH range, which could be accredited to structural interaction [57]. In addition to the electrostatic repulsion phenomena, the high BSA removal for CATP membrane was also considered as an indicator of considerable protein removal due to its structural interaction. The BSA removal phenomenon for CA and CAP membranes can be suggested due to a substantial protein deposition inside and on the membrane surfaces. Therefore, the surface depositions of BSA proteins can provide an extra hindrance to solute transportation. The high removal and comparatively lowest flux of CAT membrane, and the highest removal and improved flux results for CATP membranes after simultaneous introduction of TiO_2 and PVP were attained in this study. All the experiments in this study were performed three times. Moreover, all the experimental results were consistent with the membrane properties and agreed with each other.

Figure 4.21b presents the parameter normalized flux ratio (NFR) with filtration time (2h) of the membranes operated for three cycles. It is clearly seen in the figure that CATP displayed the highest NFR values 90%, 93%, and

90.2% for cycle 1, cycle 2, and cycle 3, respectively. Furthermore, these results indicate the lowest irreversible flux loss and less BSA deposition on the pore walls and on the surface of the membrane consistently during the three cycle fouling/rinsing operations. However, CA, CAT, and CAP revealed a considerable decline in the BSA and PW fluxes, where the results for NFR are about 28.6%, 65.7%, and 47.4%, respectively, of the initial flux during the first cycle operation.

4.4.3 Summary

In the present work, the introductions of PVP and TiO_2 into the casting solution have changed the structure of the resulting membranes during the phase inversion process. The influences of PVP and TiO_2 on the preparation of phase inverted CA ultrafiltration membrane blended with PVP and TiO_2 were explored in terms of morphology study, equilibrium water content, hydrophilicity, hydraulic resistance, permeability performance, and thermal stability. The BSA removal efficiencies and anti-fouling performances of the membranes were evaluated using ultrafiltration process. Hence, the following conclusions were made:

- After the introduction of PVP and TiO_2 to the ternary (polymer-solvent-non-solvent) system, the formations of finger-like structures and macro-voids were reduced significantly. Nevertheless, for CAP and CATP membranes, the development of micro-voids (i.e., micro porous structure) was observed.
- The introduction NPs alone into the polymer solution have formed aggregation of the NPs within the matrix of the membrane, and a small macro-void and finger-like structure was detected in the case of CAT, and these results are suggested to be related to the interference effect of NPs during the phase inversion process.
- The thermal stability of CATP membrane exhibited higher degradation temperature than CA and CAP membranes. This result was attained due to the highly stable nature of TiO_2 NPs at high temperatures.
- The improvement in the average pore size, porosity, and hydrophilic nature of the CA membranes was revealed after the introduction of PVP and TiO_2.
- The CAT blended with PVP (i.e., CATP) membrane exhibited improved BSA flux, removal efficiency, and flux recovery ratios for the three fouling/rinsing cycles.

References

1. Mulder, M., *Basic Principles of Membrane Technology*. Second ed. 1996.
2. Young, T.H., and Chen, L.W., Pore formation mechanism of membranes from phase inversion process. *Desalination*, 1995. 103(3): pp. 233–247.
3. Mohammadi, T., and Saljoughi, E., Effect of production conditions on morphology and permeability of asymmetric cellulose acetate membranes. *Desalination*, 2009. 243(1–3): pp. 1–7.
4. Strathmann, H., and Kock, K., The formation mechanism of phase inversion membranes. *Desalination*, 1977. 21(3): pp. 241–255.
5. Chakrabarty, B., Ghoshal, A.K., and Purkait, M.K., Preparation, characterization and performance studies of polysulfone membranes using PVP as an additive. *Journal of Membrane Science*, 2008. 315(1–2): pp. 36–47.
6. Zhu, Z., Xiao, J., He, W., Wang, T., Wei, Z., and Dong, Y., A phase-inversion casting process for preparation of tubular porous alumina ceramic membranes. *Journal of the European Ceramic Society*, 2015. 35(11): pp. 3187–3194.
7. Algarra, M., Vázquez, M.I., Alonso, B., Casado, C.M., Casado, J., and Benavente, J., Characterization of an engineered cellulose based membrane by thiol dendrimer for heavy metals removal. *Chemical Engineering Journal*, 2014. 253: pp. 472–477.
8. Konwarh, R., Karak, N., and Misra, M., Electrospun cellulose acetate nanofibers: The present status and gamut of biotechnological applications. *Biotechnology Advances*, 2014. 31(4): pp. 421–437.
9. Abedini, R., Mousavi, S.M., and Aminzadeh, R., A novel cellulose acetate (CA) membrane using TiO_2 nanoparticles: Preparation, characterization and permeation study. *Desalination*, 2011. 277(1–3): pp. 40–45.
10. Saljoughi, E., and Mohammadi, T., Cellulose acetate (CA)/polyvinylpyrrolidone (PVP) blend asymmetric membranes: Preparation, morphology and performance. *Desalination*, 2009. 249(2): pp. 850–854.
11. Chakrabarty, B., Ghoshal, A.K., and Purkait, M.K., Effect of molecular weight of PEG on membrane morphology and transport properties. *Journal of Membrane Science*, 2008. 309(1–2): pp. 209–221.
12. Kim, J.H., and Lee, K.H., Effect of PEG additive on membrane formation by phase inversion. *Journal of Membrane Science*, 1998. 138(2): pp. 153–163.
13. Saljoughi, E., Amirilargani, M., and Mohammadi, T., Effect of PEG additive and coagulation bath temperature on the morphology, permeability and thermal/chemical stability of asymmetric CA membranes. *Desalination*, 2010. 262(1–3): pp. 72–78.
14. Wongchitphimon, S., Wang, R., Jiraratananon, R., Shi, L., and Loh, C.H., Effect of polyethylene glycol (PEG) as an additive on the fabrication of polyvinylidene fluoride- co-hexafluropropylene (PVDF-HFP) asymmetric microporous hollow fiber membranes. *Journal of Membrane Science*, 2011. 369(1–2): pp. 329–338.

15. Yang, M., Xie, S., Li, Q., Wang, Y., Chang, X., Shan, L., Sun, L., Huang, X., and Gao, C., Effects of polyvinylpyrrolidone both as a binder and pore-former on the release of sparingly water-soluble topiramate from ethylcellulose coated pellets. *International Journal of Pharmaceutics*, 2014. 465(1–2): pp. 187–196.
16. Yoo, S.H., Kim, J.H., Jho, J.Y., Won, J., and Kang, Y.S., Influence of the addition of PVP on the morphology of asymmetric polyimide phase inversion membranes: Effect of PVP molecular weight. *Journal of Membrane Science*, 2004. 236(1–2): pp. 203–207.
17. Zhao, S., Wang, Z., Wei, X., Tian, X., Wang, J., Yang, S., and Wang, S., Comparison study of the effect of PVP and PANI nanofibers additives on membrane formation mechanism, structure and performance. *Journal of Membrane Science*, 2011. 385: pp. 110–122.
18. Liu, H., and Hsieh, Y.L., Ultrafine fibrous cellulose membranes from electrospinning of cellulose acetate. *Journal of Polymer Science Part B: Polymer Physics*, 2002. 40(18): pp. 2119–2129.
19. Tungprapa, S., Puangparn, T., Weerasombut, M., Jangchud, I., Fakum, P., Semongkhol, S., Meechaisue, C., and Supaphol, P., Electrospun cellulose acetate fibers: Effect of solvent system on morphology and fiber diameter. *Cellulose*, 2007. 14(6): pp. 563–575.
20. Paulsen, F.G., Shojaie, S.S., and Krantz, W.B., Effect of evaporation step on macrovoid formation in wet-cast polymeric membranes. *Journal of Membrane Science*, 1994. 91(3): pp. 265–282.
21. Machado, P.S.T., Habert, A.C., and Borges, C.P., Membrane formation mechanism based on precipitation kinetics and membrane morphology: Flat and hollow fiber polysulfone membranes. *Journal of Membrane Science*, 1999. 155(2): pp. 171–183.
22. Boom, R.M., Wienk, I.M., Van den Boomgaard, T., and Smolders, C.A., Microstructures in phase inversion membranes. Part 2. The role of a polymeric additive. *Journal of Membrane Science*, 1992. 73(2–3): pp. 277–292.
23. Qin, J.J., Li, Y., Lee, L.S., and Lee, H., Cellulose acetate hollow fiber ultrafiltration membranes made from CA/PVP 360 K/NMP/water. *Journal of Membrane Science*, 2003. 218(1–2): pp. 173–183.
24. Smolders, C.A., Reuvers, A.J., Boom, R.M., and Wienk, I.M., Microstructures in phase-inversion membranes. Part 1. Formation of macrovoids. *Journal of Membrane Science*, 1992. 73(2–3): pp. 259–275.
25. Reuvers, A.J., Membrane formation: Diffusion induced demixing processes in ternary polymeric systems (Doctoral dissertation, University of Twente [Host]), 1987.
26. Marchese, J., Ponce, M., Ochoa, N.A., Prádanos, P., Palacio, L., and Hernández, A., Fouling behaviour of polyethersulfone UF membranes made with different PVP. *Journal of Membrane Science*, 2003. 211(1): pp. 1–11.
27. Mosqueda-Jimenez, D.B., Narbaitz, R.M., Matsuura, T., Chowdhury, G., Pleizier, G., and Santerre, J.P., Influence of processing conditions on the properties of ultrafiltration membranes. *Journal of Membrane Science*, 2004. 231(1–2): pp. 209–224.
28. Fane, A.G., Fell, C.J.D., and Waters, A.G., The relationship between membrane surface pore characteristics and flux for ultrafiltration membranes. *Journal of Membrane Science*, 1981. 9(3): pp. 245–262.

29. Fell, C.J.D., Kim, K.J., Chen, V., Wiley, D.E., and Fane, A.G., Factors determining flux and rejection of ultrafiltration membranes. *Chemical Engineering and Processing: Process Intensification*, 1990. 27(3): pp. 165–173.
30. Idris, A., and Yet, L.K., The effect of different molecular weight PEG additives on cellulose acetate asymmetric dialysis membrane performance. *Journal of Membrane Science*, 2006. 280(1–2): pp. 920–927.
31. Chou, W.L., Yu, D.G., Yang, M.C., and Jou, C.H., Effect of molecular weight and concentration of PEG additives on morphology and permeation performance of cellulose acetate hollow fibers. *Separation and Purification Technology*, 2007. 57(2): pp. 209–219.
32. Matsuyama, H., Yamamoto, A., Yano, H., Maki, T., Teramoto, M., Mishima, K., and Matsuyama, K., Effect of organic solvents on membrane formation by phase separation with supercritical CO2. *Journal of Membrane Science*, 2002. 204(1–2): pp. 81–87.
33. Guan, R., Dai, H., Li, C., Liu, J., and Xu, J., Effect of casting solvent on the morphology and performance of sulfonated polyethersulfone membranes. *Journal of Membrane Science*, 2006. 277(1–2): pp. 148–156.
34. Lu, D., Zou, H., Guan, R., Dai, H., and Lu, L., Sulfonation of polyethersulfone by chlorosulfonic acid. *Polymer Bulletin*, 2005. 54(1–2): pp. 21–28.
35. Kruczek, B., Development and characterization of dense membranes for gas separation made from high molecular weight sulfonated poly (phenylene oxide): Effect of casting conditions on the morphology and performance of the membranes. University of Ottawa (Canada), 1999.
36. Hansen, C.M., The three dimensional solubility parameter and solvent diffusion coefficient: Their importance in surface coating formulation. Copenhagen Danish Technical Press, Kongens Lyngby, Denmark, 1967.
37. Vebber, G.C., Pranke, P., and Pereira, C.N., Calculating Hansen solubility parameters of polymers with genetic algorithms. *Journal of Applied Polymer Science*, 2014. 131(1).
38. Nair, R., Nyamweya, N., Gönen, S., Martınez-Miranda, L.J., and Hoag, S.W., Influence of various drugs on the glass transition temperature of poly (vinylpyrrolidone): A thermodynamic and spectroscopic investigation. *International Journal of Pharmaceutics*, 2001. 225(1–2): pp. 83–96.
39. Özdemir, C., and Güner, A., Solubility profiles of poly (ethylene glycol)/solvent systems, I: Qualitative comparison of solubility parameter approaches. *European Polymer Journal*, 2007. 43(7): pp. 3068–3093.
40. Kopitzke, R.W., Linkous, C.A., Anderson, H.R., and Nelson, G.L., Conductivity and water uptake of aromatic-based proton exchange membrane electrolytes. *Journal of the Electrochemical Society*, 2000. 147(5): pp. 1677–1681.
41. Shi, H., He, Y., Pan, Y., Di, H., Zeng, G., Zhang, L., and Zhang, C., A modified mussel-inspired method to fabricate TiO2 decorated superhydrophilic PVDF membrane for oil/water separation. *Journal of Membrane Science*, 2016. 506: pp. 60–70.
42. Dutta, B.K., Principles of mass transfer and separation processes. PHI Learning Private Limited, New Delhi, India, 2009.
43. Jiang, J.H., Zhu, L.P., Li, X.L., Xu, Y.Y., and Zhu, B.K., Surface modification of PE porous membranes based on the strong adhesion of polydopamine and covalent immobilization of heparin. *Journal of Membrane Science*, 2010. 364(1–2): pp. 194–202.

44. Arthanareeswaran, G., Thanikaivelan, P., Srinivasn, K., Mohan, D., and Rajendran, M., Synthesis, characterization and thermal studies on cellulose acetate membranes with additive. *European Polymer Journal*, 2004. 40(9): pp. 2153–2159.
45. Robinson, S., and Williams, P.A., Inhibition of protein adsorption onto silica by polyvinylpyrrolidone. *Langmuir*, 2002. 18(23): pp. 8743–8748.
46. Wu, J., Wang, Z., Lin, W., and Chen, S., Investigation of the interaction between poly (ethylene glycol) and protein molecules using low field nuclear magnetic resonance. Acta biomaterialia, 2013. 9(5): pp. 6414–6420.
47. Yang, Z., Galloway, J.A., and Yu, H., Protein interactions with poly (ethylene glycol) self-assembled monolayers on glass substrates: Diffusion and adsorption. *Langmuir*, 1999. 15(24): pp. 8405–8411.
48. Garcia-Ivars, J., Alcaina-Miranda, M.I., Iborra-Clar, M.I., Mendoza-Roca, J.A., and Pastor-Alcañiz, L., Enhancement in hydrophilicity of different polymer phase- inversion ultrafiltration membranes by introducing PEG/Al2O3 nanoparticles. *Separation and Purification Technology*, 2014. 128: pp. 45–57.
49. Maximous, N., Nakhla, G., Wan, W., and Wong, K., Performance of a novel ZrO2/PES membrane for wastewater filtration. *Journal of Membrane Science*, 2010. 352(1–2): pp. 222–230.
50. Shi, Q., Su, Y., Zhao, W., Li, C., Hu, Y., Jiang, Z., and Zhu, S., Zwitterionic polyethersulfone ultrafiltration membrane with superior antifouling property. *Journal of Membrane Science*, 2008. 319(1–2): pp. 271–278.
51. Tang, Y.P., Chan, J.X., Chung, T.S., Weber, M., Staudt, C., and Maletzko, C., Simultaneously covalent and ionic bridging towards antifouling of GO-imbedded nanocomposite hollow fiber membranes. *Journal of Materials Chemistry A*, 2015. 3(19): pp. 10573–10584.
52. Feng, Y., Han, G., Zhang, L., Chen, S.B., Chung, T.S., Weber, M., Staudt, C., and Maletzko, C., Rheology and phase inversion behavior of polyphenylenesulfone (PPSU) and sulfonated PPSU for membrane formation. *Polymer*, 2016. 99: pp. 72–82.
53. Luo, M.L., Zhao, J.Q., Tang, W., and Pu, C.S., Hydrophilic modification of poly (ether sulfone) ultrafiltration membrane surface by self-assembly of TiO2 nanoparticles. *Applied Surface Science*, 2005. 249(1–4): pp. 76–84.
54. Liu, Y., Koops, G.H., and Strathmann, H., Characterization of morphology controlled polyethersulfone hollow fiber membranes by the addition of polyethylene glycol to the dope and bore liquid solution. *Journal of Membrane Science*, 2003. 223(1–2): pp. 187–199.
55. Panda, S.R., and De, S., Role of polyethylene glycol with different solvents for tailor-made polysulfone membranes. *Journal of Polymer Research*, 2013. 20(7): p. 179.
56. Shi, F., Ma, Y., Ma, J., Wang, P., and Sun, W., Preparation and characterization of PVDF/TiO2 hybrid membranes with different dosage of nano-TiO2. *Journal of Membrane Science*, 2012. 389: pp. 522–531.
57. Cao, X., Ma, J., Shi, X., and Ren, Z., Effect of TiO2 nanoparticle size on the performance of PVDF membrane. *Applied Surface Science*, 2006. 253(4): pp. 2003–2010.
58. Maximous, N., Nakhla, G., Wan, W., and Wong, K., Preparation, characterization and performance of Al2O3/PES membrane for wastewater filtration. *Journal of Membrane Science*, 2009. 341(1–2): pp. 67–75.

59. Bottino, A., Capannelli, G., and Comite, A., Preparation and characterization of novel porous PVDF-ZrO2 composite membranes. *Desalination,* 2002. 146(1–3): pp. 35–40.
60. Lin, J., Ye, W., Zhong, K., Shen, J., Jullok, N., Sotto, A., and Van der Bruggen, B., Enhancement of polyethersulfone (PES) membrane doped by monodisperse Stöber silica for water treatment. *Chemical Engineering and Processing: Process Intensification,* 2016. 107: pp. 194–205.
61. Prince, J.A., Bhuvana, S., Boodhoo, K.V.K., Anbharasi, V., and Singh, G., Synthesis and characterization of PEG-Ag immobilized PES hollow fiber ultrafiltration membranes with long lasting antifouling properties. *Journal of Membrane Science,* 2014. 454: pp. 538–548.
62. Garcia-Ivars, J., Iborra-Clar, M.I., Alcaina-Miranda, M.I., Mendoza-Roca, J.A., and Pastor-Alcañiz, L., Development of fouling-resistant polyethersulfone ultrafiltration membranes via surface UV photografting with polyethylene glycol/aluminum oxide nanoparticles. *Separation and Purification Technology,* 2014. 135: pp. 88–99.
63. Ong, C.S., Lau, W.J., Goh, P.S., Ng, B.C., and Ismail, A.F., Preparation and characterization of PVDF–PVP–TiO2 composite hollow fiber membranes for oily wastewater treatment using submerged membrane system. *Desalination and Water Treatment,* 2015. 53(5): pp. 1213–1223.
64. Liu, Y., Koops, G.H., and Strathmann, H., Characterization of morphology controlled polyethersulfone hollow fiber membranes by the addition of polyethylene glycol to the dope and bore liquid solution. *Journal of Membrane Science,* 2003. 223(1–2): pp. 187–199.
65. Yuliwati, E., and Ismail, A.F., Effect of additives concentration on the surface properties and performance of PVDF ultrafiltration membranes for refinery produced wastewater treatment. *Desalination,* 2011. 273(1): pp. 226–234.
66. Garcia-Ivars, J., Iborra-Clar, M.I., Alcaina-Miranda, M.I., and Van der Bruggen, B., Comparison between hydrophilic and hydrophobic metal nanoparticles on the phase separation phenomena during formation of asymmetric polyethersulphone membranes. *Journal of Membrane Science,* 2015. 493: pp. 709–722.
67. Greenlee, L.F., and Rentz, N.S., Influence of nanoparticle processing and additives on PES casting solution viscosity and cast membrane characteristics. *Polymer,* 2016. 103: pp. 498–508.
68. Sotto, A., Boromand, A., Zhang, R., Luis, P., Arsuaga, J.M., Kim, J., and Van der Bruggen, B., Effect of nanoparticle aggregation at low concentrations of TiO2 on the hydrophilicity, morphology, and fouling resistance of PES–TiO2 membranes. *Journal of Colloid and Interface Science,* 2011. 363(2): pp. 540–550.
69. Chan, K.H., Wong, E.T., Irfan, M., Idris, A., and Yusof, N.M., Enhanced Cu (II) rejection and fouling reduction through fabrication of PEG-PES nanocomposite ultrafiltration membrane with PEG-coated cobalt doped iron oxide nanoparticle. *Journal of the Taiwan Institute of Chemical Engineers,* 2015. 47: pp. 50–58.
70. Zafar, M., Ali, M., Khan, S.M., Jamil, T., and Butt, M.T.Z., Effect of additives on the properties and performance of cellulose acetate derivative membranes in the separation of isopropanol/water mixtures. *Desalination,* 2012. 285: pp. 359–365.
71. Maria da Conceicao, C.L., de Alencar, A.E.V., Mazzeto, S.E., and de A Soares, S., The effect of additives on the thermal degradation of cellulose acetate. *Polymer Degradation and Stability,* 2003. 80(1): pp. 149–155.

72. Ali, M., Zafar, M., Jamil, T., and Butt, M.T.Z., Influence of glycol additives on the structure and performance of cellulose acetate/zinc oxide blend membranes. *Desalination*, 2011. 270(1–3): pp. 98–104.
73. Bae, T.H., and Tak, T.M., Effect of TiO2 nanoparticles on fouling mitigation of ultrafiltration membranes for activated sludge filtration. *Journal of Membrane Science*, 2005. 249(1–2): pp. 1–8.
74. Kumar, M., Gholamvand, Z., Morrissey, A., Nolan, K., Ulbricht, M., and Lawler, J., Preparation and characterization of low fouling novel hybrid ultrafiltration membranes based on the blends of GO– TiO2 nanocomposite and polysulfone for humic acid removal. *Journal of Membrane Science*, 2016. 506: pp. 38–49.
75. Mo, H., Tay, K.G., and Ng, H.Y., Fouling of reverse osmosis membrane by protein (BSA): Effects of pH, calcium, magnesium, ionic strength and temperature. *Journal of Membrane Science*, 2008. 315(1–2): pp. 28–35.
76. Lee, H.J., Jung, B., Kang, Y.S., and Lee, H., Phase separation of polymer casting solution by nonsolvent vapor. *Journal of Membrane Science*, 2004. 245(1–2): pp. 103–112.
77. Qin, J.J., Li, Y., Lee, L.S., and Lee, H., Cellulose acetate hollow fiber ultrafiltration membranes made from CA/PVP 360 K/NMP/water. *Journal of Membrane Science*, 2003. 218(1–2): pp. 173–183.
78. Yuan, Z., and Dan-Li, X., Porous PVDF/TPU blends asymmetric hollow fiber membranes prepared with the use of hydrophilic additive PVP (K30). *Desalination*, 2008. 223(1–3): pp. 438–447.
79. Lua, A.C., and Shen, Y., Preparation and characterization of polyimide–silica composite membranes and their derived carbon–silica composite membranes for gas separation. *Chemical Engineering Journal*, 2013. 220: pp. 441–451.
80. Su, Y.C., Huang, C., Pan, J.R., Hsieh, W.P., and Chu, M.C., Fouling mitigation by TiO_2 composite membrane in membrane bioreactors. *Journal of Environmental Engineering*, 2011. 138(3): pp. 344–350.
81. Saffaj, N., Younssi, S.A., Albizane, A., Messouadi, A., Bouhria, M., Persin, M., Cretin, M., and Larbot, A., Preparation and characterization of ultrafiltration membranes for toxic removal from wastewater. *Desalination*, 2004. 168: pp. 259–263.
82. Gebru, K.A., and Das, C., Effects of solubility parameter differences among PEG, PVP and CA on the preparation of ultrafiltration membranes: Impacts of solvents and additives on morphology, permeability and fouling performances. *Chinese Journal of Chemical Engineering*, 2017. 25(7): pp. 911–923.
83. Das, C., and Gebru, K.A., Preparation and characterization of CA– PEG– TiO_2 membranes: Effect of PEG and TiO2 on morphology, flux and fouling performance. *Journal of Membrane Science and Research*, 2017. 3(2): pp. 90–101.
84. Chatterjee, P.K., and Conrad, C.M., Thermogravimetric analysis of cellulose. *Journal of Polymer Science Part A: Polymer Chemistry*, 1968. 6(12): pp. 3217–3233.
85. Yu, L.Y., Shen, H.M., and Xu, Z.L., PVDF–TiO_2 composite hollow fiber ultrafiltration membranes prepared by TiO2 sol–gel method and blending method. *Journal of Applied Polymer Science*, 2009. 113(3): pp. 1763–1772.
86. Li, C.W., and Chen, Y.S., Fouling of UF membrane by humic substance: Effects of molecular weight and powder-activated carbon (PAC) pre-treatment. *Desalination*, 2004. 170(1): pp. 59–67.

5

Modification of Polymeric Membranes

5.1 Functionalization and Characterization of Cellulose Acetate Membranes for Chromium (VI) Removal

5.1.1 Introduction

Removal of heavy metals and other contaminants from water and wastewater is a very important factor with respect to environmental pollution control and human health. Heavy metal ions, such as chromium, lead, copper, cadmium, mercury, cobalt, and nickel, are some examples of water contaminants. Among these heavy metals chromium (VI) is a common effluent, such as in leather processing, alloy preparation, and metal plating industry. Chromium causes problems such as bronchogenic, bronchitis, carcinoma, ulcer formation, and liver damage [1]. Chromium toxicity is largely dependent on its oxidation states. The most common oxidation states of chromium are Cr^{+6}, Cr^{+5}, Cr^{+4}, Cr^{+3}, Cr^{+2}, and Cr^{+1}. However, Cr (III) (i.e., trivalent chromium) and Cr (VI) (i.e., hexavalent chromium) are the most hazardous forms of chromium. In an aqueous solutions, Cr (VI) exists mainly as chromate ion $\left(CrO_4^{2-}\right)$ (pH >7.0), dichromate $\left(Cr_2O7_2^-\right)$ (pH 1.0–6.0, conc. >1 g/L), hydrogen chromate $\left(HCrO_4^-\right)$ (pH 1.0–6.0), and salts of chromic acid (H_2CrO_4) (pH <1.0) [2,3]. Several methods, such as membrane processes [4–7], electrocoagulation [8–10], precipitation [11–13], ion exchange [14,15], and adsorption [2,16–20] have been used for removal of heavy metal ions from contaminated water [21]. Currently membrane-based heavy metal removal process has been getting more attention due to operational simplicity, low energy consumption, and environmental friendliness. Nanofiltration and reverse osmosis techniques are able to remove heavy metals effectively from contaminated water, but they require higher energy consumption when compared with ultrafiltration (UF) and microfiltration (MF) membrane processes. However, the conventional UF and MF membrane separation processes wouldn't remove the heavy metal ions easily due to their larger pore sizes. Therefore, modification of the conventional UF/MF membranes is required to achieve selective separation of heavy meal ions from aqueous solutions [22,23]. Now, researchers are attempting to develop UF membrane with a capability to remove heavy metals, which is intended to simplify process steps and save more energy. Therefore, modified UF membranes can offer numerous advantages for removal of heavy metal ions with higher flux, lower operating pressure, and ease of scale up [1,24]. Cellulose acetate is broadly applicable for the synthesis of membranes because of having tough, biocompatible, hydrophilic characteristics, good desalting nature, high flux, and moderately less expensive to be employed for reverse osmosis, microfiltration, ultrafiltration, and gas separation applications [25–27]. The drawback of cellulose acetate membranes is that it is susceptible to thermal and mechanical stabilities depending on the environments and conditions of application [28]. Therefore, in this study TiO_2 nanoparticles (NPs) were incorporated to CA matrix due to its stable nature, ease of availability, and other properties, such as antifouling nature, thermal stability, and mechanical stability. Moreover, TiO_2 NPs can enhance the hydrophilic nature of the polymer as well [29,30]. Amines appear to be a particular interest toward heavy metal by blending with membrane material because of their tendency to develop complex reaction between the heavy metal ions and the nitrogen atom within the functional groups [21,31,32]. Thus, in this study prior to the incorporation of TiO_2 NPs, functionalization of the NPs was an important step to improve the separation characteristics of the membrane. The high specific surface area and porous characteristics of TiO_2 NPs make the functionalization and/or impregnation of the amine groups on its surface and inside pores easily [33]. Therefore, the TiO_2 NPs were modified using three amines, namely, ethylenediamine (EDA), hexamethylenetetramine (HMTA), and tetraethylenepentamine (TEPA). After that the amine modified TiO_2 NPs were mixed with CA polymer, and flat sheet composite membranes were prepared using phase inversion process. This study can also expect to broaden the application area of the ultrafiltration membranes especially for the treatment of toxic contaminants. Moreover, the introduction of amine groups within the membranes in this work was successful.

5.1.2 Physical Blending Process

5.1.2.1 Preparation of Amine-Modified TiO_2

The amine-modified TiO_2 (i.e., Ti-EDA, Ti-HMTA, and Ti-TEPA) were prepared by impregnation of the amines into the NPs. Briefly, 3.0 mL of each amine was dissolved in 30 mL of deionized water and stirred for about 0.5 h using magnetic stirrer. After that, 3.0 g of TiO_2 was added to the solution and stirred for 6 h. Thus, the amine-modified TiO_2 NPs were then dried in a hot air oven at 70 °C for 2 h and consequently kept in a capped vial until use. The composition of amine modified TiO_2 NPs is presented in Table 5.1.

5.1.2.2 Preparation of CA/U-Ti, CA/Ti-EDA, CA/Ti-HMTA, and CA/Ti-TEPA Membranes

Phase inversion technique by immersion precipitation was used for fabricating asymmetric ultrafiltration membranes. First, uniform solutions of CA in

TABLE 5.1

Composition of Amine modified TiO_2 Nanoparticles

Sample	TiO_2 (g)	EDA (mL)	HMTA (mL)	TEPA (mL)	DI Water (mL)
Ti-EDA	3	3	0	0	30
Ti-HMTA	3	0	3	0	30
Ti-TEPA	3	0	0	3	30

TABLE 5.2

Solution Compositions of the Casting Solution: CA, PEG, U-TiO_2, and M-TiO_2 Nanoparticles

Membrane	CA (%)	U-TiO_2 (%)	M-TiO_2 (%)	PEG (%)	Ac: DMAc (70:30; v/v)(Wt.%)
CA/U-Ti	10.5	2	–	3	84.5
CA/Ti-EDA	10.5	–	2	3	84.5
CA/Ti-HMTA	10.5	–	2	3	84.5
CA/Ti-TEPA	10.5	–	2	3	84.5

acetone/DMAc (70/30; v/v) were prepared under continuous magnetic stirring at room temperature (28 ± 2 °C). After getting a uniform solution of the CA/solvent, a fixed amount of PEG additive (3 wt.%) was added and stirred continuously at least for 24 hours until the solution was fully dissolved and a homogeneous casting solution was obtained. Finally, 2 wt.% unmodified (U-TiO_2) and modified (M-TiO_2) nanoparticles were added separately and well dispersed for another one day to form a homogeneous composite solution. Therefore, the membranes with different composition were designated as CA/U-Ti, CA/Ti-HMTA, CA/Ti-EDA, and CA/Ti-TEPA. Table 5.2 shows the solution casting compositions of all the prepared membranes.

5.1.2.3 TEM Analysis of U-TiO_2 and M-TiO_2 NPs

The particle size and microstructural information of U-TiO_2 (commercial) and M-TiO_2 nanoparticles were characterized using transmission electron microscopy. First, the nanoparticle samples were prepared by dispersing in DI water (500 mg/L) and then poured on a carbon tape covered plate. Finally, the samples were dried at room temperature and ready for TEM analysis. As clearly shown in Figure 5.1, the TiO_2 NPs appeared in the form of spots. The black spots imbedded on the surface of M-TiO_2 NPs were because of the uniform dispersion of the amines. However, the highly homogeneous dispersion of the black spots was observed prominently for Ti-EDA and Ti-TEPA samples and not the Ti-HMTA, which may be due to the better interaction between the amine groups and TiO_2 NPs during the impregnation process.

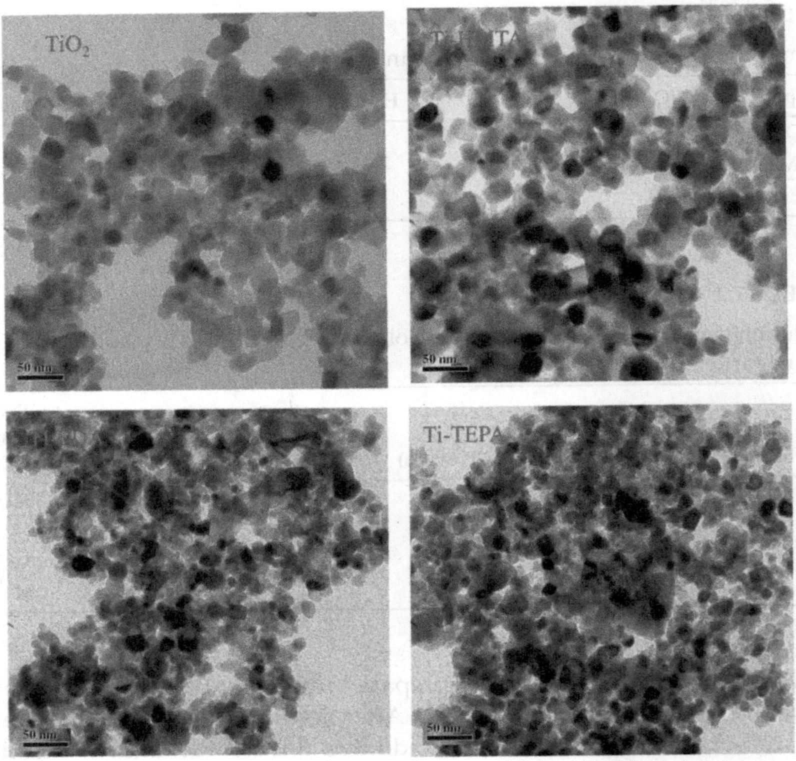

FIGURE 5.1
TEM images of U-TiO_2 and M-TiO_2 nanoparticles.

To measure the size of each nanoparticle, Image J software was employed. The sizes of the nanoparticles were not uniform, and their sizes ranged from 16 to 100 nm. The average particle sizes were approximately 29.8 nm, 30.1 nm, 29.4 nm, and 32.2 nm for TiO_2, Ti-HMTA, Ti-EDA, and Ti-TEPA, respectively; no considerable variation in average particle size was detected from this study.

5.1.2.4 ATR-FTIR and Zeta Potential (ζ) and Thermal Analysis

As presented in Figure 5.2a, in the IR spectrum of Ti-EDA, 3290 (N-H stretch), 1470 (C-H bend), the stretching vibrations of C–H bonds are present at 2953 and 2872 cm^{-1}. The peaks at 1580 and 1309 cm^{-1} are caused by the bending vibration of N–H bonds and stretching vibrations of C–N bonds, respectively. The C–O stretching vibration was observed at 1000 cm^{-1}. In the IR spectrum of Ti-TEPA, the stretching vibrations of C–H bonds are present at 2833 cm^{-1}. The peaks of N–H bonds and C–N bonds are present at 1650 and 1350 cm^{-1}, respectively. Thus, loading of the amines into the pore channels of the TiO_2 nanoparticles is

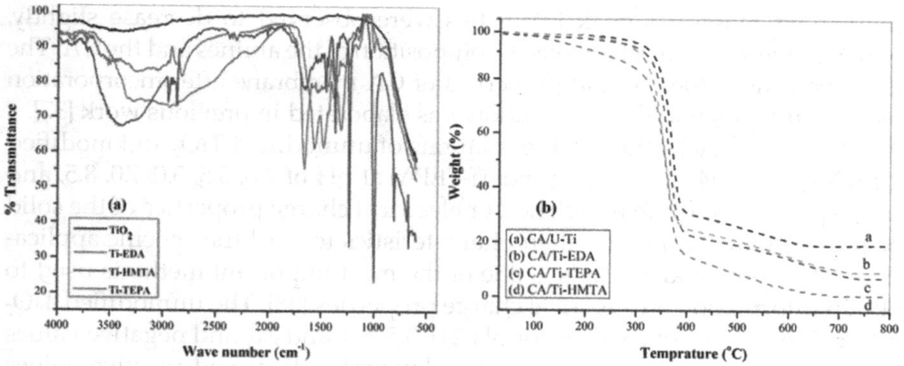

FIGURE 5.2
(a) ATR–FTIR results of unmodified TiO_2 and modified TiO_2 NPs; (b) TGA analysis CA/U-Ti, CA/Ti-HMTA, CA/Ti-EDA, and CA/Ti-TEPA membranes.

confirmed. The IR transmission spectra of Ti-HMTA reveals that the C–O stretching vibration at 1000 cm^{-1}, the C–H stretching vibration bands at 2984 cm^{-1}, the C–H bend modes at 1454 cm^{-1}, and the IR spectra for C-H rock and C-H wag $(CH_2)_4$ were observed at 1371 cm^{-1} and 1236 cm^{-1}, respectively.

The IR spectrum of TiO_2 NPs shows intense broadband in the vicinity of 600 to 800 cm^{-1} attributed to the stretching vibration of Ti–O–Ti as expected for TiO_2 samples. This result shows the successful interaction between the TiO_2 NPs and the amines. The spectrum for all the samples, except pure TiO_2 NPs, shows considerable peaks attributed to O–H group at 3290 cm^{-1}, 3388 cm^{-1}, and 3250 cm^{-1} for Ti-EDA, Ti-HMTA, and Ti-TEPA, respectively. This result is due to the sample moisture where the amines can more easily absorb moisture than the pure TiO_2 NPs.

The thermal degradation analyses (TGA graphs) of the prepared membranes before and after impregnation are presented in Figure 5.2b. The TGA results of the prepared membranes displayed two-step degradation procedures. Therefore, the start of the decomposition step for CA/U-Ti, CA/Ti-HMTA, CA/Ti-EDA, and CA/Ti-TEPA membranes are 289 °C, 135 °C, 287 °C, and 286 °C, respectively. The observed weight losses are due to the degradation of CA chains because of the pyrolysis of the backbone of the CA polymer and also followed by de-acetylation of CA [34–36]. However, the relatively weak thermal stability of CA/Ti-HMTA membrane is an indication of a relatively weak interaction between HMTA and TiO_2, where HMTA has no active groups like NH_2- or NH-, which can further influence the thermal stability of the membrane. During the last decomposition step, the degradation temperatures of CA/U-Ti, CA/Ti-HMTA, CA/Ti-EDA, and CA/Ti-TEPA are 390 °C, 384 °C, 388 °C, and 389 °C, respectively, signifying the main thermal degradation of the CA chains. From these results, it is shown that all the membranes exhibited satisfactory thermal stability performances, except a slight shift in intensity was observed. After the impregnation of the different amines, the second mass loss peak intensities were observed to decrease slightly, which is due to the simultaneous decomposition of the amines and the CA. The improvement in the thermal properties of CA membrane after incorporation of PEG and TiO_2 into the CA solution was elaborated in previous work [37].

Zeta potential (ζ) values of the solutions of unmodified TiO_2 and modified TiO_2 NPs (Ti-HMTA, Ti-EDA, and Ti-TEPA) at pH of 2.0, 3.5, 5.0, 7.0, 8.5, and 10 are presented in Table 5.3. The ζ or electrical charge properties of the solid or film polymers influence their characteristics toward the specific applications. Zeta potential analysis is one of the most important methods used to obtain information on electrical charge properties [38]. The unmodified TiO_2 NPs exhibited positive values for pH 2.0, 3.5, 5.0, and 7.0, and negative values for pH 8.5 and 10. The amine-modified samples displayed positive values for pH 2.0, 3.5, 5.0 and negative values for pH 7.0, 8.5, and 10. The high positive ζ values for the modified TiO_2 NPs at pH 2.0, 3.5, and 5.0 is due to the fact that the amines immigrated within the TiO_2 NPs tend to be protonated at lower pH [39,40]. However, the ζ values for Ti-HMTA didn't show significant changes after modification. The amines used in this study have different structures and contents of NH_2-groups. EDA has two NH_2-groups only; TEPA also has two NH_2-groups and three NH-groups; and HMTA has no NH_2- or NH- group, only N is connected with three CH_2-groups. For this reason, the protonation due to the presence of the amine group is not expected in the case of Ti-HMTA. The ζ value of the unmodified TiO_2 nanoparticles was +3.34 mV at pH of 7.0. The water molecules can occupy the oxygen vacancies and produce adsorbed –OH groups, which indicated that the nanoparticles were positively charged [41]. However, with increasing –OH concentration in the solution, adsorption of –OH on the sample increases, causing substantial ζ change with pH. Therefore, high negative ζ value (i.e., –14.22 and –43.02) at pH of 8.5 and 10 was detected for the unmodified TiO_2 nanoparticles. Moreover, high negative ζ values were attained with further increase in the pH values to 10 for all the amine modified TiO_2 NPs. This result is accredited to an increase in the amount of –OH groups in the solid barrier due to alkaline nature of the solutions [42]. The TiO_2 NPs modified with TEPA (Ti-TEPA) displayed higher negative ζ than that of the Ti-HMTA and Ti-EDA, whereas the unmodified TiO_2 NPs showed positive zeta potentials

TABLE 5.3

Zeta Potential (ζ) Results of Unmodified and Modified TiO_2 NPs

	Zeta Potential (mV)					
Samples	**pH 2.0**	**pH 3.5**	**pH 5.0**	**pH 7.0**	**pH 8.5**	**pH 10**
TiO_2	+6.23	+4.55	+4.01	+3.34	−14.22	−43.02
Ti-TEPA	+41.02	+40.76	+37.25	−32.62	−36.71	−40.86
Ti-EDA	+28.79	+28.15	+26.32	−15.66	−16.52	−20.35
Ti-HMTA	+6.39	+6.19	+6.01	−4.32	−14.01	−30.28

at pH of 7.0. Zeta potential variation as a function of the types of the amines impregnated within the TiO_2 NPs can be explained due to the structure and content of NH_2- or NH- groups. Therefore, the pH dependence of ζ for the modified and unmodified TiO_2 NPs is confirmed from this study.

5.1.2.5 Morphology and Hydrophilicity Study

The FESEM images of the U-TiO_2 and M-TiO_2 NPs (i.e., Ti-HMTA, Ti-EDA, and Ti-TEPA) are presented in Figure 5.3a. As seen from the figure, the unmodified TiO_2 NPs depicted a flower resembling structures after the amine impregnation process. Thus, a successful modification and interaction of the amines and TiO_2 NPs can be suggested from the TEM (Figure 5.1) and FESEM results. CA/U-TiO_2 and CA/M-TiO_2 casting solutions were prepared, and membranes with different amines (i.e., CA/U-Ti, CA/Ti-HMTA, CA/Ti-EDA, and CA/Ti-TEPA) were fabricated using the immersion precipitation phase inversion technique. The top surface and cross-sectional views of the FESEM images of all the prepared membranes are presented in Figure 5.3b and 5.3c. It can be seen that all the prepared membranes displayed asymmetric structures with less finger-like macro-voids and a denser top layer. The suppression or the development of finger-like structure and

(a)

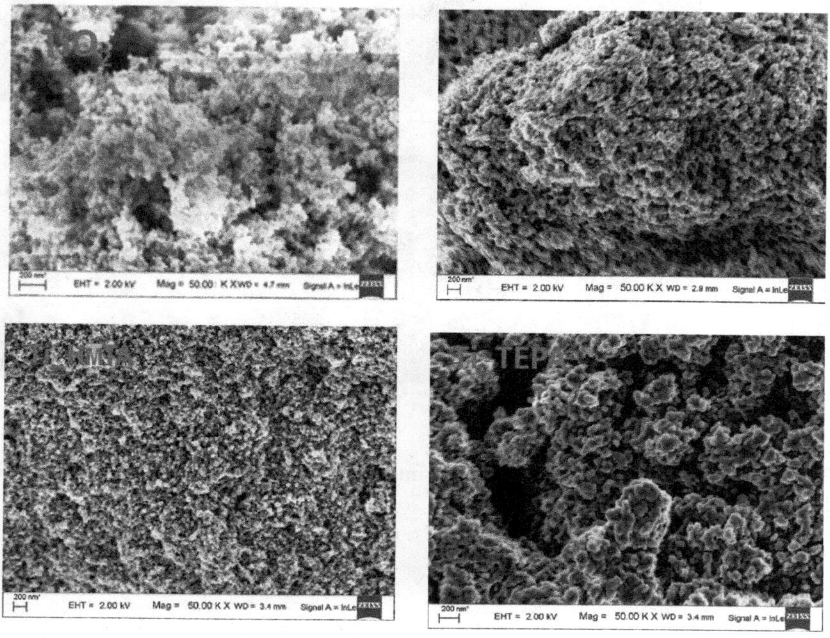

FIGURE 5.3
(a) FESEM images of U-TiO_2 and M- TiO_2 powder. *(Continued)*

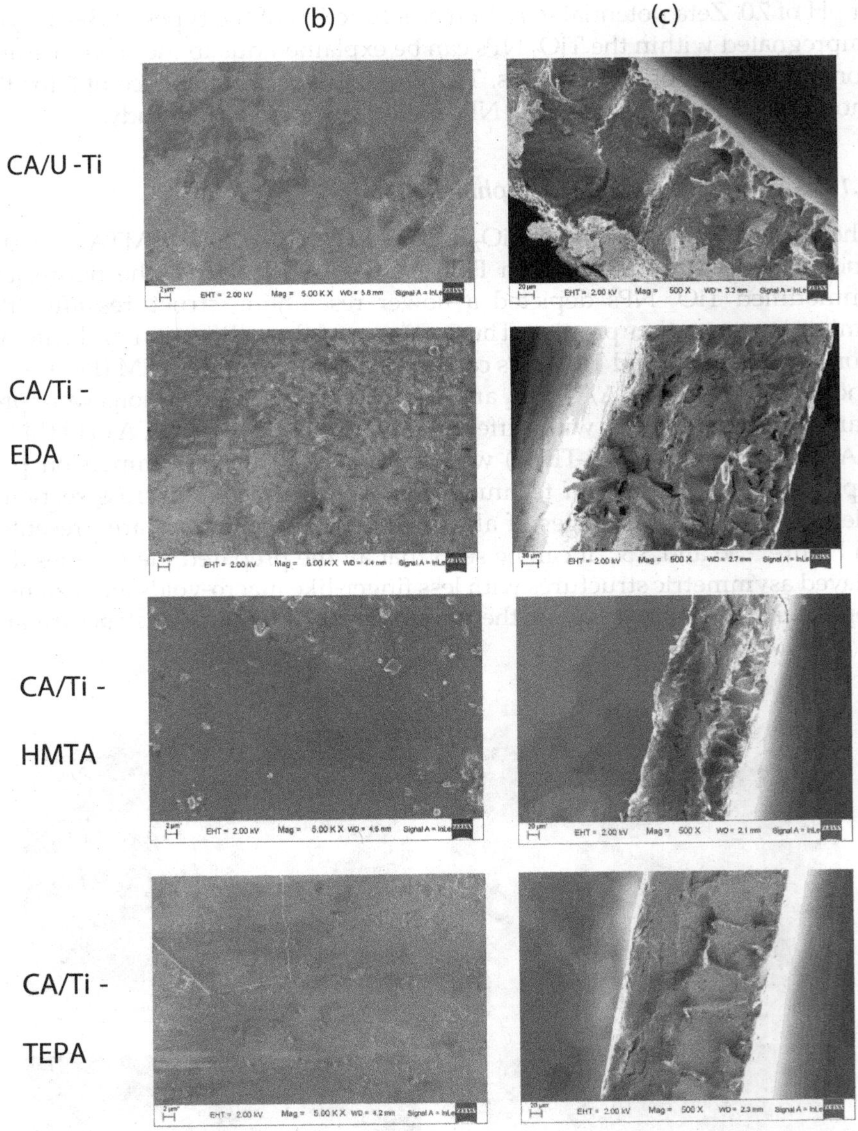

FIGURE 5.3 (CONTINUED)
(b) Top surface and (c) cross-sectional FESEM images of CA/U-Ti, CA/Ti-HMTA, CA/Ti-EDA, and CA/Ti-TEPA membranes.

macro-voids during the membrane formation is due to the presence of PEG and TiO_2 within the casting solutions. Moreover, the pore forming properties of PEG additive have also played a significant role in the development of porous sublayer structures. Therefore, the finger-like and macro-voids free structures were observed for CA/U-Ti, CA/Ti-HMTA, CA/Ti-EDA, and CA/Ti-TEPA membranes. Comparing CA/U-TiO_2 and CA/Ti-HMTA, CA/Ti-EDA, and CA/Ti-TEPA, it also can be observed that the pore size of CA/U-TiO_2 membranes was slightly decreased after modification. This result was attained because of the adsorption and attachment of the amines within the matrix of the membranes. Table 5.4 shows that all modified membranes had displayed a uniform decrease in pore size, when the type of amines was HMTA, EDA, and TEPA, respectively, and this result can be accredited to the difference in interaction level of the amines with TiO_2 NPs. However, the pore size differences of the membranes are not significant enough to influence permeation flux, which will be discussed later on.

Water contact angle measurement is an effective technique for investigating the hydrophilicity nature of the prepared membranes. The (a) water contact angles values and (b) images of the water drops on the surface of all the prepared membranes are shown in Figure 5.4. Membranes with smaller

TABLE 5.4

Compaction and Hydraulic Characteristics of Amine Modified Membranes at 250 kPa

Membrane	CF	R_m ($\times 10^{-10} m^{-1}$)	J_w (L/m^2h)	*EWC* (%)	ε (%)	r_m (nm)	Thickness (μm)
CA/U-Ti	1.50	0.50	587.8	80.1	74.8	27.0	73.5
CA/Ti-HMTA	1.41	0.42	649.2	81.5	73.6	26.5	82.4
CA/Ti-EDA	1.25	0.40	752.2	82.7	72.4	25.7	84.2
CA/Ti-TEPA	1.32	0.35	924.7	84.2	70.2	24.2	86.9

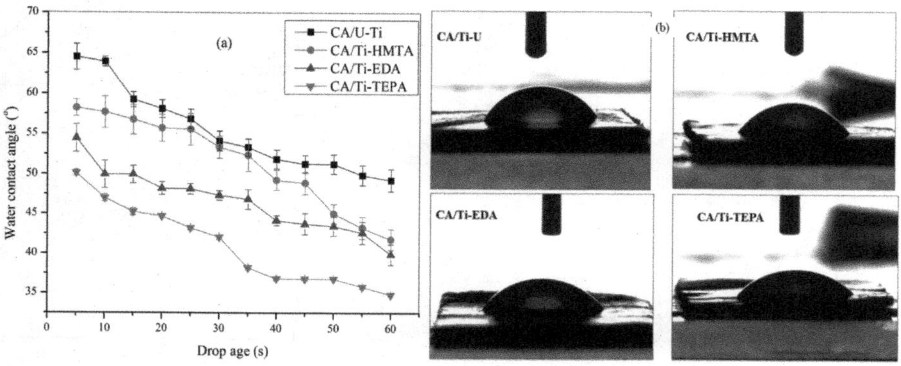

FIGURE 5.4
(a) Water contact angle values with different drop ages and (b) images of water droplets of the prepared membranes.

water contact angle values and higher water permeation rates are considered hydrophilic [43]. As seen from the figure, the water contact angle values of the amine modified membranes were reduced significantly especially for CA/Ti-EDA and CA/Ti-TEPA membranes. The average water contact angle of CA/U-Ti after 60 min was around 55.3°. The average contact angle values of the modified membranes were reduced to 51.3°, 46.6°, and 40.9° for CA/Ti-HMTA, CA/Ti-EDA, and CA/Ti-TEPA, respectively, showing an enhanced surface hydrophilicity. The significant improvement of the hydrophilicity of the CA/Ti-EDA and CA/Ti-TEPA membranes after impregnation of the amines are that the NH_2-groups are well-known polar groups having high affinity with water and hydrophilic character [44]. The relatively smallest contact angle value for CA/Ti-TEPA and improved surface hydrophilicity as compared with the other membranes were mainly attributed to the difference in structures and contents of NH_2-groups, where TEPA has two NH_2-groups and three NH-groups.

5.1.2.6 PWF Performance of Membranes

The pure water flux (PWF) results were measured to investigate the influence of the loading of the amine groups in the matrix of the prepared membranes. Therefore, the PWF results and the effect of the operating pressures are presented in Figure 5.5a and 5.5b, respectively. As can be seen from the figure, the PWF results of the prepared membranes decreased uniformly with decreasing the amount of functional amines within the matrix of the membranes. These results are attributed to the decreasing hydrophilicity (increasing water contact angle) of the membranes. As clearly presented in Table 5.4, the PWF results of CA/U-Ti, CA/Ti-EDA, CA/Ti-HMTA, and CA/Ti-TEPA at 250 kPa are 587.8, 649.2, 752.5, and 924.7 Lm^{-2} h^{-1}, respectively. However, the PWF results of all the prepared membranes are appropriate for their efficient use in filtration processes.

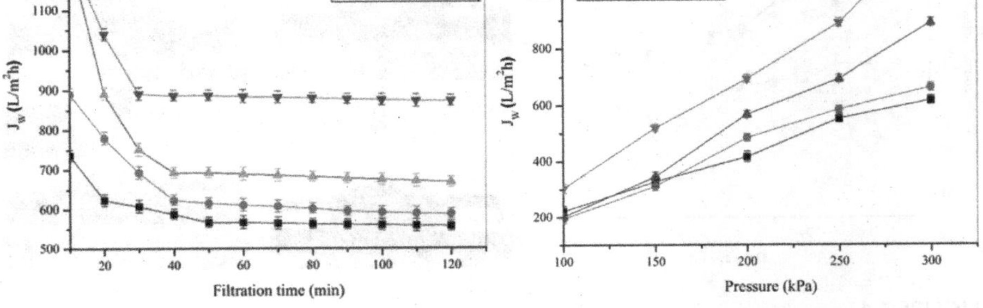

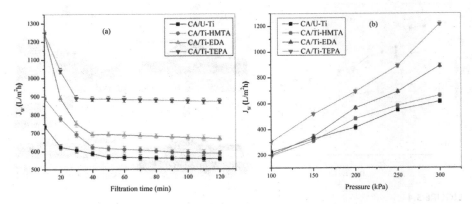

FIGURE 5.5
(a) PWF profile during compaction study (at 250 kPa) and (b) effect of pressure on PWF.

The compaction factor, EWC, porosity, average pore radius, thickness, and hydraulic resistance of all the prepared membranes in this study are presented in Table 5.4. Studying compaction factor (CF) is important to recognize the morphological structures (i.e., pore arrangements) of the membranes, especially the membrane sublayer formation. Membranes with high CF are likely to be highly compacted and may show the existence of some defective pores or macro-voids in the membrane sublayer arrangement. The CF of all the membranes shows similar results. The PWF results of all the membranes were increased almost uniformly with increases in an operating pressure from 100 to 300 kPa. Thus, the PWF for membrane results were observed to increase uniformly with varying the type of amine group added to the membrane matrix. Hence, the CA/Ti-TEPA membranes have displayed higher PWF than the other membranes due to its best hydrophilic nature. As seen from Table 5.4, no significant difference was detected in the results of hydraulic resistance, porosity, and the average pore radius of all the prepared membranes. However, a slight increase in average pore size of the membranes with varying the type of amine groups may be accredited to the number of amines involved within the matrix of the membranes. Therefore, the CA/Ti-TEPA membrane with more amine groups showed a slightly lower average pore diameter and porosity, which is due to the increase in the number of branched amines within the matrix of the membrane. The slight increase in EWC for CA/Ti-TEPA membrane when compared with other membranes is due to the decrease in water contact angle (higher hydrophilicity value). Moreover, the hydrophilicity and/or EWC were observed to increase uniformly with increasing the type of amine groups loaded to the membranes (i.e., HMTA, EDA, and TEPA). From the porosity measurement results, it is shown that all the membranes have shown satisfactory results in the range of 70 to 74%, which is accredited to lower concentration of the membrane forming polymer (10.5 wt.%) as well as due to the addition of PEG additives and TiO_2 NPs.

5.1.2.7 Cr (VI) Removal Efficiency of Membranes

The permeate flux results and efficiency of the prepared membranes for Cr (VI) ions removal was investigated using 10 ppm Cr (VI) solution. To further investigate the influence of pH on Cr (VI) ion removal in the ultrafiltration process using the prepared membranes, the solution pH was kept at 3.5 and 7.0. The flux and removal performance results of CA/U-Ti, CA/Ti-HMTA, CA/Ti-EDA, and CA/Ti-TEPA membranes are presented in Figure 5.6. As seen from the figure, all the membranes have displayed lower Cr (VI) flux values and higher Cr (VI) removal performances at pH 3.5 than at pH 7.0. The membranes that were modified with various amines (HMTA, EDA, and TEPA) have revealed considerably higher removal efficiency for Cr (VI) ions than the unmodified membrane (CA/U-Ti). The improved Cr (VI) ion removal efficiencies of CA/Ti-HMTA, CA/Ti-EDA, and CA/Ti-TEPA were

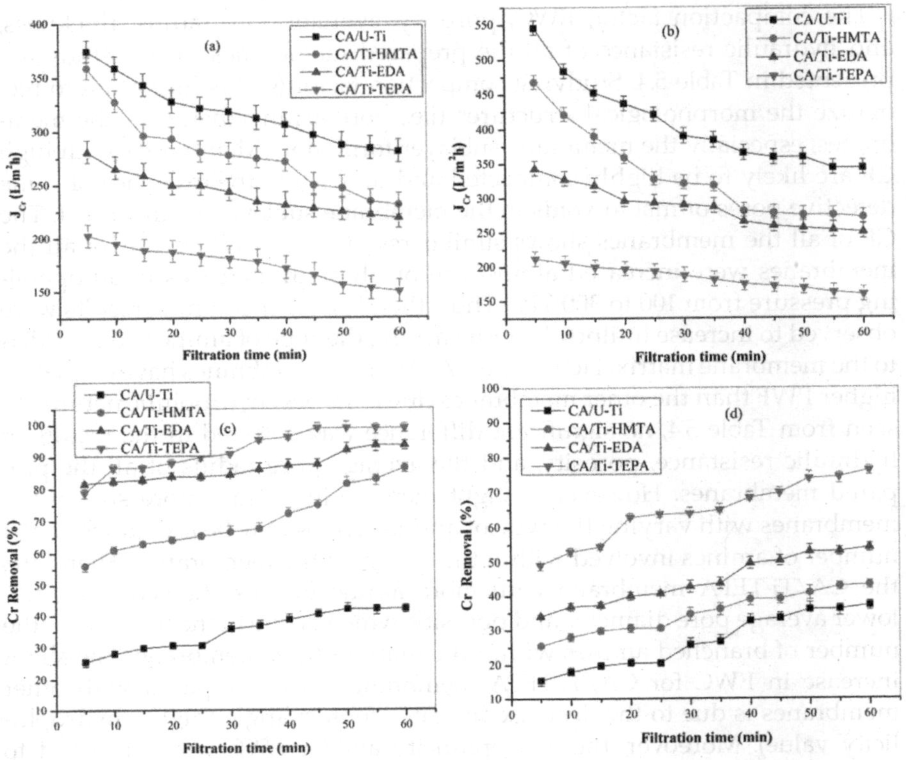

FIGURE 5.6
Permeate flux of Cr (VI) results of all the prepared membranes at (a) pH=3.5, (b) pH=7, and Cr (VI) removal % at (c) pH = 3.5, (d) pH = 7 (Cr (VI) concentration = 10 ppm, ΔP = 150 kPa, @ 200 rpm).

mainly due to the differences in content and structure of the amines within the matrix and pore walls of the membranes. This result approves that the introduction of amines within the matrix of the membranes indeed enhanced the properties of the membranes toward the removal of Cr (VI) ions. Therefore, the Cr (VI) ion removal for CA/Ti-EDA and CA/Ti-TEPA membranes showed better performances and was assumed because of the electrostatic interaction between the positively charged amines and Cr (VI) anions. Due to the de-protonation and protonation of the amine groups impregnated within the CA/Ti-EDA and CA/Ti-TEPA membrane matrixes, the amine-modified membranes could become positively charged at lower pH but not at higher pH values [40]. These amine groups impregnated within the membrane matrix can carry out protonation, and the degree of protonation depends on the pH of the solution. These results are confirmed by using zeta potential study as presented in Table 5.3. Consequently, the amine functional groups impregnated within the membrane matrix have governed the electrostatic interaction between the membrane surface and the chromium ion. Therefore, the protonated amine groups of the modified membranes led to an improved electrostatic attraction between protonated amine groups and the chromium ions at lower pH value [45]. It is also mentioned in the literature that although Cr (VI) occurs commonly as anions over the whole pH range, the proportions of the Cr (VI) species (i.e., $Cr_2O_7^{2-}$, H_2CrO_4, $HCrO_4^-$, and CrO_4^2) change with pH variation. Aroua et al. [46] have conducted speciation profile on Cr(VI), and they have found the following results. $Cr_2O_7^{2-}$ appears at pH 1.0 to 6.0 and disappears at pH 9; H_2CrO_4 (aq.) (a low concentration) appears only at pH 2.0 and 3.0; $HCrO_4^-$ occurs in low concentration at pH 1.0–8.0 and disappears at pH 10.0; and CrO_4^2 species starts to develop at pH 10 and above. Accordingly, no electrostatic attraction between Cr (VI) and the amine modified membranes was suggested, when Cr (VI) was in the form of H_2CrO_4 and CrO_4^2. The electrostatic attraction exists for both $HCrO_4^-$ and CrO_4^{2-}, though more positively charged amine groups are necessary to attract/adsorb CrO_4^{2-} because of its higher ion valence. Therefore, the variation in ionization degree of the amine modified membranes and the change in Cr (VI) species at various pH caused variation in removal efficiency of the membranes. Some of the Cr (VI) ion removal characteristic can also be described using the deposition phenomenon onto the membrane surface in this pH range, which could be accredited to structural interaction [47]. In addition to the electrostatic interaction phenomena, the Cr (VI) ion removal for the membranes was also considered as an indicator of considerable Cr (VI) ion removal due to its structural interaction. The Cr (VI) ion removal phenomenon for CA/U-Ti and CA/Ti-HMTA membranes can be suggested mainly due to a substantial Cr (VI) ion deposition inside and on the membrane surfaces due to structural interaction. Therefore, the surface depositions of Cr (VI) ion can provide an extra hindrance to solute transportation. However, lower Cr (VI) ion removal performances were attained in the case of CA/U-Ti and CA/Ti-HMTA membranes. These results are suggested due to the availability of lesser active sites where amine groups are not present; where HMTA has no any NH_2- or NH- group, only N is connected with three CH_2-groups. The amine functional groups loaded within the matrix of the Ti-EDA and Ti-TEPA membranes have lone pair of electrons from nitrogen and mainly act as an active site for the development of the complexion between chromium metal ions and amine groups. The order of the removal/adsorption capacity of the prepared membrane is CA/U-Ti < CA/Ti-HMTA < CA/Ti-EDA < CA/Ti-TEPA. The maximum removal values of the CA/U-Ti, CA/Ti-HMTA, CA/Ti-EDA, and CA/Ti-TEPA membranes at pH 3.5 are 42.2%, 86.4%, 93.6%, and 99.8%, respectively. On the other hand, the Cr (VI) removal values of the membranes at pH 7.0 are 37.1%, 43.3%, 47.2%, and 76.2%, respectively. At pH 7.0, the Cr (VI) ion removal performances of the modified membranes observed to be decreased during the ultrafiltration and these results were essentially due to the low Cr (VI) removal capability of these membranes at neutral pH. As seen from the zeta potential study in Table 5.3, all the modified membranes exhibited negative charges, and the removal of Cr (VI) in this case is due to the electrostatic repulsion between Cr (VI) ion and the membrane surfaces charges. Due to the lower negative charges for Ti-HMTA and low positive charges for unmodified TiO_2 NPs, lower removal efficiencies were attained. However, higher Cr (VI) removal/repulsion efficiency was observed for Ti-EDA and Ti-TEPA due to the availability of higher negative charges than the other membranes at the neutral pH. As observed from Figure 5.6c and 5.6d, Cr (VI) removal using the amine modified membranes is higher at low pH (3.5) and decreased with increasing the pH of the solution to 7.0. Therefore, at acidic pH the protonated amino groups of the amine impregnated Ti-EDA, and Ti-TEPA membranes have improved the removal/adsorption capabilities of the membranes to Cr (VI) ions. It was clearly observed that the removal of Cr (VI) metal ions by the membranes were in increasing order (i.e., HMTA < EDA < TEPA). This result can be accredited due to an increase in active sites for removal of Cr (VI) metal ions as the content and structure of the amine groups (NH_2- or NH-) within the membranes differs. The difference in removal of Cr (VI) metal ions behavior of the membranes was suggested due to the different electrostatic performance of the negatively charged Cr (VI) ions. Therefore, it was observed from this study that the pH of the solution was an important parameter that affects removal performances of the prepared membranes. As seen from Figure 5.6c and 5.6d, the removal of Cr (VI) using the prepared membranes was increased with filtration time and the maximum Cr (VI) ion removal was attained after 1 h filtration operation. Consequently, pH of 3.5 was considered as a suitable condition for the removal Cr (VI) ion for the modified membranes. Furthermore, comparing all the membranes, the maximum removal of Cr (VI) ion was achieved by CA/Ti-TEPA membrane. The Cr (VI) flux values for all the membranes at pH 3.5 and 7.0 are presented in Figure 5.6a and 5.6b. The uniform decrease in Cr (VI) fluxes with increasing time was suggested due to susceptible pore blocking of the membranes because of Cr (VI) ion adsorption and deposition on membrane surface and inner channels of the pores, where the effect of concentration polarization was reduced by rigorous stirring (200 rpm) on the surface of the membrane. On the other hand, flux variation with pH change was observed. At pH of 3.5 low Cr (VI) permeate flux was detected. This result was suggested due to the electrostatic attraction between the positively charged membranes and Cr (VI) ions. Moreover, lower fluxes for CA/Ti-EDA and CA/Ti-TEPA membranes were observed. The $-NH_2$ chains within CA/Ti-EDA and CA/Ti-TEPA membranes have turned into extended arrangement, which have formed a narrow effective pore size. At pH of 7.0, the $-NH_2$ chains have changed to collapsed conformation due to the de protonation of the amine groups, which caused large effective pore size and therefore resulted in high Cr (VI) ions flux results for the membranes.

Furthermore, as explained above, the flux results were observed to be dependent on the removal performances of the membranes. The schematic

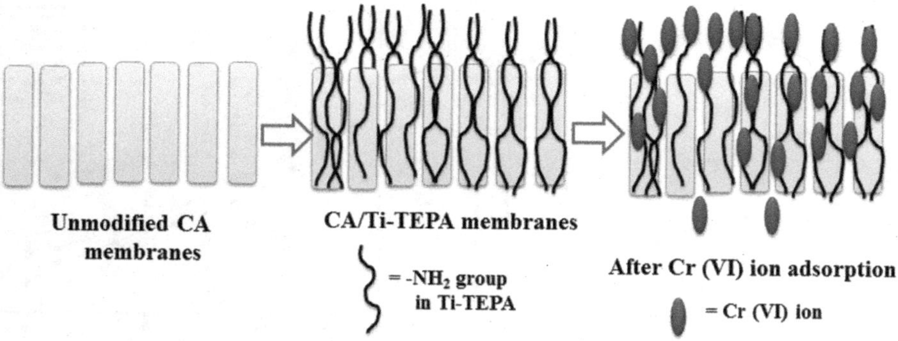

FIGURE 5.7
Schematic diagram for the impregnated amine groups and deposition of Cr (VI) ions in ultrafiltration process.

diagram for the impregnation of amine groups and deposition of Cr (VI) ions in ultrafiltration process is presented in Figure 5.7.

5.1.2.8 Effect of Cr (VI) Concentration

The removal efficiency of CA/Ti-TEPA membranes for Cr (VI) ions at different initial ion concentration was investigated. Therefore, the removal efficiency of the membranes was found to be concentration dependent. The results of permeate fluxes and removals of chromium ions for various concentrations (i.e., 10, 30, 50, 70, and 100 ppm) with time using CA/Ti-TEPA membrane were investigated and are presented in Figure 5.8. The removals of Cr (VI) ion using CA/Ti-TEPA membranes were observed to be dependent on the feed concentration of the Cr (VI) ion. As clearly depicted in Figure 5.8c and 5.8d, the removal of Cr (VI) ion of the membranes uniformly decreased with increasing the concentration of Cr (VI) ion (from 10 ppm to 100 ppm). This result can be associated with the availability of excess binding sites within the CA/Ti-TEPA membranes for removal of Cr (VI) ion. Therefore, the highest removal of Cr (VI) ion result was attained at low concentration (10 ppm) due to the availability of excess binding sites for Cr (VI) ions removal. The removal/adsorption results for 10 ppm, 30 ppm, 50 ppm, 70 ppm, and 100 ppm at pH 3.5 are 99.6%, 98.9%, 93.6%, 86.4%, and 82.2%, respectively. The flux results of CA/Ti-TEPA for both pH values, however, were observed to increase uniformly with increasing the concentration of Cr (VI) ions (Figure 5.8a and 5.8b). As already discussed in the previous sections, higher removal performances were attained at low pH (3.5) than at high pH (7.0) for all the concentrations. The removal performances of the membrane at pH 7.0 are presented in Figure 5.8d, and the maximum removal results are 80.7%, 76.2%, 69.1%, 58.3%, and 52.4% for 10 ppm, 30 ppm, 50 ppm, 70 ppm, and 100 ppm, respectively.

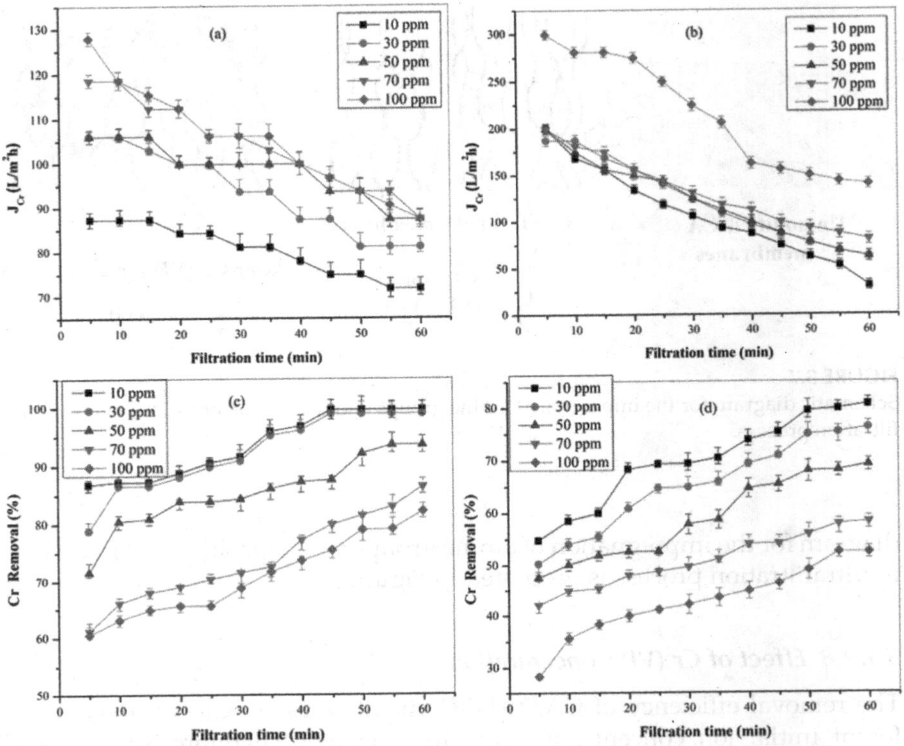

FIGURE 5.8
Effect of concentration on Cr (VI) ion permeate flux results of the CA/Ti-TEPA membrane at (a) pH = 3.5, (b) pH = 7, and Cr (VI) ion removal % at (c) pH = 3.5, (d) pH = 7 (ΔP = 150 kPa, @ 200 rpm).

5.1.2.9 Washing/Regeneration Performance Study

In this study, four washing/regeneration cycles were conducted to evaluate the fouling and removal performance of CA/Ti-TEPA membrane using KCl solution and 10 ppm Cr (VI) ion solution where the results are presented in Figure 5.9. As clearly shown in Figure 5.9a, the uniform decrease in KCl and chromium fluxes were realized to be highly noticeable, which are credited to the decreasing in porosity of the membrane due to an interior adsorption of Cr (VI) ion, which further could lead to pore blocking.

Therefore, the KCl flux results for 1st, 2nd, 3rd, and 4th cycles are 639.7, 546.4, 481.8, and 438.6 L m^{-2} h^{-1}, respectively, at pH of 7.0. The KCl flux results are 483.8, 390.4, 325.9, and 282.6 L m^{-2} h^{-1} at pH of 3.5. The Cr (VI) removal percentages for both pH values were observed to decrease with increasing the regeneration cycles. At the first cycle, the Cr (VI) removal percentage at pH 3.5 was observed to decrease slightly. The maximum Cr (VI) removal results for four washing/regeneration cycles are 99.6%, 99.5%, 98.6% and, 96.6%, respectively. As already explained previously, the pH dependence of

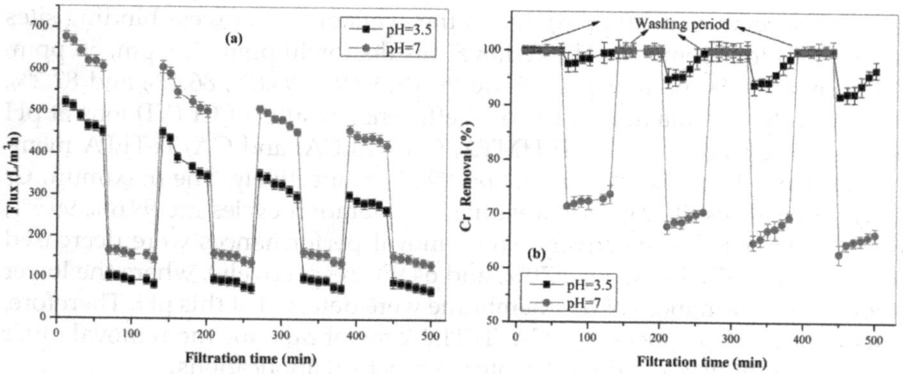

FIGURE 5.9
(a) Cr (VI) flux and (b) removal efficiency of CA/Ti-TEPA membrane during the filtration cycle (Cr (VI) concentration: 10 ppm, pressure: 150 kPa, washing operation: 2.5 g/L KCl @ 200 rpm).

ζ values for the TEPA modified membranes has influenced the chromium ion removal performances. As seen from the ζ study in Table 5.3, the TEPA modified TiO_2 NPs exhibited high positive ζ value at pH 3.5. Therefore, higher Cr (VI) removal percentage of CA/Ti-TEPA membrane at pH of 3.5 was attained due to the availability of more active sites for the adsorption of the negatively charged Cr (VI) ions. Thus, even after four washing/regeneration cycles, the Cr (VI) removal percentage remained at 96.6%. The membranes exhibited negative charges at pH of 7.0 and the removal of Cr (VI) in this case is due to the electrostatic repulsion between Cr (VI) ion and the membrane surfaces charges. The removal performances were decreased for the pH 7.0 (72.4%, 69.1%, 67.0%, and 64.9%, respectively), where the lower removal performances of the membrane were detected at this pH. These results were accredited mainly due to the low Cr (VI) removal capacity of CA/Ti-TEPA membrane at higher pH (7.0) due to the availability of lesser sites. From the above results, it can be suggested that the CA/Ti-TEPA membrane can be a good candidate for the removal of low concentration Cr (VI).

Thus, the authors strongly believe that this work will have a substantial contribution to the current state-of-the-art on the modification and enhancement of the properties of conventional cellulose acetate membranes for specific applications than the previous studies [6,27].

5.1.3 Summary

In the present study, Cr (VI) ion removal experiments were carried out using CA/U-Ti, CA/Ti-HMTA, CA/Ti-EDA, and CA/Ti-TEPA membranes. The Cr (VI) ion removal efficiency was influenced by the metal ion concentration, the pH of the solution, and type of amine. The optimized pH was 3.5 for the removal of Cr (VI) ions for all the membranes. CA/Ti-TEPA showed the best removal efficiency. The highest removal of Cr (VI) ions result was attained at low concentration (10 ppm) due to the availability of excess binding sites for Cr (VI) ions removal. The removal results for 10 ppm, 30 ppm, 50 ppm, 70 ppm, and 100 ppm at pH 3.5 are 99.6%, 98.9%, 93.6%, 86.4%, and 82.2%, respectively. The maximum removal efficiency results of Cr (VI) ions at pH of 3.5 using CA/U-Ti, CA/Ti-HMTA, CA/Ti-EDA, and CA/Ti-TEPA membranes were 47.2%, 86.4%, 93.6%, and 99.8%, respectively. The maximum Cr (VI) removal results for four washing/regeneration cycles are 99.6%, 99.5%, 98.6%, and 96.6%, respectively. The removal performances were decreased for the pH 7.0 (72.4%, 69.1%, 67.0%, and 64.9%, respectively), where the lower removal performances of the membrane were detected at this pH. Therefore, there are good prospects for CA/Ti-TEPA membrane for the removal of Cr (VI) ions from water and wastewater in practical applications.

5.2 Grafting Copolymerization of Poly Methyl Methacrylate (PMMA) onto Cellulose Acetate Modified with Amine Group for Removal of Humic Acid

5.2.1 Introduction

Recently, polysaccharides characterized by cellulose and its derivatives have fascinated researchers as environmental friendly materials. Cellulose acetate (CA) among the hydrophilic polymers is considered as the most important organic ester of cellulose due to its abundance and broad applicability for the synthesis of different products because of tough, biocompatible, hydrophilicity characteristics, and it is moderately less expensive [27,48,49]. The drawback of cellulose acetate membrane is that it is susceptible to thermal and chemical stabilities depending on the environments and conditions of application [50]. Therefore, modification of cellulose acetate using graft copolymerization process gives a substantial way to modify its chemical and physical properties [51]. Recently, the modifying of polymers has received great attention, and grafting is one of the promising approaches. Grafting co-polymerization is an attractive technique to introduce different functional groups to the backbone of a polymer [52]. Moreover, grafting is one of the chemical modification methods advanced to date and has appeared as a simple and versatile method to develop surface properties of polymers for various applications. For the graft copolymerization to be occurring, groups that can generate radicals should be introduced onto the backbone of the polymer first to activate the polymer surface. Most of the polymers having chemically inert properties can be activated via UV irradiation [53], plasma treatment [54], ion and electron beam [55–58], benzoic per oxide oxidization [59–64], and Ce^{4+} oxidization [65–68]. In most of the grafting processes, the free radicals are initiated on the backbone of polymer by numerous irradiations and free radical polymerizations of vinyl monomers or chemical initiators. Generally, cellulose and its derivatives are considered as excellent polymers for surface modification purpose because of their hydroxyl groups [69,70]. In this study, free radical polymerization was used for the synthesis of the graft copolymer of CA with PMMA using cerium sulfate (CS) as initiator. The grafting copolymer of the CA with the second polymer containing functional groups, such as PMMA, could radically increase the interaction of CA polymer with a wide range of materials. Thus, the grafting copolymerization of vinyl monomer onto the cellulose acetate using electron transfer method, a variant of living radical polymerization to achieve a controlled graft polymerization, was considered a new approach to potentially produce copolymers with well-defined structures. Furthermore, functionalization of the synthesized product was done using TEPA amine to improve the membrane selectivity. Moreover, humic acid (HA) is considered as the main component of dissolved organic matter (DOM), which can be found in both surface and groundwater. It is also known that HA contaminated water shows undesirable color and has been associated in bacterial development in wastewater [71]. The need for elimination of HA has currently become more important due to having haloacetic acids (HAA) and trihalomethane (THM) as byproducts. Carcinogenic organic compounds that are risky to human health are developed when water contaminated with HA is treated with disinfection processes such as chlorination [72,73]. Furthermore, HA has the ability to bind pesticides, heavy metals, and herbicides existing in wastewater, and complexed materials with high concentrations can be developed [74,75]. Therefore, the removal of HA is an important issue in water treatment. Currently membrane based HA removal process has been getting more attention due to their operational simplicity, low energy consumption, and environmental friendly nature. Nanofiltration and reverse osmosis techniques are able to remove HA effectively from contaminated water, but they require higher energy consumption when compared with ultrafiltration (UF) and microfiltration (MF) membrane processes. However, the conventional UF and MF membrane separation processes wouldn't remove the HA easily due to their larger pore sizes. Moreover, HA has been known as one of the main membrane fouling components during treatment wastewater by means of membrane separation process. Therefore, modification of the conventional UF/MF membranes is required to achieve selective removal of HA from aqueous solutions [71,75,76].

The main objective of this study was to prepare modified CA membranes using PMMA and amine group with improved selectivity for HA molecule removal. An effort was made to investigate the influence of polymerization time and temperature during the graft copolymerization process. Flat sheet membranes were prepared from the modified CA using phase inversion process.

5.2.2 Grafting and Amination Process

5.2.2.1 Synthesis of CA-g-PMMA and CA-g-PMMA_TEPA

Cellulose acetate powder was washed using methanol and deionized water, respectively, for three times and then was vacuum dried at 60 °C for 12 h before use. As seen in the schematic of the graft copolymerization process (Figure 5.10), the graft copolymerization of MMA onto CA took place using a three necked round bottom reactor vessel with magnetic stirrer inside the aqueous solution. For graft copolymerization, 8.0 g of CA was added into the reactor vessel comprising 100 mL of an aqueous solution of 0.006 mol/L of CS in 0.4 M sulfuric acid. After that, the reaction mixture was purged using N_2 gas approximately for 30 min prior to addition a 25 mL of 0.08 mol/L MMA monomer to start graft copolymerization. Consequently, the aqueous solution was stirred continuously at a constant rate (250 rpm) to avoid the influence of stirring on the degree of grafting copolymerization. A condenser was connected to ensure that any solvent or aqueous solution vapor cools and drips back down into the chamber containing the solid sample. Finally, the polymerization reaction was stopped with the addition of a 5.0 wt.% hydroquinone solution at a fixed time of reaction, and the mixture was discharged into a mixture of water and methanol (1:1), and the grafted CA powder was separated.

The polymerization reaction was conducted at different temperatures (i.e., 25, 45, 60, and 70 °C) in an oil-bath using heating magnetic stirrer. The pH of the polymerization reaction was adjusted at 2.0 before the reaction was

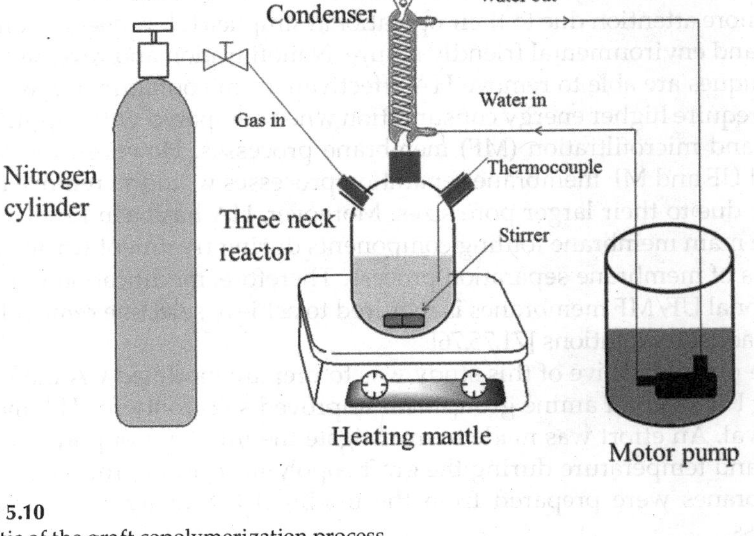

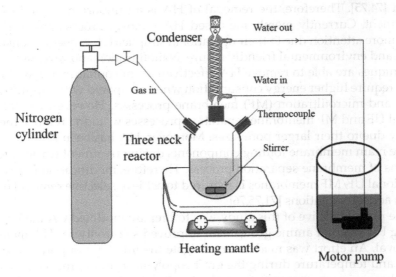

FIGURE 5.10
Schematic of the graft copolymerization process.

started. The duration of the copolymerization was varied from 2 h to 8 h to investigate its effect on the reaction. The contents of the reactor were stirred continuously throughout the reaction period to avoid the homopolymer precipitation on the substrate surfaces. The grafted samples were quickly taken out after the desired polymerization time from the reactor vessel and washed repetitively using methanol and deionized water. Table 5.5 presents the amount of CS and MMA, the variation of time and temperature on the polymerization reaction. To remove all the unreacted MMA monomer, its low MW PMMA homopolymer, and the impurities precipitated along with grafted CA powder, the PMMA grafted cellulose acetate (CA-g-PMMA) samples were Soxhlet extracted using methanol/acetone (1:1) for 24 h and repeatedly washed using deionized water. Consequently, the grafted CA powders were dried using vacuum drier at 60 °C.

The percentage of the graft was calculated by the percent increase in weight as follows [13]:

$$Gp(\%) = \left[\left(\frac{W_g - W_{ug}}{W_{ug}}\right) \times 100\right] \tag{5.1}$$

where W_g (g) and W_{ug} (g) are the weight of the grafted and the un-grafted cellulose acetate samples, respectively.

During the amination process, 6 g of CA-g-PMMA sample was dispersed in 250 mL DMF/water aqueous solution, and finally, 10%vol/vol. TEPA in DI (20 mL TEPA in 200 mL of DI) was introduced through the ring opening reaction of the methyl groups of the copolymer by reacting with 99% TEPA in DMF/water aqueous solution at 70 °C for approximately 8 h. Thus, the resultant product (i.e., CA-g-PMMA_TEPA) was filtered and washed using methanol and deionized water following drying in vacuum at 50 °C.

TABLE 5.5

The Variation of Time and Temperature on the Polymerization Reaction

Reaction	CS, mmol/L	MMA, mmol/L	Time, h	Temperature, °C
G-1	8	6	2	25
G-2	8	6	2	60
G-3	8	6	6	45
G-4	8	6	6	70
G-5	8	6	8	25
G-6	8	6	8	60

The weight gain percentage (W_{gp}) of the aminated samples were calculated using the following equation.

$$Wgp(\%) = \left[\left(\frac{W_f - W_i}{W_i}\right) \times 100\right] \tag{5.2}$$

where W_f (g) and W_i (g) are the weight of the aminated sample and the initial weight of the grafted CA, respectively.

5.2.2.2 Preparation of Un-g-CA, CA-g-PMMA, and CA-g-PMMA_TEPA Membranes

Phase inversion technique by immersion precipitation was used for fabricating asymmetric ultrafiltration membrane from modified and unmodified cellulose acetate polymer. Initially, uniform solutions of the modified and unmodified CA in acetone/DMAc (70/30; v/v) were prepared under continuous magnetic stirring at room temperature (28 ± 2 °C). After getting a uniform solution, a fixed amount of PEG additive (3 wt.%) was added and stirred continuously at least for 24 h until the solution was fully dissolved and a homogeneous casting solution was obtained. Finally, 2 wt.% TiO_2 nanoparticles were added separately and well dispersed for another one day to form a homogeneous composite solution. PEG additive and TiO_2 NPs were added to enhance the porosity, hydrophilicity, thermal/chemical stability, and anti-fouling properties of the membrane as already explained in previous studies elsewhere [14,15]. Therefore, the membranes with different composition were designated as CA, CA-g-PMMA, and CA-g-PMMA-TEPA. Table 5.6 shows the solution casting compositions of all the prepared membranes.

5.2.2.3 Graft Polymerization

The free radical locations were produced on a cellulose acetate backbone through direct oxidation by using Ce^{4+} ion. The redox potentials of the Ce^{4+} metal ion are the key factors in defining the grafting effectiveness. Metal

TABLE 5.6

Solution Casting Compositions of All the Prepared Membranes

Membrane	CA wt.%	TiO_2 wt.%	PEG wt.%	AC:DMAc (70:30; v/v)
Un-g-CA	10.5	2	3	84.5
CA-g-PMMA	10.5	2	3	84.5
CA-g-PMMA_TEPA	10.5	2	3	84.5

ions having lower oxidation potentials are chosen for enhanced grafting effectiveness. Therefore, the suggested tool for such practice has attributed to the intermediate development of a polymer chelate and metal ion complexes, where ceric ion is well-known, to develop complexes with hydroxyl groups on a CA backbone and can be dissociated through one electron transfer to provide free radicals.

$$Ce^{4+} + R_{Pol.}OH \rightarrow Complex \rightarrow R_{Pol.}O^{\bullet} + Ce^{3+} + H^{+} \tag{5.3}$$

$$R_{Pol.}O^{\bullet} + M \rightarrow R_{Pol.}OM \rightarrow R_{Pol.}OMM^{\bullet} \tag{5.4}$$

In this study, the tentative reaction mechanism during the grafting polymerization using sulfuric acid solution with the presence of ceric sulfate complex is well-known and the reaction mechanisms are suggested as follows [77]:

$$Ce_4^{+} + HSO_4^{-} \Leftrightarrow CeSO_4^{2+} + H^{+} \tag{5.5}$$

$$CeSO_4^{2+} + HSO_4^{-} \Leftrightarrow Ce\,(SO_4)_2 + H^{+} \tag{5.6}$$

$$Ce(SO_4)_2 + HSO_4^{-} \Leftrightarrow Ce(SO_4)_3^{2-} + H^{+} \tag{5.7}$$

$$HSO_4^{-} \Leftrightarrow SO_4^{2-} + H^{+} \tag{5.8}$$

During the polymerization reaction, only one species or all of these CS species are maybe reactive and have their equilibrium constants for the formation of complexes, dissociation, and termination [78,79]. The concentrations of the above species are dependent on the hydrogen and sulfate ion concentration. Therefore, the resultant covalent bond between the CA and the acetate vinyl monomer was suggested due to the formation of free radicals on the vinyl monomer and initiated hydroxyl groups of cellulose acetate. The competitive reaction for the graft of the homopolymerization can be occurred due to the active species present in the aqueous medium (i.e., –H and –OH). To investigate the optimum grafting of MMA onto cellulose acetate initiated by the redox reaction of cellulose acetate and Ce (IV) ion, factors like pH of polymerization medium, CS, and MMA concentrations were kept constant. The duration of polymerization and temperature of grafting were varied, and the effects were investigated. Therefore, initiation of graft polymerization of MMA onto cellulose acetate powder using Ce(IV) ion in sulfuric acid medium proceeds due to oxidizing of the substrate (i.e., cellulose acetate).

Consequently, the reaction complexes break down to provide CA macro radicals, H^+ and Ce(III) ion. These radicals may mainly result from sulfur–hydrogen rather than oxygen–hydrogen bonds of the CA because of the sulfur–hydrogen form weaker covalent bonds when compared with oxygen–hydrogen. Consequently, the free radical grafting copolymerization of MMA onto the cellulose acetate powder is expected to be attained. Development of CA macro radicals includes an electron transferring method from the acetate group to the Ce(IV) ions. Thus, $Ce(SO_4)_2$ is supposed to be the reactive species of CS where later has changed to the highly stable and colorless Ce(III) ion. Moreover, it is evidently observed that Ce (IV) sulfate initiator with yellow color disappears when the grafting reaction time continues.

During the modification of cellulose acetate powder, the grafting polymerization of MMA onto CA powder was proposed using CS as an initiator. The polymerization was attained in a heterogeneous system (i.e., monomer–polymer), and the commercial CA powder was first purified to achieve a better diffusion of the monomer in the peripheral layer of CA during the grafting process. For initiation of vinyl polymerizations, CS has been commonly used as a source of radicals. The general processes assumed for the synthesizing of CA-g-PMMA_TEPA are presented in Figure 5.11. It is suggested that the free radical fragments may abstract

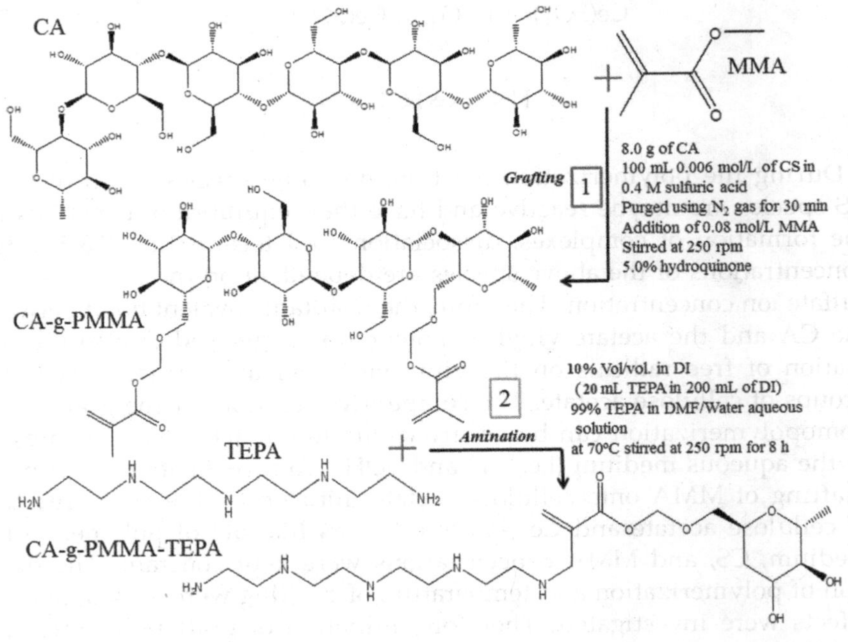

FIGURE 5.11
Scheme for preparation of PMMA, CA-g-PMMA, and CA-g-PMMA_TEPA.

hydroxyl groups of CA from the backbone to generate macro radicals, introducing graft polymerization. The other suggested reaction mechanism was that the radicals could react with the monomer to develop an increasing monomeric radical. Therefore, the increasing monomeric radical was able to transfer its radical character to the CA chain and the MMA graft copolymerization as well as homopolymerization was induced. The effective chemical grafting of MMA onto CA chains was investigated using ^{1}H NMR and the spectra of CA, MMA, and CA-g-PMMA and presented in Figure 5.12.

As clearly seen from the figures, a significant shift and splitting of the CA peaks were detected after grafting of the PMMA, and a successful graft polymerization was achieved. Subsequently, un-grafted PMMA was completely detached from the CA surface by Soxhlet extraction followed by washing using methanol and deionized water. NMR results confirmed that the CS was able to initiate graft polymerization of MMA on CA properly. Obviously, the chemical integrity of the CA and PMMA was observed after graft polymerization, as the peaks corresponding to both CA and MMA were observed in the ^{1}H NMR spectrum of the CA-g- PMMA, though the level of integrity was observed to be dependent on polymerization time and temperature.

5.2.2.4 Effect of Polymerization Time and Temperature

To examine the effects of temperature on grafting polymerization of MMA onto cellulose acetate was conducted at 25, 45, 60, and 70 °C, where the concentrations of MMA and CS were kept at 6 mmole/L and 8 mmole/L, respectively, and the pH of the polymerization medium was kept at 2.0. The reaction time was varied as 2, 6, and 8 h. The ^{1}H NMR results of CA, MMA, CA-PMMA (2/25), CA-PMMA (2/60), CA-PMMA (6/45), CA-PMMA (6/70), CA-PMMA (8/25), and CA-PMMA (8/60) are shown in Figure 5.12 and the results (Table 5.7) are ^{1}H NMR (600 MHz, DMSO) δ 3.60–3.45 (m, 35H), 2.54 (t, J = 33.2 Hz, 2H), 2.18–1.63 (m, 9H); ^{1}H NMR (600 MHz, DMSO) δ 6.05–5.97 (m, 3H), 5.60 (dd, J = 4.1, 3.5 Hz, 3H), 3.67 (s, 10H), 3.32–3.27 (m, 1H), 1.86 (d, J = 1.2 Hz, 10H); ^{1}H NMR (600 MHz, DMSO) δ 4.05–3.72 (m, 121H), 2.56–2.43 (m, 1H), 2.10–1.76 (m, 10H); ^{1}H NMR (600 MHz, DMSO) δ 4.04–3.55 (m, 127H), 2.58–2.41 (m, 1H), 2.10–1.78 (m, 14H); ^{1}H NMR (600 MHz, DMSO) δ 3.78–3.38 (m, 118H), 2.55–2.45 (m, 2H), 2.12–1.75 (m, 24H); ^{1}H NMR (600 MHz, DMSO) δ 3.89–3.44 (m, 84H), 2.50 (s, 1H), 2.14–1.73 (m, 12H); ^{1}H NMR (600 MHz, DMSO) δ 3.63 (d, J = 9.6 Hz, 64H), 2.50 (s, 1H), 1.94 (dd, J = 66.6, 49.2 Hz, 10H); and ^{1}H NMR (600 MHz, DMSO) δ 3.96–3.48 (m, 150H), 2.50 (s, 2H), 2.13–1.76 (m, 23H), respectively. From these results, the CA signals at 2.54 and 2.18–1.63 ppm were significantly improved after the grafting of the PMMA. However, the chemical shifts and the peak intensities were varied with the time and temperature. In this study, the combined effects of the time and temperature were observed, and the appearances of the peaks

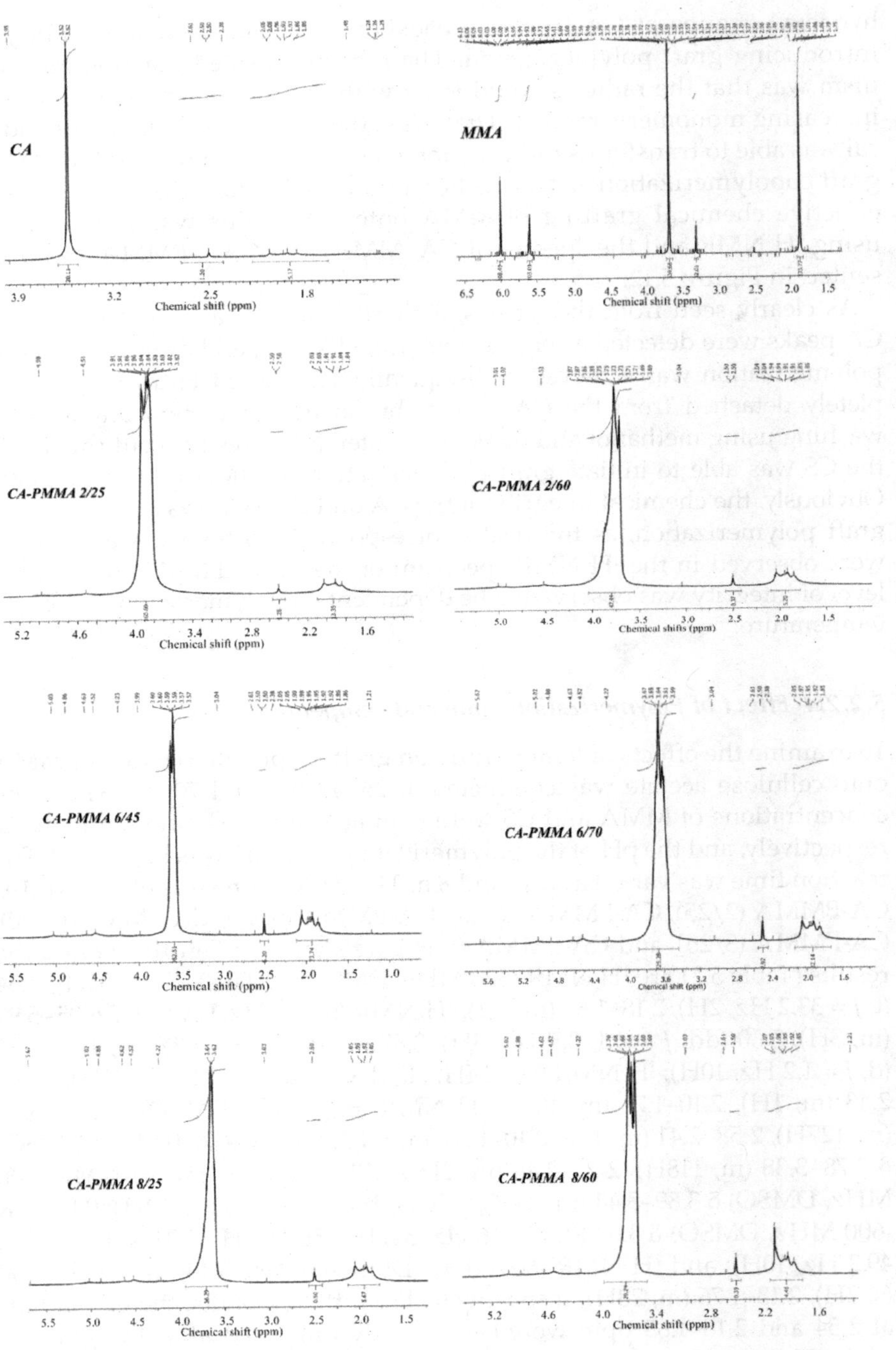

FIGURE 5.12
^{1}H NMR spectra of CA, MMA, and CA-g-PMMA dissolved in DMSO-d6 at different polymerization time and temperature.

TABLE 5.7
^{1}H NMR Results of Prepared Samples

Sample	Shift	H's	Integral
CA	1.95	9	8.90
	2.54	2	2.09
	3.52	35	35.01
MMA	3.67	10	10.00
	1.86	10	9.83
	5.60	3	3.03
	6.02	3	3.19
CA-PMMA (2/25)	1.93	10	10.06
	2.50	1	0.96
	3.85	121	120.98
CA-PMMA (2/60)	1.94	14	13.93
	2.50	1	0.98
	3.77	127	127.09
CA-PMMA (6/45)	1.95	24	24.00
	2.50	2	2.02
	3.59	118	117.98
CA-PMMA (6/70)	1.95	12	11.98
	2.50	1	1.09
	3.63	84	83.94
CA-PMMA (8/25)	1.94	10	10.02
	3.63	64	64.10
	2.50	1	0.88
CA-PMMA (8/60)	2.50	2	1.99
	1.97	23	22.91
	3.64	150	150.10

having higher intensities were detected for CA-PMMA 6/45, CA-PMMA 6/70, and CA-PMMA 8/60. It is obvious that the grafting is time and temperature dependent, where the combination of time and temperature at 8 h and 60 °C (i.e., CA-PMMA, 8/60) seems to be the optimum temperature and time for the grafting of MMA onto CA by the CA–Ce (IV) sulfate redox system. Moreover, the peak for CA at around 3.60–3.45 ppm was split after the graft polymerization reaction of PMMA onto the CA backbone. The chemical shift and splitting of the proton at 3.6 ppm are accredited to protons of $-OCH_3$ in PMMA, which is an indication of the successful attachment of the MMA monomers on the CA backbone [80]. Raising the temperature while keeping the time constant enhances the effectiveness of the redox instigation system and in turn improves the degree of the grafting reaction. If we observe the graphs for CA-PMMA, 8/25 and CA-PMMA, 8/60, new peaks with higher intensities were detected at 60 °C than 25 °C. Nonetheless, the contribution of the higher polymerization time, the higher temperature, has played a great role for the enhancement of the grafting process. The polymerization temperature is one of the key factors that govern the kinetics of grafting copolymerization. The grafting efficiency improves with higher temperature until the optimum bound is achieved. The reason for this result can be due to increasing of the monomeric diffusional process in the polymeric backbone with rising temperature and then simplifying the grafting process. During the grafting of MMA on CA, the grafting efficiency increased considerably as the temperature was raised due to a larger swelling of the CA backbone and a resultant improved degree of the monomers' diffusion within the locality of CA backbone. The results can be explained due to an increased thermal decomposition rate of CS and its capacity in generating free radicals on the polymer backbone with increasing temperature. Therefore, this incident can cause increased CA polymer macro radicals concentration, and consequently improved the graft polymerization. The effect of variation of reaction time was also investigated during the grafting polymerization. A better grafting efficiency was observed when the reaction time was increased. Regardless of the effect of temperature, an increase in the reaction time increases the rate of grafting. As clearly observed from the graphs presented in Figure 5.12, CA-PMMA, 2/60 and CA-PMMA, 8/60, peaks with higher intensities were detected for the polymerization time of 8 h. This result is explained as the polymerization time increases the CA backbone, MMA monomer, and CS initiator interactions increase. Moreover, compared with other efforts to chemical modification of CA, the functionalization method would keep the main-chain structures of CA integral without altering much its unique physical properties. Moreover, the graft percentage (*Gp*) of CA-g-PMMA and the weight gain percentage (W_{gp}) of CA-g-PMMA_TEPA were calculated using Equations 5.1 and 5.2. The *Gp* results for CA-PMMA (2/25), CA-PMMA (2/60), CA-PMMA (8/25), CA-PMMA (6/45), CA-PMMA (6/70), and CA-PMMA (8/60) were found to be 19.4, 20.2, 23.1, 25.8, 28.3, and 36.3%, respectively. It was observed that the effects of time and temperature were indicated clearly from the *Gp* results of the samples and the highest grafting percentage was attained for CA-PMMA (8/60). Accordingly, CA-PMMA (8/60) sample was selected for the amination process. Furthermore, the weight gain percentage (W_{gp}) of CA-g-PMMA-TEPA was calculated as 33.3%. The results indicate that the grafting and amination process were successful in this study.

5.2.2.5 ATR-FTIR Spectroscopy and Zeta Potential Analysis

To examine the chemical structure of the synthesized samples, the ATR-FTIR analysis was completed. Therefore, the functional groups of CA, MMA, CA-g-PMMA, and CA-g-PMMA_TEPA were characterized and presented in Figure 5.13a. In the spectra of pure CA, the broadband detected at 3419 cm^{-1} corresponded to –OH stretching because of the strong hydrogen

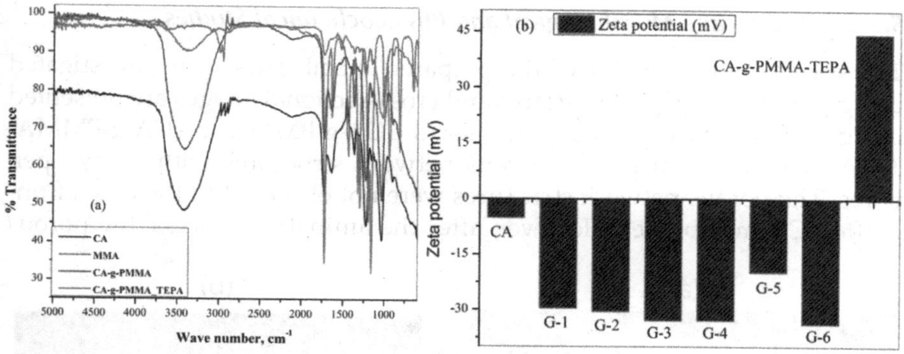

FIGURE 5.13
(a) FTIR spectra and (b) zeta potential (ζ) results of the unmodified and modified CA.

bond of intermolecular and intramolecular kinds. The characteristic band of 2960 cm^{-1} attributed to the C–H stretching [81]. In the IR spectrum of CA-g-PMMA_TEPA, the stretching vibrations of C–H bonds are present at 2890 cm^{-1}. The peaks of N–H bonds and C–N bonds are present at 1720 and 1250 cm^{-1}, respectively. Thus, the results show that TEPA was effectively introduced onto the surface of the CA-g-PMMA. The spectrum for all the samples, except MMA, show considerable peaks attributed to O–H group at 3419 cm^{-1}, 3423 cm^{-1}, and 3396 cm^{-1} for CA, CA-g-PMMA, and CA-g-PMMA_TEPA, respectively. This result is due to the interaction of CA with MMA and TEPA successfully. Comparing the FTIR spectrum of the CA-g-PMMA and CA-g-PMMA_TEPA to that of the unmodified CA, the broadening peak of hydroxyl groups in the glucose rings turn smaller after the reaction of the CA with MMA and PMMA_TEPA. This result indicates that some of the hydroxyl groups were replaced during the synthesizing process. Furthermore, the zeta potential (ζ) values of the modified and unmodified CA at pH of 7.0 are presented in Figure 5.13b. Zeta potential analysis is one of the most important methods used to obtain information on the electrical charge properties [38]. Therefore, the ζ or electrical charge properties of the powder or film polymers impact their performances toward specific applications. As seen from the figure, the un-grafted CA powder displayed low negative ζ value (i.e., –5.1 mV). The ζ values of the grafted powders at different temperature and time displayed higher negative values than the un-grafted powder. The high negative ζ values for the PMMA grafted samples is due to an increase in the extent of –COOH groups on the surface of CA [71]. A variation in zeta potential results as a function of time and temperature was confirmed by this study. High positive ζ value (+44.4 mV) was detected for the CA-g-PMMA modified with TEPA. This positive ζ result was attained due to the introduction of the amine group (TEPA) through the ring opening reaction of the methyl groups of the copolymer in the polymer amination process [82,83].

5.2.2.6 Membrane Morphological and Physicochemical Studies

The surface morphologies of the prepared membranes were investigated using FESEM, and the top surfaces and cross-sectional images are presented in Figure 5.14. As seen from Figure 5.14a, the CA-g-PMMA and CA-g-PMMA-TEPA membranes displayed porous network structures with many open pores. The porous network structures were not observed in the case of un-grafted CA membranes. However, after the amination process, the porous

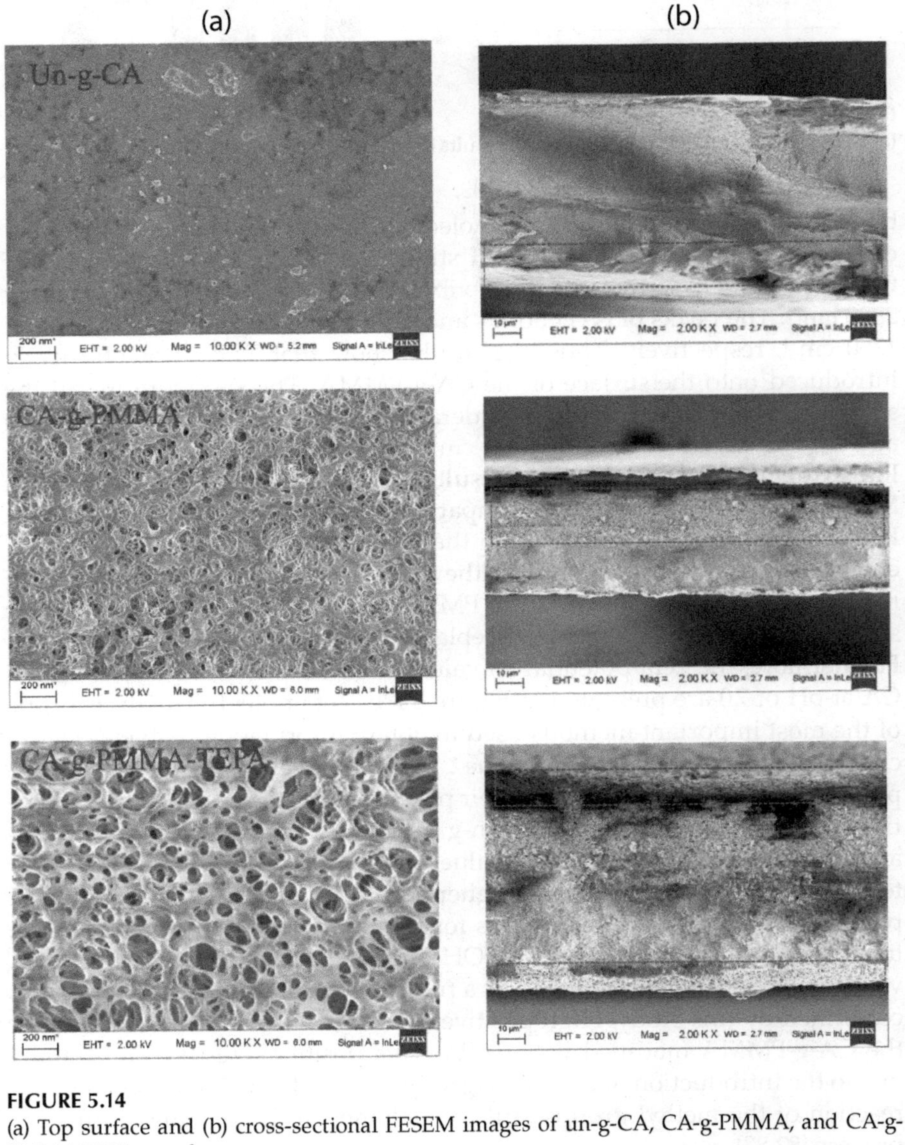

FIGURE 5.14
(a) Top surface and (b) cross-sectional FESEM images of un-g-CA, CA-g-PMMA, and CA-g-PMMA-TEPA membranes.

network structures on the membrane were observed to be smooth without showing major microscopic defects. The membranes were having an asymmetric structure consisting of a thin dense top layer and porous sublayer and without the development of macro-voids were observed. The formation of macro-voids occurs under rapid precipitation condition, and the precipitations are faster at high coagulation temperature [84]. In the present study, the coagulation bath temperature and evaporation time (i.e., before immersion) were fixed to 28 ± 2 °C and 30 s, respectively, during the preparation of all the membranes. As explained in detailed in the previous study, an additional important factor that can influence the development or suppression of macro-voids is the de-mixing that happened during the phase separation process, where the instantaneous de-mixing and the delayed de-mixing depend on the mutual affinity of the solvent and non-solvent in the ternary system [37]. The suppression or the formation of macro-voids may be due to adding of the PEG and TiO_2 additives to the ternary (polymer/solvent/non-solvent) system. The presence of theses additives can increase the concentration/viscosity of the casting solution, which may diminish the diffusional exchange rate of the solvent and non-solvent during the membrane formation process. Consequently, this may hinder the instantaneous liquid-liquid de-mixing process that suppresses the formation of macro-voids [85]. As observed from Figure 5.14b, when the modified CA solution was cast, finger-like structure-free and porous cross-section were attained, unlike the unmodified CA, where some finger-like structures with less porous cross-section were detected. Therefore, the grafting of PMMA onto the CA surface made the membrane structure more porous. Since the top layer is highly porous to form the resistance layer that can limit nuclei development in the sublayer, the porous membrane with no finger-like structures can be developed under the less stable system [82]. Therefore, the porous top layer is believed to suppress the formation of finger-like structures in the sublayer and result in porous, finger-like structure-free cross-section. Furthermore, a highly porous membrane (CA-g-PMMA-TEPA) was obtained with the addition of TEPA to the grafted polymer due to the development of quaternary aminated polymer. Unlike the CA-g-PMMA membrane, a porous structure was observed below the top layer in the cross-section in the case of CA-g-PMMA-TEPA membrane. Therefore, with the addition of TEPA to the grafted polymer, the top layer can lead the liquid–liquid phase separation instantly, causing the development of many pores in the top layer of the membrane. Moreover, the top layer of the membrane was porous enough to lead the large amounts of the non-solvent into the sublayer to start a number of nuclei creations. The open pore structures developed in the membranes are formed by nucleation and development of the polymer-lean phase in the metastable area between the bi-nodal and the spinodal curve [86,87]. Nevertheless, the top layer structure of the ultrafiltration membrane frequently doesn't reveal an open pore structures and not a completely uniform gel-layer, which is stated as a nodular structure. The development of nodular structures can't be described by nucleation of the polymer lean part. It is also not common that nucleation of the polymer-rich part happens because this only occurs at initially low polymer concentrations, below the critical point. A possible explanation for the formation of a nodular structure on the top surface of the membranes could be due to the spinodal de-mixing because the diffusion process throughout the development of the top layer is faster for the homogenous system to develop greatly unstable and crosses the spinodal curvature [88]. The top surface with improved interconnected pores structures is observed more prominently in the case of CA-g-PMMA-TEPA membrane, which is achieved due to the spinodal decomposition.

Studying of compaction factor (CF) of the prepared membranes is important to distinguish the morphological structures (i.e., pore arrangements) of the membranes, particularly the membrane sublayer configuration. Therefore, membranes having high CF indicate that these are highly compacted and indicate the existence of defective pore structures in the membrane sublayer. The compaction factor and hydraulic characteristics of all prepared membranes are presented in Table 5.8. It was observed that for both CA-g-PMMA and CA-g-PMMA-TEPA membranes, the CF decreases to 1.4 ± 0.5 and 1.2 ± 0.7, respectively, from 2.2 ± 0.2 (i.e., un-g-CA). These results were attained due to the modification process; membranes appear to be free of finger-like structures. Furthermore, the CA-g-PMMA-TEPA membrane exhibited the smallest compaction factor result (i.e., 1.2 ± 0.7) because of less microscopic defects as compared with CA-g-PMMA membrane.

The hydraulic resistance (R_m), EWC, average pore radius, porosity, and thickness of all the prepared membranes are reported in Table 5.8. The average pore size and pore size distributions were determined using water filtration velocity method and using Image J software from FESEM images, and the results are presented in Figure 5.15 and Table 5.5. The measured average pore size results for un-g-CA, CA-g-PMMA, and CA-g-PMMA-TEPA membranes were 42.4 nm, 56.7 nm, and 62.8 nm, respectively.

As seen from these results the hydraulic resistance of un-g-CA (i.e., $0.95 \times 10^{-10} m^{-1}$) was greater than CA-g-PMMA (i.e., $0.80 \times 10^{-10} m^{-1}$) and CA-g-PMMA-TEPA (i.e., $0.79 \times 10^{-10} m^{-1}$) membranes. This higher resistance of un-g-CA is because of its less porous nature than the other membranes, which can resist the water flux better than the other membranes. From the porosity

TABLE 5.8

Compaction and Hydraulic Characteristics of the Prepared Membranes at 250 kPa

Membrane	*CF*	R_m ($\times 10^{-10} m^{-1}$)	J_w (L/m^2h)	*EWC* (%)	ε (%)	r_m (nm)	Thickness (μm)
Un-g-CA	2.2 ± 0.2	0.95	292.7	81.2 ± 4.1	74.9 ± 3.5	42.4 ± 4.4	82.7
CA-g-PMMA	1.4 ± 0.5	0.80	312.2	85.4 ± 3.2	77.2 ± 2.3	56.7 ± 3.2	71.8
CA-g-PMMA_TEPA	1.2 ± 0.7	0.79	334.5	87.7 ± 2.7	82.4 ± 1.5	62.8 ± 2.3	73.5

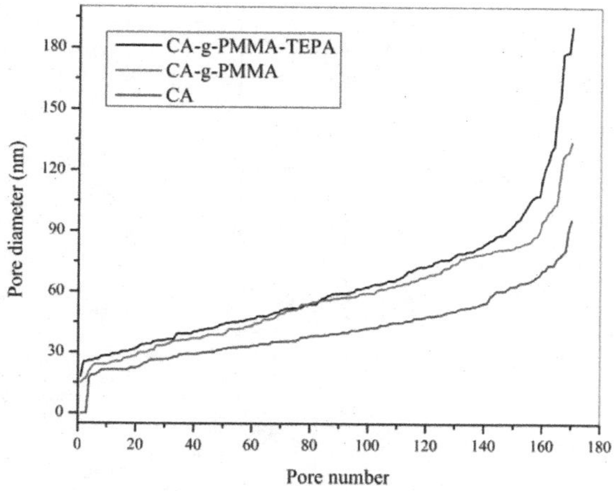

FIGURE 5.15
Pore size distribution of CA, CA-g-PMMA, and CA-g-PMMA-TEPA membranes.

measurement results it is shown that all the membranes have shown satisfactory results (i.e., 74.9%, 77.2%, and 82.4% for un-g-CA, CA-g-PMMA, and CA-g-PMMA-TEPA, respectively), which is accredited to the lower concentration of the membrane forming polymer (10.5 wt.%) as well as due to the addition of PEG additives and TiO_2 NPs together with the type of solvent used and type of de-mixing occurred [37].

5.2.2.7 Ultrafiltration of PW and HA Solutions

The pure water flux (PWF) studies were performed to investigate the effect of the grafting and amination of the prepared membranes, and the PWF results and the effect of the operating pressures are presented in Figure 5.16a and 5.16b, respectively.

As clearly presented in Table 5.8, the PWF results of un-g-CA, CA-g-PMMA, and CA-g-PMMA-TEPA membranes at 250 kPa are 292.7, 312.2, and 334.5 Lm^{-2} h^{-1}, respectively. The pure water flux results of the membranes are consistent with their pore size and porosity results. Nevertheless, the PWF results of all the prepared membranes are suitable for their efficient use in filtration processes. The PWF results at different operating pressures are presented in Figure 5.16b. The results of PWF for all the membranes were increased almost uniformly with increasing in an operating pressure from 100 to 300 kPa. The permeate flux results and HA removal efficiencies of the prepared membranes were investigated using 10 ppm HA solution at pH of 7.0, and the results are presented in Figure 5.16c and 5.16d, respectively. The uniform decrease in HA fluxes with increasing time was suggested due to susceptible pore blocking of the membranes because of HA molecules adsorption and deposition on the membrane surface and inner channels

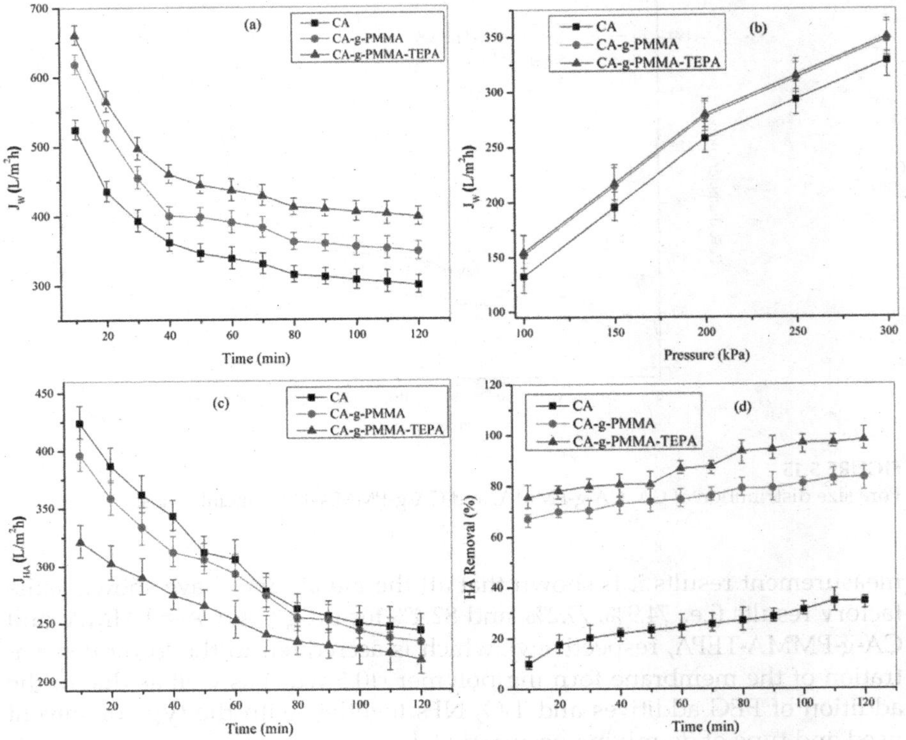

FIGURE 5.16
(a, b) Pure water flux, (c) HA flux, and (d) removal % results of un-g-CA, CA-g-PMMA, and CA-g-PMMA-TEPA membranes at pH = 7, HA concentration = 10 ppm, ΔP = 150 kPa, @ 200 rpm.

of the pore, where the effect of concentration polarization was reduced by rigorous stirring (200 rpm) on the surface of the membrane. As seen from the figure, the un-g-CA, CA-g-PMMA, and CA-g-PMMA-TEPA membranes exhibited lower HA flux values as compared with PWF results.

Furthermore, the grafted and aminated membranes (CA-g-PMMA and CA-g-PMMA-TEPA, respectively) have revealed considerably higher removal efficiency for HA than the unmodified membrane (un-g-CA). The enhanced HA removal efficiencies of the modified membranes were mainly due to the introduction of PMMA and TEPA through grafting and amination processes, respectively. These results confirmed that the introduction of PMMA and TEPA on the surface and within the matrix of CA indeed improve d the characteristics of the membranes toward the removal of HA. Therefore, the HA removal was suggested due to the electrostatic interaction between the membranes and HA molecules. As already explained in the previous section in Figure 5.13b, the results of zeta potential study showed that the surface charge value of un-g-CA (−5.1 mV) was increased to −33.7 mV (i.e., CA-g-PMMA).

After the amination process of CA-g-PMMA using TEPA, the surface charge result for CA-g-PMMA-TEPA was +44.4 mV. The measured zeta potential value of HA solution at pH of 7.0 was −60 mV [89]. As seen from Figure 5.16d, the HA molecule removal performance for un-g-CA membrane was lower as compared with modified membranes. This result is clear that more negative charges are required to remove (repulsion, in this case) the HA molecules with high negative charge. However, the negative charges of un-g-CA (−5.1 mV) were not enough to attain higher repulsion of the HA molecules with −60 mV. Higher removal efficiencies were observed for CA-g-PMMA and CA-g-PMMA-TEPA membranes. Therefore, the HA molecules removal phenomenon for CA-g-PMMA membrane is due to electrostatic repulsion as both the high negatively surface charged membrane (−33.7 mV) and the negatively charged of the HA aggregates would result in high electrostatic repulsion [90]. Furthermore, the highest removal efficiency was achieved for CA-g-PMMA-TEPA membranes due to the availability of high positive charges (+44.4 mV) for the adsorption of negatively charged HA molecules. Therefore, the maximum removal efficiencies for un-g-CA, CA-g-PMMA, and CA-g-PMMA-TEPA membranes at pH of 7.0 were 34.5%, 83.3%, and 99.1%, respectively. The main assumption of this study was that the surface charges of the modified membranes were altered after the grafting and amination process of the CA polymer. Grafting method has numerous benefits, such as the capacity of modifying the polymer surfaces to have distinctive properties by choosing the various monomers. The introduction of grafting chains with a high density and an exact localization on the substrate surface can be done easily and well-regulated without changing the properties of the bulk polymer. Moreover, graft chains onto a substrate surface with covalent attachments and promising long-term chemical stability, in contrast to physically coated polymer chains, can be introduced. Initially, -OCH_3 of PMMA was covalently attached by replacing the –H atom of the –OH groups in CA (Figure 5.11). Second, the TEPA was introduced through the ring opening reaction between the methyl groups of the PMMA and –NH_2 of the TEPA. As confirmed by the zeta potential study, the negative and positive charges were introduced on the layer of the CA polymer alternatively. The schematic diagram for the membrane grafting, amination process, and adsorption of HA molecules in UF process is presented in Figure 5.17.

5.2.2.8 Membrane Regeneration Performance Study

To investigate the fouling and HA removal performances of the un-g-CA, CA-g-PMMA, and CA-g-PMMA-TEPA membranes, four washing/regeneration cycles were conducted using 10 ppm HA solution, and the results are shown in Figure 5.18. After filtration of HA solutions, fouled membranes were cleaned with 0.05 M NaOH solution followed by rinsing using DI water for 30 min, and the DI water fluxes of the tested membranes were measured. As clearly shown in Figure 5.18a, the uniform decrease in

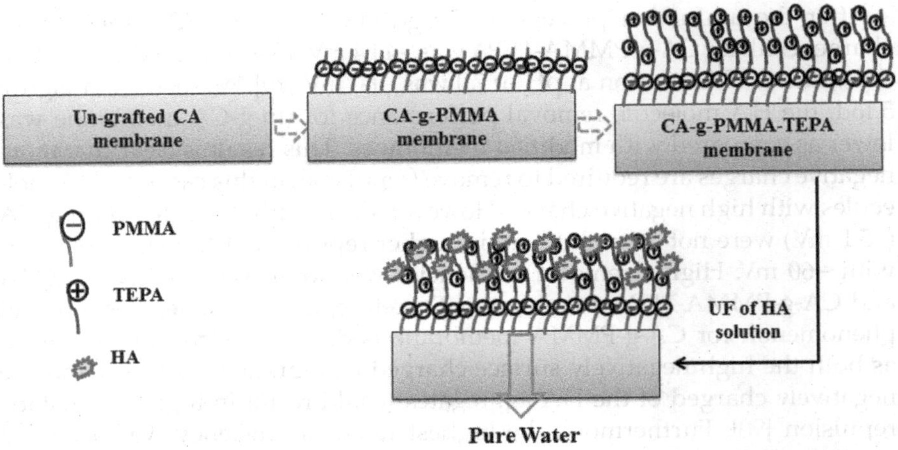

FIGURE 5.17
Schematic diagram for the membrane grafting and amination process and adsorption of HA in UF process.

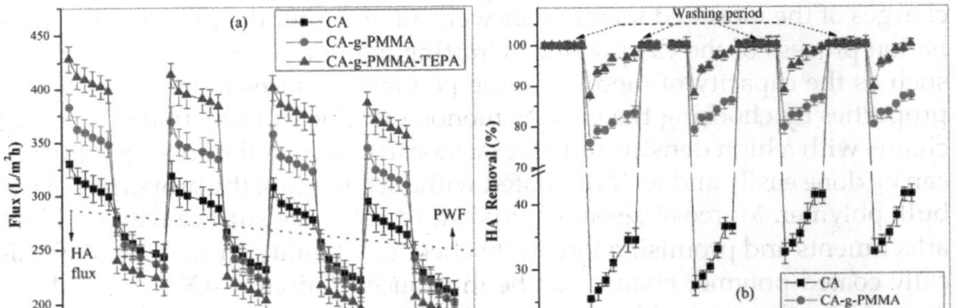

FIGURE 5.18
Washing/regeneration study of un-g-CA, CA-g-PMMA, and CA-g-PMMA-TEPA membranes at pH = 7, HA concentration = 10 ppm, ΔP = 150 kPa, @ 200 rpm.

DI water and HA fluxes were understood to be highly noticeable, which are attributed to the reduction in porosity (pore blocking) of the membranes due to an interior adsorption of HA molecules. Consequently, the DI water flux results for the four filtration cycles at pH of 7.0 for un-g-CA membrane are 311, 299, 288, and 274 L m^{-2} h^{-1}, respectively; for CA-g-PMMA membrane are 360, 346, 334, and 321 L m^{-2} h^{-1}, respectively; and for CA-g-PMMA-TEPA membrane are 409, 394, 381, and 366 L m^{-2} h^{-1}, respectively. The HA flux results for the four washing/regeneration cycles are 255, 243,

233, and 221 L m^{-2} h^{-1}, respectively, for un-g-CA membrane; 249, 236, 227, and 216 L m^{-2} h^{-1}, respectively, for CA-g-PMMA membrane; and 229, 214, 206, and 195 L m^{-2} h^{-1}, respectively, for CA-g-PMMA-TEPA membrane. The HA removal efficiencies of all the prepared membranes at pH of 7.0 are presented in Figure 5.18b. The removal percentage results were detected to increase with increasing of the regeneration cycles. At the end of the fourth cycle, the HA removal percentages were 41.6%, 87.4%, and 99.9% for un-g-CA, CA-g-PMMA, and CA-g-PMMA-TEPA membranes, respectively. The increasing in the removal performances of the membranes with increasing the regeneration cycles can be attributed due to the permanent deposition of the HA molecules onto the membrane surface and within the membrane pores, which could be accredited to structural interaction [47]. In addition to the electrostatic interaction phenomena the HA removal for the membrane with increasing the regeneration cycle was also considered as an indicator of considerable removal due to its structural interaction. Therefore, the surface and pore channel depositions of HA molecules can provide extra resistance to solute transport. From this study, it can be suggested that the CA-g-PMMA-TEPA membrane can be a good candidate for the removal of HA molecules from aqueous solutions.

5.2.3 Summary

The feasibility of grafting of PMMA onto CA polymer using CS as the redox initiator was investigated in the present work. The polymerization reaction conditions, such as initiator concentration and monomer concentration, were kept constant. The polymerization reaction time and temperature were varied and had a great impact on grafting copolymerization. Moreover, the functionalization of the synthesized product was done using TEPA amine. The study of ^{1}H NMR and FTIR spectra results indicated that the graft copolymerization and functionalization were successfully achieved. A highly porous membrane (CA-g-PMMA-TEPA) was obtained with the addition of TEPA to the grafted polymer due to the development of quaternary aminated polymer. Unlike the CA-g-PMMA membrane, a porous structure was observed below the top layer in the cross-section in the case of CA-g-PMMA-TEPA membrane. The maximum removal efficiencies for un-g-CA, CA-g-PMMA, and CA-g-PMMA-TEPA membranes at pH of 7.0 were 34.5%, 83.3%, and 99.1%, respectively. The removal percentage results were detected to increase with increasing in the regeneration cycles. At the end of the fourth cycle, the HA removal percentages were 41.6%, 87.4%, and 99.9% for un-g-CA, CA-g-PMMA, and CA-g-PMMA-TEPA membranes, respectively. Therefore, the grafted and aminated membrane, CA-g-PMMA-TEPA, has shown a higher HA removal efficiency and can be used for various industrial applications.

References

1. Bao, Y., Yan, X., Du, W., Xie, X., Pan, Z., Zhou, J., and Li, L., Application of amine-functionalized MCM-41 modified ultrafiltration membrane to remove chromium (VI) and copper (II). *Chemical Engineering Journal*, 2015. 281: pp. 460–467.
2. Mohan, D., and Pittman Jr, C.U., Activated carbons and low cost adsorbents for remediation of tri- and hexavalent chromium from water. *Journal of Hazardous Materials*, 2006. 137(2): pp. 762–811.
3. Yao, Z., Du, S., Zhang, Y., Zhu, B., Zhu, L., and John, A.E., Positively charged membrane for removing low concentration Cr(VI) in ultrafiltration process. *Journal of Water Process Engineering*, 2015. 8: pp. 99–107.
4. Jayalakshmi, A., Rajesh, S., Senthilkumar, S., and Mohan, D., Epoxy functionalized poly(ether-sulfone) incorporated cellulose acetate ultrafiltration membrane for the removal of chromium ions. *Separation and Purification Technology*, 2012. 90: pp. 120–132.
5. Labanda, J., Khaidar, M.S., and Llorens, J., Feasibility study on the recovery of chromium (III) by polymer enhanced ultrafiltration. *Desalination*, 2009. 249(2): pp. 577–581.
6. Muthukrishnan, M., and Guha, B.K., Effect of pH on rejection of hexavalent chromium by nanofiltration. *Desalination*, 2008. 219(1–3): pp. 171–178.
7. Ortega, L.M., Lebrun, R., Noël, I.M., and Hausler, R., Application of nanofiltration in the recovery of chromium (III) from tannery effluents. *Separation and Purification Technology*, 2005. 44(1): pp. 45–52.
8. Bhatti, M.S., Reddy, A.S., and Thukral, A.K., Electrocoagulation removal of Cr(VI) from simulated wastewater using response surface methodology. *Journal of Hazardous Materials*, 2009. 172(2–3): pp. 839–846.
9. Gao, P., Chen, X., Shen, F., and Chen, G., Removal of chromium(VI) from wastewater by combined electrocoagulation-electroflotation without a filter. *Separation and Purification Technology*, 2005. 43(2): pp. 117–123.
10. Heidmann, I., and Calmano, W., Removal of Cr(VI) from model wastewaters by electrocoagulation with Fe electrodes. *Separation and Purification Technology*, 2008. 61(1): pp. 15–21.
11. Chen, S.S., Cheng, C.Y., Li, C.W., Chai, P.H., and Chang, Y.M., Reduction of chromate from electroplating wastewater from pH 1 to 2 using fluidized zero valent iron process. *Journal of Hazardous Materials*, 2007. 142(1–2): pp. 362–367.
12. Fu, F., Ma, J., Xie, L., Tang, B., Han, W., and Lin, S., Chromium removal using resin supported nanoscale zero-valent iron. *Journal of Environmental Management*, 2013. 128: pp. 822–827.
13. Němeček, J., Lhotský, O., and Cajthaml, T., Nanoscale zero-valent iron application for in situ reduction of hexavalent chromium and its effects on indigenous microorganism populations. *Science of The Total Environment*, 2014. 485–486: pp. 739–747.
14. Owlad, M., Aroua, M. K., Daud, W.A.W., Baroutian, S., Removal of hexavalent chromium-contaminated water and wastewater: A review. *Water Air and Soil Pollution*, 2009. 200: pp. 59–77.
15. Lin, S.H., and Kiang, C.D., Chromic acid recovery from waste acid solution by an ion exchange process: Equilibrium and column ion exchange modeling. *Chemical Engineering Journal*, 2003. 92(1–3): pp. 193–199.

16. Hatano, T., and Tsuruta, T., Removal and recovery of chromium(III) from aqueous chromium(III) using arthrobacter nicotianae cells. *Advances in Microbiology,* 2017. 7: pp. 487–497.
17. Bayramoğlu, G., and Arica, M.Y., Adsorption of Cr(VI) onto PEI immobilized acrylate-based magnetic beads: Isotherms, kinetics and thermodynamics study. *Chemical Engineering Journal,* 2008. 139(1): pp. 20–28.
18. Cheng, Q., Li, C., Xu, L., Li, J., and Zhai, M., Adsorption of Cr(VI) ions using the amphiphilic gels based on 2-(dimethylamino)ethyl methacrylate modified with 1-bromoalkanes. *Chemical Engineering Journal,* 2011. 173(1): pp. 42–48.
19. Li, C., Zhang, Y., Peng, J., Wu, H., Li, J., and Zhai, M., Adsorption of Cr(VI) using cellulose microsphere-based adsorbent prepared by radiation-induced grafting. *Radiation Physics and Chemistry,* 2012. 81(8): pp. 967–970.
20. Qiu, J., Wang, Z., Li, H., Xu, L., Peng, J., Zhai, M., Yang, C., Li, J., and Wei, G., Adsorption of Cr(VI) using silica-based adsorbent prepared by radiation-induced grafting. *Journal of Hazardous Materials,* 2009. 166(1): pp. 270–276.
21. Heidari, A., Younesi, H., and Mehraban, Z., Removal of Ni(II), Cd(II), and Pb(II) from a ternary aqueous solution by amino functionalized mesoporous and nano mesoporous silica. *Chemical Engineering Journal,* 2009. 153(1-3): pp. 70–79.
22. Diallo, M.S., Christie, S., Swaminathan, P., Johnson, J.H., and Goddard, W.A., Dendrimer enhanced ultrafiltration. 1. Recovery of Cu (II) from aqueous solutions using PAMAM dendrimers with ethylene diamine core and terminal NH_2 groups. *Environmental Science & Technology,* 2005, 39(5), pp. 1366–1377.
23. Bessbousse, H., Rhlalou, T., Verchère, J.F., and Lebrun, L., Removal of heavy metal ions from aqueous solutions by filtration with a novel complexing membrane containing poly(ethyleneimine) in a poly(vinyl alcohol) matrix. *Journal of Membrane Science,* 2008. 307(2): pp. 249–259.
24. Ghaee, A., Shariaty-Niassar, M., Barzin, J., and Matsuura, T., Effects of chitosan membrane morphology on copper ion adsorption. *Chemical Engineering Journal,* 2010. 165(1): pp. 46–55.
25. Algarra, M., Vázquez, M.I., Alonso, B., Casado, C.M., Casado, J., and Benavente, J., Characterization of an engineered cellulose based membrane by thiol dendrimer for heavy metals removal. *Chemical Engineering Journal,* 2014. 253: pp. 472–477.
26. Konwarh, R., Karak, N., and Misra, M., Electrospun cellulose acetate nanofibers: The present status and gamut of biotechnological applications. *Biotechnology Advances,* 2013. 31(4): pp. 421–437.
27. Abedini, R., Mousavi, S.M., and Aminzadeh, R., A novel cellulose acetate (CA) membrane using TiO_2 nanoparticles: Preparation, characterization and permeation study. *Desalination,* 2011. 277(1–3): pp. 40–45.
28. Arthanareeswaran, G., Thanikaivelan, P., Srinivasn, K., Mohan, D., and Rajendran, M., Synthesis, characterization and thermal studies on cellulose acetate membranes with additive. *European Polymer Journal,* 2004. 40(9): pp. 2153–2159.
29. Cao, X., Ma, J., Shi, X., and Ren, Z., Effect of TiO_2 nanoparticle size on the performance of PVDF membrane. *Applied Surface Science,* 2006. 253(4): pp. 2003–2010.
30. Saffaj, N., Younssi, S.A., Albizane, A., Messouadi, A., Bouhria, M., Persin, M., Cretin, M., and Larbot, A., Preparation and characterization of ultrafiltration membranes for toxic removal from wastewater. *Desalination,* 2004. 168: pp. 259–263.
31. Hao, S., Zhong, Y., Pepe, F., and Zhu, W., Adsorption of Pb2+ and Cu2+ on anionic surfactant-templated amino-functionalized mesoporous silicas. *Chemical Engineering Journal,* 2012. 189–190: pp. 160–167.

32. Najafi, M., Yousefi, Y., and Rafati, A.A., Synthesis, characterization and adsorption studies of several heavy metal ions on amino-functionalized silica nano hollow sphere and silica gel. *Separation and Purification Technology*, 2012. 85: pp. 193–205.
33. Song, F., Zhao, Y., Cao, Y., Ding, J., Bu, Y., and Zhong, Q., Capture of carbon dioxide from flue gases by amine-functionalized TiO2 nanotubes. *Applied Surface Science*, 2013. 268: pp. 124–128.
34. Shieh, J.J., and Chung, T.S., Effect of liquid-liquid demixing on the membrane morphology, gas permeation, thermal and mechanical properties of cellulose acetate hollow fibers. *Journal of Membrane Science*, 1998. 140: pp. 67–79.
35. Maria da Conceicao, C.L., de Alencar, A.E.V., Mazzeto, S.E., and de A Soares, S., et al., The effect of additives on the thermal degradation of cellulose acetate. *Polymer Degradation and Stability*, 2003. 80(1): pp. 149–155.
36. Chatterjee, P.K., and Conrad, C.M., Thermogravimetric analysis of cellulose. *Journal of Polymer Science Part A: Polymer Chemistry*, 1968. 6: pp. 3217–3233.
37. Gebru, K.A., and Das, C., Preparation and characterization of CA–PEG–TiO_2 membranes effect of PEG and TiO_2 on morphology, flux and fouling performance. *Journal of Membrane Science and Research*, 2017. 3(2): pp. 90–101.
38. Kato, K., Uchida, E., Kang, E.T., Uyama, Y., and Ikada, Y., Polymer surface with graft chains. *Progress in Polymer Science*, 2003. 28: pp. 209–259.
39. Dong, Z., Wei, H., Mao, J., Wang, D., Yang, M., Bo, S., and Ji, X., Synthesis and responsive behavior of poly(N,N-dimethylaminoethyl methacrylate) brushes grafted on silica nanoparticles and their quaternized derivatives. *Polymer*, 2012. 53(10): pp. 2074–2084.
40. Yao, Z., Du, S., Zhang, Y., Zhu, B., Zhu, L., and John, A.E., Positively charged membrane for removing low concentration Cr(VI) in ultrafiltration process. *Journal of Water Process Engineering*, 2015. 8: pp. 99–107.
41. Liao, D.L., Wu, G.S., and Liao, B.Q., Zeta potential of shape-controlled TiO_2 nanoparticles with surfactants. *Colloids and Surfaces A: Physicochemical and Engineering Aspects*, 2009. 348(1–3): pp. 270–275.
42. Gebru, K.A., and Das, C., Removal of Pb (II) and Cu (II) ions from wastewater using composite electrospun cellulose acetate/titanium oxide (TiO_2) adsorbent. *Journal of Water Process Engineering*, 2017. 16: pp. 1–13.
43. Zarrabi, H., Yekavalangi, M.E., Vatanpour, V., Shockravi, A., and Safarpour, M., Improvement in desalination performance of thin film nanocomposite nanofiltration membrane using amine-functionalized multiwalled carbon nanotube. *Desalination*, 2016. 394: pp. 83–90.
44. Vatanpour, V., Esmaeili, M., and Farahani, M.H.D.A., Farahani, Fouling reduction and retention increment of polyethersulfone nanofiltration membranes embedded by amine-functionalized multi-walled carbon nanotubes. *Journal of Membrane Science*, 2014. 466: pp. 70–81.
45. Bayramoğlu, G., and Arica, M.Y., Adsorption of Cr(VI) onto PEI immobilized acrylate-based magnetic beads: Isotherms, kinetics and thermodynamics study. *Chemical Engineering Journal*, 2008. 139(1): pp. 20–28.
46. Aroua, M.K., Zuki, F.M., and Sulaiman, N.M., Removal of chromium ions from aqueous solutions by polymer-enhanced ultrafiltration. *Journal of Hazardous Materials*, 2007. 147(3): pp. 752–758.
47. Cao, X., Ma, J., Shi, X., and Ren, Z., Effect of TiO_2 nanoparticle size on the performance of PVDF membrane. *Applied Surface Science*, 2006. 253(4): pp. 2003–2010.

48. Algarra, M., Vázquez, M.I., Alonso, B., Casado, C.M., Casado, J., and Benavente, J., Characterization of an engineered cellulose based membrane by thiol dendrimer for heavy metals removal. *Chemical Engineering Journal*, 2014. 253: pp. 472–477.
49. Konwarh, R., Karak, N., and Misra, M., Electrospun cellulose acetate nanofibers: The present status and gamut of biotechnological applications. *Biotechnology Advances*, 2013. 31(4): pp. 421–437.
50. Arthanareeswaran, G., Thanikaivelan, P., Srinivasn, K., Mohan, D., and Rajendran, M., Synthesis, characterization and thermal studies on cellulose acetate membranes with additive. *European Polymer Journal*, 2004. 40(9): pp. 2153–2159.
51. Shen, D., and Huang, Y., The synthesis of CDA-g-PMMA copolymers through atom transfer radical polymerization. *Polymer*, 2004. 45(21): pp. 7091–7097.
52. Bhattacharya, A., and Misra, B.N., Grafting: A versatile means to modify polymers: Techniques, factors and applications. *Progress in Polymer Science*, 2004. 29(8): pp. 767–814.
53. Khayet, M., Seman, M.A., and Hilal, N., Response surface modeling and optimization of composite nanofiltration modified membranes. *Journal of Membrane Science*, 2010. 349(1–2): pp. 113–122.
54. Khayet, M., Seman, M.A., and Hilal, N., Surface nano-structuring of reverse osmosis membranes via atmospheric pressure plasma-induced graft polymerization for reduction of mineral scaling propensity. *Journal of Membrane Science*, 2010. 354(1–2): pp. 142–149.
55. Linggawati, A., Mohammad, A.W., and Ghazali, Z., Effect of electron beam irradiation on morphology and sieving characteristics of nylon-66 membranes. *European Polymer Journal*, 2009. 45(10): pp. 2797–2804.
56. Linggawati, A., Mohammad, A.W., and Leo, C.P., Effects of APTEOS content and electron beam irradiation on physical and separation properties of hybrid nylon-66 membranes. *Materials Chemistry and Physics*, 2012. 133(1): pp. 110–117.
57. Mukherjee, P., Jones, K.L., and Abitoye, J.O., Surface modification of nanofiltration membranes by ion implantation. *Journal of Membrane Science*, 2005. 254(1–2): pp. 303–310.
58. Wanichapichart, P., Bootluck, W., Thopan, P., and Yu, L.D., Influence of nitrogen ion implantation on filtration of fluoride and cadmium using polysulfone/chitosan blend membranes. *Nuclear Instruments and Methods in Physics Research Section B: Beam Interactions with Materials and Atoms*, 2014. 326: pp. 195–199.
59. Sacak, M., and Oflaz, F., Benzoyl-peroxide-initiated graft copolymerization of poly (ethylene terephthalate) fibers with acrylic acid. *Journal of Applied Polymer Science*, 1993. 50: pp. 1909–1916.
60. Sacak, M., and Pulat, E., Benzoyl-peroxide-initiated graft copolymerization of poly(ethyleneTerephthalate) fibers with acrylamide. *Journal of Applied Polymer Science*, 1989. 38: pp. 539–546.
61. Pulat, M., and Babayigit, D. Graft copolymerization of PU membranes with acrylic acid and crotonic acid using benzoyl peroxide initiator. *Journal of Applied Polymer Science*, 2001. 80: pp. 2690–2695.
62. Sacak, M., Sertkaya, F., and Talu, M., Grafting of poly (ethylene Terephthalate) fibers with methacrylic acid using benzoyl peroxide. *Journal of Applied Polymer Science*, 1992. 44: pp. 1737–1742.
63. Pulat, M., and Isakoca, C., Chemically induced graft copolymerization of vinyl monomers onto cotton fibers. *Journal of Applied Polymer Science*, 2006. 100(3): pp. 2343–2347.

64. Sanli, O., and Ünal, H.I., Graft copolymerization of 2-hydroxy ethyl methacrylate on dimethyl sulfoxide pretreated poly(ethylene terephthalate) films using benzoyl peroxide. *Journal of Macromolecular Science, Part A*, 2002. 39(5): pp. 447–465.
65. Kulkarni, A.Y., and Mehta, P.C., Ceric ion induced redox polymerization of acrylonitrile in presence of cellulose. *Journal of Applied Polymer Science*, 1965, 9(7): p. 2633.
66. Kulkarni, A.Y., and Mehta, P.C., Ceric ion-induced redox polymerization of acrylonitrile on cellulose. *Journal of Applied Polymer Science*, 1968. 12: pp. 1321–1342.
67. Hebeish, A., and Mehta, P.C., Cerium-initiated grafting of acrylonitrile onto cellulosic materials. *Journal of Applied Polymer Science*, 1968. 12: pp. 1625–1647.
68. Joshi, J.M., and Sinha, V.K., Ceric ammonium nitrate induced grafting of polyacrylamide onto carboxymethyl chitosan. *Carbohydrate Polymers*, 2007. 67(3): pp. 427–435.
69. Habibi, Y., Lucia, L.A., and Rojas, O.J., Cellulose nanocrystals: Chemistry, self-assembly, and applications. *Chemical Reviews*, 2010. 110: pp. 3479–3500.
70. Yu, X., Tong, S., Ge, M., Wu, L., Zuo, J., Cao, C., and Song, W., Synthesis and characterization of multi-amino-functionalized cellulose for arsenic adsorption. *Carbohydrate Polymers*, 2013. 92(1): pp. 380–387.
71. Kumar, M., Gholamvand, Z., Morrissey, A., Nolan, K., Ulbricht, M., and Lawler, J., Preparation and characterization of low fouling novel hybrid ultrafiltration membranes based on the blends of GO–TiO2 nanocomposite and polysulfone for humic acid removal. *Journal of Membrane Science*, 2016. 506: pp. 38–49.
72. Lowe, J., and Hossain, M.M., Application of ultrafiltration membranes for removal of humic acid from drinking water. *Desalination*, 2008. 218(1–3): pp. 343–354.
73. Shao, J., Zhao, L., Chen, X., and He, Y., Humic acid rejection and flux decline with negatively charged membranes of different spacer arm lengths and charge groups. *Journal of Membrane Science*, 2013. 435: pp. 38–45.
74. Ma, B., Yu, W., Jefferson, W.A., Liu, H., and Qu, J., Modification of ultrafiltration membrane with nanoscale zerovalent iron layers for humic acid fouling reduction. *Water Research*, 2015. 71: pp. 140–149.
75. Shao, J., Hou, J., and Song, H., Comparison of humic acid rejection and flux decline during filtration with negatively charged and uncharged ultrafiltration membranes. *Water Research*, 2011. 45(2): pp. 473–482.
76. Bessbousse, H., Rhlalou, T., Verchère, J.F., and Lebrun, L., Removal of heavy metal ions from aqueous solutions by filtration with a novel complexing membrane containing poly(ethyleneimine) in a poly(vinyl alcohol) matrix. *Journal of Membrane Science*, 2008. 307(2): pp. 249–259.
77. Zahran, M.K., Grafting of methacrylic acid and other vinyl monomers onto cotton fabric using Ce (IV) ion–cellulose thiocarbonate redox system. *Journal of Polymer Research*, 2005. 13(1): pp. 65–71.
78. Hintz, H.L., and Johnson, D.C., The mechanism of oxidation of cyclic alcohols by cerium(1V). *The Journal of Organic Chemistry*, 1967. 32(3): pp. 556–564.
79. Kaliyamurthy, K., Elayaperumal, P., Balakrishnan, T., and Santappa, M., Kinetics of polymerization of vinyl monomers initiated by manganese (III) acetate (II). *Polymer Journal*, 1982. 14: pp. 107–113.
80. Zhong, J.F., Chai, X.S., and Fu, S.Y., Homogeneous grafting poly (methyl methacrylate) on cellulose by atom transfer radical polymerization. *Carbohydrate Polymers*, 2012. 87(2): pp. 1869–1873.

81. Jin, X., Xu, J., Wang, X., Xie, Z., Liu, Z., Liang, B., Chen, D., and Shen, G., Flexible TiO2/cellulose acetate hybrid film as a recyclable photocatalyst. *RSC Advances*, 2014. 4(25): p. 12640.
82. Tang, B., Xu, T., Gong, M., and Yang, W., A novel positively charged asymmetry membranes from poly(2,6-dimethyl-1,4-phenylene oxide) by benzyl bromination and in situ amination: Membrane preparation and characterization. *Journal of Membrane Science*, 2005. 248(1–2): pp. 119–125.
83. Tongwen, X., and Weihua, Y., A novel positively charged composite membranes for nanofiltration prepared from poly(2,6-dimethyl-1,4-phenylene oxide) by in situ amines crosslinking. *Journal of Membrane Science*, 2003. 215(1–2): pp. 25–32.
84. Saljoughi, E., Amirilargani, M., and Mohammadi, T., Effect of PEG additive and coagulation bath temperature on the morphology, permeability and thermal/chemical stability of asymmetric CA membranes. *Desalination*, 2010. 262(1–3): pp. 72–78.
85. Qin, J.J., Li, Y., Lee, L.S., and Lee, H., Cellulose acetate hollow fiber ultrafiltration membranes made from CA/PVP 360 K/NMP/water. *Journal of Membrane Science*, 2003. 218: pp. 173–183.
86. Boom, R.M., Wienk, I.M., Van den Boomgaard, T., and Smolders, C.A., Microstructures in phase inversion membranes. Part 2. The role of a polymeric additive. *Journal of Membrane Science*, 1992. 73: pp. 277–292.
87. Smolders, C.A., Reuvers, A.J., Boom, R.M., and Wienk, I.M., Microstructures in phase-inversion membranes. Part 1. Formation of macrovoids. *Journal of Membrane Science*, 1992. 73: pp. 259–275.
88. Gebru, K.A., and Das, C., Effects of solubility parameter differences among PEG, PVP and CA on the preparation of ultrafiltration membranes: Impacts of solvents and additives on morphology, permeability and fouling performances. *Chinese Journal of Chemical Engineering*, 2017. 25: pp. 911–923.
89. Esfahani, M.R., Tyler, J.L., Stretz, H.A., and Wells, M.J., Effects of a dual nanofiller, nano-TiO_2 and MWCNT, for polysulfone-based nanocomposite membranes for water purification. *Desalination*, 2015. 372: pp. 47–56.
90. Jones, K.L., and O'Melia, C.R., Protein and humic acid adsorption onto hydrophilic membrane surfaces: Effects of pH and ionic strength. *Journal of Membrane Science*, 2000. 165: pp. 31–46.

6

Polymeric Membranes for Industrial Effluent Treatments Applications

6.1 Industrial Effluents

6.1.1 Introduction

In recent years, the contamination of freshwater is a major concern due the direct discharge of industrial wastewater such as textile, leather, pulp and paper, printing, and distillery industries, etc. to the environment, which contains a huge number of toxic inorganic or organic materials. The chemical oxygen demand (COD) becomes high for the release of untreated industrial effluents due to the non-biodegradable chemicals, residual dyes, and raw materials that are also responsible for increasing the color and odor of freshwater bodies. The direct discharge of the untreated effluents has already created severe environmental issues for both flora and fauna [1]. To minimize the concentration level of toxicity of the pollutants from the freshwater bodies, treatment of the industrial effluents is currently a serious concern [2].

6.1.2 Behavior of Industrial Wastewater

The sources of industrial effluents are mainly raw material preparation section, cooling tower region, production lines, washing of instruments, etc. The properties of the wastewater depend on the industry due to the use of different kind of raw materials and organic and inorganic chemicals during the manufacturing of final products. Generally, highly contaminated effluent contains proteins, fat, oil compounds, heavy metal ions, inorganic salts, toxic reagents, phenols such as organic materials, etc. based on the types of the industries. Thus, the suspended materials, dissolved organic compounds, and the non-biodegradable components enhance the chemical oxygen demand of the industrial effluents severely [3]. According to literature, if the COD level of any effluent is less than 2000 mg L^{-1}, it is considered as low strength level wastewater, though it is mentioned that the strength of toxicity of the effluents depends on the biodegradability of the dissolved materials and type of the industry. The chemical oxygen demand (COD) level varies widely from 2000 to 16,000 mg L^{-1} dissolved O_2 for the tannery industry effluents as the variation of raw effluents is found for the above industry. Not only that, the textile industrial effluent contains a high amount of inorganic salts and organic dye reagents resulting in high range of chemical oxygen demand (COD) starting from 1500 to 6000 mg L^{-1} of dissolved O_2. The analysis of biodegradability based on biological oxygen demand (BOD) also helps to recognize the strength of wastewater. The analytical ratio of BOD_5/COD is an important parameter to categorize the type of wastewater. Some of the effluents, such as dairy, wheat and starch, food and beverages, and phenolic industries contain biodegradable components resulting in a high ratio of BOD_5/COD. The organic components, present in the dairy, food, and beverages industries, are significantly broken down due the activities of microorganisms present in the environment. The ratio of BOD_5/COD is low for the effluents generated from tannery, paper and pulp, textile, and dyeing industries, indicating the proper and effective physical and chemical treatment before discharge in freshwater bodies. The amount of suspended materials is high for starch (more than 13,000 mg L^{-1}), food and beverages (54,000 mg L^{-1}), and vegetable oil industries (approximately 24,000 mg L^{-1}), which need proper treatment to minimize the water pollution [4,5]. Table 6.1 shows the characteristics of high strength effluents of different industries. The treated effluents can be released into the freshwater bodies, whereas some of the above industries reuse the effluents after treatment in the field of sanitation purposes or landscaping. However, the regulation and standard levels of the toxic compounds made by environmental pollution control board should be maintained before discharging the effluents in the environment [6].

6.2 Industries Generating Hazardous Effluents

6.2.1 Food-Processing Industry

The dairy industry consumes a huge amount of freshwater for the manufacturing of milk products, and during the washing of machineries, cleaning of milk containers, packaging of products, wastewater is generated every day. It contains high concentration of whey proteins, fat, lactose, minerals, and edible chemicals whose biological oxygen demand (BOD) level is high. Due to the stinking in nature, and substantial black flocculated sludges, dairy industry effluents cause discomfort to the surrounding environment that requires proper and significant treatment process [8]. Olive oil mill wastewater also contains high amounts of volatile solids, high concentration of phenolic components, and a huge amount of inorganic materials that increases the COD and BOD load of the untreated effluent. The untreated low molecular weight phenolic components, present in the olive oil mill wastewater, are mainly responsible

TABLE 6.1
High Strength Wastewater Characteristic

Industry	BOD_5 (mg L^{-1})	COD (mg L^{-1})	BOD_5/COD	TSS (mg/L)	SO_4^{2-} (mg/L)	PO_4^{3-} (mg/L)	NH_4-N (mg/L)	Oil (mg/L)
Textile	700	1500–6000	0.117			120	20	
Tannery	250–5000	2500–16,000	0.1–0.313				450	
Wheat starch	16,000	35,000	0.457	13,300	–	–	–	–
Dairy	2200	3500	0.629	–	–	–	120	–
Beverage	1000	1800	0.556	–	–	–	–	–
Palm oil	34,000	67,000	0.507	24,000	–	–	50	100,000
Pet food	10,000	21,000	0.476	54,000	–	200	110	–
Pharmaceutical	3225	6300	0.512	1679	–	–	–	–

Sources: Mutamim, N.S.A. et al., *Chemical Engineering Journal*, 225: pp. 109–119, 2013; Sivagami, K. et al., *Journal of Environmental Chemical Engineering*, 6: pp. 3656–3663, 2018.

for increasing the toxicity level of freshwater bodies and can hinder aquatic life [9]. The fish processing industry generated wastewater is a major source of pollution at the coastal regions of the world. The fish processing industries contains a wide range of inorganic and organic pollutants in different forms, such as soluble materials, suspended particles, colloidal components, and particulate forms. The intensity of the pollutant generated from this industry mainly depends on the particular operations involved in fish processing such as washing, packaging, cutting, etc., which enhances the total organic carbon content, biological, and chemical oxygen demand level of the effluents [10].

6.2.1.1 Characterization of the Food Industrial Effluents

The main compositions of the dairy industrial effluent are whey proteins and milk sugar, which helps to maintain the pH of the wastewater as neutral. Due to the rapid fermentation of the milk products to lactic acid, the pH level changes to acidic in nature resulting in precipitation of casein in the effluents. According to the literature, the high electrical conductivity value indicates the presence of different kinds of minerals in the dairy industry effluents. Not only that, the high total dissolved solids (TDS) materials can hinder the aquatic life after decreasing the oxygen level from the freshwater bodies during degradation of organic compounds [8]. The nitrates and phosphate compounds can be found with high range in the dairy industry effluents. The use of additives and detergent during the cleaning process can change the pH level of wastewater significantly. As a result, the high amount of BOD (300–1400 mg L^{-1} of dissolved O_2) and COD (650–3000 mg L^{-1} of dissolved O_2) levels have been reported for the dairy industry wastewater [11]. A significant amount of metal ions such as sodium (Na), potassium (K), calcium (Ca), iron (Fe), magnesium (Mg), nickel (Ni) are also present in any kind of food and beverages industrial effluents like the dairy industry. It is reported in the literature that the potassium and cadmium content have been found higher in the dairy industry wastewater due to different activities during final product processing. These metal components can easily show adverse effects in the environment [8].

6.2.1.2 Electrocoagulation of Effluents

The effects of electrocoagulation process have already been performed to treat the dairy industry wastewater successfully. Due to the process simplicity, electrochemical process like electrocoagulation has gained consideration during the removal of organic components and metal ions to treat the food industry wastewater. The electrocoagulation process occurs as follows:

i. At the beginning of the process, electrolytic reactions occur at the electrode surfaces made by iron or aluminum.
ii. The formation of coagulant reagents happens in the aqueous phase.

iii. Finally, the adsorption of soluble and collided components on the formed coagulants helps to remove the untreated inorganic or organic particles from the. Aluminum and iron anodes produce the aluminum and iron hydroxide flocks at the anode section followed by the hydrolysis reaction, which helps to adsorb the organic components from the wastewater.

Though the electrochemical processes are efficient and simple methods, the initial cost of this process is the major drawback on use of this process in industrial level effluents [8].

6.2.1.3 Powdered Activated Charcoal Treatment

Activated charcoal treatment is a significant process to purify industrial wastewater. The adsorption of dissolved organic materials happens on the surface of the adsorbent compounds, like activated charcoal. The use of powdered activated charcoal has been increased to treat the toxic contaminants present in industrial effluents. The adsorption of pollutants happens when the attractive forces at the activated carbon surface are higher than the attractive forces of the liquid phase. According to previous studies, activated charcoal has been considered as a significant and efficient adsorbent for the removal of different kind of toxic materials from industrial effluents. Activated charcoals made by different sources have already been studied to treat a wide range of organic and inorganic materials, ions, and metals from simulated and real wastewater. However, the commercialized production of activated carbon is expensive and time consuming. Beside the activated carbons, the removal of toxic pollutants has also been reported using activated alumina and bauxite as adsorbents successfully. As a result, the uses of activated carbons and alumina in the field of the removal of toxic materials from the industrial effluents have been considered one efficient technology recently [8].

6.2.1.4 Membrane Applications in Food Industry

The effluents generated from the food and beverages industries such as dairy, vegetable and fruit processing, palm oil processing, meat, and fishery industry contain high toxic compounds that depend on the types of processing operations. The membrane technology shows various advantages on the treatment of inorganic and organic pollutants from the industrial effluents. The organic and inorganic components present in the wastewater can be separated and recovered using different kinds of membrane filtration such as microfiltration (MF), ultrafiltration (UF), nanofiltration (NF), and reverse osmosis (RO) depending on the molecular size of the materials. The treated water that is known as produced water can be recycled for different purposes. The membrane technology can reduce the process material cost after reusing the organic and inorganic compounds that are separated by membrane filtration. Not only that, this technology helps to minimize waste disposal cost significantly [12]. Membrane technology has special credit to treat the food industry effluents and recycle water to minimize the consumption of freshwater daily. For the treatment of dairy industry effluents, research has already been performed conveniently using different membrane filtration techniques. The significant reflection on the recovery of organic compounds and the production of reusable water has been found for the nanofiltration and the reverse osmosis process during dairy industry wastewater treatment. To characterize the membrane materials, solvent permeability and the selectivity of separation are important factors toward membrane filtration [13].

The membrane filtration depends on the transmembrane pressure drops (TMP), membrane molecular weight cutoff, feed, and retentate flow rate, feed concentration to achieve an efficient treatment of industrial wastewater. The efficiency of the polymeric membrane filtration can be calculated based on the observed rejection of the solute particles. Membrane materials play an important role to achieve significant amount of rejection during filtration process. However, microporous membranes such as NF and RO have the capacity to reject solute particles at their molecular level. The selectivity of the membranes mainly varies based on the chemical nature of the solute molecules. Several researchers have focused on matter of the choice of membrane materials to describe the selectivity of the separation of the organic compounds from food industry wastewater. Though, the dairy industry effluent treatments using RO and NF polymeric membranes are already described successfully, a strong and efficient progress and growth of various membrane filtration can be performed during the treatment of the other food industries effluents [14].

Due to water scarcity and the increasing demands on freshwater preservation, the reuse of effluents in food industry is an essential activity. In recent years, the reuse of wastewater generated from the food industry has been started successfully toward water sustainability. The recycled water can be used in cooling towers, washing, and for the process water [15]. Though the food industrial waste effluents contain less toxic materials than other industries such as textile, leather, petroleum industries, etc., the chemical and biological oxygen demands, total dissolved solids materials are significantly high and require proper treatments before discharge. Table 6.2 describes the efficiency of the membrane technology to treat the different industrial effluents and their reutilization.

TABLE 6.2

The Efficiency of the Membrane Technology to Treat the Different Industrial Effluents and Their Reutilization

Industry/Wastewater	Combined Membrane Treatments	Water Recycling
Dairy/Vapor condensates from concentration and drying steps	Cartridge filtration-NF-RO-UV oxidation	Water use in boilers Drinking
Dairy/Flash coolers	Cartridge filtration-NF-UV	Boiler make up water
Milk/Bottles machines, chess processing	MF, UF, NF, and RO	Unspecified
Beverage/bottle rinsing, brewing room, bright beer reservoir	MBR-NF, RO	Unspecified
Fruit and vegetable processing/rinsing beans, cereal processing	MF, UF, NF, RO	Rinsing beans
Tomato/cleaning, sorting and moving the processed	NF	Unspecified
Fruit juices/bottle washing, fruit processing, juice production and cleaning of tanks, pipes	NF	Drinking
Vegetable oil/olive mill, washing	MF, UF, NF, RO	Drinking
Meat and seafood/slaughterhouse fish and crustaceans and tuna cooking	Two NF steps-UV SBR, MBR, UF and RO in different combinations	Drinking

Source: Muro, C. et al., *Food Industrial Processes – Methods and Equipment*, 14: pp. 254–280, 2012.

6.2.2 Leather Industry

6.2.2.1 Introduction

The tanning industry, one of the oldest industries in the world, has been categorized as a severe pollution generating industry after releasing highly contaminated waste effluents in the environment recently. The effluent coming from a leather factory is categorized as a critical environmental hazard due to the high toxic inorganic and organic chemical levels, high level of salinity, high biological and chemical oxygen demand, low BOD_5/COD level, inorganic and organic dissolved matters, high amount of suspended materials, presence of sulphide, chromium and lead ions, other heavy metals, total kjeldahl nitrogen (TKN), chlorine content, ammonia, and other specific pollutants such as toxic dye reagents, etc. [16]. A huge amount of freshwater is used in the leather industry for the tanning process. The release of untreated effluents in the environment is also increasing, which requires proper treatment to control the water pollution. Not only that, gaseous and solid wastes are generated during the tanning process that also need appropriate treatment before discharge into the environment. In the field of the chrome tanning process in a leather industry, more than 40% of the residual chromium salts (hexavalent chromium salts) are directly released in the wastewater, resulting in a severe threat to the living environment. Lung cancer, ulcer, skin cancer, and nasal septum perforation are critical diseases due to the heavy release of untreated pentachlorophenol, chromium salts, and other different toxic pollutants. A huge number of project works to treat the tannery industry effluents applying a numerous number of technologies such as coagulation, flotation, adsorption, electrochemical processes, and membrane technologies have been reported previously. Among all the processes, membrane processes such as ultrafiltration and nanofiltration have been selected as efficient, eco–friendly, and cost effective processes to treat leather and tanning processing wastewater [17].

The use of chromium salts during tanning process is a common method in the leather industry. The chrome tanning is performed based on the chemical reaction of the cattle hides and a basic chromium sulfate [17]. This process is performed with the increasing of pH level from 3 to above. After completing the tanning process, the important steps such as piled down of chrome-tanned leather, wrung, grading of the leather for the quality and thickness, fragmented into flesh and grain sheets, and shaved to the preferred thickness is performed and require a huge amount of freshwater for mainly washing purposes and generate spent tanned liquor.

Wide studies have been performed to treat and recycle the spent tanned liquor in the leather industry. The recovering of chromium salts from the spent tanned liquor can reduce the fresh chromium uptake during the leather tanning process. Though the electrodialysis system may be beneficial to recover the neutral salts from the spent tanned liquor, the high cost estimation is the major demerit for this process. The significant effects of the traditional method to recover chromium salts have already been studied based on the precipitation of target components such as chromium salts with sodium hydroxide (NaOH) and after that the dissolution of Chromium(III) hydroxide ($Cr(OH)_3$) in sulfuric acid (H_2SO_4). However, the low quality of the recovered materials due to the presence of different impurities, such as lipid substances and other metals, is the drawback of acid-alkaline recovery of the chromium salts. The membrane technology has already shown its significant ability to recover used chromium from tannery wastewater. Membrane-based separation process is known as a cleaner technology due to its various advantages to purify the tannery, textile, paper, and paint industrial effluents. In present years, membrane-based technologies have been emerging quickly and the economic cost is continuing to decrease while the application potentials toward strong effluents treatment are spreading. The main advantage of a membrane-based process is that concentration and separation is achieved without a change of state and without use of chemicals or thermal energy, thus making the process energy-efficient and ideally suited for recovery applications. The possibility of applying membrane processes in the treatment of the exhausted bath of a single step offers interesting perspectives for the survival of this industry and for recovering and recycling of primary resources. The pilot scale study for the removal of unreacted chromium ions from spent tannery liquid has also been carried out using reverse osmosis (RO). However, the use of integrated membrane systems such as MF followed by NF, UF followed by RO, and NF followed by RO in the field of the treatment of spent tannery liquid and the recovery of chromium and other materials is critically challenged due to the presence of high BOD load and the unreacted parts of protein molecules resulting in fouling of the polymeric membrane surfaces and the subsequent process failure temporarily or forever [17].

6.2.2.2 Membrane Application in the Treatment of Tannery Effluents

The effects of membrane microfiltration (MF), ultrafiltration (UF), nanofiltration (NF), and reverse osmosis (RO) to recapture the chromium (III) ions from the leather industry effluents have already been discussed successfully. Previously, the application of conventional technologies to decrease the polluted materials from the wastewater were mainly based on end-pipe treatments. The use of the conventional technologies is time consuming, space-requiring, high cost estimating, and does not certify the recuperating and the reuse of water and chemicals. Subsequently, the produced sludge is generally discharged into landfills even though it contains major chemicals that could be recycled at the industrial level. To reduce the effective polluting parameters, and to recover the primary resources, membrane technologies have represented an efficient and significant effect over the treatment of tannery effluents. With the help of membrane technology, the cleaning-up of remaining wastes is simplified and is reused in the field of agricultural purposes. Figure 6.1 represents the experimental applications of membrane systems for the recycling of tanning effluents, and the discussions of the all processes are summarized below [18].

6.2.2.2.1 Soaking

During the soaking process, the collected raw skins are washed and cleaned with freshwater to remove the unwanted materials and salts. To hydrate and to solubilize the skin proteins, the treatment of cleaned skins was performed using water. As a result, the effluents coming from the soaking process contains a huge amount of organic materials whose biological oxygen demand is high and requires proper treatment. Ultrafiltration of soaking effluent is an attractive and prominent technology to remove the organic materials from the soaking effluents. Preliminary treatments such as sedimentation and steel spring filters (200–300 mm size of net) are required to remove the suspended materials to control the clogging occurrences of membranes.

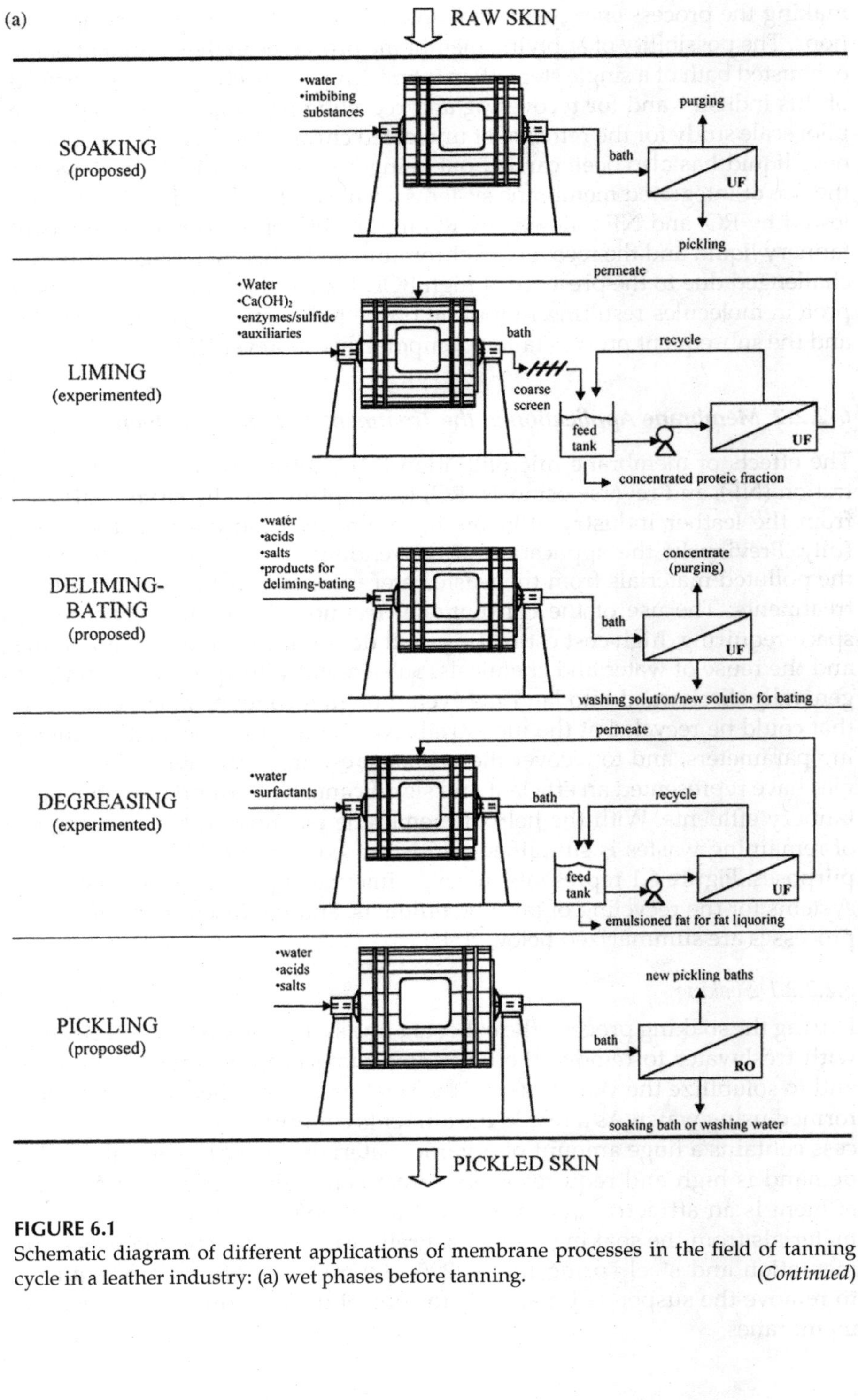

FIGURE 6.1
Schematic diagram of different applications of membrane processes in the field of tanning cycle in a leather industry: (a) wet phases before tanning. *(Continued)*

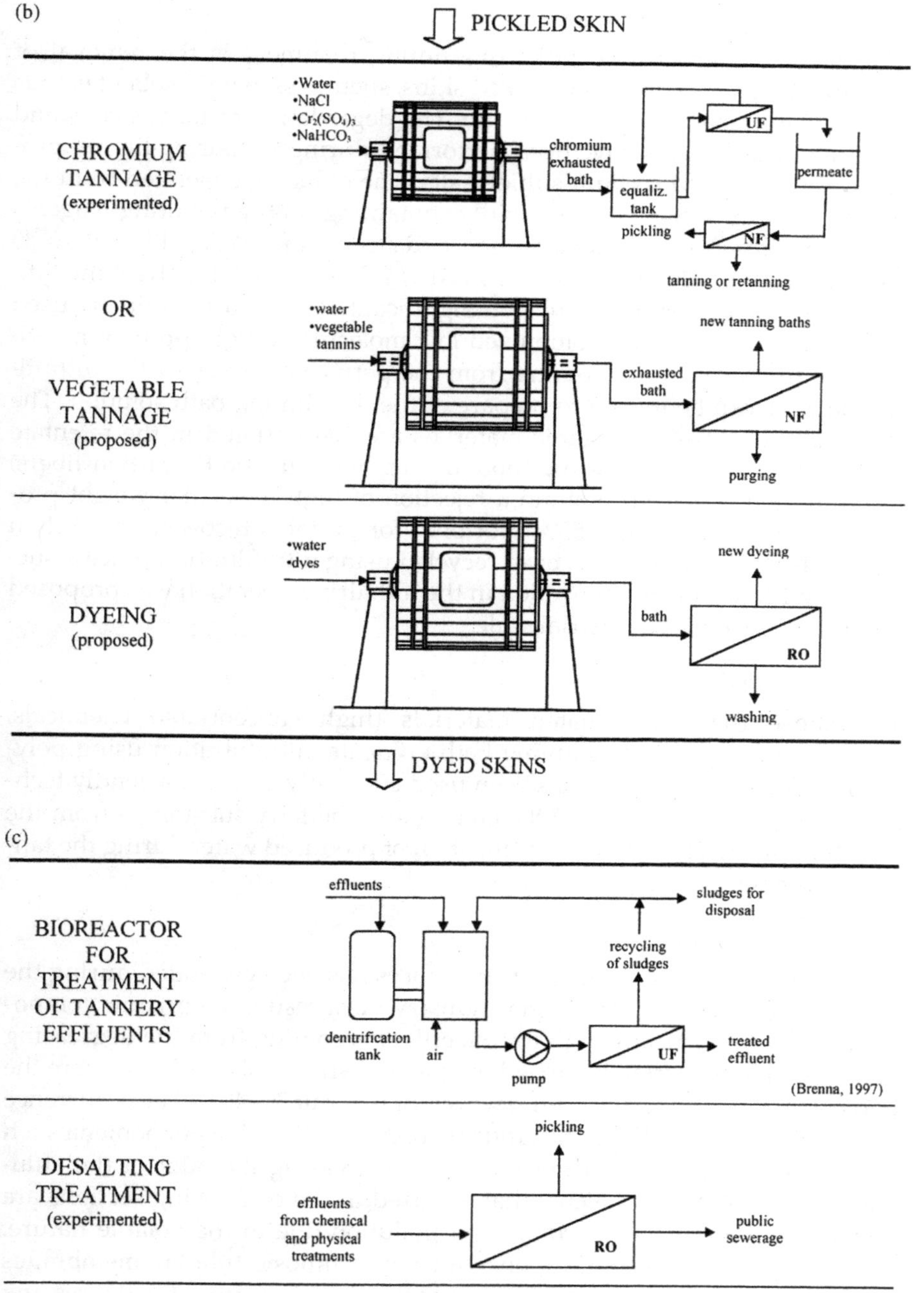

FIGURE 6.1 (CONTINUED)
Schematic diagram of different applications of membrane processes in the field of tanning cycle in a leather industry: (b) tanning of skin and dyeing; and (c) the treatment of final effluents. (From Cassano, A. et al., *Journal of Membrane Science*, 181(1): pp. 111–126, 2001.)

6.2.2.2.2 *Unhairing*

The major concern of the unhairing-liming treatment is the removal of unwanted materials from the soaked skins, such as skin hair, subcutaneous adipose layer, and fat substances, etc. The degradation of mucus cells and swelling of the dorms have been performed during unhairing-liming process using lime and sodium sulfide salts. The effluents generated from the liming-unhairing sections are mainly containing highly polluting reagents due to the presence of the toxic sulfide, and amines materials with high COD value (20,000–40,000 mg L^{-1} of dissolved O_2). The use of UF to treat the liming effluents has already delivered significant results to recover the used sulfide materials and the solubilized low molecular weight protein materials. The produced water coming from the permeate section of the ultrafiltration zone can be reused to prepare of the next liming bath solution. The high molecular weight organic materials are concentrated in the retentate section. It is reported that the tubular membrane (made by carbon fibers) ultrafiltration achieved to attain a rejection of high molecular weight proteins approximately 60 to 85%. Not only for proteins recovery, more than 60% of the initial sulfide has been recycled using ultrafiltration process successfully. Moreover, an innovation in the unhairing operation was proposed and tested on a pilot industrial scale.

6.2.2.2.3 *Deliming-Bating*

To remove the contaminated materials (high concentrated chemicals, enzymes, etc.) from the deliming bath effluents, ultrafiltration using polymeric membrane materials has been used efficiently. This eco-friendly technology helps to reduce the COD loading and the fatty substances from the tannery effluents toward the reutilization of produced water during the tanning processes.

6.2.2.2.4 *Degreasing*

The excess amount of natural fat substances has been generated during the degreasing process of the leather industry. As a result, the organic component loading is critically high for the effluents coming from the degreasing sections. This operation is carried out particularly for sheepskins where the percentage of fat substances on raw weight is about 30–40%. The occurrence of high quantities of fat in the effluent creates undesirable phenomena such as loss of physical strength of the water, increasing the odor of the effluents, etc. The organic solvents that are used in the degreasing sections are increasing the risk of environmental pollution due to the volatile nature. The application of ultrafiltration with non-cellulosic tubular membranes has been performed to treat and recover the chemicals from the degreasing effluents with the observed rejection of 91%. It is reported that the polymeric membrane ultrafiltration of tannery effluents permitted the high removal capacity of fatty and lipid materials, protein substances, organic components from the different sections of the leather industry. Water recycling has also been performed successfully with the help of membrane ultrafiltration technology.

6.2.2.2.5 Pickling

The pickling process is the final operation to eliminate the remaining lime from the tanned skin samples with the help of acidification and dehydration and washing. As a result, the effluent generated from this section contains high concentrated sulfuric, chloridic, formic, lactic acids, adequate amount of sodium chloride and sodium sulfate salts, and chromium ions. The reverse osmosis process can be performed to recover the salts that can be reused in the pickling section for further treatment of tanned skins.

6.2.2.2.6 Chromium Recovery

Chromium sulfate is known as the most commonly used tanning material today. The chromium tannage wastewater contains approximately 30% of initial chromium salts, which requires proper treatment before discharge in freshwater. The serious health problems have been increased rapidly due to the untreated chromium release in the environment. The recovery of chromium ions using polymeric membrane nanofiltration from tanning exhausted effluents represents an effective and significant economic benefit for the leather tanning industry after its reusing the waste materials and toward the minimization of water pollution. It is reported that the quality of the recovered chromium salts during acid alkaline treatment is not always optimal due to the presence of impurities such as different metals and unreacted lipid and fat substances. To increase the quality of the recycled chromium salts, membrane filtration has been chosen as an effective and alternative method that has already been studied widely. The pilot scale spiral wound ultrafiltration (UF) module with flat sheet polymeric membranes were used with varying different operating conditions such as membrane materials, transmembrane pressure drops (TMP), chromium ions initial concentration, temperature of the feed solution, feed, and retentate flow rates, etc. (Figure 6.1b). It is reported that only 28% removal of chromium ions has been observed using spiral wound membrane ultrafiltration with the chromium (III) initial concentration of 4343 mg L^{-1}, the maximum transmembrane pressure drop of 3.8 bar, membrane active surface area of 3 m^2, and molecular weight cutoff (MWCO) of 15–25 kDa. Suspended materials, the fat contents, and organic nitrogen compounds were found 84%, 98%, and 40%, respectively.

To increase the chromium removal efficiency, the UF module was integrated with the nanofiltration system with the membrane active area of 5.5 m^2, MWCO of 150 kDa, maximum TMP drop of 16 bar, and the axial flow rate of 1200–4500 L h^{-1}. According to the analytical results, up to 99% rejection (Figure 6.2) of chromium ions was obtained using spiral wound

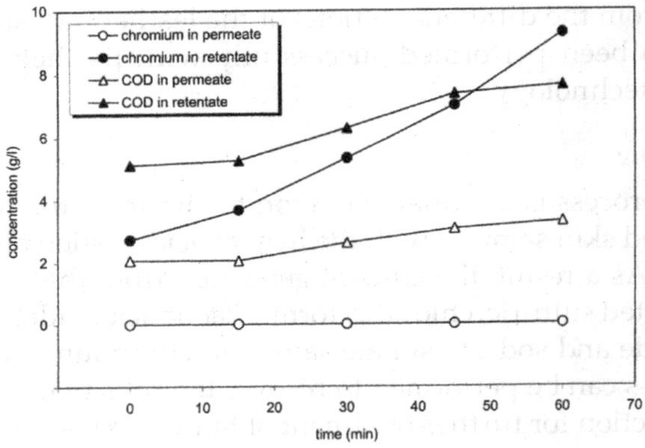

FIGURE 6.2
Concentration of chromium ions and the chemical oxygen demand (COD) with respect to process time in a spiral wound nanofiltration unit (parametric conditions: temperature: 25°C; TMP drop: 16 bar; axial flow rate of feed: 2200 L h^{-1}; and permeate flux: 3.63–36.0 L m^{-2} h^{-1}). (From Cassano, A. et al., *Journal of Membrane Science*, 181(1): pp. 111–126, 2001.)

nanofiltration membrane module. However, less rejection of organic substances has been observed as the chemical oxygen demand (COD) was increased in the permeate samples with increasing process time due to fouling. Finally, the recovered chromium (III) salts was reused during the tanning processes and compared with the fresh chromium (III) solution. The physical and the chemical analysis for the both the chromium (III) salts revealed that the values were significant and similar according to the tensile strength, shrinkage temperature, breaking load, etc.

6.2.2.3 Fenton's Reaction Followed by Membrane Filtration

The leather and tanning industry is considered one of the vital cost-effective industrial revenues in several countries. Meanwhile, it is one of the major consumers of water, which is largely converted to wastewater by soaking, liming, de-liming, pickling, bating, tanning, and wet-finishing operations. This industry is commonly associated with aggressive odors coming from animal hide, raw substances, solids, liquids, and gaseous wastes springing from such industrial and the medieval methods of processing [19]. The generated effluents from this process are highly loaded with toxic organic hazardous wastes that are categorized as the most intolerable materials in the tanning industry. Such effluents contain various chemicals such as sulfide, excess lime during lime and delime section, and organic substances that are generated due to the hydrolytic degradation of hair, keratins, and inter-fibrillar proteins with high BOD and COD loading. It is reported that the more than 60% of the total BOD, COD, and the total solids (TS) are contributed during the liming, unhairing, and deliming processes. Previously, several studies were performed to control the water pollution due to the tanning industry. The discharge of liming/unharing effluents to the natural water bodies could also cause a serious environmental contamination and health risks. The appearances of wastewater vary extensively depending on the nature of the liming/unharing adopted process as well as the amount of water and used chemicals. It is reported that the conventional treatment of the tannery wastewater, including primary sedimentation, biological oxidation, and secondary sedimentation, was not efficient to meet the required limits for the treated of tannery effluents at least for some parameters such as sulfides, COD, ammonia, salinity, and surfactants. Furthermore, biological treatment of wastewaters containing resisting and toxic compounds requires a long period of time to reach the required limit. This is mainly due to the presence of several organic compounds applied in leather tanning processes that resist the conventional chemical and/or biological methods. The anaerobic process and the combination of anaerobic/aerobic system were examined for the treatment of the leather tanning industrial wastewater. Further study was carried out using aerobic treatment process to decrease the level of pollutants in the treated effluent. Other technologies are studied as an alternative to biological and classical physico-chemical treatment. These technologies depend mainly on the conventional phase separation techniques such as adsorption processes and microfiltration. Other methods depend on destroying the contaminants (using chemical oxidation/reduction) or pretreatment followed by biological oxidation techniques. Fenton technique is one of the most active systems for the oxidation of organics in water. This reactivity is due to the in situ generation of highly oxidation species (hydroxyl radicals) as a result of the dissociation of H_2O_2 molecule. Hydroxyl radicals are extremely reactive, and membrane filtration technology (MF), with its different applications, has proven to be a reliable technique for the treatment of water and wastewater. During the last several years, membrane filtration has received great development by researchers and manufacturers. Therefore, the materials and techniques of membranes are improved greatly, which provide higher fluxes, longer life time, partly refining the fouling as well as reducing the operation and maintenance cost. Figure 6.3 describes the industrial leather preparation steps, and the pollution problems as well as the environmental impact of preservation and beam house operations.

An integrated system based on using of the Fenton oxidation process followed by membrane microfiltration for the treatment of tannery wastewater was conducted. Figure 6.4 illustrates the diagram of the treatment system. The raw tannery liming/unhairing wastewater was submitted to the first sedimentation tank where the effluent was directed by gravity to the mixing tank. Addition of Fenton reaction, namely, H_2O_2 and Fe^{2+} catalyst, was carried out in the mixing tank. The wastewater flow was slowed down for the formation of flocks and finally to three baffled tanks for sedimentation and separation. The effluent of this tank was directed to the microfiltration process (Figure 6.4).

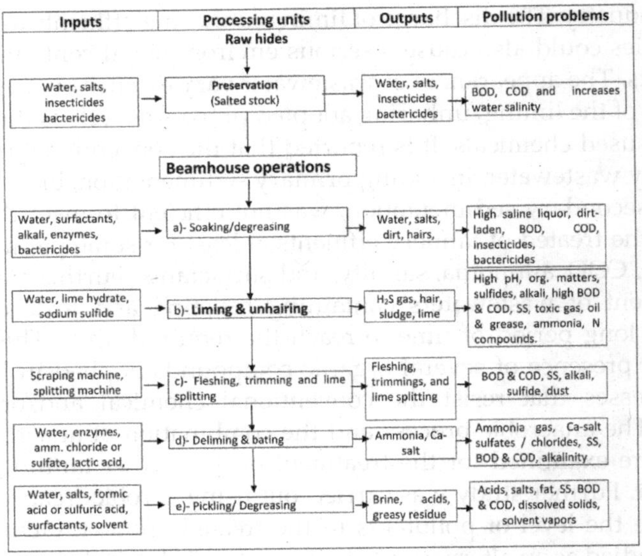

FIGURE 6.3
Schematics of the industrial leather preparation steps; pollution problems and environmental impact of preservation and beam house operations. (From Abdel-Shafy, H.I. et al., *Water Science and Technology*, 74(3): pp. 586–594, 2016.)

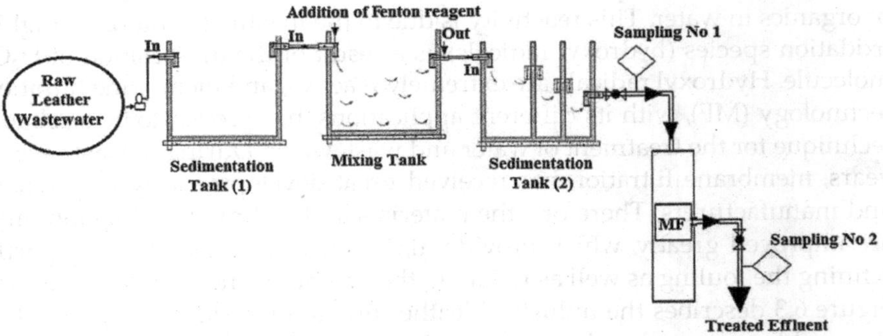

FIGURE 6.4
Schematic diagram of the integrated treatment system using Fenton oxidation followed by the microfiltration processes. (From Abdel-Shafy, H.I. et al., *Water Science and Technology*, 74(3): pp. 586–594, 2016.)

6.2.2.3.1 *Microfiltration Membrane (MF)*

The microfiltration process was designed as a semi-pilot scale unit for the treatment of Fenton's treated effluent (Figure 6.4). The MF system is supplied with a continuous aeration that flows from the bottom of the filter. Tangential flow along the membrane surface was maintained to minimize the fouling problem on the surface of the plate and frame module. A continuous aeration was provided by a compressor to the MF system to limit the problem of fouling or clogging. Meanwhile, it could maintain the aerobic condition of the reactor. It is reported that the necessary transmembrane pressure difference could be realized by 1.4 m of water head above the used membrane (gravity flow) without consumption of any energy as recommended by [20]. Such gravity flow of wastewater was able to save between 30 to 40% of the total required energy. Consequently, the aeration system of this MF consumed only 60% of the total required energy. In addition, the membrane filter allowed the mixed liquor suspended materials concentrations to be by far exceeding the usual 2–4 g L^{-1} in conventional treatment systems. The aim is to obtain a final effluent free of particles, organic compounds, and possible germs. The membrane filtration was operated under aerobic conditions in which the oxygen concentration in the tank was maintained between 1 to 4 mg L^{-1} of O_2. The technical characteristics of the studied MF were as follows:

- Membrane material is (PEC)
- Number of membranes are (8)
- Membrane surface = m^2 (0.6)
- Resistance to hydrogen peroxide (H_2O_2) and (NaOCl), ppm (3,000–5,000 (normal 500))
- Resistance/pH range from (1.5 to 10)
- Resistance/temperature (WC) is <50
- Resistance/pressure as mWS (Max. 1 to 3 (1.02 mWS = 10 kPa))

6.2.2.3.2 *Continuous Treatment System of Fenton's Reaction Followed by Membrane Filtration*

In a continuous treatment system (Figure 6.4), a semi-pilot scale was operated via two steps, namely Fenton oxidation process followed by membrane filtration. In the first treatment system the Fenton oxidation process was employed at the predetermined optimum dose of the hydrogen peroxide and ferrous sulfates. The average removal rate of COD, BOD, TSS, oil and grease, S_2, TKN and T.P. reached to 97.1, 85.2, 99.3, 99.9, 99.9, 84.5, and 94.3%, respectively (Table 6.3). The corresponding residual concentrations are shown in Table 6.3. The above results indicate that the biodegradability of lime/unhair wastewater was enhanced by employing the Fenton reaction. After the Fenton's oxidation, the BOD/COD ratio of the studied wastewater was increased from 0.155 to 0.80 by about 5.16 times as an indication of effluent improvement, which is in a good accordance with the results obtained by. Such achievement confirmed that the treatment with Fenton's oxidation does not only improve the biodegradability but also reduces the toxicity of the wastewaters by decreasing the level of sulfides and TKN. Meanwhile, the concentrations of the pollution parameters, namely, COD and BOD5, TSS (Table 6.3) were still over the

TABLE 6.3

Physicochemical Analysis of Lime/Unhair Effluent and the Removal Percentage after Treatment Using Combined Treatment Process

		Fenton Oxidation Treated Effluent		Membrane Filtration Treated Effluent		Over all	Advanced Wastewater Treatment*
Parameter	Raw	Concentration	%R	Concentration	%R	%R	
COD (mg L^{-1} of O_2)	23,400	679.0	97.1	19.0	97.2	99.9	40
BOD_5 (mg L^{-1} of O_2)	3,650	540.2	85.2	16	97.0	99.6	20
TSS (mg L^{-1})	3,680	25.76	99.3	1.8	93.0	99.95	20
Oil and grease (mg L^{-1})	4,900	5	99.9	0.63	87.4	99.98	5
Sulfide (mg L^{-1})	3,400	2.1	99.9	0.08	96.2	99.99	–
TKN (mg L^{-1})	614	95.2	84.5	23.8	75.0	96.1	–
TP (mg L^{-1})	67	3.8	94.3	2.25	40.8	96.6	–
Ratio of BOD_5/COD	0.15	0.80	–	0.84	–	–	–

Source: Abdel-Shafy, H.I. et al., *Water Science and Technology*, 74(3): pp. 586–594, 2016.

* Egyptian regulation: Egyptian Environmental Affairs Authority, Law 48, No. 61–63, Permissible values for wastes in River Nile (1982) and Law 9, Law of the Environmental Protection (2009).

permissible level of the National Regulatory Standards. Many governments over the world developed more suitable effluent standards to protect the environment. The standard of sulfide is restricted, especially for lime/unhairing effluents, and it is in the range of 1 to 2 mg L^{-1} [21]. The effluent of advanced oxidation was directed to an aerated membrane filtration (Figure 6.4). The characteristics of the final effluent are shown in Table 6.3. The achieved removal rate of the COD, BOD, TSS, oil and grease, S2, TKN and total phosphorous (TP) was 97.2, 97.0, 93.0, 87.4, 96.2, 75.0, and 40.8% respectively (Table 6.3). The corresponding residual concentrations are given in Table 6.3. Meanwhile, slight increase in the BOD/COD ratio was reached namely from 0.80 to 0.84 as an indication of effluent improvement. The physical and chemical analysis of the final treated effluent was found to successfully meet the permissible limits of the unrestricted water reuse according to the standards. Therefore, the treated water can be used for irrigating vegetables and other plants (i.e., the treated water meets the national water reuse guidelines that are related to quality of water including public health issues).

The overall results indicated that the raw leather industries generate a large amount of waste effluent that are critically loaded with organic toxic materials. These wastewaters are characterized by high value of the pollution parameters including TSS, COD, BOD, ammonia, and sulfides. The presence of high level of sulfides is attributed to the use of sodium sulfides in the unhairing tanning process as well as the high organic load of this wastewater. Such wastewater should be treated before discharge. Otherwise, serious environmental pollution and health hazards could be developed. The presence of high sulfides could cause corrosion to the sewer systems. It was reported that the conventional treatment such as chemical and biological treatment of waste effluents is not highly efficient due to the presence of resisting and toxic compounds [21]. In this respect, several investigators confirmed that oxidation and/or aerobic treatment are effective in the treatment of leather wastewater. In addition, the combination of anaerobic/aerobic treatment proved to be more efficient. The present extensive study deals with the treatment of the wastewater coming from the leather lime, de-lime, and unhair sections using the integration of Fenton reaction followed by membrane filtration. Bench scale jar-test experiments were employed to determine the effect of Fenton's reaction. The optimum dose of hydrogen peroxide in combination with ferrous sulfate was determined. When a semi-pilot plant continuous system was employed the results showed that the biodegradability of lime/unhair wastewater was enhanced by addition of the Fenton reaction. This biodegradability is attributed to the oxidation of the organic pollutants in the wastewater as well as the oxidation of the toxic elements such as sulfides and ammonia [21]. It was also confirmed that the treatment with Fenton's oxidation not only improves the biodegradability but also reduces the toxicity of the wastewaters by decreasing the level of sulfides and TKN. When the effluent of advanced oxidation was directed to an aerated membrane filtration (MF) further achievement in the removal of COD, BOD, TSS, oil and grease, S^{2-}, TKN and T.P. was achieved. Meanwhile, slight increase in the BOD/COD ratio was reached as an indication of effluent improvement. This achievement can be attributed to the fact that the combination of the aerated MF reactor and the filtration capacity are the main factors of such success [20]. The characteristics of the final treated effluent were found to meet the permissible limits of the unrestricted water reuse successfully according to the literatures.

6.2.3 Petroleum Industry

6.2.3.1 Introduction

Crude oil is a composite mixture that comprises generally alicyclic, aliphatic, and aromatic hydrocarbons. The refining process needs de-emulsifiers and the production water generated from the decantation of the oil and water emulsion offers a wide range of salinity up to three times higher than the salinity of seawater [22]. The removal of oil and gas (O&G) resources from the production water can be performed using conventional and unconventional technologies. The conventional process of the extraction of hydrocarbons, after the well drilling process, is based on the natural pressure of the well supporting the next operations such as pumping or compression. After the reduction of the well's natural pressure drop, the different techniques such as water and gas injection and other depletion followed by compression techniques are applied to increase the production rate. The oil and gas reservoir section is still considered as a conventional resource in the world. The heavy oil, oil shale, and oil sands are the primary sources of the unconventional oil that are found in the low-permeability rock areas.

6.2.3.2 Water and the Petroleum Industry

Most sections of the oil industry (Figure 6.5), including natural gas production, liquefied natural gas (LNG) generation, production of gas to liquid (GTL), and other oil fields, consume a huge amount of freshwater and produce contaminated effluents. Basically, the water to be achieved is either during the co-production of the hydrocarbons, which is generated as a byproduct from oil and gas dispensation, or water applied at the facility to control the production processes. For the cooling and oil processing all the conventional oil production industries also use freshwater during downstream processes. It is observed that the groundwater, seawater, and surface water typically show an imperative role for all the cases, including upstream and downstream operations. Thus, profitable water management is an essential part of the oil and gas industries to confirm the optimized supportable actions and attain the license to work significantly in the world. The highly

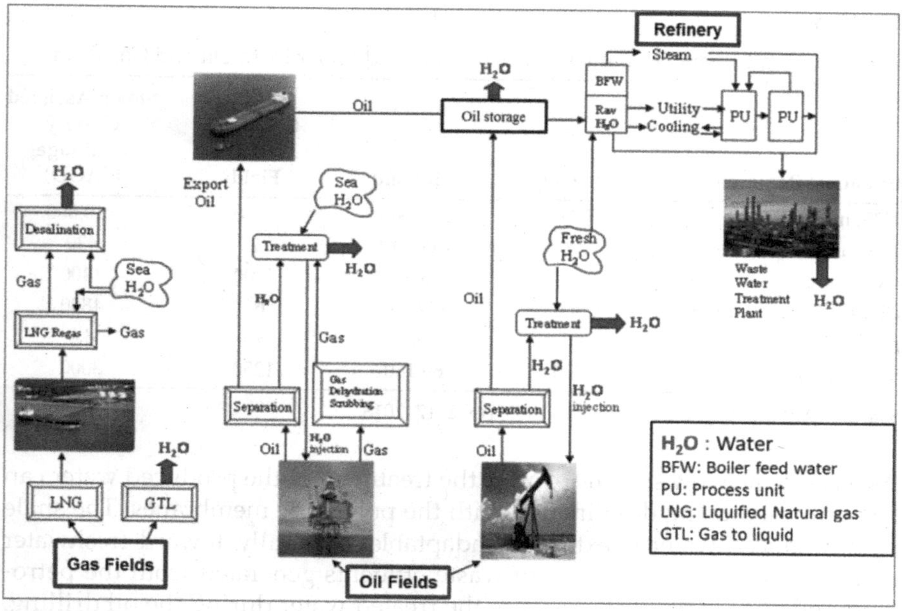

FIGURE 6.5
The use of water in a conventional petroleum industry. (From Adham, S. et al., *Desalination*, 440: pp. 2–17, 2018.)

contaminated effluents are generated from the upstream processes containing biodegradable hydrocarbons. As a result, the BOD load is high for this kind of waste effluents [22]. The effluents generated from both the operations such as upstream and downstream processes require proper treatment before discharge. The reutilization of treated water can be done in the oil and petroleum industry after the significant treatment.

6.2.3.2.1 *Chemical Characteristics*

The chemical characteristics such as salinity of the oil reservoir and refinery generated wastewater vary with the geographical location, the types of hydrocarbons, and especially the age of the reservoirs. Table 6.4 shows the common chemical characteristics of the various produced water generated from the different oil and gas field sections. The variation of the organic and inorganic parameters is the two major concerns to categorize the oil field generated wastewater. The range of the total organic carbon (TOC) concentration is reported typically 500 to 1000 mg L^{-1}, whereas coal bed methane comprises low organic carbon load of approximately 2 mg L^{-1}. The total dissolved solids (TDS) of the produced water in terms of salinity is found critically high and above 247,000 mg L^{-1} in the oil field produced water, which requires immediate attention, whereas the total dissolved solids (TDS) from the different kinds of hydrocarbons is comparatively low (less than 10,000 mg L^{-1}) and is much lower than the seawater salinity (35,000 to 40,000 mg L^{-1}).

TABLE 6.4
The Common Chemical Characteristics of Produced Water in an Oil and Gas Field

Chemical Parameters	Gas Field	Oil Field	Coal Bed Methane Field	Steam Assisted Gravity Drainage (SAGD)
TDS, mg L^{-1}	5200	247,000	2510	<10,000
TOC, mg L^{-1}	500	500–2000	2	430
HCO_3, mg L^{-1}	–	310	1700	1400
Chloride content, mg L^{-1}	2300	152,750	62	4800
pH	4.3	5.6	8.4	8.8
Sodium, mg L^{-1}	1030	69,160	1350	3000

Source: Adham, S. et al., *Desalination*, 440: pp. 2–17, 2018.

As a result, at a low specific energy, the treatment of the produced water can be performed using desalination with the polymeric membranes. The shale play produced water is extremely adaptable. Generally, toward freshwater management, the purification of waste effluents generated from the petroleum industry is required to reuse the treated water during the oil drilling, hydrocarbons co-production purposes, etc.

6.2.4 Textile Industry

6.2.4.1 Introduction

The adverse impacts of the various industries have been reflected in the world as the socioeconomic revolution has spread worldwide. Inappropriately, the growth of the various industrial sectors has beaten up certain accidental effects, causing an unavoidable balance between industrial development and environmental poverty. The textile industry consumes a huge amount of freshwater and discharges highly contaminated effluents from the various stages of the textile processing containing various dye reagents, heavy metals, different inorganic salts, etc. [23]. However, it is reported that some parts of the substantial quantities of wastewater are not suitable to reuse due to the presence of non-degradable components. Thus, textile industrial effluents are likely to cause environmental difficulties if proper treatment is not completed. The wastewater found from the various sections of the textile industry is mostly rich in color due to the use of dye reagents, contains high chemical oxygen demand (COD), low BOD_5/COD ratio, complex chemicals, organic and inorganic salts, high total dissolved solids (TDS), normally high pH, high salinity, and turbidity [24]. According to the literature, the textile industry generated wastes can be classified into four different principal groups: dispersible materials, hard-to-treat wastes due to non-degradable matters, high sludge volume, and hazardous and toxic components [25]. Among the different complex chemicals found in the common textile effluents, the dye reagents have been inarguably selected as the most unconditional source of pollution. The direct release of the colored dye reagents into the environment unfavorably affects aquatic life, freshwater transparency, change of the taste of the drinking water, and disturbed dissolved oxygen content present in the freshwater due to the highly complex structure of the dyes, high molecular weight, and tendency of low biodegradability [26]. Further, the mutagenic and the carcinogenic effects of the dye reagents have been reported previously [27]. The presence of these comparatively recalcitrant dye reagents along with organic and the inorganic salts materials, various acids and alkali compounds, and other remaining chemicals in the wastewater directly released into the sewage systems hinders the biological treatment of the effluents [28]. Possibly, the utmost risk to environmental sustainability is modelled by the offensive consumption of water by the textile industries, resulting in the exhaustion of the availability of freshwater resources. The shortage in the accessibility of freshwater can be assessed by the example of textile industries in India, which consume approximately 0.2 m^3 of freshwater per kg of textile products fabricated, and generate above 200–350 m^3 of effluents per ton of final product [29]. According to the recent study, the water consumption by the industrial sector is rising with time due to the massive industrial growth and a substantial increase in population, which will possibly account for around 8.5 and 10.1% of the total freshwater extraction in the years 2025 and 2050, respectively [23]. The declining supply of freshwater is hence an associated result of the development of enormous industrial sectors and is sure to bring about a decrement in the performance of the various industrial sectors such as the textile industry due to the serious scantiness of freshwater resources or degradation in the quality of available drinking water. The various conventional methods to treat effluents have been incorporated to control the risk of textile industrial effluents. The wastewater treatment process should be equally skilled in the field of reclaiming the treated water using in textile processing to a large extent; such a preparation is crucial for the sustainable progress in the industrial section in our society. Previously, different treatment methods were in practice to diminish the contaminant levels of the textile effluents. However, all the methods suffer from certain severe problems. The eco-friendly biological methods, such as conventional activated sludge processes, an anaerobic treatment of textile industry generated wastes, and bioremediation processes are also performed for the treatment of textile industry wastewater. Their respective efficiencies are unfavorably affected by the biologically determined constitution of the impurities present in the textile effluents as well as by the diurnal variation in the atmosphere by means of the variation in effluent pH, temperature, and the concentration differences of pollutants in the textile effluents [30]. Furthermore, the conventional biological methods do not perform the complete mineralization of the target components such as dye reagents. Later, the toxicity of the discharged wastewater from the textile industry remains almost unchanged due to the presence of dye contaminants. This problem harshly obstructs the scale-up of the biological methods for the resulting reactor unpredictability [31]. The composite rheology of the textile industry discharge therefore involves either particular or integrated application of the physicochemical processes, such as coagulation-flocculation chlorination of effluents, adsorption process, and advanced oxidation processes, mainly ozonation of wastewater, Fenton oxidation process, electro-Fenton combined methods, photo-Fenton oxidation processes, and photo electro catalytic process to degrade the toxic materials from the textile wastewater completely [32]. The potential of the adsorption process is mainly unexploited due to the restrictions postured by the eco-friendly discard of the disbursed adsorbents, and the difficulty in restoration of the used adsorbents process efficiency and high chemical costs [33]. Advanced oxidation processes suffer from an annoyingly short half-life period and hence the process usually displays insufficient decolorization productivities for insoluble particles. The chemicals used in some operations such as chlorination and coagulation not only increase the cost of treatment but also produce a huge amount of residues that require secondary treatment. Furthermore, the degradation products present in the treated water may hamper the further production of fibers. Also, the efficiency of operations such as flocculation is restricted due to the high electrolytic strength generally detected for the textile effluents. These disadvantages can be acceptably overcome using membrane-based treatments including microfiltration, ultrafiltration, nanofiltration, reverse osmosis, and hybridization or integration of two or more of these processes for the textile industry effluents treatment. The membrane technology is normally known as clean and eco-friendly technology. The characteristic simplicity of the membrane separation process, the facility of modular design for treatment of feed samples in a large-scale operation, application of moderate temperature conditions, with no physical and chemical phase changing, no waste byproducts, and the insignificant use of additives are major key points of the membrane-based effluent treatment techniques. Besides, the considerable retention efficacies and steadiness characterizing are also significant factors for the membrane-based processes under different experimental environments [34]. These benefits account for the rising attention to membrane-based technology. However, a major weakness of any membrane filtration is severe membrane fouling. The appropriate selection of membrane separation process and the regular membrane cleaning using different chemical and mechanical techniques can minimize the membrane fouling problems and decrease used membrane replacement costs [35].

6.2.4.2 Characteristics of Wastewaters Generated from Textile Industry

The processing methods used in the different textile industries can be extensively classified, such as wet and dry processing, in accordance with the effluents properties produced therein [24]. Effluents found in textile industries, mainly in the wet processing, differ in the composition and the degree of toxicity due to the use of various chemicals. This toxicity depends on the mainly raw materials preparations, exact processes in operation, the machineries used in the processes, the quality of the water used for fiber processing, and the predominant management theory applied to control the water use. However, the dry processing areas mainly generate solid wastes containing chemicals and fabric rejects. Water is mostly used in a textile industry during the scrubbing, flushing of the raw materials, desizing of the cleaned materials, scouring or kiering process, bleaching of the scrubbed materials, mercerizing process, dyeing of the polished products, washing of the products, neutralization, and finally salt bath washing. Figure 6.6 reveals the entire process steps involved in a textile manufacturing industry where freshwater is consumed on a regular basis. It is reported that the use of freshwater varies with each process involved in the textile industry. Among all the steps, the dyeing and the printing sections are more water exhaustive processes than the others. The effluents generated from the dyeing and printing section contains high amount of toxic contaminants resulting in high biochemical oxygen demand (BOD), chemical oxygen demand (COD), total dissolved solids (TDS), and the total suspended solids (TSS) of the wastewater (Table 6.5). Hence, diversity has been found in the composition of the textile industry wastewater. The eco-toxicological effects of the textile industry generated waste materials released from different sources can be credited to a number of factors. The treatment of the toxic components using membrane-based technology is a major concern today.

6.2.4.3 Membrane-Based Treatment Processes

The use of membrane technology in such cases mainly depends on the materials involved to prepare the membrane. The mechanical, chemical, thermal resistance, and the membrane exposure are the major parameters to prepare the polymeric membranes to minimize the fouling behavior. Moreover, the organization of the concentrate streams containing target compounds from the feed solution and the recovery of other supplementary toxic chemicals during membrane filtration delivers the fruitful application of membrane-based separation processes. Thus, the additional membrane concentrate treatment processes have to be evaluated based on cost estimation, energy-intensiveness, final product efficiency, and environmental gentleness prior to the release of the residuals to the environment after membrane filtration [36]. For example, the direct discharge of the nanofiltration (NF) and reverse osmosis (RO) rejected materials after the treatment of dye bath effluents is not an environmentally accepted action due to the composite rheology of the concentrated materials. However, the bioremediation of the rejected materials through an activated sludge process is often not efficient technology to degrade the recalcitrant materials, whereas an anaerobic degradation process used in association with the membrane separation process can be a feasible option to treat textile industry generated effluents.

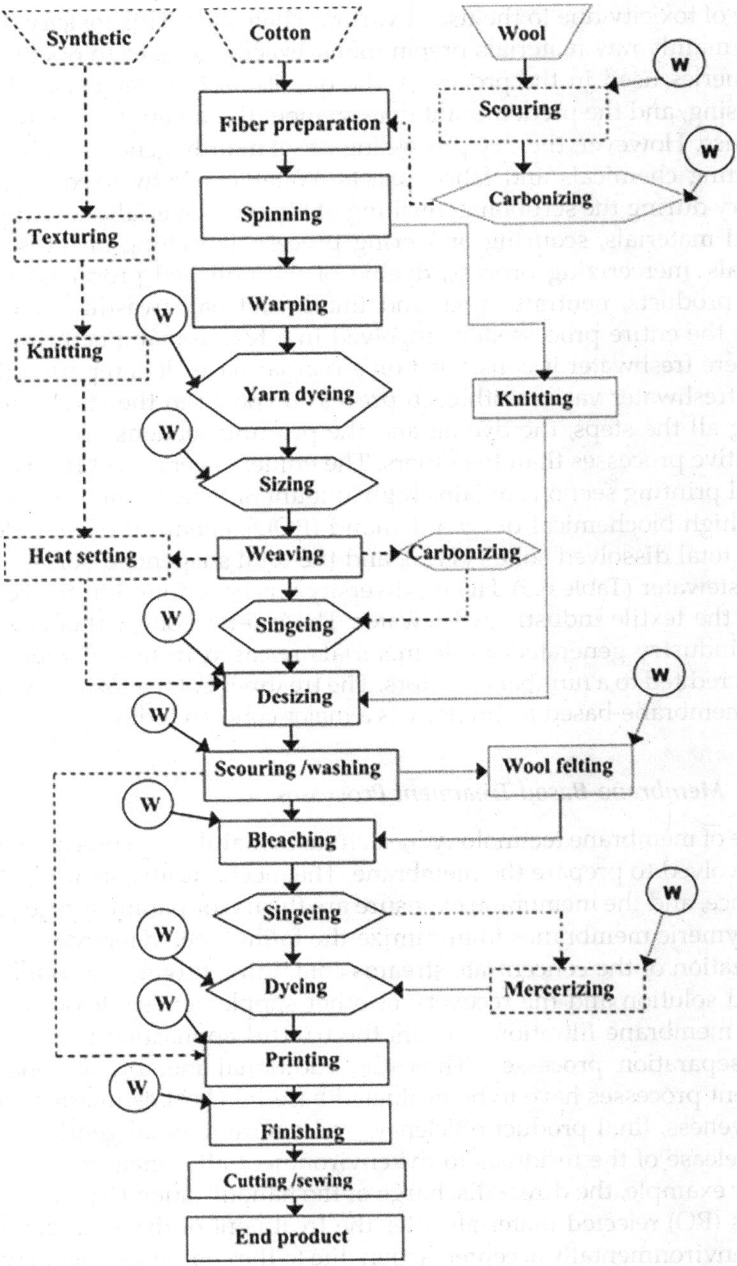

FIGURE 6.6
The process flowchart for the general steps involved in textile fabrication. (From Dasgupta, J. et al., *Journal of Environmental Management*, 147: pp. 55–72, 2015.)

TABLE 6.5

Physicochemical Analysis of the Textile Effluents Collected from Various Stages of the Textile Processing

Characteristics	Scouring	Bleaching	Mercerizing	Dyeing	Composite	Discharge Limit into Public Sewage	Maximum Permissible Limit for Water Reutilization
Color	–	–	Extremely colored	Highly colored	Highly colored (>14,000 Pt-Co units)	Colorless water	Colorless water
TDS (mg L^{-1})	12,000–30,000	2500–11,000	2000–2600	1500–4000	2900–10,000	2100	–
TSS (mg L^{-1})	1000–2000	200–400	600–1900	50–350	100–700	100	–
COD (mg L^{-1} of dissolved O_2)	10,000–20,000	1200–1600	250–400	400–1400	250–8000	250–500	–
BOD (mg L^{-1} of dissolved O_2)	2500–3500	100–500	50–120	100–400	50–550	30	–
Sulphates (mg L^{-1})	–	–	100–350	–	50–300	1000	400
Chlorides (mg L^{-1})	–	–	350–700	–	100–500	600–1000	600
pH	9–14	8.5–11	8–10	1.5–10	1.9–13	5.5–9.0	6.5–9.2

Source: Dasgupta, J. et al., *Journal of Environmental Management*, 147: pp. 55–72, 2015.

6.2.4.3.1 Microfiltration

Microfiltration has inadequate application during the treatment of textile industry generated effluents because of its adjacent similarity to basic filtration process, like sand filtration and mechanical sieve analysis [37]. The pore size of a typical microfiltration membrane varies in the range 0.1–10 mm. The separation of materials using microfiltration is naturally performed at a low transmembrane pressure (TMP) difference within 2 bar. The microfiltration is primarily applied to remove the suspended and colloidal dye particles from the dye bath of a textile industry. Hence, the microfiltration is rarely used as an unassisted independent treatment process for the remediation of toxic and complex industrial effluents such as textile industry wastewater. As such, it is frequently employed as a significant pretreatment process in a hybrid system to remove the suspended materials present in the effluents. For an example, a comparative study of textile effluents has been performed between the microfiltration followed by the nanofiltration, and the coagulation and flocculation process followed by the nanofiltration. The results showed that more significant amounts of suspended materials have been removed using microfiltration process than coagulation and flocculation as the high permeate flux has been obtained during the nanofiltration of the microfiltration pretreated wastewater. This observation was credited to the significant retention of COD, color of the raw effluent, turbidity, and the salinity of wastewater as the substantial removal of suspended materials was performed using the microfiltration of the textile wastewater resulting in less fouling characteristics during the nanofiltration process [38]. The study thus recognized the advantage of the use of microfiltration over the coagulation and flocculation process as the pretreatment step prior to the final treatment, like nanofiltration process to purify the textile wastewater. However, microfiltration can also be performed as a final treatment process while purifying industrial effluents. Previously, the electrooxidation reaction using an anode made of titanium oxide followed by ceramic membrane microfiltration has been studied to remove toxic soluble organic components and suspended materials, respectively, from a model textile wastewater prepared by azo dye [39].

The main factor for all the cases was to assimilate two appropriate treatment techniques by which the toxic materials and suspended components can be easily removed. Recently, an attempt using asymmetric tubular carbon microfiltration of textile industry wastewater has been performed successfully. Mineral coal powder was used to prepare the tubular carbon based microfiltration membrane. Figure 6.7 demonstrates the experimental apparatus accepted for the treatment of textile effluent. It is reported that more than 50% of COD removal was obtained in a single-handed process such as microfiltration, whereas approximately 30% of the existing salinity has been retained and significant removal of colored component has been found using tubular carbon microfiltration membrane in the field of textile industry effluent treatment [40]. In other investigation, to remove the methylene blue dye

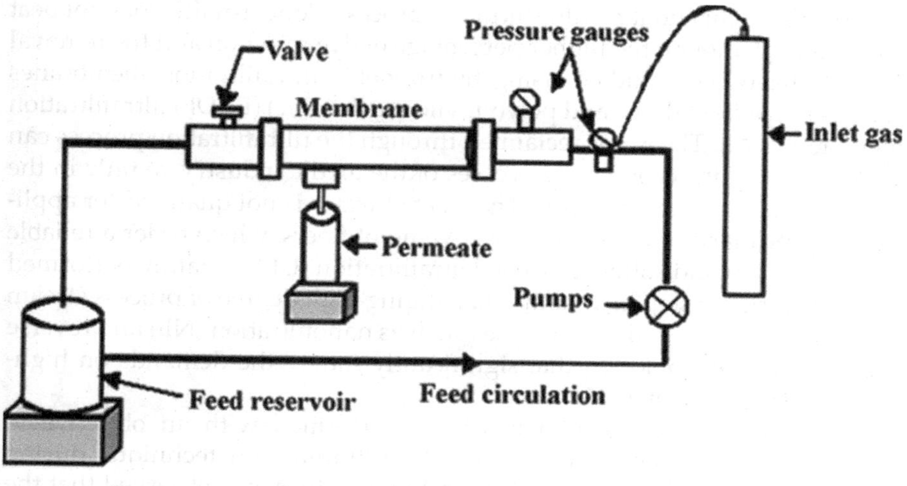

FIGURE 6.7
A schematic diagram of the pilot scale setup for the cross-flow mode of microfiltration process; temperature: 25°C, transmembrane pressure drop (TMP): 1–5 bars. (From Dasgupta, J. et al., *Journal of Environmental Management*, 147: pp. 55–72, 2015.)

from aqueous solutions the performance of a thin film composite (TFC) membrane made of organoclay/chitosan nanocomposite materials coated on the commercial polyvinylidene fluoride (PVDF) microfiltration membrane has been observed significantly [41]. The novel composite membranes were prepared using polyamide microfiltration membranes by merging two cationic materials, poly(ethyleneimine)-PEI and chitosan-CHI, and an anionic (poly (acrylic acid)-PAA) polyelectrolyte materials through the layer-by-layer (LbL) assembly. Two illustrative textile effluents, namely, methylene blue (MB) and Coomassie brilliant blue (CBB) solutions, were prepared to estimate the removal capacity of the above said membranes. Approximately, 95% COD, 79.9% MB removal, and 87.1% CBB retention were achieved using the PAA/CHI multilayers microfiltration membrane [42].

6.2.4.3.2 *Ultrafiltration*

The ultrafiltration is a process frequently used in the field of the separation of macromolecules and colloidal particles from industrial effluents. The solutes rejected in a common ultrafiltration process have molecular weights of a few thousands of Daltons [37]. Although hugely fruitful in handling pollutants present in the effluents released from the various chemical, pharmaceutical, and food industries, the ultrafiltration methods have partial applications during the treatment of textile industry wastewater, and this is mostly because the molecular weights of the dye reagent present in the toxic effluents are generally much lower than the molecular weight cutoff (MWCO) of the ultrafiltration membranes [43]. Consequently, the dye rejection brought about by ultrafiltration process alone usually does not beat 90% removal, though the higher percentage of dye rejection and the removal of COD have been studied using hydrophobic ultrafiltration membranes such as polyethersulfone and poly(vinylidene fluoride) (PVDF) ultrafiltration membranes [44]. The water reclaimed through the ultrafiltration process can be reused only in secondary processes of the textile industry, mainly in the rinsing and washing processes. The treated water is not qualified for application in primary processes such as dyeing of fibers, which order a reliable supply of clean and softened water. Ultrafiltration (UF) is mainly performed as a pretreatment step in systems challenging high degree of process stream cleansing; it is followed by processes such as nanofiltration (NF) and reverse osmosis (RO) technologies that significantly satisfy the demands on high-quality produced water [45].

Several innovative research has been performed with an objective to recover the performance showed by the ultrafiltration technique during the treatment of textile wastewaters. In the literature, it is observed that the ultrafiltration membranes were prepared using polysulfone materials at different evaporation temperatures by phase inversion process and studied the textile effluents removal efficiencies [46].

The other research described the novel mechanisms containing polymer and polyelectrolyte enhanced ultrafiltration (PEUF), which includes the complexation of dye reagents with the high molecular weight polymers, followed by the ultrafiltration process [47], and the micellar enhanced ultrafiltration (MEUF) technology, where some surfactants at a concentration beyond their critical micelle concentration (CMC) have been added to an aqueous solution containing toxic dye reagents to procedure micelles that solubilize the organic materials, which are later separated using ultrafiltration processes [48].

However, the progressive membrane fouling and the consequent decrease in permeate flux are the major demerits for the ultrafiltration process. This requires the outline of the novel hybrid processes, where the ultrafiltration process is preceded by an appropriate feed pretreatment process such as coagulation and flocculation processes. Continuous development of the membrane technology has been simplified by the expansion of modified membranes such as modified poly(vinylidene fluoride) membranes using styrene-acrylonitrile (SAN) composites [49] that have been rationalized to exhibit acceptable resistance to fouling.

6.2.4.3.3 Nanofiltration

Nanofiltration (NF) membrane technology is usually placed between the ultrafiltration and the reverse osmosis process. Its rising acceptance over the years as an attractive and easy textile effluents treatment process can be credited to the several aids it delivers in terms of environmental contamination abatement, retention, reutilization of textile dyes, divalent inorganic and organic salts, various auxiliary chemicals, and reuse of brine solution [23].

Additionally, the achieving of quality permeate permits the reuse of treated effluents in the major processes such as dyeing and finishing in the textile industry. Nanofiltration process operates at a comparatively low pressure drop, ranges from 500 to 1000 kPa, and it allows low retention of monovalent ion materials, which boosts the scope for low brine retention and reuse while allowing approximately 100% retention of multivalent ion materials, thus ensuring high solute selectivity. The rejection of solutes in the nanofiltration process is governed with the help of the steric and charge repulsion activities. Other beneficial characteristics of nanofiltration comprise its solvent permeability with high rate, rejection of dissolved uncharged materials like organic matters, with molecular weight of 150 Da (approximately), significant scale-up facilities, comfort chemical cleaning process, and the capability of NF membranes to perform at high temperature of 70°C, which decreases the energy consumption [23].

Numerous investigations have been explored to purify the textile effluents using the novel nanofiltration process [50]. It is found that the advantageous amalgamation of hollow fiber membrane configuration and submerged membrane filtration process has been obtained using thin film composite hollow fiber membranes made of sodium carboxymethyl cellulose (CMCNa)/polypropylene (PP) during the treatment of anionic dye solution. The hollow fiber membranes are often favored to flat-sheet membranes, due to their superior energy efficiency attached with high surface to volume ratio; submerged membrane filtration process, however, brings advantages such as relatively lower energy utilization and cleaning necessities than the other tangential filtration modes.

It is observed that the negatively charged CMC Na/PP complex hollow fiber membrane with a MWCO of 700 Da, approximately, was reasonably active in the field of the rejection of anionic dyes such as Congo red and Methyl blue from synthetic solution with the neutral pH. Results showed that the percentage of dye rejection, salt retention rate, and the water flux for a simulated solution containing 2000 mg L^{-1} Congo red dye agent and 10,000 mg L^{-1} NaCl solution were approximately, 99.8%, less than 2.0%, and 7.0 L m^{-2} h^{-1} bar^{-1}, respectively. The electrostatic repulsion between the dye particles and the negatively charged active membrane surface of the modernly invented membrane was observed as the vital and significant mechanism leading the submerged nanofiltration process of a saline anionic dye aqueous solutions [51]. In the other study, the performance of the different spiral wound NF membranes has been evaluated during the treatment of the secondary textile effluents. The behavior of all nanofiltration membranes was explored over an extensive range of volume concentration factors (VCF) (Figure 6.8). The resulting dissimilarity in the membrane fouling characteristics and the permeate flux decline nature was studied successfully [52]. Table 6.6 describes the different investigations performed by the different researchers during the treatment of textile effluents using nanofiltration.

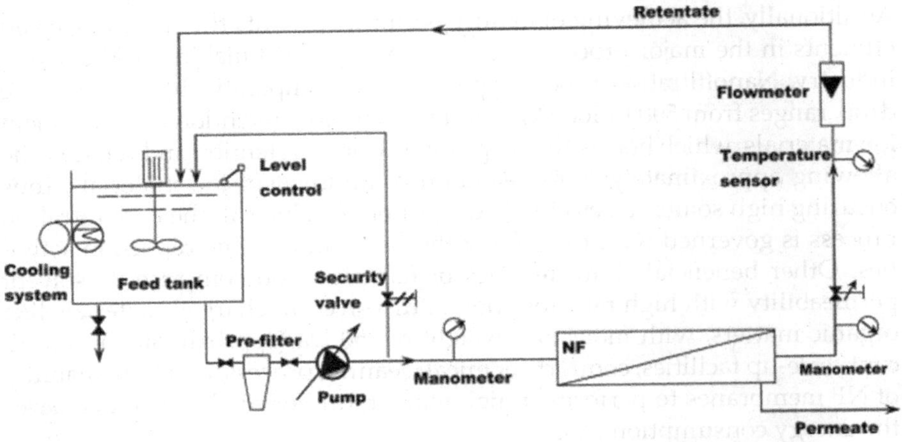

FIGURE 6.8
Schematic diagram of the experimental pilot plant setup for NF experiments. (From Bes-Piá, A. et al., *Journal of Hazardous Materials*, 178(1–3): pp. 341–348, 2010.)

6.2.4.3.4 *Reverse Osmosis*

Reverse osmosis (RO) is active in recovering and removing macromolecules as well as ionic compounds from the textile effluents. The treated effluent obtained from the reverse osmosis process is generally colorless and has less salinity. The usage of the dense polymeric membranes in the field of reverse osmosis and the high osmotic pressure generation due to presence of high inorganic salt concentrations significantly restrict the permeate flux increment, and at times severe fouling occurs, which distresses the membrane performance characteristics [53]. However, in RO process, the transmembrane pressures drop (TMP) greater than 2000 kPa are essential to preserve the reasonable permeate flux gaining, which again deals a critical blow to the method economics. Cross-flow filtration processes were carried out using a reverse osmosis and nanofiltration membranes while treating the textile effluents. It is reported that the treated products, for both cases, met the recovery criteria, and recyclable water of significant quality generated with each membrane filtrations could hence be reused to the textile manufacturing processes such as washing and dyeing sections, thereby saving on freshwater utilization and energy consumption and additional costs involved in the downstream purification of water [54].

6.2.4.3.5 *Electrodialysis*

Electrodialysis is performed, rather rarely, in industries such as textile industries for reduction of textile effluents contaminants. Literature review exposes that there is a lack of quality studies dealing with the utilization of electrodialysis based textile effluent purification techniques. Electrodialysis is extremely functional to remove chlorides components and hence is predominantly effectual for the significant remediation of wastewaters released

TABLE 6.6

List of the Different Investigations Performed by Different Researchers during the Treatment of Textile Effluents Using Nanofiltration

Process Description	Membrane Specification	Effluents Present	Component(s) Removed	Permeate Flux
Textile wastewater reclamation	Three flat sheet NF membranes: Desal-5, NE-70, and TS-40	Textile effluent (dyeing facility)	>99% dye removal Rejection (NE-70) > rejection (Desal-5) Turbidity, hardness, TOC and color removal: <0.2 NTU, 60 mg/L as $CaCO_3$, 10 mg/L and 5 HU, respectively	Flux (NE-70) twice flux (Desal-5)
Direct NF and UF/NF	Membranes (NF): NF90, NF200, and NF270 Set up: pilot plant Flat-sheet module	Secondary textile effluent (cotton thread factory) Effluent COD: 200 mg O_2/L TDS: 5000 mg/L	99% COD reduction Maximum (95–97%) salt rejection (NF90) Permeate conductivity (NF90) < 500 mS/cm UF pretreatment: 40% COD reduction	Flux trend: J (NF270) > J (NF200) > J (NF90) UF pretreatment: NF permeate flux increase (50%)
Nanofiltration using novelly fabricated membranes	UV-photografting (sodium p-styrene sulfonate) monomer on polysulfone UF membrane MWCO: 1200e1300 Da	Dyes: Acid red 4, Acid orange 10, Direct red 80, Disperse blue 56, Reactive orange 16 Salts: Na_2SO_4, NaCl	97% dye retention (0.4 MPa) Fouling tendency (photografted membrane) < fouling tendency (commercial polyamide membrane Desal SDK	0.23–0.28 m^3/m^2 day (0.4 MPa)

(Continued)

TABLE 6.6 (CONTINUED)

List of the Different Investigations Performed by Different Researchers during the Treatment of Textile Effluents Using Nanofiltration

Process Description	Membrane Specification	Effluents Present	Component(s) Removed	Permeate Flux
Cross flow nanofiltration	Flat sheet polysulfone based thin film composite (TFC-SR2) nanofilter	CI reactive black 5 (Bayer, Sydney), Salt: NaCl	Average dye rejection: 98% Average NaCl rejection: <14%	Average flux at 500 kPa: 59.58–78.4% Mean water flux recovery: 99%
Nanofiltration	UV-photo grafted nanofiltration membrane MWCO increases with increasing hydraulic permeability	Direct red 80 (DR80), disperse blue 56 (DB56), acid red 4 (AR4), reactive orange 16 (RO16) and basic blue 3 (BB3)	Dye retention: >96% at 0.4 MPa	Hydraulic permeability: 0.48–0.56 m^3/m^2 day
Dye wastewater reuse	Nanofiltration polyamide (PA) composite membranes MWCO: 500 Da	Separate aqueous solutions of 5 different dyes: Direct Red 75, 80, and 81 Direct Yellow 8 and 27 Model dyeing wastewater: Direct Red 75, PVA, NaCl, and Na_2SO_4	Almost 100% dyes rejection Retention efficiency improvement after coagulation (alum) pretreatment	20% flux improvement after coagulation (alum) pretreatment

Source: Dasgupta, J. et al., *Journal of Environmental Management*, 147: pp. 55–72, 2015.

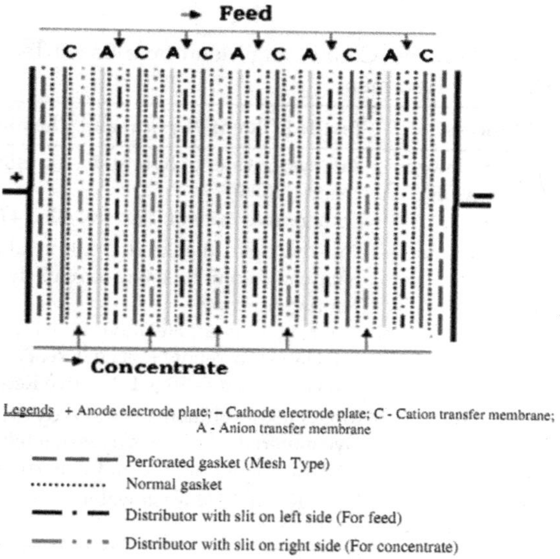

FIGURE 6.9
The schematic diagram of an electrodialysis process. (From Dasgupta, J. et al., *Journal of Environmental Management*, 147: pp. 55–72, 2015.)

from textile industries containing a huge amount of sodium chloride (NaCl), or salt. The electrodialysis process using the bipolar membrane materials (EDBM) has currently concerned attention due to its high energy efficacy and cost effectiveness than reverse osmosis (RO) technologies [23].

The treatment of textile effluents using electrodialysis process has been investigated in terms of total dissolved solids (TDS) reduction in a batch mode and continuous mode operation under constant current with sodium chloride solution (7500 mg L^{-1}) and sodium sulphate solution (5000 mg L^{-1}). The operating current density, in the range 3.6 to 4.8 mA cm^{-2}, was recommended for the treatment of textile CETP effluent with TDS, approximately 7000 mg L^{-1} [55].

The electrodialysis process is able to remove the successive volume load on evaporators through concentration of retained molecules obtained from a reverse osmosis (RO) process. The efficiency of an electrodialysis membrane system in achieving the anticipated concentration of the reverse osmosis treated textile discharge was also investigated (Figure 6.9) successfully with the current densities of 2.15–3.35 A m^{-2}, feed flow rate of 18–108 L h^{-1} [56]. The electrodialysis process successfully concentrated the reverse osmosis rejected solution during the treatment of textile industrial effluents.

6.2.4.3.6 Integrated Process

Today, many researchers have reported about the use of integrated membrane processes in the field of the various industrial effluents treatment. The significant effects of the integrated membrane technology have been

TABLE 6.7

The Different Integrated Process Connecting Membrane-Based Technologies for the Textile Wastewater Treatment

Hybrid or Integrated Process	Textile Effluent Physical and Chemical Characteristics
Fenton oxidation process followed by membrane bioreactor (MBR) system	Raw wastewater of hybrid dyeing effluent treatment plant (HDWTP) Primary dye reagent: Reactive Blue 4 (RB4)
Nanofiltration (NF) process followed by an anoxic biodegradation system	Post-dyeing textile (knitted cotton fabric) wastewater: simulated solution using various dyes Temperature: low Dye reagent used: Helaktyn Blue F-R (CI Reactive Blue 4), Helaktyn Yellow F-5G (CI Reactive Yellow 1), and Helaktyn Red Fe5B (CI Reactive Red 2)
Sequencing batch reactor (SBR) followed by nanofiltration (NF) process	Model textile effluents: various dye reagents solution Dye materials: 3 various dyes, namely, Remazol Yellow RR, Remazol Blue RR, and Remazol Red RR
Coagulation of wastewater followed by ultrafiltration process	Feed: Raw textile wastewater

Source: Dasgupta, J. et al., *Journal of Environmental Management*, 147: pp. 55–72, 2015.

obtained with favorable results during the purification of highly contaminated wastewater generated from the textile industries containing dyes, salts, and auxiliary chemicals [23]. However, many of the integrated systems are focused on the potential of conjugate with biological, chemical, and membrane separation technologies such as integrated ozone biological aerated filtrations followed by membrane reverse osmosis processes resulting in recyclable process stream [57]. For an example, a low-pressure microfiltration with hollow fiber membranes was performed [58] to separate and recover the anatase titanium dioxide (TiO_2) photocatalyst previously used to reduce the effects of organic azo dye reagent. The electro-catalytic oxidation using titanium and silicon based anode materials (Ti/SnO_2–Sb_2O_3–Y) followed by nanofiltration process was significantly carried out during another dye reagent, namely Acid red 73. The substantial effects of the oxidation process prior to the membrane filtration helps to reduce the concentration polarization and irreversible fouling of active membranes [59]. Table 6.7 delivers the different integrated membrane-based technologies, concerning two or more treatment methods, which simultaneously address the different features of textile effluents purification with pollutant removal, water recycling, and the recovery of dyes and chemicals for further use.

6.2.4.4 *Economic Evaluation: Textile Wastewater Treatment Using Membrane Separation*

The applicability of the different technologies in the industrial possibility can be determined and analyzed only after evaluating the realism of the process from the economical point of view. Hence, a huge number of assessments were directed by various scientists to confirm the economic viability of membrane-based separation processes to certify the successful service of these technologies in the field of textile effluents treatment [60]. For instance, the techno-economical assessment in terms of filtration cost effectiveness, energy consumption of an ultrafiltration, and reverse osmosis based effluent treatment of a secondary effluents from dyeing and finishing plants was carried out on a pilot scale mode [61]. The economic estimation based on payback period, the net present value, and the internal return rate during the ultrafiltration of textile waste effluent has been reported significantly [44]. The economic viability of the zero liquid discharge (ZLD) method was also evaluated during the treatment of the highly contaminated textile industry generated effluents using integrated membrane process [62]. Therefore, the explicit utilization of increasingly evolving membrane separation processes in the field of textile wastewater treatment can positively bring about the process growth in textile facilities in terms of the initial capital cost estimation through the conservation of water and a wide reuse of energy, process materials, and produced water [23].

6.2.5 Summary

In the present study, satisfactory diminution of various parameters, mainly of turbidity, COD, EC, TDS, and hardness, were observed. From the initial characterization of the effluents, ice-cream effluent was observed with more pollution potential but varies with type of process used in the industry. Toxic chromium, lead, and cadmium were observed in all food industrial effluents and may cause damage to aquatic and other life forms through bioaccumulation. Electrocoagulation with aluminum rods is a convenient route for the treatment of three food industrial effluents. After electrocoagulation, COD, turbidity, EC, and hardness showed decrease in values that also indicate that the technique is best and more efficient to treat such type of effluents. Powdered activated charcoal treatment proved to be better as compared with alum dosages. We also recommend that all the three effluents should be used for irrigation purpose as they have good fertilizing value. Useful nutrients from these effluents can be recycled from the sludge. Electrocoagulation should be promoted for such type of food effluents and other highly complicated liquid wastes. Wastewaters generated in the food and beverages industries depend on the specific site activity. Animal processors and rendering plants will generate effluents with different characteristics to those from fruit/vegetable washers and edible oil refiners (suspended/colloidal and dissolved solids, organic pollution, and oil and greases as well as microbial contamination). MF and UF systems can reduce suspended solids and microorganisms, while UF/RO combinations can also remove dissolved solids and provide a supply of process water and simultaneously reducing waste streams. UF systems can get more than 90% reduction in BOD and less than 5 mg L^{-1} in residual solids and less than 50 mg L^{-1} in grease and oil. NF systems are being used in a number of applications thanks to the quick development in new membrane materials. In case of RO process, BOD removal rate of 90 to 99% is possible providing a low cost controlled source of bacteria-free water. The favorable characteristics (modular) of membrane technologies allow different techniques as it has been seen all along this chapter. These hybrid processes can include traditional techniques as centrifugation, cartridge filtration, disinfection, and different membrane techniques building a "cascade design" used in many of the applications reviewed. The risk of membrane damage due to the contact with particles, salt conglomerates, chemicals, or others substances must be minimized to prevent short membrane life. Operation parameters must be carefully selected to obtain good results, especially not to overpass maximum temperature and transmembrane pressures recommended by membrane manufacturers. From the point of view of each particular process, to work at permeate flow rates below critical flux will assure longer runs. Membrane operating optimization is another aspect of paramount importance. It seems likely that the application of membrane systems in the food industry will continue growing rapidly. In particular, wastewater treatments will become more important in the next years because of the increasing cost of mains water and effluent sewer disposal. A membrane wastewater treatment system can be a major contribution to a food sector, and its introduction may feature as part of the continuous improvement plans within an environmental management system. Water touches most segments of the petroleum industry. Water managed by the industry is either co-produced with the hydrocarbons, generated as a byproduct from oil/gas processing, and/or used to support production operations. An overview of the various case studies of the oil and gas industry are provided where membrane processes were installed or considered and a review of applied research projects that included membrane technology evaluations. The current critical evaluation significantly highlights the advantages of the membrane-based treatment methods in generating regained textile effluents is quite tangible. The careful choice of the suitable membrane-based method is an important factor to achieve a high-quality permeate with low energy consumption. For an example, the quality of water recovered using microfiltration and ultrafiltration processes generally does not meet the criteria for reusing the reclaimed water in the serious processes such as dyeing of fibers. As a result, nanofiltration and reverse osmosis processes are therefore essential to produce the best quality treated water that can be reutilizes in the primary textile steps such as the dying processes. Furthermore, the concentrated solutes from the nanofiltration and reverse osmosis processes can be treated again using a comparatively energy effective process such as membrane crystallization and membrane distillation systems to regain the materials toward the concept of zero liquid discharge (ZLD).

6.3 Polyelectrolyte Membranes for Treatment of Industrial Effluents

6.3.1 Introduction

Industrial wastewater is considered a major source of water pollution due to the presence of various toxic contaminants. With the rapid growth of industrialization, the water scarcity and the degradation of water quality are also increasing due to the huge amount of freshwater consumption by industries such as paper, textile, leather, carpet, printing, and distillery and the discharge of highly contaminated effluents into rivers and other water bodies [63]. The complex characteristics of the contaminants due to the different features of raw materials, intermediates products, auxiliary inorganic and organic chemicals, and some residual dye agents, which are not reduced efficiently by the various biological processes, creates toxicity, high color, chemical, and biological oxygen demand [64]. This poses critical environmental contaminations, as this effluent increases the oxygen demand of the freshwater that is highly toxic to both fauna and flora. The high color is also an unwanted feature to the visual nature of the living environment. The dye reagents used in most of the industries, like textile, leather tanning, printing industries, etc., are steady to the light oxidation and aerobic ingestion resulting in decrement of dissolved oxygen in the water bodies. A few dye effluents (less than 1 mg L^{-1}) report color to the water bodies. The treatment of the industrial effluent is essential to minimize the concentration of toxic impurities in effluent to the level of allowable limits before discharge [65]. Several technologies have been established to eliminate the color components from the industrial effluents. However, there is still research to evaluate a cost effective, efficient, and eco-friendly process to remove or recover the toxic components from the industrial wastewater [66]. The activated sludge process can mainly minimize the biological oxygen demand with the help of bacterial oxidation; however, this system is not sufficient to remove the entire toxic materials responsible for the generation of color in the effluents. Hydrogen peroxide treatment of wastewater has also been carried out to degrade the toxic elements [67]. The Fenton oxidation process using H_2O_2 and Fe (II) salts can decolorize and degrade the soluble and insoluble dye reagents; however, the generation of sludge materials through flocculation is the demerit for this process. H_2O_2-activation in the presence of ultraviolet radiation is also an efficient process to degrade the pollutants but can produce hazardous derivatives [68]. Therefore, a significant and economically viable method should be developed to remove the toxic color components from the industrial effluent before being released into freshwater bodies. Membranes have been beneficial to treat wastewater generated from textile, paper, and other industries and to purify water. Membrane technologies such as ultrafiltration, nanofiltration, and reverse osmosis are effectively used in the field of groundwater and seawater purification. The naturally occurring

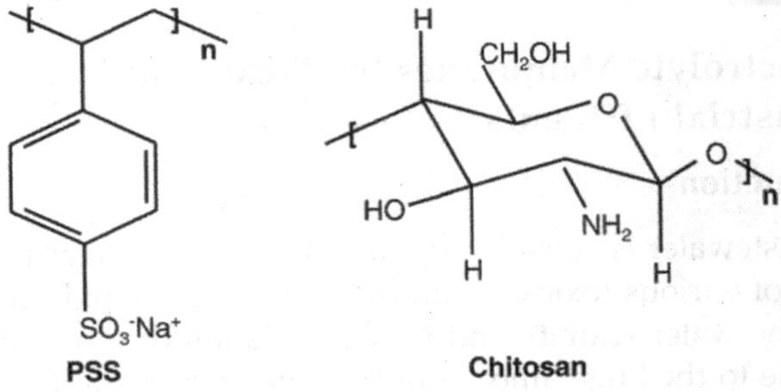

FIGURE 6.10
The polyelectrolytes used for deposition [1]. (From Aravind, U.K. et al., *Desalination*, 252(1–3): pp. 27–32, 2010.)

biodegradable materials can be incorporated economically to modify the polymeric multilayered membranes for the effective treatment of industrial wastewater [69]. The multilayered membranes can effectively encapsulate the different toxic molecules. The characterization of the effluent has been addressed in terms of the conventional parameters such as chemical oxygen demand (COD), color of the effluents, total dissolved solids (TDS), electrical conductivity of the wastewater, and pH. The membrane performance has been calculated for the textile and paper industry effluents at the various dilutions.

The polyethersulfone supporting membrane was cleaned with the double distilled water and kept in deionized water for the time period of 24 h. Polycationic and polyanionic solutions were used to rinse the clean polyethersulfone membrane. pH of the fresh polyelectrolyte solutions was adjusted to 1.7 by using HCl solution. The characteristic structure of the polyelectrolytes used for deposition is shown in Figure 6.10. The multilayers were prepared with the help of the consecutive adsorption of polystyrene sulfonate and chitosan on the polyethersulfone membrane [70]. Films were accumulated with a total number of 5.5, 10.5, 15.5, and 20.5 bilayers. The membrane used was rounded in shape with an effective diameter of 4.8 cm^2. The prepared chitosan–polystyrene sulfonate is designated as the polyelectrolyte pair. Chitosan components contain free amino groups that are a fragile polyelectrolyte. However, polyelectrolytes with the low charge density are proficient to form a less cross-linked multilayer, resulting in swollen membranes. Polystyrene sulfonate materials are the most extensively used polyanion in the research of the preparation of multilayers. The development of the multilayers on the surface of polyethersulfone membrane has been evaluated using scanning electron microscope (SEM), which is shown in Figure 6.11 [1].

6.3.2 Treatment of Paper Mill Effluent

Paper industry also generates a huge number of byproducts that create serious discarding problems and affect the ecosystem [71]. The paper industry

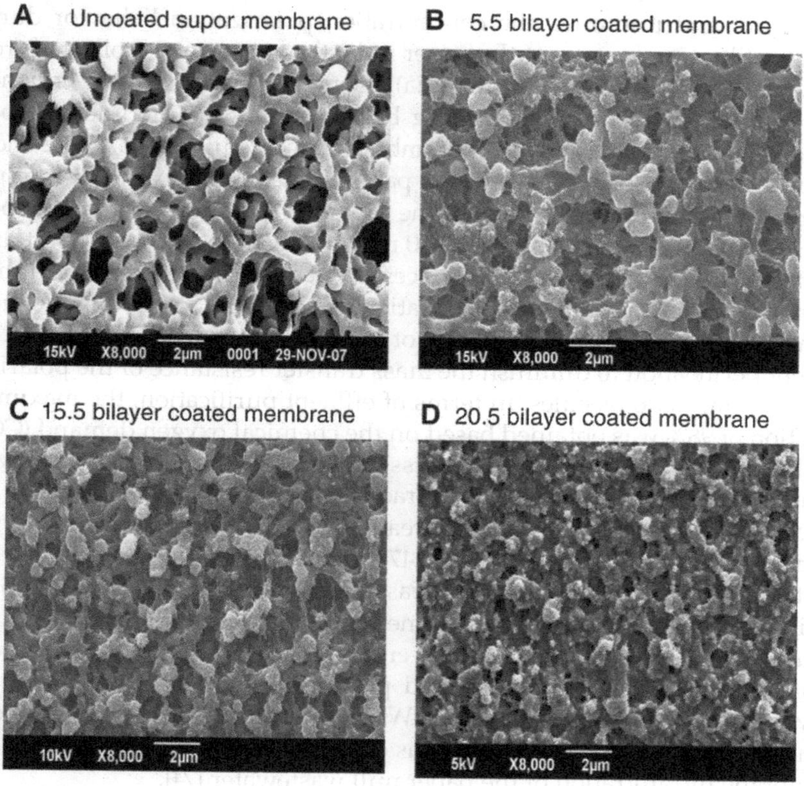

FIGURE 6.11
Scanning electron microscope (SEM) images of different multilayer membranes (CHI/PSS). (From Aravind, U.K. et al., *Desalination*, 252(1–3): pp. 27–32, 2010.)

generated black liquor was already used to remove the Pb^{2+} and Zn^{2+} ions from model wastewater. The adsorption capacity of 1865 and 95 mg g^{-1} was found for the $Pb^{2+}$and Zn^{2+}, respectively, at 40°C [72]. The adsorption of phenolic compounds such as 2-nitrophenol, 2-chlorophenol, 4-chlorophenol, phenol, 2,4-dichlorophenol, 3,5-dichlorophenol, etc. using paper industry sludges was also evaluated significantly with a high removal rate by [73]. The major disadvantage of the adsorption process is the disposal problems of used adsorbents.

The paper industry generated effluents contain fatty and resins acids, tannin and lignin derivative materials with high biochemical and chemical oxygen demands (BOD and COD). The treatment of paper industry effluents has been performed previously using conventional processes such as adsorption of toxic components using activated carbon from the wastewater, coagulation technology, biological treatment using aerobic and anaerobic digestions, sedimentation process, etc. However, the membrane technology delivers as an attractive method due to its cost effectiveness, the reuse of produced water, and the recovery of valuable products from the industrial effluents.

To control and minimize the concentration polarization behavior during the ultrafiltration process of the paper industry wastewater (collected from, M/s. Nagaon Paper Mill, Assam, India), the newly designed shear-enhanced membrane system, namely spinning basket membrane module, has been explored using polyethersulfone membrane successfully. Figure 6.12 shows the representation of the treatment of paper industry wastewater using spinning basket membrane module. In the presence of various rotational speed of the membrane basket (10.47 to 73.30 rad s^{-1}), the deposition of the retained solutes on the membrane active surfaces was less resulting from a minimum membrane resistance during the filtration process. The rotation of the membrane basket helped to rise the rate of membrane shear that enhanced the solvent permeation to diminish the mass transfer resistance of the polarized layer of retained molecules. In terms of effluent purification, the maximum rejection of 98% was obtained based on the chemical oxygen demand (COD) at a high applied transmembrane pressure drop (TMP drop) of 414 kPa with a membrane basket rotation of 52.36 rad s^{-1}. The initial COD load of the collected wastewater and after the pretreatment of the raw effluents using vacuum filtration were approximately 8470 and 6430 mg L^{-1}, respectively. The electrical conductivity was also decreased significantly from 4.8 to 0.58 S m^{-1} during the spinning basket membrane ultrafiltration of the paper industry effluent. Not only that, the total power supply during the filtration process was calculated by energy consumed per unit volume and the maximum power supply was obtained as 0.32 kW. Thus, at the cost of less energy consumption, good rejection capacity was observed during the spinning basket membrane ultrafiltration of the paper mill wastewater [74].

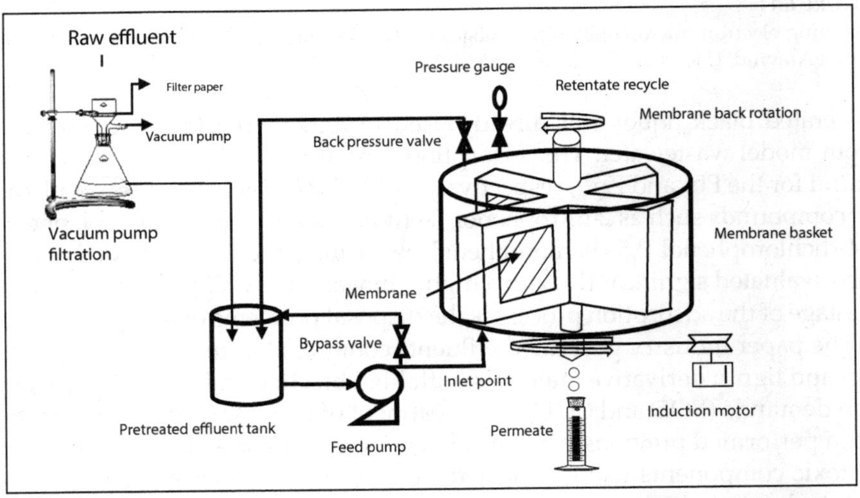

FIGURE 6.12
The schematic representation of the treatment of paper industry wastewater using spinning basket membrane module. (From Saha, S., and Das, C., *Journal of Environmental Chemical Engineering*, 5(5): pp. 4583–4593, 2017.)

6.3.2.1 Color

The characterization of the paper industry effluent can be performed based on the analysis of the effluents' color, clarity, total dissolved solids content, COD and BOD load, pH, electrical conductivity, etc. The color of the wastewater is generally analyzed using potassium chloroplatinate and cobaltous chloride solutions. The analytical results are described in Table 6.8. The removal of the colored components from the effluent having pH 7.3 was evaluated with different numbers of bilayers membranes. The high molecular weight materials such as tannins, lignin derived materials, and associated compounds are mostly responsible for the concentrated color of the paper industry effluents [75]. The interaction between the negatively charged compounds and the polyelectrolyte bilayer supported polymeric membranes has been observed in the literature resulting in a successful removal of the toxic chemicals during the treatment of the paper industry effluents. The assembled polyelectrolyte bilayers deposited are arranged alternatively, like positive (chitosan–CHI) and negative (polystyrene sulfonate–PSS) charges, respectively, and the membrane surface layer is always finished with the chitosan materials. As a result, the active interaction and restriction of the organic compounds present in the paper effluent to the bilayers can occur and the efficiency of the immobilization of the toxic materials raised with increased bilayers starting from 5.5 to 20.5. The reduction in color components with the varying pH range of 6 to 9 has been obtained during the membrane treatment process [1]. The control of the fouling characteristics was also reported for the newly composite membranes formed by the polyelectrolytes bilayers. The UV–visible absorption results for the untreated and the treated paper mill effluents has been revealed in Figure 6.13. It is observed that the lignin like materials are principally accountable to generate the strong colors in the wastewater and the absorbance range of the lignin compounds is approximately 250–300 nm. A reduced absorbance in this region has confirmed the removal of lignin compounds during multilayered membrane treatment of the paper industry effluents.

6.3.2.2 pH

pH is used to show the acidity and alkalinity condition of a prepared solution. The pH range of the natural water is in the range of 4 to 9. Due to the presence of carbonates and bicarbonate compounds, most of the waters are somewhat alkaline in nature. For the drinking water, the range of the pH is between 7.0 and 8.5, generally. The pH of the untreated effluents varies with the raw materials, chemical, and other parameters involved during the production. For example, the initial pH of a paper mill wastewater was reported about to 10. With the help of the spinning basket membrane module, the pH of the treated effluents has been changed to 7.45, which became invariant for all the conditions [74]. However, after the treatment of an industrial effluents

TABLE 6.8
The Physicochemical Properties of a Paper Mill Effluent

Properties	Total Dissolved Solids (TDS) (ppt)		pH		Electric Conductivity (m S)		Turbidity Standard Suspension (40 NTU)		COD mg L^{-1}		Color Hazen Units	
Effluent	Paper	Textile	Paper	Textile	Paper	Textile	Paper	Textile	Paper	Textile	Paper	Textile
Raw effluent	1.62	6.85	7.37	11.68	3.71	10.4	22.1	2480	12,500	2750	3000	–
After treatment 5.5 bl	1.53	3.24	8.75	10.17	3.34	7.20	7.7	2200	4000	2400	2000	–
10.5 bl	1.51	3.2	8.68	10.08	3.29	7.06	8.1	2110	2500	2200	1600	–
15.5 bl	1.51	3.15	8.65	9.75	3.26	6.97	8	1600	2000	2000	1100	–
20.5 bl	1.53	3.15	8.63	9.75	3.32	7	7.7	720	1500	1700	1000	–

Source: Aravind, U.K. et al., *Desalination*, 252(1–3): pp. 27–32, 2010.

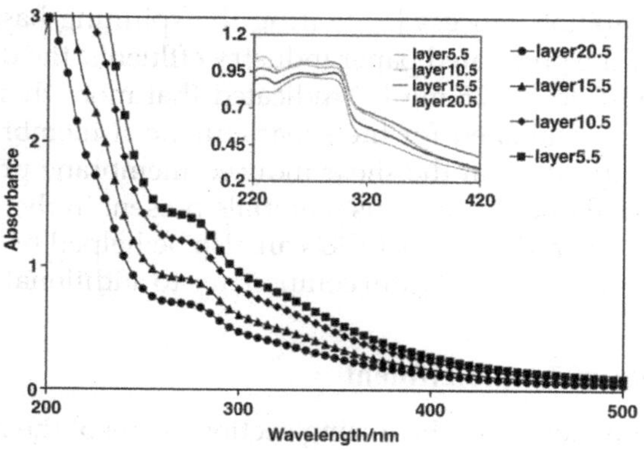

FIGURE 6.13
UV–visible absorption results of the paper industry effluents after membrane treatment. (From Aravind, U.K. et al., *Desalination*, 252(1–3): pp. 27–32, 2010.)

with a polyelectrolyte bilayer membrane, the pH of the sample was increased from 7.45 to 8.5 due to the decrease in the concentration of hydrogen ions after the adsorption of the lower carboxylic acids [1].

6.3.2.3 Total Dissolved Solids (TDS), Electrical Conductivity, and Turbidity

Electrical conductivity of natural water is directly proportional to the dissolved minerals present in the water, since the electrical conductivity differs directly with the water temperature. The difference in the dissolved organic and inorganic solid contents is specified by conductivity measurements. The variation of the electrical conductivity has not been varied significantly during the treatment of paper industry effluent using membrane bilayers [1]. With the help of various rotational speeds of a spinning basket membrane module, the electrical conductivity decreased from 4.80 to 1.52 S m^{-1} using polyethersulfone membrane with molecular weight cutoff (MWCO) of 50 kDa during the shear-enhanced treatment of paper mill wastewater [74]. However, the suspended materials are unwanted in treated waters because that shrinks the brightness, distresses the color, inhibits texture and consistency, and favor progress in slimes [1].

6.3.2.4 Chemical Oxygen Demand

Chemical oxygen demand (mg L^{-1} of dissolved O_2) is the amount of oxygen consumed during the oxidation of organic and inorganic materials under some specified conditions, modified for the influence of the chlorides. COD test is extensively employed to verify the pollution strength of industrial and domestic wastes. The COD of an untreated paper industry effluent decreased from 12,500 to 1500 mg L^{-1} using polyelectrolyte bilayer membrane filtration indicating that approximately 88% removal of COD has been obtained during this separation process [1]. During the spinning basket membrane ultrafiltration of the pretreated paper industry effluents, the decreased value of COD from 6340 to 160.75 mg L^{-1} indicated that more than 97% removal capacity has been evaluated for the shear-enhanced membrane ultrafiltration [74]. This indicates that the shear-induced membrane process is operative in the removal of carbonaceous materials present in the paper industry waste effluents. The reduction of COD can also be helped by the conversion of some functional groups of lignin components to additional oxidized form.

6.3.3 Treatment of Textile Effluent

The effluent generated from the dyeing section is one of the major contributors to the textile wastewater, and this colored effluent has a severe influence on the living environment. The effluent is mainly comprised of secondary chemicals and the residual dye components. The most common dyes used in the textile industry are Brown R, Red 6B, Orange RR, Black BB, etc. The dye removal rate can be evaluated by investigating the color and COD level of the effluents. The pH, TDS, and electrical conductivity analysis are also important parameters to evaluate the toxicity level of the textile industry wastewater. The physicochemical parameters help to compare the treated effluent with raw wastewater during membrane-based purification. Table 6.8 displays some typical observation of the physicochemical characteristics of textile industry generated effluent [1].

6.3.3.1 Color and COD

During the bilayer membrane treatment of textile industry effluent, the color is found to reduce with the rise in the number of bilayers in the membrane setup. The pH control plays an important role during the polyelectrolyte membrane-based wastewater treatment. The maximum removal of the toxic colored materials is found at lower pH. The COD reduction efficiency for the textile wastewater is in the range of 70%. Chitosan being a fragile polyelectrolyte material, the molecular construction of the film can be organized by varying the different charge density with varying the polymer chains. The multilayer composition is showed to varying with pH conditions of the effluent, and as a result, the degree of ionization deviates that disturbs the degree of interpenetration between the active layers, and the number of unbound chemical functional groups. Thus, the optimization of the pH is required to control the interaction between the bilayer and the molecules. The neutral pH plays a significant role to remove the color and COD level significantly. The removal degree depends on the structure of the used chemicals during the industrial process and the molecular size of the dye reagent. It is reported that it is the important reason that the COD reduction of the textile effluent is lower than the paper effluent as the textile effluent contains highly complexed dye reagents [1].

6.3.3.2 pH, TDS, and Electrical Conductivity

pH of the textile industry wastewater is comparatively higher than other effluents. Most of the manufacturing processes in the textile industry are performed in an alkaline pH. The high alkaline pH of above 11 indicates that the textile effluent contains soluble anionic dyes. After the treatment with polyelectrolyte bilayer membrane, the decreasing rate of the pH has been noticed. The lower removal rate of the COD has been indicated due to the lower rejection of the organic compounds during the bilayer membrane separation. It is highly possible that the unabsorbed dye materials still convey the high pH of the treated effluent. The reduction of COD and color components can be improved with the adjustment of pH of the raw effluent. The remarkable variation of total dissolved solids (TDS) and the electrical conductivity has been observed after membrane separation of the textile industry effluent. Electrical conductivity was decreased from 10.4 m s to 7 m s representing the significant absorbance of small molecular weight materials [1].

6.3.4 Summary

The treatment of paper industry and textile mill wastewater using chitosan-based polyelectrolyte membranes has been demonstrated in this section. The comparative study using spinning basket membrane module during the treatment of paper industry effluent has also been discussed here. Chitosan is predominantly important due to its biodegradable characteristics. Thus, a considerable reduction of COD and color has been observed during the paper industry effluent using chitosan-based membrane. This method is estimated to deliver an effectual route for the purification of industrial effluents such as paper and textile mill wastewater. Chitosan is previously recognized as a decent adsorbent for the removal of dye reagent. During the paper mill effluent, the chemical oxygen demand (COD) reduction was more effective than textile effluent using bilayer membrane separation. Spinning basket membrane ultrafiltration using polyethersulfone membrane delivered higher COD removal than chitosan-based membranes. However, during the textile effluent treatment, the pH adjustment played an important role to remove the color materials and COD. The composite membrane can be reused for further purification after the removal of the adsorbed components with the help of pH variation.

6.4 Membrane Bioreactor for Industrial Wastewater Treatment

6.4.1 Introduction

Application of membrane bioreactor (MBR) in water and wastewater treatment is not a new technology. Research on membrane bioreactor (MBR) is ongoing to make more efficient and healthy treatment of industrial wastewater containing severe toxic materials [5]. In the industrial area, the presentation of membrane bioreactor has been extensively deliberated since the early 1990s when the first large connection of membrane bioreactor was performed in the United States through General Motors at their plant situated in Mansfield, Ohio [76]. The first continuous scale internal membrane bioreactor process to treat effluents from the food and beverages industry was installed in North America in 1998 [77]. In the 1990s, the submerged membrane bioreactor was successfully commercialized and it was observed to have low working cost compared to other types of membrane bioreactor [78]. In membrane bioreactor, the biological systems perform a vital role where solutes as well as contaminants in effluent are transformed into the final products before filtration is performed by the active membrane [79]. Membrane bioreactor (MBR) is also recognized as a different process for the conventional activated sludge (CAS) treatment where clarifier is substituted and recovered with the membrane system to overcome the settling problem during the formation of undesired biomass. Membrane bioreactor (MBR) also can produce high performance during the treatment of water besides having small tracks compared to the conventional activated sludge process where clarifiers are removed. MBR also gives high quality effluent [80], is decent in eliminating the organic and inorganic pollutants, proficient of deterring high organic loading [81], and produces less slurry [78]. With all the merits of membrane bioreactor (MBR), various industries have already started the installation of the MBR to diminish the cost of water by recycling the treated water for other processes. For an example, the treated water can be used in landscape and industrial sanitary purposes. Not only that, the high-quality treated water from MBR can be reused for the heat integration and processing by confirming that the treated water has small amounts of impurities to avoid the failure of complex equipment or pipes [82]. However, membrane fouling is a critical factor for the MBR and study on how to diminish the fouling behavior is still continuing [83,84]. Other restrictions of concern are the restrictions in pH, temperature, pressure, and some corrosive chemicals [78,82,85]. Fouling infects not only the microorganisms in the reactor but also abolishes the membrane morphology. Recently, some researchers have carried out alterations and incorporation of MBR to decrease the restrictions [86,87]. Some of the most common manufacturers of membrane bioreactor (MBR) are M/s. Kubota from Japan, M/s. Zenon from Canada, and M/s. Mitsubishi. M/s. Kubota has conquered the installation; in the meantime, M/s. Zenon has installed more capacity water treated through their membrane process [82]. In industries, the merits during the management of MBR are to generate the optimum condition to operate MBR for high strength industrial effluents, shock loading rate, to regulate the biofouling effect, and diminish the energy consumption during operational [78]. Recently, the fouling effect by soluble extracellular polymeric substances (EPS), soluble microbial products (SMP), and inorganic compound dominates in MBR research.

6.4.1.1 MBR Configuration

Large clarifying basins are needed in conventional activated sludge (CAS) treatment to ensure the complete settlement of the flocs. High power for diffuser is utilized in the aeration basin to make sure that the nutrients are totally transformed to the final products. In industries, membrane bioreactor is performed as a secondary treatment process to decrease the biodegradable and non-biodegradable compounds in the final product or in advance treatment to recover the remaining nutrients that are not fully used during secondary treatment. The basic view of MBR configuration is imperative before the MBR alteration for improvement is made. Basic schematic diagram of MBR configuration is shown in Figure 6.14. Figure 6.14a shows a side-stream or external membrane module while Figure 6.14b shows an immersed membrane bioreactor (iMBR) or submerged membrane bioreactor (sMBR) module [5]. For sMBR system, the feed wastewater is directly in contact with biomass. Wastewater and biomass are both pumped through the recirculation loop consisting of membranes. The concentrated sludge is recycled back to the reactor while the water effluent is discharged. The idea of separating the membrane and bioreactor is to ease the membrane maintenance, but it will increase the operational cost due to recirculation loop installation [88]. The iMBR system has less operational cost because there is no recirculation loop compared to the sMBR system and a biological process occurs around

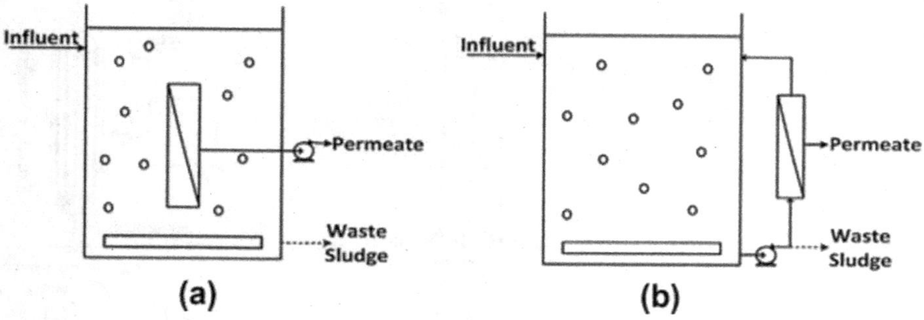

FIGURE 6.14
Basic schematic of MBR: (a) immersed MBR, (b) side-stream MBR. (From Mutamim, N.S.A. et al., *Desalination*, 305: pp. 1–11, 2012.)

the membrane in iMBR. Both iMBR and sMBR need to pump out the excess sludge to maintain sludge age. The mode of membrane transportation could be pressure driven or vacuum driven. Radjenovic et al. stated that pressure-driven filtration is used in sMBR and vacuum-driven is used for iMBR, which operates in dead-end mode [82]. The air bubbles are supplied to both the systems for aeration besides scrubbing, especially for the immersed system to reduce membrane fouling in cross-flow effect through the membrane surface [80,89]. There are also aerobic and anaerobic MBR where oxygen acts as an important medium for the microbial growth in the aerobic process while anaerobic is done without oxygen. Anaerobic MBR is less efficient in removing COD and takes a long time for startup [5]. Usually, anaerobic treatment is used for treating high-strength wastewater at low temperature, which is suitable for microbial growth. Moreover, it is difficult to adjust low temperature for the waste feed, and it causes high fouling compared with aerobic at low flux.

According to the natures of the effluent, membrane bioreactors have been modified into three types. Figure 6.15a MBR shows biomass separation, Figure 6.15b is a membrane aeration bioreactor (also called membrane aerated biofilm reactor [MABR]), and Figure 6.15c is an extractive MBR (EMBR). However, MABR and EMBR were applied in a pilot scale for industrial wastewater. Purified oxygen is directly used by MABR without bubble formation for biofilm growth on the external side of the membrane. Within the biofilm under aerobic condition, the organic matters were biodegraded and the entire oxygen supplied was almost utilized for biodegradation. Its difficulty is to maintain the optimum thickness of biofilm to get sufficient oxidation; the growth of excessive biofilm can hinder the liquid flow problem [90]. EMBR is typically operated in the presence of high concentration of inorganic materials such as high salinity of compounds and extreme pH

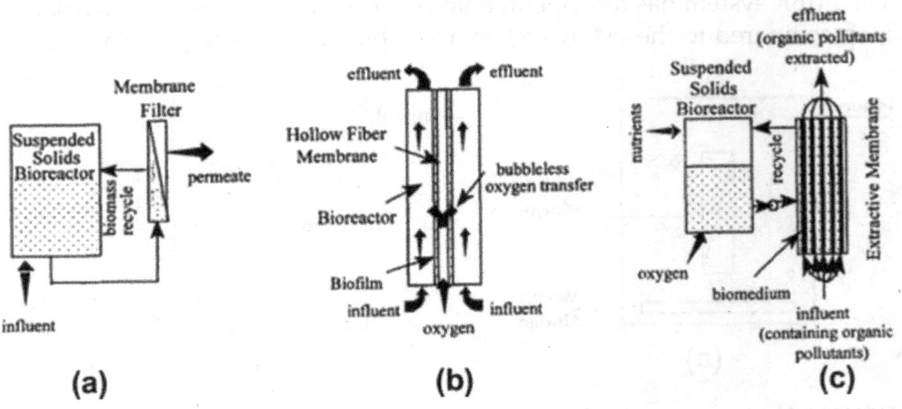

FIGURE 6.15
Types of MBR processes: (a) MBR biomass separation, (b) membrane aeration bioreactor, (c) extraction MBR.

value that might hinder the biodegradation process. EMBR selectively recovers the specific organic contaminants, namely, phenol, hydrogen sulphide, and some inorganic that can be treated in a separated bioreactor [91].

6.4.1.2 Membrane Behavior

The importance of studying membrane behavior is to select good quality membrane in treating high strength industrial wastewater. High strength wastewater consists of diverse contaminants that could possibly corrode the membrane and lead to operational failure [5]. The efficiency of the membrane also depends on the size of pores, types of materials, types of wastewater to be treated, solubility, and retention time. Retention is observed due to the mix liquor suspended solid (MLSS) concentration change between the retentate (a part of solution that cannot cross over the membrane) and permeate (solution after filtration). Permeability, flux, transmembrane pressure (TMP), and resistance are the parameters that also need to be considered. Permeability is flux per transmembrane pressure drop ($J/\Delta P$). Flux (L/m^{2}h) is the flow of permeate per unit of membrane (component accessibility to the membrane) and it is related to hydraulic resistance, thickness of the membrane, or cake layer and driven force. Driving force is the gradient of membrane potential area (unit area of the membrane) of mass transport that involve pressure and concentration of particles. The mass transport mechanism for the membrane depends on the structure and materials of the membrane. Membrane structure plays an important role in transport mechanism whether the structure is parallel or in series. Diffusion and solubility of the component are related to the kinetic ability of mass transport for membrane. Membrane pore size participates in kinetic mass transport for the membrane itself [92]. Depending on the size of the impurities in the treatment process, the types of membranes used are different. Generally, two types of membrane materials are widely used, namely, polymeric and ceramic. Ceramic membranes are usually used for industrial wastewater treatment as these have good performance in filtration compared to polymer because of high chemical resistance, and they excessively inert and easy to clean [93,94]. Chemical stability of the membrane does not only depend on the materials used but also on the pore size; it reduces when the structure is very fine. Ceramic also has advanced hydrophilic capability due to the significant water contact angle. However, the main hindrance of ceramic membrane is it is exclusive to fabricate and fragile [95]. However, in the current membrane separation process, various polymers have been used commercially in the form of PVDF, PES, PE, and PP due to the respectable physical and chemical resistance. The porous polymer membrane has its own weaknesses where it can foul easily due to the hydrophobic characteristic. The separation of materials is performed using hydrophobic membrane because the pore size can easily be fabricated. Weakness of hydrophobic membrane is improved by providing hydrophobic polymer coating [96]. PE is more quickly fouled compared to PVDF [78]. Membrane

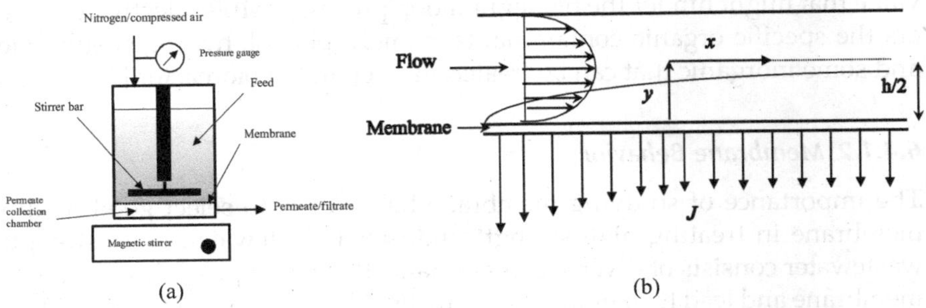

FIGURE 6.16
(a) Dead-end filtration, (b) cross-flow filtration.

configuration also plays an important role as every configuration has its own merit and demerit based on the cost, capability to withstand turbulence and back-flushing, etc. [88]. From the operation point of view, membrane processes are of two types, namely, dead end and cross flow. These are shown schematically in Figure 6.16a and 6.16b. In dead end filtration, solutes are deposited on the membrane surface and permeate is collected from the bottom of the cell. In a cross flow system, the direction of feed flow is normal to the direction of the permeate flow, hence defined as a cross flow system, and the feed is allowed to flow tangentially over the membrane surface and the growth of concentration boundary layer is arrested. Cross flow patterns are observed in plate and frame module, tubular membrane module, hollow fiber membrane module, spiral wound membrane module, spinning basket membrane module, and rotating disk membrane module.

6.4.2 Application of MBR in Industrial Wastewater Treatment

During MBR operation, there are different operating conditions depending on the level of constituents of high strength wastewater. The operating conditions cover the sludge behaviors such as MLSS, dissolved oxygen (DO), hydraulic retention time (HRT), and solid retention time (SRT) and membrane behaviors in terms of membrane configuration and pore size. The operational parameters for textile and food industries are tabulated in Table 6.9. Table 6.10 shows the characteristics as high strength wastewaters, but these are different in terms of biodegradability. Food industries have high biodegradability due to the presence of slow biodegradable organic/toxic matters compared to food industries [97]. In food industry wastewater, the level of biodegradability is high due to the presence of readily biodegradable organic matters [98].

TABLE 6.9

MBR Operational Parameters for Industrial Wastewater

	Textile			Food		
	Textile [99]	**Textile [100]**	**Textile [101]**	**Pet Food [102]**	**Palm Oil [87]**	**Dairy Product [103]**
Reactor volume (L)	500	20	230	20	20	20
Reactor type	Aerobic, side-stream	Aerobic, side-stream	Aerobic, submerged	Aerobic, submerged	Aerobic, submerged	Aerobic, submerged
Membrane configuration	UF (7 tubular modules), PVDF	UF, external tubular cross-flow, PVDF	HF	2 modules	FS, 1 module	MF, 34 strands of a HF
Membrane surface area (m^2)	–	0.28	–	0.047	0.1	0.00162
Pore size (μm)	0.025	–	0.04	0.04	0.4	0.4
Flux (L/m^2h)	–	30	20	–	10	Horizontal: 5.03 Vertical: 2.27
MLSS (mg/L)	5000–15,000	–	13,900	–	5000	4000–10,000
MLVSS (mg/L)	–	–	–	–	47,000	–
DO (mg/L)	1–3	2–3	–	3	8	–
HRT (day)	2	0.7–4	0.58	2.9	0.8	–
SRT (day)	–	11	25	50	–	–
COD removal (%)	97	90	97	97	94	–
Colour removal (%)	70	98	98	–	–	–
TSS removal (%)	–	–	99	–	–	99

TABLE 6.10
High Strength Wastewater Characteristic

Industry	COD (mg/L)	BOD_5 (mg/L)	BOD_5/COD	NH_4–N (mg/L)	TSS (mg/L)	SO_4^{2-}	PO_4^{3-} (mg/L)	Oil (mg/L)
Tannery	16,000	5000	0.313	450	–	–	–	–
Textile	6000	700	0.117	20	–	–	120	–
Textile	4000	500	0.125	4.8	–	200	2	–
Dyeing	1300	250	0.192	100	200	–	–	40
Textile	1500	500	0.333	50	140	–	7	–
Wheat starch	35,000	16,000	0.457	–	13,300	–	–	–
Dairy	3500	2200	0.629	120	–	–	–	–
Beverage	1800	1000	0.556	–	–	–	–	–
Palm oil	67,000	34,000	0.507	50	24,000	–	–	100,000
Pet food	21,000	10,000	0.476	110	54,000	–	200	–
Dairy product	880	680	0.773	–	2480	–	–	–
Pharmaceutical	6300	3225	0.512	–	1679	–	–	–

6.4.2.1 Textile Industries

In textile industries, the principal source of wastewater is spent dye and washing water, contain huge color, salt, toxicity, low biodegradability. These industries apply adsorption, biological treatment, chemical precipitation, and membrane technology [101]. Hai et al. used hollow fiber (HF) and flat sheet (FS) membranes at the same condition with 50–200 μm pore of membrane for treating high strength synthetic textile industry wastewater comprised of various chemicals, organic loading, and color. Mutamin et al. have used white-rot fungi *Coriolus versicolor,* NBRC 9791 to remove specifically color and other nutrients with a reactor volume of 12.5 L and each of two HF-FS membranes with same surface area of 0.2 m^2 and size of 0.4 μm (Mitsubishi Rayon) [5]. Badani et al. have shown that the best performance by mix liquor suspended solid (MLSS) was at 15,000 mg/L with average color removal of 70%. The flux decline was observed due to membrane fouling with trans-membrane pressure (TMP). For a long-term performance, the hydraulic retention time (HRT) of 2 days was sufficient for almost 97% COD degradation [99]. In another study, Brik et al. specified that at the beginning of the process, the sludge 5000 mg/L of MLSS was treated with municipal wastewater before being acclimatized with textile effluent. The MLSS was found to increase to 10,000 mg/L and continued to increase up to 15,000 mg/L before sludge withdrawal. It was also observed that the effect of nutrient addition could not improve COD removal. However, adding the nutrient contributed to conductivity due to the inorganic content and causing severe fouling [100]. A report by Yigit et al. revealed that colonies of biomass with a wide spectrum of degradation capability were able to degrade as readily biodegradable, slow biodegradable, and biorecalcitrant for a highly concentrated mixed textile wastewater with BOD_5/COD ratio of 0.32. [101].

6.4.2.2 Food Industries

Food processing industries generate highly contaminated wastewater containing various organic materials. A wide range of protein molecules and various fatty acids are present in this kind of wastewater. As a result, the BOD and COD load is high for food industry generated effluents. However, the primary treatment of the food industry effluent is generally performed before discharging the wastewater to municipal effluent treatment plants [104]. According to literature, the removal of organic materials from the food industry wastewater was found about 37% using membranes due to the rejection of particles larger than the active membrane pore size. The mechanism of nitrogen removal rate was also evaluated into two stages for membrane bioreactor (MBR) such as nitrification–denitrification. The removal rate of the nitrogen at different treatment states was 21% of influent nitrogen was removed by SND; after that, 31% was removed using cell synthesis (N_{cell}), and 13% was purified by stripped process ($N_{stripped}$) during the first stage of

MBR treatment [105]. Another study showed the performance of the different configurations of membrane such as horizontal and vertical at the various concentrations of MLSS and their significant effect on the performance of MBR, although the COD and BOD values were found less, and the amount of total suspended solids (TSS) was high. The result described that horizontal flux was declined slowly than vertical flux and by rising the MLSS concentration, the permeate flux was reduced. TSS and turbidity removal was good when the MLSS concentration was low. This study also delivered that the pH performance was significant at low with high MLSS, which raised to 7–9 after treatment [106].

The treatment of palm oil mill wastewater was performed successfully using hybrid membrane bioreactor system coupled with ultrafiltration and reverse osmosis. An anaerobic expanded granular sludge bed (EGSB) and aerobic biofilm reactor (ABR) were used during biological treatment of oily wastewater. Results showed that 93% COD removal and 43% organic matters removal were found using EGSM system. The recovered organic materials can be further used in the field of the generation of bioenergy. In ABR, the COD removal rate was observed at only 27%. During the reverse osmosis treatment, near 99.99% COD removal rate was noticed. All the suspended solids were removed by the UF and the dissolved solids and inorganic salts were filtered significantly using RO [107].

6.4.2.3 MBR: Fouling, Limitation, and Mitigation

The cost estimation is a major restriction to membrane bioreactor (MBR) technology due to the preparation of membrane, which leads to increase in the maintenance and operational costs. Severe membrane fouling is mainly responsible for the increase in the cleaning cost. The analysis of fouling needs deliberation when it originates to membrane. During the treatment of high strength effluents containing various amounts of contaminants, it will generate high clogging of the active membrane surface area. Some major factors that effect the membrane fouling during membrane bioreactor (MBR) process are biomass concentration, suspended materials, dissolved solids, particle size, MBR configuration, operating parameters (transmembrane pressure drop, cross-flow velocity, hydraulic retention time), etc. With the help of the variation of cross-flow velocity, TMP drop, and aeration time, fouling can be checked and controlled. Figure 6.17 demonstrates the different irreversible pore clogging behavior. Fouling is mainly the physicochemical interface between the bio fluid and the membrane to generate a cake or gel layer and the severe adsorption of the dissolved organic materials into the active membrane pores resulting in flux decline continuously. In a general membrane bioreactor system, the side streams have higher fouling affinity than a submerged MBR system. The high energy of pumping requires the side stream of MBR that generates high permeate flux, which will lead to reiterating the fouling characteristics compared to the submerged MBR system

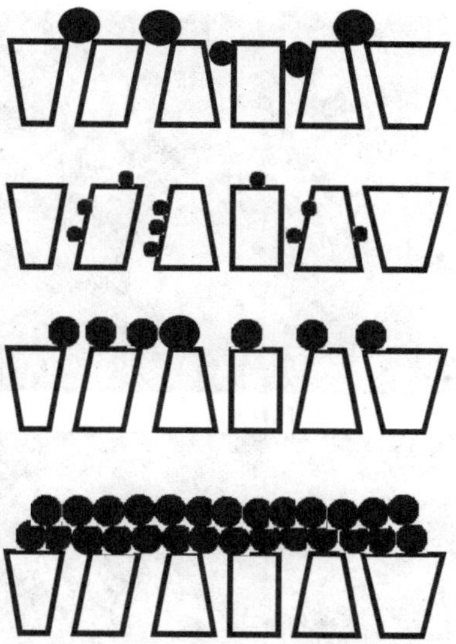

FIGURE 6.17
Diagrammatic representation of membrane fouling mechanisms: (a) complete blocking, (b) standard blocking, (c) intermediate blocking, (d) cake filtration.

[108]. The deposition of the biopolymers including proteins and polysaccharides is responsible for the organic fouling in a MBR system. The deposition of biopolymers is critical to remove [109]. Figure 6.18 describes the scanning electron microscopic demonstration of the fouled membranes. Figure 6.19 shows the energy-dispersive X-ray spectroscopy results of the fouled membranes [110].

The insoluble forms of the extracellular polymeric substance and the soluble microbial products lead to severe membrane fouling. The extracellular polymeric substance (EPS) is found at the outside of the cell surface and the soluble microbial product is the organic materials that is released during the substrate metabolism or due to the biomass decay. Both of the compounds contain proteins, short chain polysaccharides, nucleic acids, fat and lipid, humic substances, etc. The connection between the EPS and SMP with membrane material is unfavorably difficult [111]. From literature, the EPS and SMP overlapping mechanisms have been analyzed by the unified theory. The EPS and the SMP compounds overlap with each other and the active cells use the electrons from the electron-donor materials to form active biomass and generate bound EPS at the same time. In count, some of the SMP are adsorbed by the biomass flocks to produce bound EPS. The different fouling mechanisms depend on the characteristics of the sludge materials. The excess growth of filaments controlled by low dissolved oxygen concentration produces better filtration than normal sludge treatment because of the large particle distribution. The extracellular polymeric substance (EPS) [112] is nearly related with the formation of specific cake

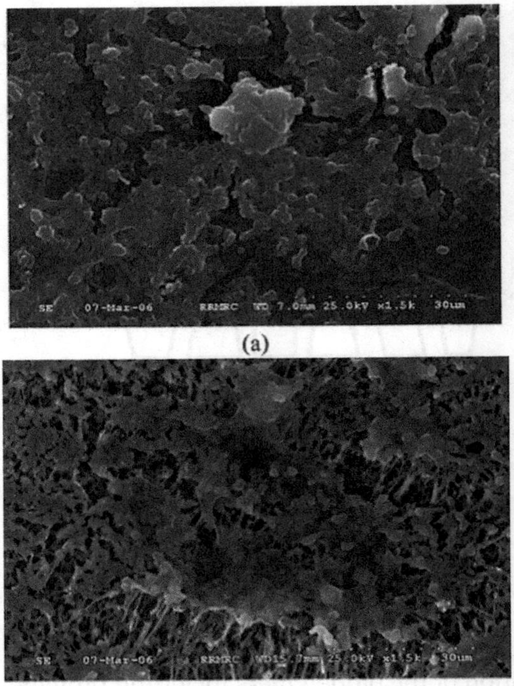

(a)

(b)

FIGURE 6.18
The SEM of fouled membrane (a) outer surface and (b) inner surface. (From Chang, C.Y. et al., *Desalination*, 234(1–3): pp. 393–401, 2008.)

resistance. The filamentous bacteria make bulking problems leading to production of high extracellular polymeric substance (EPS) concentration [113].

With the rising of heavy metal concentration, the membrane permeability is decreased due to high formation of extracellular polymeric substance (EPS) [114]. It is reported that the soluble microbial products content is considered as an indicator of the fouling control level since with the rising in soluble microbial products concentrations in membrane bioreactor (MBR) tends to decrease the membrane permeability. The high irreversible membrane fouling happens due to the less sludge age. Thus, the fouling mitigation operational parameters are the part of limitation during the treatment of the wastewater using membrane bioreactor [5]. As a result, the operating parameters require judicious control to reduce the severe fouling effects. Sludge retention time (SRT) is a valuable operational parameter that influences membrane bioreactor (MBR) performance, particularly in the control of irreversible fouling problem. With the long sludge retention time normally recovers filtration performance and decreases EPS and SMP production with the help of starvation conditions. The severe fouling can happen with the long SRT.

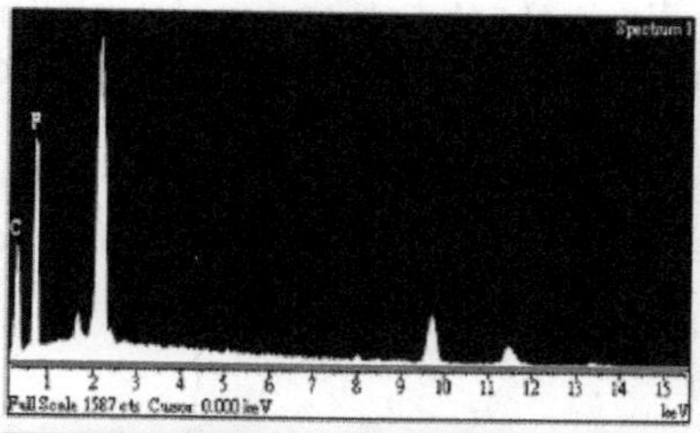

Element	Wt.%	At%
C	43.84	55.25
F	56.16	44.75

(a)

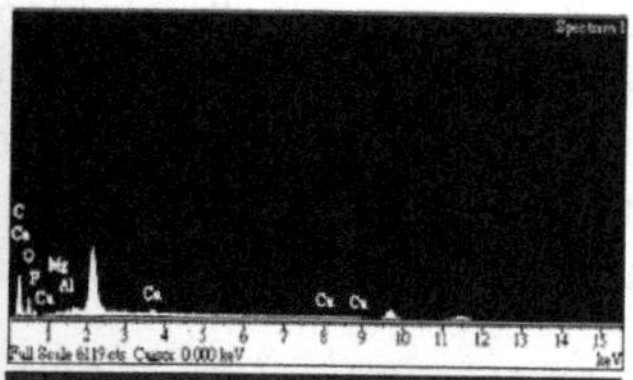

Element	Wt.%	At%
C	47.24	57.59
O	31.33	28.68
F	14.85	11.45
Mg	0.72	0.43
Al	0.88	0.48
Ca	1.66	0.61
Cu	3.32	0.76

(b)

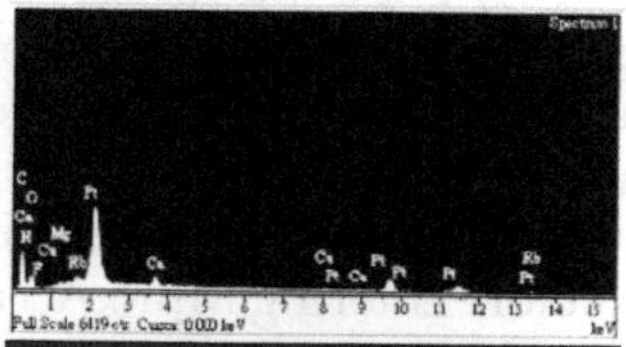

Element	Wt.%	At%
C	35.49	45.57
N	15.92	17.53
O	31.51	30.38
F	4.09	3.32
Mg	0.85	0.54
Ca	3.60	1.39
Cu	2.79	0.68
Rb	1.36	0.24
Pt	4.40	0.35

(c)

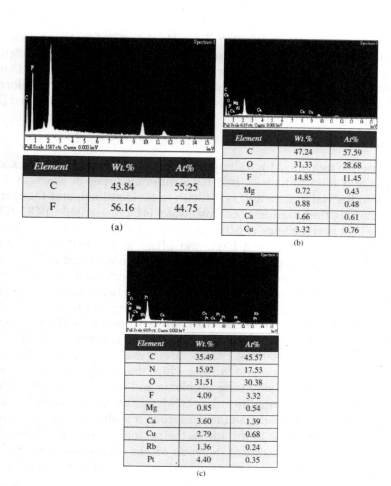

FIGURE 6.19
EDX spectra and elemental composition of nascent and fouled membranes: (a) nascent membrane, (b) inner surface of fouled membrane, (c) outer surface of fouled membrane. (From Chang, C.Y. et al., *Desalination*, 234(1–3): pp. 393–401, 2008.)

Correspondingly, if the sludge retention time (SRT) is too small, there will be a chance to reduce the performance of membrane bioreactor (MBR) due to less biomass generation. Further, for the operational control, the membrane cleaning is required during the decline of the permeate flux, slightly. There are generally three types of membrane cleaning during filtration: physical, chemical, and combination of the physical and chemical cleaning. Physical cleaning comprises backwashing, and the water is sent in the reverse direction of the filtration. The backwashing is a speedy process but is less operative than chemical cleaning [115]. Fundamentally, the physical cleaning removes the coarse solids or cake formation on the active surface of the membranes, whereas the chemical cleaning removes the flocs and also the deposited materials. It can also eliminate the strong materials that stick on the active membrane surface. However, the chemical cleaning, except the fouling elimination, can hinder the membrane surface morphology. During the industrial persistence, the in situ cleaning process is typically performed if the fouling is not critical; otherwise, the ex-situ cleaning is performed [116].

6.4.3 Summary

MBR is used to treat most high strength wastewaters successfully and shown in textile and food industries that have different types of strength. High strength industrial wastewater is difficult to be identified, and the method used to identify the "hardness" of wastewater is biodegradable (BOD_5/COD) ratio. The best performance of MBR can be produced by controlling parameters such as SRT, HRT, TMP, Flux, and MLSS to an optimum condition. Besides, the control of operating parameters is demonstrated to be effective in dropping fouling characteristics. Permeability performance is affected by the recent parameters such as SMP and EPS that need more investigation in the industrial sectors. To avoid membrane fouling, wastewater with high loading has to be treated before entering MBR. According to the MBR operation condition, physiological characteristics (MLSS concentration, EPS, SMP, organic and inorganic matters) change, and these make it critical to regulate and forecast the membrane fouling. Thus, there are some techniques to eliminate or minimize the fouling problems and enhance the performance of MBR to produce high-quality effluents. Membrane is the heart of this system but besides pH, pressure, and temperature, it is also sensitive to unusual chemicals that contribute to minimize the performance of MBR. Since the quality of wastewater treatment depends on the involvement of processes and the environment, these problems need to be considered. To prolong the life span of the membrane, besides the maintaining of microbial growth, adjustments of the wastewater, dilution for the high strength and toxic materials contained effluents, and pretreatment or neutralization of acidic or alkaline nature of effluents are sometimes required.

6.5 Polymer-Enhanced Ultrafiltration Membranes

6.5.1 Introduction

Due to the endless growth of polluted wastewaters discharged in the environment by urban and industrial communities and to their potential noxious impact on human health, the international regulations have been steadily toughened in the last decades. Remediation of waters polluted by heavy metals, and especially in their ionic form, has therefore become a major concern.

Hence, industries have been compelled to find novel alternatives to treat their effluents before they are released into the environment [116–118]. In this context, the implementation of membrane technology appears to be a potential option for the treatment of contaminated effluents [119–121]. In particular, nanoporous membranes used for nanofiltration or ultrafiltration have demonstrated remarkable abilities for the removal of ionic contaminants in terms of performances but also in terms of economic [122] and ecological [123–125] benefits. They allow separation of ionic components according to size and electric mechanisms [126]. Although nanoporous membranes have demonstrated several benefits, they unfortunately exhibit drawbacks for metal ion removal due to the size of the membrane pores. Large pores of UF membranes allow considerable permeation fluxes but low ion rejections, whereas the tight pores of NF membranes lead to high rejections but with low volume fluxes. Consequently, many recent studies were devoted to improve rejection performances without compromising on flux [127–129]. Among the viable options, the use of polymers from natural resources or synthetic procedures has demonstrated outstanding achievements to this end. Polymers can be used to modify the physicochemical properties of the membrane to increase its rejection performances. This modification can be implemented either by surface deposition, for instance, via coating [130–131] or grafting [132], or by including additives during the membrane manufacture [133–135]. Besides membrane modification, a polymer can also be introduced in the feed solution to increase the effective size of the pollutant ions by complexation. After this preliminary step, the feed solution containing complexed ions is then filtered by ultrafiltration and ion rejection is therefore potentially enhanced (compared to non-complexed ions) by steric or electrostatic mechanisms [136]. This hybrid process, which is usually called polymer-enhanced ultrafiltration (PEUF), is much easier to implement compared to chemical membrane modification, especially in view of a potential application for the treatment of urban or industrial wastewaters [137]. PEUF is mostly used to remove metal ions by adding synthetic chelating agents such as polyacrylic acid (PAA) [138] or polyethylenimine (PEI) [139]. The complexation of metal ions has already been studied in literature, but this section aims at accurately investigating the enhancement of performances induced by the addition of two eco-friendly polymers, namely chitosan and carboxy methyl cellulose (CMC), in several conditions. More specifically, the influence of metal or polymer concentrations, salt addition, or solution pH value on ion removal performances is deeply studied for Ni(II). Among metals that may cause disorders, nickel is not necessarily the most toxic, but its accumulation in the environment may represent a critical hazard to human health [140]. Moreover, this ion is commonly present in industrial effluents, especially those of surface treatment industry, and its removal is revealed to be essential [141]. Finally, industrial discharge water from industry has also been treated by ultrafiltration assisted by chitosan to establish if performances can be extrapolated to complex mixtures containing several ions.

6.5.1.1 Ion and Polymer Solutions

All synthetic solutions studied were made with demineralized water with a residual conductivity lower than 0.1 mS/cm. Sodium nickel salt provided by Acros Organics was used as reference pollutant. Nickel concentrations investigated in this study were chosen to cover a wide range of values commonly observed for industrial effluents, from wastewater before any treatment to discharge water released in the environment. For this reason, concentrations were varied from 1 to 100 ppm corresponding to molar concentrations from 1.7×10^{-5} to 1.7×10^{-2} mol L^{-1}. The real effluent was collected before discharge (i.e., after usual physicochemical treatment to comply with safety standards) from a surface treatment industry of Franche-Comte. Chitosan is a biodegradable and biocompatible polysaccharide, which is proved to be a perfect substitute for synthetic polymers from petrochemical industries [142]. Chitosan used in this study is provided by France Chitine (Orange, France) [136]. It exhibits a molecular weight of 1.8×10^{-5} g mol^{-1} and a degree of acetylation (DA) close to 15%, which means that it contains 85% of amino groups (NH_2) and 15% of acetamide groups ($NHCOCH_3$), as can be seen in Figure 6.20a. The pKa value of chitosan is around 6.3, which means that it is positively charged at natural pH (5.4 in our case) but it can become mainly neutral for pH above 6.3 [137]. Carboxymethyl cellulose (CMC) is a cellulose derivative that contains carboxymethyl groups (CH_2COOH) bound to some of the hydroxyl groups of the glucopyranose monomers (Figure 6.20b). It is provided by Sigma Aldrich in the form of sodium salt with a degree of substitution of 0.7 and a molecular weight of 9×10^{-5} g mol^{-1}. CMC contains anionic groups with an approximate pKa value of 3.5–4.0 depending on the conditions [143]. Above this value, CMC is thus negatively charged. Quantities of chitosan and carboxymethyl cellulose were chosen so that 12–1200 mol of repeating monomer unit can interact with each mole of Ni^{2+}, that is, from 2×10^{-4} to 2×10^{-2} mol L^{-1} of repeating unit.

6.5.1.2 Polymer Addition and Filtration Experiments

Firstly, CMC is dissolved in 5 L of demineralized water, chitosan is dissolved in 5 L of diluted acetic acid solution 0.1 M, and both are then stirred

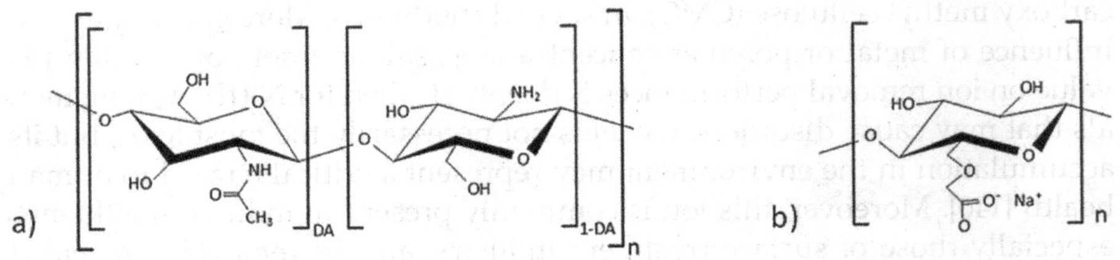

FIGURE 6.20
Chemical structure of (a) chitosan, (b) carboxymethyl cellulose. (From Lam, B. et al., *Journal of Cleaner Production*, 171: pp. 927–933, 2018.)

during 1 h at room temperature to obtain complete dissolution. After that, 30 L of feed solution were prepared by mixing solution of dissolved polymers and solution containing Ni^{2+} ions [136]. This solution is stirred once again during one night in a tank to optimize interactions between nickel and polymers. This solution containing both polymer and ions is used as feed solution for filtration experiments within the framework of the PEUF. Filtration experiments were implemented for feed solutions containing Ni^{2+} ions with and without addition of polymers. Solutions containing only ions are filtered without pretreatment whereas solutions containing polymers are introduced in the feed tank after the step of polymer addition described in the previous paragraph. Filtration experiments are carried out with a setup at the semi-industrial scale. As seen from Figure 6.21, 30 L of feed solution are pumped in the cross-flow unit containing a flat-sheet membrane before being recycled in the feed tank to keep concentrations constant. Filtration cell is equipped with a polyamide thin film composite flat-sheet membrane (Desal GK, Sm = 14×10^{-3} m^2, MWCO = 3.5 kDa) supplied by M/s. GE Water & Process Technologies (Trevose, USA). The membrane is washed before experiments by an acid/base cycle before being equilibrated by water filtration until permeability remains constant. Once the stabilization phase is done, solutions are filtered at various applied pressures at maximum flow-rate possible with this cell (i.e., 800 L/h corresponding to a cross-flow velocity close to 4 m/s) to minimize concentration polarization phenomenon at the membrane wall thanks to turbulence [144].

Thermoregulation of the solutions is ensured by a cooling unit that maintains temperature at 25 ± 1°C. After a stabilization time during which

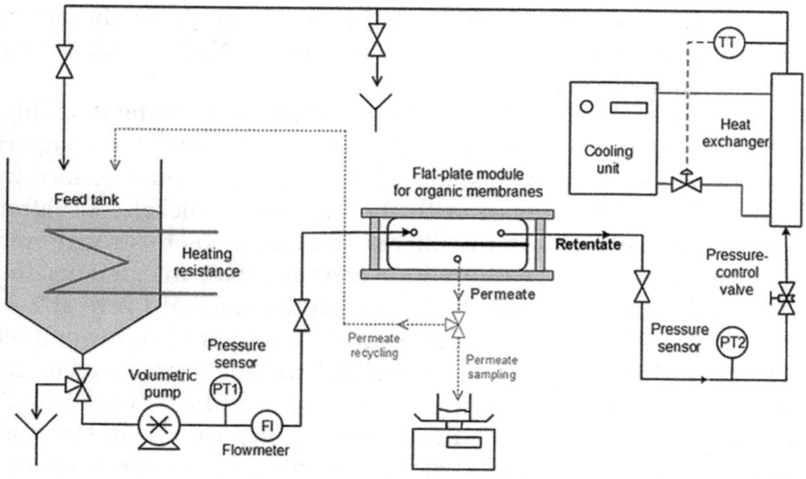

FIGURE 6.21
Scheme of the setup used for filtration experiments. (From Lam, B. et al., *Journal of Cleaner Production*, 171: pp. 927–933, 2018.)

permeate stream is recycled into the feed tank, samples are collected for analyses. Each experiment consists in filtrating a feed solution at six given pressures from 5 to 28 bar, which corresponds to one rejection curve ($R_i = f(Jv)$). For each applied pressure, the mass of permeate stream (m_p) is collected during a given time (D_t) and the concentrations of retentate ($C_{i,r}$) and permeate ($C_{i,p}$) solutions are determined to calculate the permeation flux J_v and rejections R_i with Equations 6.1 and 6.2, respectively.

$$J_v = \frac{m_p}{\Delta t_p S_m} \tag{6.1}$$

$$R_i = 1 - \frac{C_{i,p}}{C_{i,r}} \tag{6.2}$$

6.5.2 Effect of Ion Concentration on Nickel Rejection

6.5.2.1 Effect of Nickel Concentration and Salt Addition

Filtration of solutions containing nickel ions was performed for four various values of concentration (from 1.7×10^{-5} to 1.7×10^{-2} mol L^{-1}) to analyze its influence on Ni^{2+} rejection. It is reported that the increase in nickel concentration tends to decrease rejection, probably due to the screening of membrane charge [136]. Indeed, the increase of ionic strength induces a collapse of the Electrical Double Layer (EDL), which leads to a decline of the electrostatic interactions between ions and membrane fixed charges. This means that performances of metal removal are optimized at the low concentrations that are found in usual industrial discharge waters. However, the latter usually contains a large amount of mineral salts, such as NaCl, which can hinder the high rejection of ions.

The presence of salts in typical effluents is investigated by studying the rejection curves of Ni^{2+} at two NaCl concentrations and the comparison with performances obtained without salt addition has been described in the literature. In the same way as with the increase of nickel concentration, the presence of NaCl within the filtered solution tends to strongly reduce the Ni^{2+} rejection due to the diminution of electrostatic interactions. In particular, it can be observed that Ni^{2+} rejection tends toward zero at 2×10^{-1} mol L^{-1}. This concentration thus corresponds to a limit value beyond which it can be considered that the influence of electrostatic interactions (between ions and membrane fixed charges) on rejection are almost negligible. This result is consistent with observations found in literature [145]. This conclusion is primordial for extrapolation to industrial waters since it means that removal performances of ionic pollutants will necessarily be low due to the large amounts of mineral salts usually present in wastewaters. In this condition, the step of polymer addition before filtration is of particular interest since mechanisms governing the ion complexation are not supposed to be impacted by the increase of ionic strength [146]. For this reason, results provided thereafter in the presence of polymers were all obtained in the presence of 2×10^{-1} mol L^{-1} of NaCl, so that electrostatic interactions are minimized.

6.5.2.2 Effect of Polymer on Nickel Rejection

6.5.2.2.1 Polymer Rejection

Before investigating the influence of polymer addition on removal performances, it is of prime importance to check that polymers are sufficiently rejected by the membrane, so that the step of polymer addition is pertinent for rejection enhancement. The rejection of solutes obtained at natural pH with chitosan and CMC for a polymer concentration of 2×10^{-4} mol L^{-1} has been stated. As expected due to their large molecular weight, both chitosan and CMC are highly retained by the membrane and rejections are higher than 90%. It can be seen that CMC is slightly more rejected than chitosan ($R > 95\%$) but nickel removal is unlikely to be influenced by this difference [136]. However, it should be noted that permeation flux obtained in the same pressure conditions is lower for solution containing chitosan than for that containing CMC. This difference may be due to a higher viscosity in the former case or a lower accumulation of CMC at the membrane surface due to electrostatic repulsion between the negative charge of the membrane and the negative carboxyl groups of CMC. It should also be noted that the membrane hydraulic permeability was also different between these two experiments, which can explain such a marked discrepancy.

6.5.2.2.2 Effect of the Amount of Polymer

Various quantities of polymer were added to solutions containing nickel ions before filtration and the behavior of the rejection has been described significantly. According to the literature, both chitosan and CMC do not prove any removal enhancement when 12 (2×10^{-4} mol L^{-1}) and 120 (2×10^{-3} mol L^{-1}) mol of monomer units per mol of nickel ion are added. Oppositely, the addition of 1200 (2×10^{-2} mol L^{-1}) mol of monomer units per mol of Ni^{2+} leads to a noticeable increase of nickel rejection. This increase is similar irrespective of the polymer added in preliminary step, even though rejection with CMC is slightly higher at low flux. It is worthwhile to mention that this removal improvement is associated to a flux decline when polymer concentration increases. Such a decrease can be partially attributed to the rise of viscosity of the filtered solution induced by polymer addition. Additionally, it is also probable that polymer is progressively accumulated at the membrane surface by the convective flux. This additional layer could therefore enhance the hydraulic resistance, leading to a flux decline [136]. This enhancement of nickel removal by polymer addition emerges as a promising result, but the quantity required to treat large amounts of metal ions could be tremendous if a minimal molar ratio of 1200 between monomer units and ions had to be met for effective performances.

6.5.2.2.3 *Influence of the Amount of Nickel*

It was previously highlighted that a ratio of 1200 monomer units per Ni^{2+} ion is required. However, it is necessary to determine if removal enhancement is governed by this ratio or the concentration of polymer. For this purpose, filtration experiments were implemented for a constant concentration of monomer units (2×10^{-2} mol L^{-1}) but for various nickel concentrations (1.7×10^{-5} to 1.7×10^{-3} mol L^{-1}) corresponding to molar ratios (monomer units to Ni^{2+}) from 1200 to 12. It is obtained that the nickel rejection is not significantly impacted by the quantity of nickel, and thus by the concentration ratio between monomer units and nickel ions. This conclusion can be drawn for the two polymers considered, both exhibiting rejections close to 40% regardless of the molar ratio. This means that the ion rejection enhancement is mainly governed by the polymer quantity and not necessarily by the relative quantity compared to that of nickel. Indeed, a significant enhancement of nickel rejection requires a minimal concentration of polymer, but this amount can enhance performances regardless of the nickel concentration.

6.5.2.2.4 *Influence of pH*

The pH value is known to have a strong influence on ion rejection since it strongly affects the membrane charge. Moreover, in the case of PEUF, pH value may also affect the physical or chemical form of the polymer and therefore its electrical charge. Consequently, it is obvious that the rejection of metal ion/polymer is partially governed by the membrane charge. The nickel rejection was investigated for solutions containing 1.7×10^{-5} mol L^{-1} of Ni^{2+} and 2×10^{-2} mol L^{-1} of polymer at various pH values. It is clearly demonstrating that the pH value has a huge impact on metal rejection. Indeed, a higher pH value enables a higher rejection and consequently a better metal removal and permeate quality. At pH = 7.4, outstanding rejection is achieved after chitosan addition, probably because this pH value is above the chitosan pKa [136]. In this condition, a part of the amino groups (NH^{3+}) are deprotonated (NH_2) and the lone pair of nitrogen is thus available for complexation of nickel ions. Moreover, it is likely that a slight fraction of chitosan may have precipitated since the chitosan solubility decreases for pH values higher than pKa. However, no precipitate has been observed from turbidity measurements and precipitation cannot thus be invoked to explain such a rejection increase at this pH value. For acid pH values, the accumulation of a positive polymer at the membrane surface may also induce electrostatic repulsion with metal cation, which could enhance their rejection. The collapse of rejection performances with CMC for pH = 3 can probably be explained by the progressive protonation of the carboxyl groups (COOH) when pH value is lower than pKa, which tends to decrease the ionic bonding between these groups and the metal cations (Ni^{2+} for instance) [136]. It is worthwhile to mention that the permeation flux of solutions containing chitosan strongly declines with pH increase, probably due to an increase of solution viscosity or polymer accumulation. On the contrary, permeation flux of solution containing CMC seems not to be influenced by pH value except for pH = 3. The flux decrease in the presence of CMC at pH = 3 can probably be explained by the change in the form of carboxyl groups. At pH 5, electrostatic repulsion between the negative charge of carboxylate groups COO and the negative charge of the membrane limits the polymer accumulation at the membrane surface and thus the flux decline. Oppositely, carboxyl groups are in the neutral form COOH at pH = 3 and the accumulation of CMC at the surface is probably increased by the lack of electrostatic repulsion with the positive membrane charge. Indeed, the membrane charge is strongly influenced by the pH value [147]. This means that above this value, the membrane is negatively charged, whereas it becomes positively charged when pH is less than 3.5. It should be stressed that turbidity measurements have also shown that no precipitation (or gel formation) was detected and solution viscosity was not noticeably affected by the presence of CMC.

6.5.3 Treatment of Industrial Wastewater Using Chitosan-Enhanced Ultrafiltration Membrane

In this section, industrial discharge water from the surface treatment industries was filtered with and without chitosan addition. The concentrations of the various species are detailed before and after treatment (UF or PEUF) in Table 6.11. It is revealed from the literature that the rejection of metal ions slightly increases due to polymer enhancement, and especially at low permeation flux. For a potential application, the magnitude of this improvement can be considered to be weak since all the ions are competing with each other [148]. It is possible that other species (such as organic matter) present in the effluent interact with ions or polymers leading to a decrease of performances. Similarly, the presence of other polymers in the process may perhaps increase the rejection of ions depending on the kind of polymer. Competition between ions tends to diminish the impact of chitosan for each ion and a more pronounced improvement of each ion rejection could perhaps be achieved with a larger amount of chitosan [136]. It seems that chitosan addition has a positive impact on most heavy metals. However, a conclusion is difficult to draw since the concentration of the various ions in the feed solution differs considerably, and their impact on rejection will be different according to the concentration of the ion considered.

Moreover, it is worth mentioning that the rejection performances without polymer enhancement are already relatively high compared to rejection obtained with synthetic solutions with a large salt content [136]. This can be explained by the relatively low amount of sodium (the potassium content being negligible) in the effluent, 0.06 mol L^{-1} compared to 0.2 mol L^{-1} for synthetic solutions. A more intensive enhancement would have perhaps been

TABLE 6.11

Concentration of Various Elements within the Industrial Effluent before and after Treatment by UF and PEUF and the Corresponding Rejections

Ions	Al	B	Ca	Co	Fe	K	Mg	Mn	Na	Ni	S	Si	Sr	Zn
Concentrations in effluent (mg/L)	1.58	1.50	752	1.52	0.59	37.4	3.46	0.19	1311	0.20	125	0.83	0.26	0.72
Concentrations after Uf[a] (mg/L)	0.35	1.42	677	1.08	0.14	35.5	3.14	0.12	1194	0.10	60.1	0.34	0.23	0.42
Concentrations after PEUF[a] (mg/L)	0.21	1.36	533	0.76	0.09	30.9	9.27	0.11	1035	0.06	44.4	0.61	0.22	0.28
Rejections after UF[a] (%)	78	5	10	29	76	5	9	38	9	50	52	59	12	41
Rejections after PEUF[a] (%)	91	6	22	45	87	11	29	39	12	69	59	34	24	64

Source: Lam, B. et al., *Journal of Cleaner Production*, 171: pp. 927–933, 2018.

[a] Performances are given for the highest pressure investigated (28 bar).

observed if salt content had been higher and thus rejections without polymer lower. Finally, it appears that a more comprehensive study that would deeply investigate this phenomenon of competitive interaction between ions and polymers could be relevant for drawing pertinent conclusions about the polymer-enhanced ultrafiltration of complex mixtures such as industrial wastewaters.

6.5.4 Summary

Ultrafiltration of solutions containing nickel ions with or without a preliminary step of polymer addition has demonstrated a potential efficiency to remove heavy metals such as nickel ions from contaminated solutions. The large amount of salt present in industrial wastewaters tends to strongly decrease the rejection of heavy metals by UF membranes due to screening of electrostatic interactions. For this reason, a preliminary step of polymer addition shows an overriding interest to restore the high removal performances. In this view, additions of natural chitosan or synthetic carboxymethyl cellulose have both led to a substantial enhancement of nickel rejection, provided that an adequate amount of polymer is used. It has also been highlighted that the pH value has a strong influence on removal performances and the higher value, the better removal performances. Among the two polymers investigated, both have shown similar performance enhancement at natural pH value. However, chitosan has proved outstanding rejection performances in basic conditions (above its pKa), whereas CMC has demonstrated weak rejection performances when pH is below its pKa. In terms of metal rejection, chitosan is thus probably the better option for basic or strongly acid ($pH < 4$) conditions. Nevertheless, CMC should probably be preferred to chitosan in natural conditions ($4 < pH < 8$) since its impact on the permeation flux is less significant. Moreover, chitosan dissolution and use is much more complicated for a potential application at the industrial scale. Finally, it was shown that, although polymer addition noticeably improves rejection of single salt solutions, extrapolation to complex effluents is still a major challenge due to competing effect between the various species.

References

1. Aravind, U.K., George, B., Baburaj, M.S., Thomas, S., Thomas, A.P., and Aravindakumar, C.T., Treatment of industrial effluents using polyelectrolyte membranes. *Desalination*, 2010. 252(1–3): pp. 27–32.
2. Freire, R.S., Kunz, A., and Duran, N., Some chemical and toxicological aspects about paper mill effluent treatment with ozone. *Environmental Technology*, 2000. 21(6): pp. 717–721.

3. Summer (2003). "Pipeline, in: National Small Flows Clearinghouse."
4. Ganesh, P.S., Ramasamy, E.V., Gajalakshmi, S., Sanjeevi, R., and Abbasi, S.A., Studies on treatment of low-strength effluents by UASB reactor and its application to dairy industry wash waters. *Indian Journal of Biotechnology*, 2007. 6: pp. 234–238.
5. Mutamim, N.S.A., Noor, Z.Z., Hassan, M.A.A., and Olsson, G., Application of membrane bioreactor technology in treating high strength industrial wastewater: A performance review. *Desalination*, 2012. 305: pp. 1–11.
6. Mutamim, N.S.A., Noor, Z.Z., Hassan, M.A.A., Yuniarto, A., and Olsson, G., Membrane bioreactor: Applications and limitations in treating high strength industrial wastewater. *Chemical Engineering Journal*, 2013. 225: pp. 109–119.
7. Sivagami, K., Sakthivel, K.P., and Nambi, I.M., Advanced oxidation processes for the treatment of tannery wastewater. *Journal of Environmental Chemical Engineering*, 2018. 6: pp. 3656–3663.
8. Qasim, W., and Mane, A.V., Characterization and treatment of selected food industrial effluents by coagulation and adsorption techniques. *Water Resources and Industry*, 2013. 4: pp. 1–12.
9. Fiorentino, A., Gentili, A., Isidori, M., Lavorgna, M., Parrella, A., and Temussi, F., Olive oil mill wastewater treatment using a chemical and biological approach. *Journal of Agricultural and Food Chemistry*, 2004. 52(16): pp. 5151–5154.
10. Dhanke, P., Wagh, S., and Kanse, N., Degradation of fish processing industry wastewater in hydro-cavitation reactor. *Materials Today: Proceedings*, 2018. 5(2): pp. 3699–3703.
11. Cecconet, D., Molognoni, D., Callegari, A., and Capodaglio, A.G., Agro-food industry wastewater treatment with microbial fuel cells: Energetic recovery issues. *International Journal of Hydrogen Energy*, 2018. 43(1): pp. 500–511.
12. Muro, C., Riera, F., and del Carmen Díaz, M., Membrane separation process in wastewater treatment of food industry. In Food Industrial Processes-Methods and Equipment. *Food Industrial Processes – Methods and Equipment*, 2012. pp. 254–280.
13. Vourch, M., Balannec, B., Chaufer, B., and Dorange, G., Treatment of dairy industry wastewater by reverse osmosis for water reuse. *Desalination*, 2008. 219(1–3): pp. 190–202.
14. Turano, E., Curcio, S., De Paola, M.G., Calabrò, V., and Iorio, G., An integrated centrifugation–ultrafiltration system in the treatment of olive mill wastewater. *Journal of Membrane Science*, 2002. 209(2): pp. 519–531.
15. Casani, S., Rouhany, M., and Knøchel, S., A discussion paper on challenges and limitations to water reuse and hygiene in the food industry. *Water Research*, 2005. 39(6): pp. 1134–1146.
16. Chowdhury, M., Mostafa, M.G., Biswas, T.K., and Saha, A.K., Treatment of leather industrial effluents by filtration and coagulation processes. *Water Resources and Industry*, 2013. 3: pp. 11–22.
17. Das, C., Patel, P., De, S., and Das Gupta, S., Treatment of tanning effluent using nanofiltration followed by reverse osmosis. *Separation and Purification Technology*, 2006. 50(3): pp. 291–299.
18. Cassano, A., Molinari, R., Romano, M., and Drioli, E., Treatment of aqueous effluents of the leather industry by membrane processes: A review. *Journal of Membrane Science*, 2001. 181(1): pp. 111–126.

19. Abdel-Shafy, H.I., El-Khateeb, M.A., and Mansour, M.S., Treatment of leather industrial wastewater via combined advanced oxidation and membrane filtration. *Water Science and Technology*, 2016. 74(3): pp. 586–594.
20. Abdel-Shafy, H.I., Schories, G., Mohamed-Mansour, M.S., and Bordei, V., Integrated membranes for the recovery and concentration of antioxidant from olive mill wastewater. *Desalination and Water Treatment*, 2015. 56(2): pp. 305–314.
21. Abdel-Shafy, H., and Aly, R., Water issue in Egypt: Resources, pollution and protection endeavors. *Navigation*, 2002. 49(3.1): pp. 4–6.
22. Adham, S., Hussain, A., Minier-Matar, J., Janson, A., and Sharma, R., Membrane applications and opportunities for water management in the oil & gas industry. *Desalination*, 2018. https://doi.org/10.1016/j.desal.2018.01.030
23. Dasgupta, J., Sikder, J., Chakraborty, S., Curcio, S., and Drioli, E., Remediation of textile effluents by membrane based treatment techniques: A state of the art review. *Journal of Environmental Management*, 2015. 147: pp. 55–72.
24. Verma, A.K., Dash, R.R., and Bhunia, P., A review on chemical coagulation/flocculation technologies for removal of colour from textile wastewaters. *Journal of Environmental Management*, 2012. 93(1): pp. 154–168.
25. Foo, K.Y., and Hameed, B.H., Decontamination of textile wastewater via TiO_2/activated carbon composite materials. *Advances in Colloid and Interface Science*, 2010. 159(2): pp. 130–143.
26. Duarte, F., Morais, V., Maldonado-Hódar, F.J., and Madeira, L.M., Treatment of textile effluents by the heterogeneous Fenton process in a continuous packed-bed reactor using Fe/activated carbon as catalyst. *Chemical Engineering Journal*, 2013. 232: pp. 34–41.
27. Wang, Q., Luan, Z., Wei, N., Li, J., and Liu, C., The color removal of dye wastewater by magnesium chloride/red mud (MRM) from aqueous solution. *Journal of Hazardous Materials*, 2009. 170(2–3): pp. 690–698.
28. Arslan-Alaton, I., Gursoy, B.H., and Schmidt, J.E., Advanced oxidation of acid and reactive dyes: Effect of Fenton treatment on aerobic, anoxic and anaerobic processes. *Dyes and Pigments*, 2008. 78(2): pp. 117–130.
29. Ranganathan, K., Karunagaran, K., and Sharma, D.C., Recycling of wastewaters of textile dyeing industries using advanced treatment technology and cost analysis—Case studies. *Resources, Conservation and Recycling*, 2007. 50(3): pp. 306–318.
30. Kapdan, I.K., Kargia, F., McMullan, G., and Marchant, R., Effect of environmental conditions on biological decolorization of textile dyestuff by C. versicolor. *Enzyme and Microbial Technology*, 2007. 26(5–6): pp. 381–387.
31. EIDefrawy, N.M.H., and Shaalan, H.F., Integrated membrane solutions for green textile industries. *Desalination*, 2007. 204(1–3): pp. 241–254.
32. Álvarez, M.S., Moscoso, F., Rodríguez, A., Sanromán, M.A., and Deive, F.J., Novel physico-biological treatment for the remediation of textile dyes-containing industrial effluents. *Bioresource Technology*, 2013. 146: pp. 689–695.
33. Robinson, T., McMullan, G., Marchant, R., and Nigam, P., Remediation of dyes in textile effluent: A critical review on current treatment technologies with a proposed alternative. *Bioresource Technology*, 2001. 77(3): pp. 247–255.
34. Dasgupta, J., Chakraborty, S., Sikder, J., Kumar, R., Pal, D., Curcio, S., and Drioli, E., The effects of thermally stable titanium silicon oxide nanoparticles on structure and performance of cellulose acetate ultrafiltration membranes. *Separation and Purification Technology*, 2014. 133: pp. 55–68.

35. Cheng, S., Oatley, D.L., Williams, P.M., and Wright, C.J., Characterisation and application of a novel positively charged nanofiltration membrane for the treatment of textile industry wastewaters. *Water Research*, 2012. 46(1): pp. 33–42.
36. Chelme-Ayala, P., Smith, D.W., and El-Din, M.G., Membrane concentrate management options: A comprehensive critical review. *Canadian Journal of Civil Engineering*, 2009. 36(6): pp. 1107–1119.
37. Mulder, J., *Basic principles of membrane technology*. Springer Science & Business Media, 2012.
38. Ellouze, E., Tahri, N., and Amar, R.B., Enhancement of textile wastewater treatment process using nanofiltration. *Desalination*, 2012. 286: pp. 16–23.
39. Juang, Y., Nurhayati, E., Huang, C., Pan, J.R., and Huang, S., A hybrid electrochemical advanced oxidation/microfiltration system using BDD/Ti anode for acid yellow 36 dye wastewater treatment. *Separation and Purification Technology*, 2013. 120: pp. 289–295.
40. Tahri, N., Jedidi, I., Cerneaux, S., Cretin, M., and Amar, R.B., Development of an asymmetric carbon microfiltration membrane: Application to the treatment of industrial textile wastewater. *Separation and Purification Technology*, 2013. 118: pp. 179–187.
41. Daraei, P., Madaeni, S.S., Salehi, E., Ghaemi, N., Ghari, H.S., Khadivi, M.A., and Rostami, E., Novel thin film composite membrane fabricated by mixed matrix nanoclay/chitosan on PVDF microfiltration support: Preparation, characterization and performance in dye removal. *Journal of Membrane Science*, 2013. 436: pp. 97–108.
42. Baburaj, M.S., Aravindakumar, C.T., Sreedhanya, S., Thomas, A.P., and Aravind, U.K., Treatment of model textile effluents with PAA/CHI and PAA/PEI composite membranes. *Desalination*, 2012. 288: pp. 72–79.
43. Arthanareeswaran, G., Thanikaivelan, P., Jaya, N., Mohan, D., and Raajenthiren, M., Removal of chromium from aqueous solution using cellulose acetate and sulfonated poly (ether ether ketone) blend ultrafiltration membranes. *Journal of Hazardous Materials*, 2007. 139(1): pp. 44–49.
44. Simonič, M., Efficiency of ultrafiltration for the pre-treatment of dye-bath effluents. *Desalination*, 2009. 245(1–3): pp. 701–707.
45. Barredo-Damas, S., Alcaina-Miranda, M.I., Bes-Piá, A., Iborra-Clar, M.I., Iborra-Clar, A., and Mendoza-Roca, J.A., Ceramic membrane behavior in textile wastewater ultrafiltration. *Desalination*, 2010. 250(2): pp. 623–628.
46. Koseoglu-Imer, D.Y., The determination of performances of polysulfone (PS) ultrafiltration membranes fabricated at different evaporation temperatures for the pretreatment of textile wastewater. *Desalination*, 2013. 316: pp. 110–119.
47. Mondal, S., Ouni, H., Dhahbi, M., and De, S., Kinetic modeling for dye removal using polyelectrolyte enhanced ultrafiltration. *Journal of Hazardous Materials*, 2012. 229: pp. 381–389.
48. Zaghbani, N., Hafiane, A., and Dhahbi, M., Removal of Safranin T from wastewater using micellar enhanced ultrafiltration. *Desalination*, 2008. 222(1–3): pp. 348–356.
49. Srivastava, H.P., Arthanareeswaran, G., Anantharaman, N., and Starov, V.M., Performance of modified poly (vinylidene fluoride) membrane for textile wastewater ultrafiltration. *Desalination*, 2011. 282: pp. 87–94.

50. Khouni, I., Marrot, B., Moulin, P., and Amar, R.B., Decolourization of the reconstituted textile effluent by different process treatments: Enzymatic catalysis, coagulation/flocculation and nanofiltration processes. *Desalination*, 2011. 268 (1–3): pp. 27–37.
51. Yu, S., Chen, Z., Cheng, Q., Lü, Z., Liu, M., and Gao, C., Application of thin-film composite hollow fiber membrane to submerged nanofiltration of anionic dye aqueous solutions. *Separation and Purification Technology*, 2012. 88: pp. 121–129.
52. Bes-Piá, A., Cuartas-Uribe, B., Mendoza-Roca, J.A., and Alcaina-Miranda, M.I., Study of the behaviour of different NF membranes for the reclamation of a secondary textile effluent in rinsing processes. *Journal of Hazardous Materials*, 2010. 178(1–3): pp. 341–348.
53. Kołtuniewicz, A.B., and Drioli, E., Membranes in clean technologies: Theory and practice, WILEY-VCH Verlag GmbH & Co. KGaA, Weinheim, Germany, 2008.
54. Liu, M., Lü, Z., Chen, Z., Yu, S., and Gao, C., Comparison of reverse osmosis and nanofiltration membranes in the treatment of biologically treated textile effluent for water reuse. *Desalination*, 2011. 281: pp. 372–378.
55. Chandramowleeswaran, M., and Palanivelu, K., Treatability studies on textile effluent for total dissolved solids reduction using electrodialysis. *Desalination*, 2006. 201(1–3): pp. 164–174.
56. Praneeth, K., Manjunath, D., Bhargava, S.K., Tardio, J., and Sridhar, S., Economical treatment of reverse osmosis reject of textile industry effluent by electrodialysis–evaporation integrated process. *Desalination*, 2014. 333(1): pp. 82–91.
57. He, Y., Wang, X., Xu, J., Yan, J., Ge, Q., Gu, X., and Jian, L., Application of integrated ozone biological aerated filters and membrane filtration in water reuse of textile effluents. *Bioresource Technology*, 2013. 133: pp. 150–157.
58. Kertèsz, S., Cakl, J., and Jiránková, H., Submerged hollow fiber microfiltration as a part of hybrid photocatalytic process for dye wastewater treatment. *Desalination*, 2014. 343: pp. 106–112.
59. Xu, L., Zhang, L., Du, L., and Zhang, S., Electro-catalytic oxidation in treating CI Acid Red 73 wastewater coupled with nanofiltration and energy consumption analysis. *Journal of Membrane Science*, 2014. 452: pp. 1–10.
60. He, Y., Li, G., Jiang, Z., Wang, H., Zhao, J., Su, H., and Huang, Q., Diafiltration and concentration of Reactive Brilliant Blue KN-R solution by two-stage ultrafiltration process at pilot scale: Technical and economic feasibility. *Desalination*, 2011. 279(1–3): pp. 235–242.
61. Ciardelli, G., Corsi, L., and Marcucci, M., Membrane separation for wastewater reuse in the textile industry. *Resources, Conservation and Recycling*, 2001. 31(2): pp. 189–197.
62. Vergili, I., Kaya, Y., Sen, U., Gönder, Z.B., and Aydiner, C., Techno-economic analysis of textile dye bath wastewater treatment by integrated membrane processes under the zero liquid discharge approach. *Resources, Conservation and Recycling*, 2012. 58: pp. 25–35.
63. Taseli, B.K., and Gokcay, C.F., Biological treatment of paper pulping effluents by using a fungal reactor. *Water Science and Technology*, 1999. 40(11–12): pp. 93–99.
64. Shawwa, A.R., Smith, D.W., and Sego, D.C., Color and chlorinated organics removal from pulp mills wastewater using activated petroleum coke. *Water Research*, 2001. 35(3): pp. 745–749.

65. Cecen, F., Urban, W., and Haberl, R., Biological and advanced treatment of sulfate pulp bleaching effluents. *Water Science and Technology*, 1992. 26(1–2): pp. 435–444.
66. Haroun, M., and Idris, A., Treatment of textile wastewater with an anaerobic fluidized bed reactor. *Desalination*, 2009. 237(1–3): pp. 357–366.
67. Kansal, S.K., Singh, M., and Sud, D., Effluent quality at kraft/soda agro-based paper mills and its treatment using a heterogeneous photocatalytic system. *Desalination*, 2008. 228(1–3): pp. 183–190.
68. Slokar, Y.M., and Le Marechal, A.M., Methods of decoloration of textile wastewaters. *Dyes and Pigments*, 1998. 37(4): pp. 335–356.
69. Krasemann, L., and Tieke, B., Composite membranes with ultrathin separation layer prepared by self-assembly of polyelectrolytes. *Materials Science and Engineering: C*, 1999. 8: pp. 513–518.
70. Aravind, U.K., Mathew, J., and Aravindakumar, C.T., Transport studies of BSA, lysozyme and ovalbumin through chitosan/polystyrene sulfonate multilayer membrane. *Journal of Membrane Science*, 2007. 299(1–2): pp. 146–155.
71. Bhatnagar, A., and Sillanpää, M., Utilization of agro-industrial and municipal waste materials as potential adsorbents for water treatment—A review. *Chemical Engineering Journal*, 2010. 157(2–3): pp. 277–296.
72. Srivastava, S.K., Singh, A.K., and Sharma, A., Studies on the uptake of lead and zinc by lignin obtained from black liquor–a paper industry waste material. *Environmental Technology*, 1994. 15(4): pp. 353–361.
73. Calace, N., Nardi, E., Petronio, B.M., and Pietroletti, M., Adsorption of phenols by papermill sludges. *Environmental Pollution*, 2002. 118(3): pp. 315–319.
74. Saha, S., and Das, C., Spinning basket membrane ultrafiltration of paper industry waste effluent: Experimental and theoretical aspects. *Journal of Environmental Chemical Engineering*, 2017. 5(5): pp. 4583–4593.
75. Garg, A., Mishra, I.M., and Chand, S., Thermochemical precipitation as a pretreatment step for the chemical oxygen demand and color removal from pulp and paper mill effluent. *Industrial & Engineering Chemistry Research*, 2005. 44(7): pp. 2016–2026.
76. Sutton, P.M., Membrane bioreactors for industrial wastewater treatment: Applicability and selection of optimal system configuration. *Proceedings of the Water Environment Federation*, 2006(9): pp. 3233–3248.
77. Sutton, P.M., Membrane bioreactors for industrial wastewater treatment: The state-of-the-art based on full scale commercial applications. *Proceedings of the Water Environment Federation*, 2003(6): pp. 23–32.
78. Le-Clech, P., Chen, V., and Fane, T.A., Fouling in membrane bioreactors used in wastewater treatment. *Journal of Membrane Science*, 2006. 284(1–2): pp. 17–53.
79. Widjaja, T., and Soeprijanto, A.A., Effect of powdered activated carbon addition on a submerged membrane adsorption hybrid bioreactor with shock loading of a toxic compound. *Journal of Mathematics and Technology*, 2010. pp. 139–146.
80. Chang, J.S., Chang, C.Y., Chen, A.C., Erdei, L., and Vigneswaran, S., Long-term operation of submerged membrane bioreactor for the treatment of high strength acrylonitrile-butadiene-styrene (ABS) wastewater: Effect of hydraulic retention time. *Desalination*, 2006. 191(1–3): pp. 45–51.
81. Zhang, Q., Performance evaluation and characterization of an innovative membrane bioreactor in the treatment of wastewater and removal of pharmaceuticals and pesticides (Doctoral dissertation, University of Cincinnati), 2009.

82. Radjenović, J., Matošić, M., Mijatović, I., Petrović, M., and Barceló, D., Membrane bioreactor (MBR) as an advanced wastewater treatment technology. In *Emerging Contaminants from Industrial and Municipal Waste*. 37–101. Springer Berlin Heidelberg, 2008. pp. 37–101.
83. Bouhabila, E.H., Aïm, R.B., and Buisson, H., Fouling characterisation in membrane bioreactors. *Separation and Purification Technology*, 2001. 22: pp. 123–132.
84. Choo, K.H., and Lee, C.H., Membrane fouling mechanisms in the membrane-coupled anaerobic bioreactor. *Water Research*, 1996. 30(8): pp. 1771–1780.
85. Kurian, R., and Nakhla, G., Performance of aerobic MBR treating high strength oily wastewater at mesophilic–thermophilic transitional temperatures. *Proceedings of the Water Environment Federation*, 2006(9): pp. 3249–3255.
86. Zhang, Y., Li, Y.A.N., Xiangli, Q.I.A.O., Lina, C.H.I., Xiangjun, N.I.U., Zhijian, M.E.I., and Zhang, Z., Integration of biological method and membrane technology in treating palm oil mill effluent. *Journal of Environmental Sciences*, 2008. 20(5): pp. 558–564.
87. Yuniarto, A., Ujang, Z., and Noor, Z.Z., Performance of bio-fouling reducers in aerobic submerged membrane bioreactor for palm oil mill effluent treatment. *Journal Teknologi UTM*, 2008. 49: pp. 555–566.
88. Judd S., *Principles and Applications of Membrane Bioreactors in Water and Wastewater Treatment*, First ed. Elsevier, U.K., 2006.
89. Sombatsompop, K.M., Membrane fouling studies in suspended and attached growth membrane bioreactor systems, Asian Institute of Technology School of Environment, Resources & Development Environmental Engineering & Management, Thailand, 2007.
90. Frederickson, K.C., The application of a membrane bioreactor for wastewater treatment on a northern Manitoban Aboriginal Community, Department of Biosystems Engineering, University of Manitoba, Winnipeg, Manitoba, Canada, 2006.
91. Jianga, T., Characterization and Modelling of Soluble Microbial Products in Membrane Bioreactors, Institute for Water Education, University Gent, 2007.
92. Koltuniewicz, A.B., and Drioli, E., Membranes in Clean Technologies, Wiley-VCH, Weinheim, 2008.
93. Jin, L., Ong, S.L., and Ng, H.Y., Comparison of fouling characteristics in different pore-sized submerged ceramic membrane bioreactors. *Water Research*, 2010. 44(20): pp. 5907–5918.
94. Hofs, B., Ogier, J., Vries, D., Beerendonk, E.F., and Cornelissen, E.R., Comparison of ceramic and polymeric membrane permeability and fouling using surface water. *Separation and Purification Technology*, 2011. 79(3), pp. 365–374.
95. Huang, H., Schwab, K., and Jacangelo, J.G., Pretreatment for low pressure membranes in water treatment: A review. *Environmental Science & Technology*, 2009. 43(9): pp. 3011–3019.
96. Hanif, S.H.M., Fabrication of Chitosan Membrane: The Effects of Different Polyethylene Glycol Compositions on Membrane Performance in Oily Wastewater Treatment, Faculty of Chemical Engineering and Natural Resource, University College of Engineering and Technology, Malaysia, 2008.
97. Durai, G., and Rajasimman, M., Biological treatment of tannery wastewater: A review. *Journal of Environmental Science and Technology*, 2011. 4(1): pp. 1–17.
98. Buenrostro-Zagal, J.F., Ramirez-Oliva, A., Caffarel-Mendez, S., Schettino-Bermudez, B., and Poggi-Varaldo, H.M., Treatment of a 2, 4-dichlorophenoxyacetic acid (2, 4-D) contamined wastewater in a membrane bioreactor. *Water Science and Technology*, 2000. 42(5–6): pp. 185–192.

99. Badani, Z., Ait-Amar, H., Si-Salah, A., Brik, M., and Fuchs, W., Treatment of textile waste water by membrane bioreactor and reuse. *Desalination*, 2005. 185(1–3): pp. 411–417.
100. Brik, M., Schoeberl, P., Chamam, B., Braun, R., and Fuchs, W., Advanced treatment of textile wastewater towards reuse using a membrane bioreactor. *Process Biochemistry*, 2006. 41(8): pp. 1751–1757.
101. Yigit, N.O., Uzal, N., Koseoglu, H., Harman, I., Yukseler, H., Yetis, U., Civelekoglu, G., and Kitis, M., Treatment of a denim producing textile industry wastewater using pilot-scale membrane bioreactor. *Desalination*, 2009. 240(1–3): pp. 143–150.
102. Rosenberger, S., Krüger, U., Witzig, R., Manz, W., Szewzyk, U., and Kraume, M., Performance of a bioreactor with submerged membranes for aerobic treatment of municipal waste water. *Water Research*, 2002. 36(2): pp. 413–420.
103. Katayon, S., Noor, M.M.M., Ahmad, J., Ghani, L.A., Nagaoka, H., and Aya, H., Effects of mixed liquor suspended solid concentrations on membrane bioreactor efficiency for treatment of food industry wastewater. *Desalination*, 2004. 167: pp. 153–158.
104. Cicek, N., A review of membrane bioreactors and their potential application in the treatment of agricultural wastewater. *Canadian Biosystems Engineering*, 2003. 45: pp. 6–37.
105. Acharya, C., Nakhla, G., and Bassi, A., Operational optimization and mass balances in a two-stage MBR treating high strength pet food wastewater. *Journal of Environmental Engineering*, 2006. 132(7): pp. 810–817.
106. Meng, F., Chae, S.R., Drews, A., Kraume, M., Shin, H.S., and Yang, F., Recent advances in membrane bioreactors (MBRs): Membrane fouling and membrane material. *Water Research*, 2009. 43(6): pp. 1489–1512.
107. Zhang, Y., Li, Y.A.N., Xiangli, Q.I.A.O., Lina, C.H.I., Xiangjun, N.I.U., Zhijian, M.E.I., and Zhang, Z., Integration of biological method and membrane technology in treating palm oil mill effluent. *Journal of Environmental Sciences*, 2008. 20(5): pp. 558–564.
108. Chang, I.S., Le Clech, P., Jefferson, B., and Judd, S., Membrane fouling in membrane bioreactors for wastewater treatment. *Journal of Environmental Engineering*, 2002. 128(11): pp. 1018–1029.
109. Viero, A.F., de Melo, T.M., Torres, A.P.R., Ferreira, N.R., Sant'Anna Jr, G.L., Borges, C.P., and Santiago, V.M., The effects of long-term feeding of high organic loading in a submerged membrane bioreactor treating oil refinery wastewater. *Journal of Membrane Science*, 2008. 319(1–2): pp. 223–230.
110. Chang, C.Y., Chang, J.S., Vigneswaran, S., and Kandasamy, J., Pharmaceutical wastewater treatment by membrane bioreactor process: A case study in southern Taiwan. *Desalination*, 2008. 234(1–3): pp. 393–401.
111. Laspidou, C.S., and Rittmann, B.E., A unified theory for extracellular polymeric substances, soluble microbial products, and active and inert biomass. *Water Research*, 2002. 36(11): pp. 2711–2720.
112. Meng, F., Chae, S.R., Drews, A., Kraume, M., Shin, H.S., and Yang, F., Recent advances in membrane bioreactors (MBRs): Membrane fouling and membrane material. *Water Research*, 2009. 43(6): pp. 1489–1512.
113. Ahmed, Z., Cho, J., Lim, B.R., Song, K.G., and Ahn, K.H., Effects of sludge retention time on membrane fouling and microbial community structure in a membrane bioreactor. *Journal of Membrane Science*, 2007. 287(2): pp. 211–218.

114. Amiri, S., Mehrnia, M.R., Azami, H., Barzegari, D., Shavandi, M., and Sarrafzadeh, M.H., Effect of heavy metals on fouling behavior in membrane bioreactors. *Iranian Journal of Environmental Health Science & Engineering*, 2010. 7(5): p. 377.
115. Zsirai, T., Buzatu, P., Aerts, P., and Judd, S., Efficacy of relaxation, backflushing, chemical cleaning and clogging removal for an immersed hollow fibre membrane bioreactor. *Water Research*, 2012. 46(14): pp. 4499–4507.
116. Blöcher, C., Noronha, M., Fünfrocken, L., Dorda, J., Mavrov, V., Janke, H.D., and Chmiel, H., Recycling of spent process water in the food industry by an integrated process of biological treatment and membrane separation. *Desalination*, 2002. 144(1–3): pp. 143–150.
117. Fu, F., and Wang, Q., Removal of heavy metal ions from wastewaters: A review. *Journal of Environmental Management*, 2011. 92(3): pp. 407–418.
118. Martín-Lara, M.A., Blázquez, G., Trujillo, M.C., Pérez, A., and Calero, M., New treatment of real electroplating wastewater containing heavy metal ions by adsorption onto olive stone. *Journal of Cleaner Production*, 2014. 81: pp. 120–129.
119. Fane, A.G., and Fane, S.A., The role of membrane technology in sustainable decentralized wastewater systems. *Water Science and Technology*, 2005. 51(10). 317–325.
120. Jeppesen, T., Shu, L., Keir, G., and Jegatheesan, V., Metal recovery from reverse osmosis concentrate. *Journal of Cleaner Production*, 2009. 17(7): pp. 703–707.
121. Peters, T., Membrane technology for water treatment. *Chemical Engineering & Technology*, 2010. 33(8): pp. 1233–1240.
122. Chew, C.M., Aroua, M.K., Hussain, M.A., and Ismail, W.M.Z.W., Evaluation of ultrafiltration and conventional water treatment systems for sustainable development: An industrial scale case study. *Journal of Cleaner Production*, 2016. 112: pp. 3152–3163.
123. Efligenir, A., Déon, S., Fievet, P., Druart, C., Morin-Crini, N., and Crini, G., Decontamination of polluted discharge waters from surface treatment industries by pressure-driven membranes: Removal performances and environmental impact. *Chemical Engineering Journal*, 2014. 258: pp. 309–319.
124. Otero-Fernández, A., Otero, J.A., Maroto, A., Carmona, J., Palacio, L., Prádanos, P., and Hernández, A., Concentration-polarization in nanofiltration of low concentration Cr (VI) aqueous solutions. Effect of operative conditions on retention. *Journal of Cleaner Production*, 2017. 150: pp. 243–252.
125. Petrinic, I., Korenak, J., Povodnik, D., and Hélix-Nielsen, C., A feasibility study of ultrafiltration/reverse osmosis (UF/RO)-based wastewater treatment and reuse in the metal finishing industry. *Journal of Cleaner Production*, 2015. 101: pp. 292–300.
126. Déon, S., Escoda, A., Fievet, P., Dutournié, P., and Bourseau, P., How to use a multi-ionic transport model to fully predict rejection of mineral salts by nanofiltration membranes. *Chemical Engineering Journal*, 2012. 189: pp. 24–31.
127. Du, C., Ma, X., Li, J., and Wu, C., Improving the charged and antifouling properties of PVDF ultrafiltration membranes by blending with polymerized ionic liquid copolymer P (MMA-b-MEBIm-Br). *Journal of Applied Polymer Science*, 2017. 134(17): 44751.
128. Kaveh, R., Shariatinia, Z., and Arefazar, A., Improvement of polyacrylonitrile ultrafiltration membranes' properties using decane-functionalized reduced graphene oxide nanoparticles. *Water Science and Technology: Water Supply*, 2016. 16(5): pp. 1378–1387.

129. Razmjou, A., Resosudarmo, A., Holmes, R.L., Li, H., Mansouri, J., and Chen, V., The effect of modified TiO2 nanoparticles on the polyethersulfone ultrafiltration hollow fiber membranes. *Desalination*, 2012. 287: pp. 271–280.
130. Ilyas, S., Abtahi, S.M., Akkilic, N., Roesink, H.D.W., and de Vos, W.M., Weak polyelectrolyte multilayers as tunable separation layers for micro-pollutant removal by hollow fiber nanofiltration membranes. *Journal of Membrane Science*, 2017. 537: pp. 220–228.
131. Laakso, T., Kallioinen, M., Pihlajamäki, A., Mänttäri, M., and Wong, J.E., Polyelectrolyte multilayer coated ultrafiltration membranes for wood extract fractionation. *Separation and Purification Technology*, 2015. 156: pp. 772–779.
132. Bolto, B., Tran, T., Hoang, M., and Xie, Z., Crosslinked poly (vinyl alcohol) membranes. *Progress in Polymer Science*, 2009. 34(9): pp. 969–981.
133. Nayak, V., Jyothi, M.S., Balakrishna, R.G., Padaki, M., and Deon, S., Novel modified poly vinyl chloride blend membranes for removal of heavy metals from mixed ion feed sample. *Journal of Hazardous Materials*, 2017. 331: pp. 289–299.
134. Rahimpour, A., and Madaeni, S.S., Improvement of performance and surface properties of nano-porous polyethersulfone (PES) membrane using hydrophilic monomers as additives in the casting solution. *Journal of Membrane Science*, 2010. 360(1–2): pp. 371–379.
135. Zhao, S., Wang, Z., Wei, X., Zhao, B., Wang, J., Yang, S., and Wang, S., Performance improvement of polysulfone ultrafiltration membrane using PANiEB as both pore forming agent and hydrophilic modifier. *Journal of Membrane Science*, 2011. 385: pp. 251–262.
136. Lam, B., Déon, S., Morin-Crini, N., Crini, G., and Fievet, P., Polymer-enhanced ultrafiltration for heavy metal removal: Influence of chitosan and carboxymethyl cellulose on filtration performances. *Journal of Cleaner Production*, 2018. 171: pp. 927–933.
137. Crini, G., and Badot, P.M., Application of chitosan, a natural aminopolysaccharide, for dye removal from aqueous solutions by adsorption processes using batch studies: A review of recent literature. *Progress in Polymer Science*, 2008. 33(4): pp. 399–447.
138. Choo, K.H., Han, S.C., Choi, S.J., Jung, J.H., Chang, D., Ahn, J.H., and Benjamin, M.M., Use of chelating polymers to enhance manganese removal in ultrafiltration for drinking water treatment. *Journal of Industrial and Engineering Chemistry*, 2007. 13(2): pp. 163–169.
139. Aroua, M.K., Zuki, F.M., and Sulaiman, N.M., Removal of chromium ions from aqueous solutions by polymer-enhanced ultrafiltration. *Journal of Hazardous Materials*, 2007. 147(3): pp. 752–758.
140. Denkhaus, E., and Salnikow, K., Nickel essentiality, toxicity, and carcinogenicity. *Critical Reviews in Oncology/Hematology*, 2002. 42(1): pp. 35–56.
141. Charles, J., Sancey, B., Morin-Crini, N., Badot, P.M., Degiorgi, F., Trunfio, G., and Crini, G., Evaluation of the phytotoxicity of polycontaminated industrial effluents using the lettuce plant (Lactuca sativa) as a bioindicator. *Ecotoxicology and Environmental Safety*, 2011. 74(7): pp. 2057–2064.
142. Chandra, R., and Rustgi, R., Biodegradable polymers. *Progress in Polymer Science*, 1998. 23(7): pp. 1273–1335.

143. Hoogendam, C.W., De Keizer, A., Cohen Stuart, M.A., Bijsterbosch, B.H., Smit, J.A.M., Van Dijk, J.A.P.P., Van der Horst, P.M., and Batelaan, J.G., Persistence length of carboxymethyl cellulose as evaluated from size exclusion chromatography and potentiometric titrations. *Macromolecules*, 1998. 31(18): pp. 6297–6309.
144. Déon, S., Dutournié, P., Fievet, P., Limousy, L., and Bourseau, P., Concentration polarization phenomenon during the nanofiltration of multi-ionic solutions: Influence of the filtrated solution and operating conditions. *Water Research*, 2013. 47(7): pp. 2260–2272.
145. Déon, S., Dutournié, P., and Bourseau, P., Transfer of monovalent salts through nanofiltration membranes: A model combining transport through pores and the polarization layer. *Industrial & Engineering Chemistry Research*, 2007. 46(21): pp. 6752–6761.
146. Labanda, J., Khaidar, M.S., and Llorens, J., Feasibility study on the recovery of chromium (III) by polymer enhanced ultrafiltration. *Desalination*, 2009. 249(2): pp. 577–581.
147. Déon, S., Deher, J., Lam, B., Crini, N., Crini, G., and Fievet, P., Remediation of solutions containing oxyanions of selenium by ultrafiltration: Study of rejection performances with and without chitosan addition. *Industrial & Engineering Chemistry* Research, 2017. 56(37): pp. 10461–10471.
148. Evans, N.D.M., Gascón, S.A., Vines, S., and Felipe-Sotelo, M., Effect of competition from other metals on nickel complexation by α-isosaccharinic, gluconic and picolinic acids. *Mineralogical Magazine*, 2012. 76(8): pp. 3425–3434.

7

Polymeric Membranes for Biomedical and Biotechnology Applications

7.1 Electrospun Polymeric Fibers for Biomedical Applications

7.1.1 Introduction

Biological systems, basically, are comprised with hierarchically complex structures [1–4], and these constructions convey the miscellaneous functionalities to the organisms such as various nanoscale viruses, microscale cells, and different macroscale tissues. The bone tissue parts take an unusual mixture of remarkable toughness and strength, credited to the exactly systematized hierarchical structures such as the nanometer-sized hydroxyapatite crystals sometimes deposited into the gap zones of collagen fibrils in the bones, which has excellent nano mechanical heterogeneities characteristics and creating high energy dissipation that may be the reason for fracture [3]. To control the fracture of the bones, different kinds of hierarchical structures such as zero-dimensional (0-D) particles, one-dimensional (1-D) fibers, two-dimensional (2-D) substrates, and the three-dimensional (3-D) supports have been developed in the field of biomedicine and other purposes [5]. The hierarchically structured fibers signify a major group and have involved a great deal of consideration in the field of biomedical and other scientific research areas, including optical areas, electronics, various sensors, catalysis, energy storage, oil and water separation, and air purification, due to the larger active area and extra heterogeneous boundaries, which may play a significant role in improving the nano size effect [6–12]. Abundant methods have been established to fabricate the one dimensional (1-D) fibers in current years. In this section, an inclusive summary of the current research on the fabricating of the electrospun hierarchically structured polymer fibers in the field of biomedical applications including tissue engineering, drug delivery process, and diagnostics are discussed. Electrospinning is a superficial and multifunctional process used to make ultrafine and continuous fibers with manageable diameters in the range of nanometers to micrometers, from both the organic or inorganic compounds. These electrospun fibers have a wide specific surface area, high aspect ratio, high porosity value, a manageable small pore size, and the capability to look like an extracellular matrix's construction. The fibers are widely used in different fields such as water filtration and solutes adsorption, optical sensing, catalysis, self-cleaning, food and beverages engineering, nanofiber-reinforced mixtures, energy storage, and especially in the area of biomedical applications [13–16].

The electrospun fibers are generally required in a small quantity, and the electrospinning method is a resourceful process as the fibers can be rotated into any kind of shape using various polymers [17,18]. Electrospun nanofibers are largely applied in the field of biomedical applications, such as tissue engineering platforms, for the wound remedial, drug delivery system, plasma filtration, affinity membrane, during immobilization of essential enzymes, small diameter vascular graft implants, healthcare, biotechnology, environmental engineering for water treatment, defense and security, energy storage and generation, and in different research purposes [19–23]. The electrospun polymer nanofibers have already been projected for many biomedical purposes including vascular and breast prostheses applications since the 1980s. Different U.S. patents have been allotted on fabrication and characterization methods and techniques for the vascular prostheses such as 4044404, 4552707, 4689186, 4878908, 4965110, 5866217, and breast prosthesis (5376117). Reviewing the number of patents, it is clear that about two-thirds of all the applications regarding electrospinning processes are in the field of medical purposes. However, the filtration application has different patents [24,25]. Owing to application of these nanofibers in various fields, different research is being completed now. A schematic illustration of the applications of electrospinning process in different fields is shown in Figure 7.1.

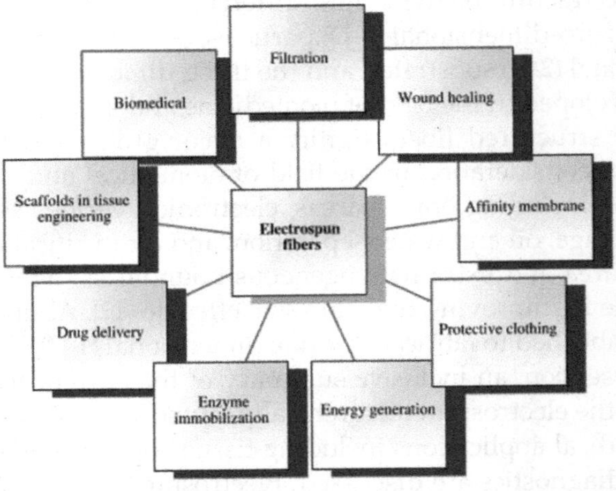

FIGURE 7.1
Applications of electrospun fibers in different sectors. (Reprinted from *Biotechnology Advances*, 28(3), Bhardwaj, N., and Kundu, S.C., Electrospinning: A fascinating fiber fabrication technique, pp. 325–347, Copyright 2010, with permission from Elsevier.)

7.1.2 Electrospun Polymeric Fibers for Biomedical Applications

The electrospun polymeric fibers can be widely used in different fields due to their high surface area, structural diversity, and adjustable porosity [27]. Particularly in the biomedical field, properties like structural diversity and high active surface area are helpful for the loading and subsequent discharge of different drugs during delivery. Moreover, the adjustable porosity and the capability of electrospun fibers to mimic the fibrillary nature of the extracellular matrix (ECM) allow its utilization as promising two-dimensional (2-D) or three-dimensional (3-D) skeletons for cell support and development in tissue engineering. The adjustable porosity and high surface area of the electrospun fibers also provide more active surface area to immobilize the functional groups in diagnostics and different applications.

7.1.2.1 Drug Delivery

Electrospun polymeric fibers have performed as drug transporters during drug delivery due to the high surface area and functional characteristics. The drug delivery process depends on the principle that the dissolution rate of a particulate drug rises with increasing the active surface area of both the drugs and the consistent carrier. Therefore, the electrospun polymeric fibers with biocompatible delivery matrices can release a controlled drug at a definite rate over a fixed period of time. Thus, biodegradable polymeric materials are typically used during drug delivery process to transport healing agents due to their easier design for programmed distribution in a controlled style [28,29]. The large active surface area connected with nanospun fibers allows for fast and effective solvent evaporation, which delivers the combined drug limited time to recrystallize, which helps the development of amorphous distributions or solid solutions [30]. The pharmaceutical dosage releasing technology can be deliberated as instantaneous, rapid, and modified dissolution process based on the polymer carrier used. A variety of solutions comprising low molecular weight drugs have been processed for electrospinning, including lipophilic drugs such as ibuprofen, cefazolin, rifampin, paclitaxel, Itraconazol, and mefoxin and tetracycline hydrochloride like hydrophilic drugs [29–33]. Few researchers have captured proteins in electrospun polymer fibers [34,35]. Moreover, the efficiency of electrospun nonwoven bio absorbable poly (lactide-co-glycolide) (PLGA) impregnated with antibiotics (mefoxin) in reducing post-surgery adhesion on an in vivo rat model have also been studied [36]. Encapsulation of a model protein, fluorescein isothiocyanate conjugated bovine serum albumin, along with poly(ethylene glycol) (PEG) in poly (ε-caprolactone) (PCL) fibers, has been performed using a coaxial formation and demonstrated a relatively smooth discharge of the drug over a retro of five days from the electrospun nanofibrous mats [37].

Drug release mechanism is related with polymer deprivation and complex diffusion path along nano-void spaces within nanofiber network. It has been found that the drug release outlines can be personalized by different formulation conditions such as polymer properties, combination of various polymers, superficial coating, and especially the state of drug molecules in a solid phase [38]. Hollow nanofibrous tubes by coaxial electrospinning also delivered an auspicious structure for the encapsulation of target drug materials [39]. This approach prospered in attaining high loading of drug and simplification of the solubilization of some stubborn and insoluble drugs. The formulation of a rich variety of therapeutic materials such as antibiotics, anti-cancer drugs, polysaccharides, proteins, and growth factors have been performed within the bulk segment of electrospun nanofibers or on their active surface for achieving controlled topical release of drug within the distinct period of time [40–43].

7.1.2.1.1 Controlled Delivery of a Single Drug

The selection of polymer is a key feature for attaining a continued release of drugs from electrospun fibers. Different processes can be performed to attain a sustained release of a single drug by electrospun fibers, containing matching drugs with appropriate polymers, using core-shell electrospun fibers or covered electrospun fiber mats as drug transporters, or by capturing drug-loaded components into electrospun fibers [44]. Then, after effective loadings, drugs need to cross the obstacles created by the fiber matrix and then diffuse farther into the release medium. In the case of biodegradable polymers, the release of drugs is mainly dependent on the degradation process of the polymer fibers because the degradation of the polymer matrix results in variations in the electrospun fiber morphology, changing the drug diffusion pathway and affecting the drug release. Moreover, drugs in the degraded regions of the fiber matrix may also dissolve directly into the release medium.

In the case of non-swelling non-degradable polymers, the drug diffusion path would not change because there is no variation in fiber morphology over time. The conditions controlling the release of the drugs are drug partitioning in the polymer, drug solubility, and drug diffusivity in the polymer. Generally, the release of drugs from the fiber matrix may be influenced by hydrophilicity, morphology, degradability, drug diffusivity, and drug partitioning. Currently, some researchers blended drugs with polymers to electrospin into drug-loaded fiber mats such as a successful incorporation of a hydrophilic antibiotic drug (MefoxinR, cefoxitin sodium) into a poly PLGA (poly(lactic-co-glycolic acid)-based electrospun nanofiber mat has been done to attain a controlled release of drugs over the course of 1 h. They have detected no loss in fiber structure or bioactivity, and they also studied that the accumulation of an amphiphilic block copolymer (PEG-b-PLA) decreased the cumulative amount of discharged drug during the previous release time, extending the drug release time up to 7 days [32]. A hydrophobic polymer dopant, namely, poly (glycerol monostearate-co-caprolactone) with polycaprolactone (PCL), was used to make a super hydrophobic electrospun fibrous mesh, resulting in a mesh with an extraordinary apparent contact angle [45]. It is believed that the air in the porous electrospun network may also perform as a barrier component and further control the drug release rate. The types of drugs that were positively encapsulated into electrospun fibers, such as antibiotics, plasmid DNA, growth factors, antimicrobial components (e.g., silver nanoparticles), proteins, bacteria, and phages, were also summarized by researchers [46]. However, choosing a great diversity of drugs may have its own influence, such as the chemical, physical characteristics, and the hydrophobicity of various drugs will disturb the controlled release procedure. The release of various compounds such as azidothymidine, maraviroc, raltegravir, and tenofovir disoproxil fumarate was also reported, and it was much quicker than tenovir in corresponding PCL/PLGA fiber formulations, signifying that even subtle differences in the chemical structures of drugs would distress the release rates [47]. In general, the application of electrospun fibers as a drug delivery method needs a cautious match between the drug and the polymer [27]. Moreover, changing the drug diffusion path is an additional effective way to regulate the release of drugs.

Electrospun fibers with various structures can be invented to control the drug release process by changing the electrospinning parameters. One method is to study coaxial electrospinning for the assembly of core-shell electrospun fibers as a drug delivery process. When drugs are injected into the core layers of these electrospun fibers, the outer shell helps as an operative barrier for averting the diffusion of water molecules and the dispersion of drugs [48]. Precisely, the configuration, thickness, and penetrability of the outer shell effects the discharge of drugs in the core area of the fibers [49]. Additionally, emulsion electrospinning was also performed to make core-shell fibers for regulating drug release [50]. Recently, a drug-loaded layered electrospun mesh was fabricated successfully with the chemotherapeutic agent, namely, SN-38 incorporated into a central "core" layer between the two super hydrophobic "shield" layers of electrospun mesh without any drug. It was found that the "shield" layers efficiently hindered the release of SN-38, and by rising the thickness of the "shield" layers, one could obviously extend the discharge time of the drug. With the 300 μm thick layers on both edges of the drug-loaded core, there was a layer of metastable air still present that could be evacuated with an ethanol usage after 100 days of soaking, resulting in increase in drug release [51]. Moreover, the encapsulation of drug-loaded micro-/nanoparticles into electrospun fibers has been described as an emerging system for regulating drug release. Thus, the grouping of micro-/nanoparticles with a large capability for holding drugs and the electrospun fibers can change the drug diffusion way. Initially, the drug molecules have to first diffuse out of the main particles; second, the drugs must overcome the barrier made by the electrospun fibers; and finally, they must reach the discharge medium, which may result in a low initial eruption monitored by a longer-term drug release technique [27].

As indicated in Figure 7.2a–e, electrospun composite fibers by incorporating ibuprofen (IBU)-loaded modified mesoporous silica (MMS) nanoparticles into electrospun poly(L-lac-tide) (PLLA) fibers (PLLA-MMS-IBU fibers) was fabricated to attain a long-term (over 100 days) controlled drug release [52]. As shown in Figure 7.2f, embedded drug-loaded liposomes into electrospun core-shell fibers with PVA using as an inner core and PCL as the outer shell of the fibers was obtained, with the potential for this structure to be used to regulate the release of TGF-, IGF-4, bFGF and other growth-enhancing drugs [53]. Electrospun drug-loaded microspheres (approximately, 10 μm to 20 μm in diameter) into sacrificial polyethylene oxide (PEO) fibers combined with slow-degrading poly (ε-caprolactone) (PCL) fibers was studied [54]. As seen from Figure 7.2g, after the elimination of the sacrificial component, the microspheres remained steadily trapped in the gap of poly (ε-caprolactone) (PCL) fibers. It is clear from Figure 7.2h that the use of a combined microspheres system with electrospun fibers electro sprayed vascular endothelial

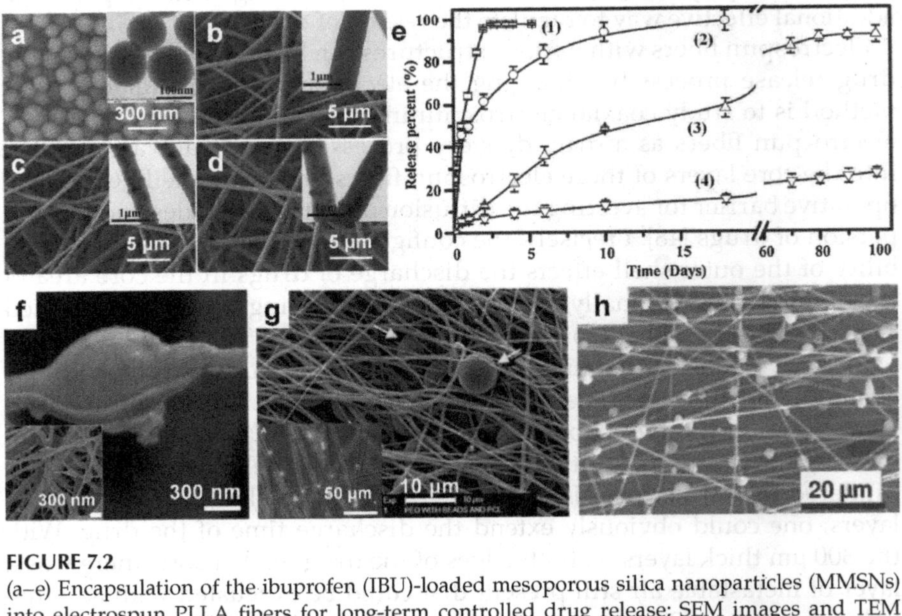

FIGURE 7.2
(a–e) Encapsulation of the ibuprofen (IBU)-loaded mesoporous silica nanoparticles (MMSNs) into electrospun PLLA fibers for long-term controlled drug release; SEM images and TEM images (inset of a) MMS-IBU and (b–d) electrospun PLLA-MMS-IBU composite fibers with different drug content; (e) in vitro IBU release profiles of (1) MMSNs, (2) PLLA-IBU9, (3) PLLA-MMS10-IBU5, and (4) PLLA-MMS15-IBU15; (f) SEM images of lipsome-embedded core-shell (PVA-PCL) nanofibers; (g) SEM image and fluorescent image (inset) of an anisotropic nanofiber/microsphere composite; in the inset, blue represents microsphere, green represents PCL fibers and black represents PEO fibers; (h) the SEM image of an angiogenic microfiber/microparticle composite. (Reprinted from *Progress in Polymer Science*, Yang, G., From nano to micro to macro: Electrospun hierarchically structured polymeric fibers for biomedical applications, Copyright 2018, with permission from Elsevier.)

growth factor (VEGF)-encapsulating PLGA micro particles with positively charged surfaces onto negatively charged electrospun PLA microfibers has been reported. Furthermore, the formation of an angiogenic microfiber/microparticle composite patch was also found [55]. Furthermore, some inorganic nanomaterials, such as carbon nanotubes and Fe_3O_4 nanoparticles, with the capability to adsorb the essential drugs and decrease the rate of diffusion of drugs in the polymer matrix were also encapsulated into the electrospun fibers to regulate the release of drug [56,57].

7.1.2.1.2 Controlled Delivery of Multiple Drugs

Multi drug loading and the programmable discharge of each drug types are key goals to attain effective drug delivery systems in electrospun fibrous. The treatment of diseases such as combined chemotherapy for a malignant growth of cells and the restoration of some defective tissues such as regeneration of bone defect need a staggered release of various therapeutic drugs or growth factors to accomplish the best results [27]. The most efficient way to realize a controlled discharged of each species in a fibrous system requires loading of various drugs in different areas in the individual fiber or within the fiber mats. Therefore, the differences in the spatial distribution of drugs may accomplish variations in release time or may program the discharge of each drug species. Numerous methods such as embedding drugs with various hydrophobicity in the same polymeric nanofiber matrix, loading different drugs into the inner or outer layer of core-shell fibers, incorporating different drug loaded particles into electrospun fibers, or changing the configuration of different drug loaded fibers may aid in achieving the targeted goals above.

To realize sequential and distinct release of drugs, inserting the drugs at different hydrophobicity in the same polymeric nanofiber matrix is a simple pathway. For instance, both hydrophobic dexamethasone (DEX) and hydrophilic green tea polyphenols (GTP) into electrospun PLGA fibers have loaded, previously. Therefore, the hydrophobic dexamethasone (DEX) molecules inside the fiber matrix only discharged from the channels formed after the early release of the hydrophilic green tea polyphenols (GTP) molecules. The technique of integrating different drug loaded molecules into the electrospun fibers is an operative and reasonably modest way to include the different drugs into the same fiber matrix where the different hydrophilic and hydrophobic drugs can be injected in the same fibers [58]. A successful encapsulation of two distinct compounds as separate domains within a single electrospun fiber was achieved by first directly synthesizing two sets of PVA nanoparticles, one set containing Texas-Red labeled BSA protein and the other set containing Alexa Fluor 488-labeled epidermal growth factor [59]. These nanoparticles were entrenched in a biocompatible polymer (PU or PLGA) through a single emulsion method. Furthermore, the mixing of these nanoparticle comprising solutions followed by electrospinning of this combination achieved separate drug domains within the fibers.

The drug release of these particle loaded fibers mixed colloidal particles with PCL was effectively electrospun the mixture into fibers. As seen in Figure 7.3a and 7.3b, all the electrospun fibers comprised of a linear array of colloidal particles at the core and the PCL as a shell. Then, a cross-linked poly (N-isopropyl acrylamide) (PNIPAM) and PMMA molecules containing the model drugs (fluorescein or Rhodamine B) was prepared, followed by

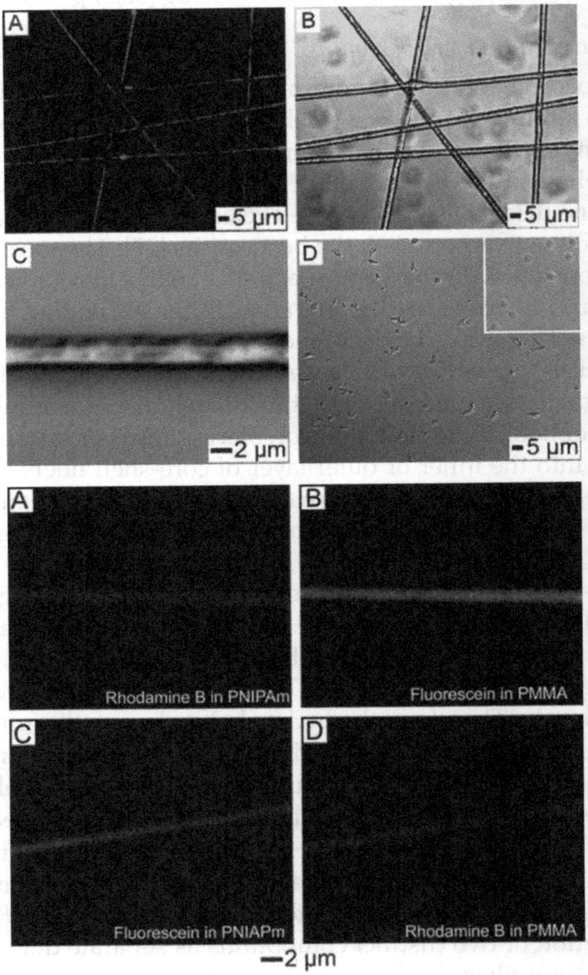

FIGURE 7.3
(a–b) CLSM images of colloids loaded PCL core-shell fibers; (c–f) CLSM images of electrospun fibers containing fluorescein or RhodaminB loaded crosslinked PMMA or PNIPAm colloids; (g) release profiles of fluorescein and Rhodamin B from different samples. (Jo, E. et al.: Core-Sheath Nanofibers Containing Colloidal Arrays in the Core for Programmable Multi-Agent Delivery. *Advanced Materials*. 2009. 21(9). pp. 968–972. Copyright Wiley-VCH Verlag GmbH & Co. KGaA. Reproduced with permission.) *(Continued)*

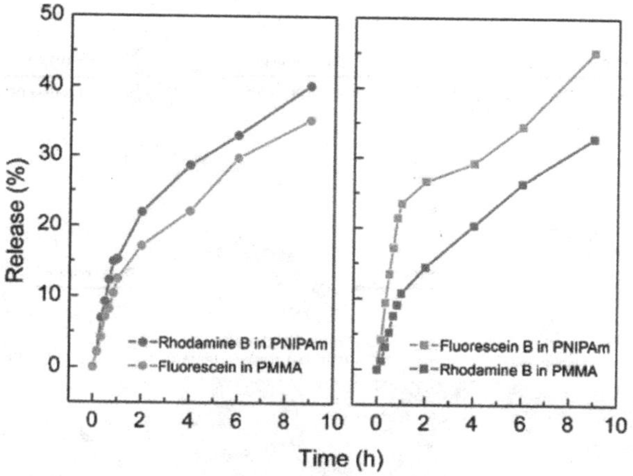

FIGURE 7.3 (CONTINUED)
(a–b) CLSM images of colloids loaded PCL core-shell fibers; (c–f) CLSM images of electrospun fibers containing fluorescein or RhodaminB loaded crosslinked PMMA or PNIPAm colloids; (g) release profiles of fluorescein and Rhodamin B from different samples. (Jo, E. et al.: Core-Sheath Nanofibers Containing Colloidal Arrays in the Core for Programmable Multi-Agent Delivery. *Advanced Materials*. 2009. 21(9). pp. 968–972. Copyright Wiley-VCH Verlag GmbH & Co. KGaA. Reproduced with permission.)

electrospinning the mixture of these two different materials comprising different drugs and the PCL [60]. Finally, core-shell fibers loaded with two different drugs was attained as detected in Figure 7.3c–f. Moreover, the two model drugs followed two distinct rate of release due to the differences in swelling performance of the two colloids, resulting in discrete release pathways for each drug (Figure 7.3g). Moreover, it is clear from the curves of drug release that the exchanging of carrier of the two different drugs would lead to an exchange of the rate of release, revealing that these particle containing fibrous drug delivery systems can realize the programmed release of multiple agents.

Furthermore, the different drugs can also be introduced in the particles and the prepared fiber matrix at the same period of time [61]. Because of different release procedures of the drug reagents in the fiber matrix versus the drug in the particles, there should be separate release methodologies. As indicated in Figure 7.4a, doxorubicin loaded in silica nanoparticles and domethacin into PCL/gelatin composite nanofibers and successful release of the two drugs were achieved [62]. Moreover, as seen from Figure 7.4b, polymeric micelles were also assimilated into electrospun nanofibers to appreciate time-programmed multi agent release of drug. Because of using a water soluble polymer, namely, PVA fiber matrix, the drug reagent and the drug-loaded micelles can instantaneously discharge from the nanofibers in an aqueous solution. Ultimately, the drug agent, present in the micelles, will release gradually. Thus, the electrospun micelle/drug-loaded nanofiber process provides multiple drugs safely entrenched within the fiber matrix and allows the independent control of each drug discharge [63]. Furthermore,

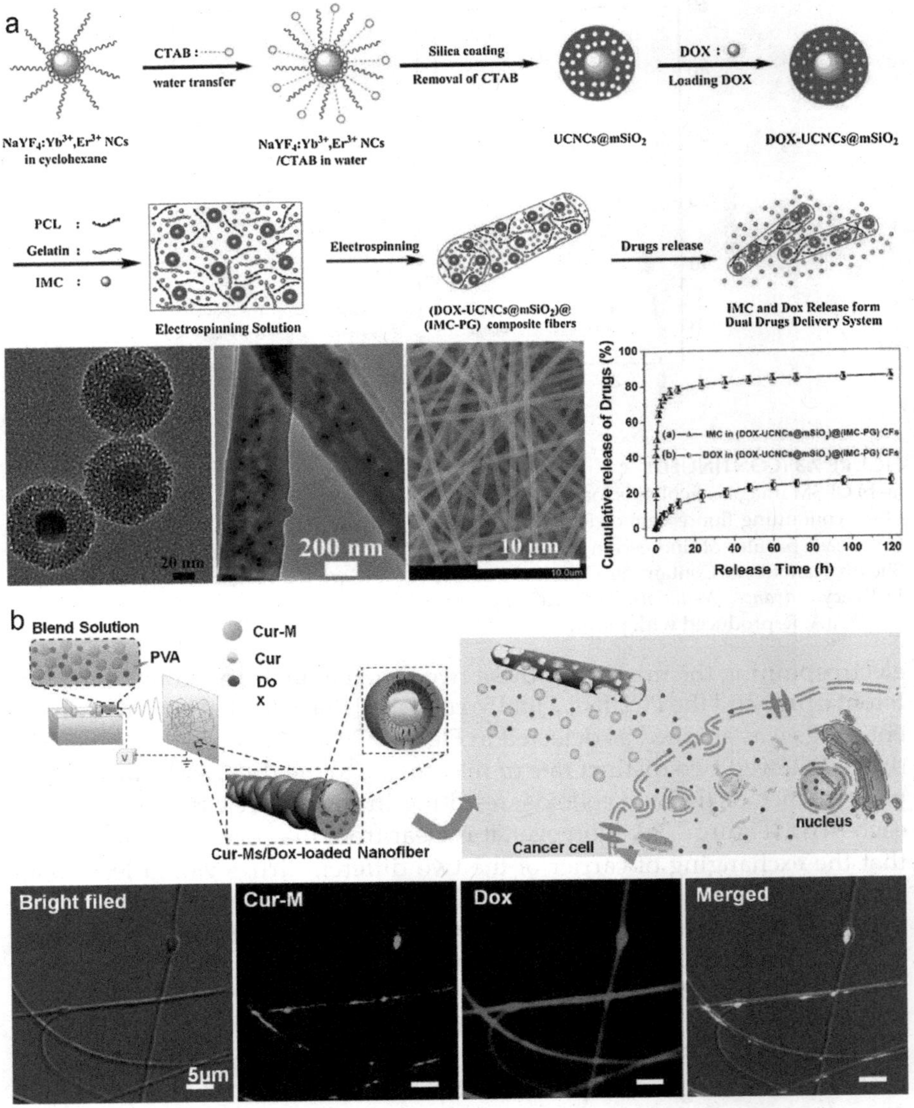

FIGURE 7.4

(a) The preparation and characterization of PCL/gelatin composite electrospun fibers containing doxorubicin-loaded silica nanoparticles and indomethacin and the release profiles of the two drugs; (b) the fabrication of the electrospun micelles/drug-loaded nanofibers, the time-programmed release of micelles and Dox from the nanofibers, the delivery of released micelles and Dox into cancer cells. The lower panel is the confocal laser scanning microscopy images of the micelles/drug-loaded nanofibers. (a. Reprinted with permission from Hou, Z. et al., *Langmuir*, 29(30): 9473–9482, 2013. Copyright 2013 American Chemical Society; b. Yang, G. et al.: Electrospun micelles/drugloaded nanofibers for time-programmed multi-agent release. *Macromolecular Bioscience*. 2014. 14(7). pp. 965–976. Copyright Wiley-VCH Verlag GmbH & Co. KGaA. Reproduced with permission.)

core-shell fibers are another active process for controlling the discharge of different drugs from fiber matrix, due to the easy encapsulation of various drugs in the shell and core layers of the fiber. Rhodamine B and bovine serum albumin (BSA) were generally used as model drugs and equated the profile of the rate of release under various parametric situations, such as loading condition of both drugs into the same layer (core or shell) and loading different drugs into the two different layers [64]. In this study, the authors observed that when either drug was in the shell layer clear initial burst release would occur with relatively fast release rate.

When a drug was injected in the core layer, the initial eruption release was blocked and the release rate was decreased. Moreover, electrospun layered fiber mats can also competently regulate the discharge of various drugs. As seen from Figure 7.5, fabrication of a multilayered drug-loaded poly (L-lactide-co-caprolactone) (PLCL) nanofiber mesh was reported and used for time-programmed dual release. They have indicated that the carrier for drug ChroB was present on the top layer of the mesh; barrier mesh was connected in the second position; after that, another drug, namely, 5, 10, 15, 20-tetraphenyl-21H, 23H-porphinetetrasulfonic acid disulfuric acid (TPPS), was carried by the third layer, and basement mesh was present in the bottom. An understandable difference in the rate of release profiles of the two different drugs was noticed in part to the barrier mesh. The chromazurol B (ChroB) from the top layer confirmed an obvious initial burst discharge, while the weak and slow initial burst release was observed in the third layer

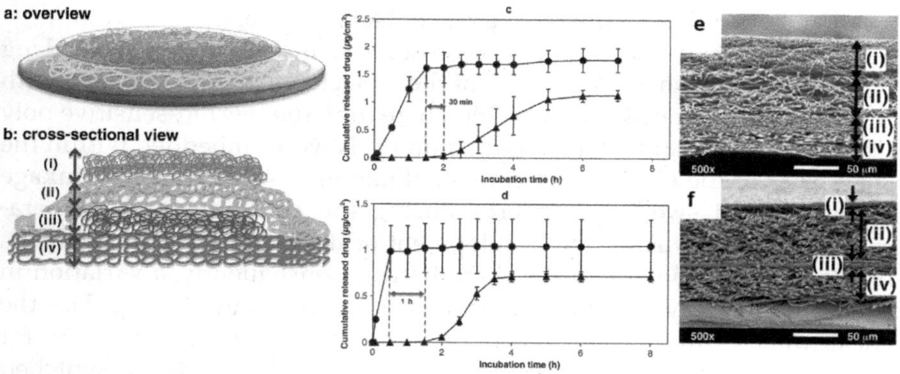

FIGURE 7.5
(a–b) Schematic diagram of multilayered drug-loaded nanofiber meshes; (c) the drug release profiles of samples; (d) the drug release profiles of samples in (e–f) cross-sectional SEM observations of multilayered nanofiber meshes with different layer thickness, in which (i) stands for ChroB-loaded fiber mat, (ii) stands for barrier fiber mat, (iii) stands for TPPS-loaded fiber mat, (iv) stands for basal fat mat. (Reprinted from *Journal of Controlled Release*, 143(2), Okuda, T. et al., Time-programmed dual release formulation by multilayered drug-loaded nanofiber meshes, pp. 258–264, Copyright 2010, with permission from Elsevier.)

during TPPS release. As seen from Figure 7.5c and 7.5d, thinner drug-loaded fibers controlled to faster drug discharge rates, while a thicker barrier mesh instigated a slower rate of drag release [65].

7.1.2.2 Stimuli-Responsive Release of Drugs

The stimuli-responsive polymeric fibers reply to peripheral stimulation, accompanied by corresponding changes in the physical and chemical properties of the polymers [27]. Based on the various types of stimulation, the stimuli-responsive fibers can be categorized as physical stimulus-responsive electrospun fibers such as temperature-responsive fibers, magnetic field-responsive fibers, light-responsive fibers, electro-responsive fibers; chemical stimulus-responsive electrospun fibers like pH-responsive fibers, humidity-responsive fibers, gas-responsive fibers, ethanol-responsive fibers, reduction-responsive fibers, glucose-responsive and protein-responsive fibers and multi stimuli-responsive fibers [66–68]. In theory, the application of stimuli-responsive electrospun fibers in a drug delivery system may give a more accurately controlled drug release than the other fibrous drug delivery systems explained in the previous sections. As seen from Figure 7.6a–c, dual-responsive electrospun fibers containing nanoparticles such as doxorubicin and magnetic within a fiber matrix were made from a temperature sensitive polymer (poly N-isopropylacrylamide) [69]. By applying an interchanging magnetic field to the fiber mats, the periodic heat has been produced by the magnetic nanoparticles inside the fiber matrix. This causes the polymer to shrink or swell and efficiently regulates the rate of DOX release.

Figure 7.6d and 7.6e shows an innovative nano gel in microfiber drug delivery system where the release of the molecules can be controlled with the help of temperature change [70]. Thereafter, the thermo-sensitive poly (N-isopropylacrylamide) (PIPAAm) nano gels were embedded within the shell layer of the electrospun core-shell nanofibers due to the shrinkage in nature and swelling property during the environmental temperature change, producing the development or departure of nano channels between the nano gel and the shell matrix. Consequently, a variation in shell penetrability was noticed and assisted as a valve to regulate the encapsulated drug release in the core layer. As shown in Figure 7.6e, for the nano gel in microfiber device (E2), a clearly temperature switched adjustable discharge of the model drug, namely, methyl orange (MO), was studied, where the concentration of MO raised abruptly at 40°C and increased marginally at 20°C during the all cycles. Then, the collective drug release of E2 extended up to 92.72% after three cycles. For the material without nano gel device (E1), a comparatively slow and continued release pattern was observed due to the direct dispersal of MO from the hydrophobic PCL shell, and the cumulative drug discharge of E1 was approximately 40%.

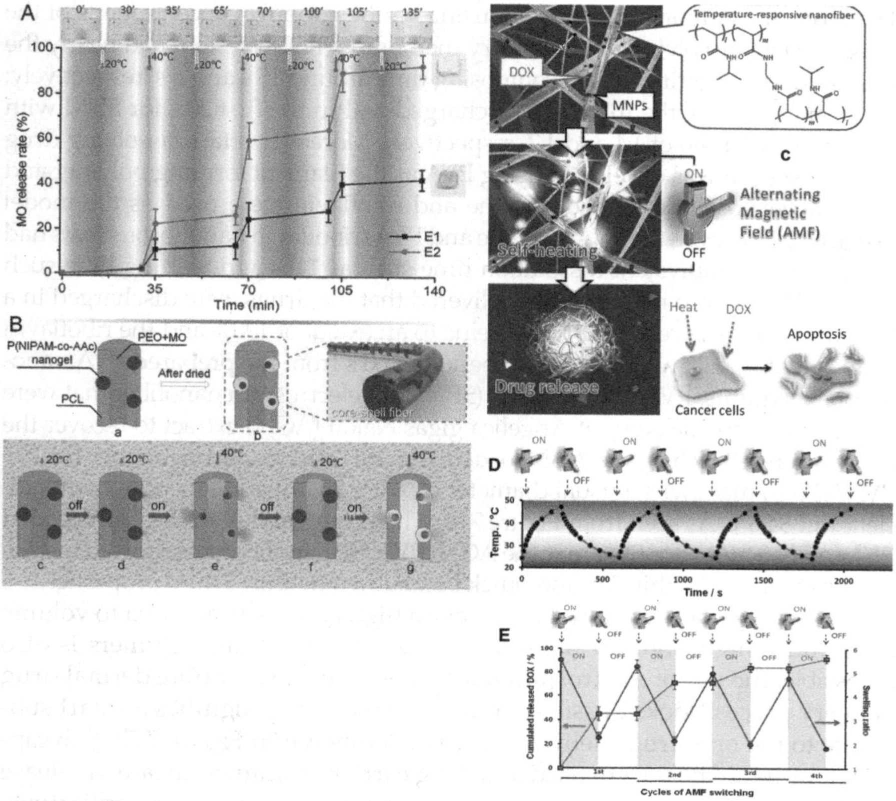

FIGURE 7.6
(a) The illustration of thermally switched drug release mechanism of electrospun nano gel-in-microfiber device (E2) under lower-to-higher temperature conversion; (b) the drug release performances of no nano gel fiber device (E1) and E2 under lower-to-higher temperature conversion for three cycles; (c) the schematic diagram of the drug release mechanism of a smart hyperthermia nanofiber system; (d) temperature changes of the nanofibers under the stimulation of alternating magnetic field; (e) the drug release profile and the swelling ratio of the nanofiber system under the stimulation of alternating magnetic field. (a–b. Li, L. et al.: Thermally Switched Release from a Nanogel-in-Microfiber Device. *Advanced Healthcare Materials*. 2015. 4(11). pp. 1658–1663. Copyright Wiley-VCH Verlag GmbH & Co. KGaA. Reproduced with permission; c–e. Kim, Y.J. et al.: A smart hyperthermia nanofiber with switchable drug release for inducing cancer apoptosis. *Advanced Functional Materials*. 2013. 23(46). pp. 5753–5761. Copyright Wiley-VCH Verlag GmbH & Co. KGaA. Reproduced with permission.)

7.1.2.3 Oral, Transdermal, and Implantable Drug Delivery Systems

Electrospun nanofibers were also observed to be beneficial during the oral delivery of some partially soluble drugs; meanwhile, the high active surface area to volume ratio improves the rate of dissolution of the nanosized drugs installed in the nanofibers [71]. The ultrafine fibers, mainly PVP loaded with IBU, were used for the development of a fast dissolving drug delivery membrane (FDM) [72].

Therefore, the authors reported that using various doping concentrations of the drug, the fast dissolving drug delivery membrane (FDM) system had nearly the similar kind of wetting and decomposing time around 15 s and 8 s, respectively; 84.9% and 58.7% of the IBU were discharged within the first 20 s for FDM with a drug-to-PVP ratio of 1:4 and 1:2, respectively. Moreover, a fast dissolving drug delivery system was developed using PVA as the filament forming polymer and drug carrier to fabricate, and caffeine and riboflavin were used as the model drugs [73]. However, the PVA/caffeine and PVA/riboflavin nanofibrous mats had practically the equivalent dissolution time such as 1.5 s and wetting time such as 4.5 s. The in vitro release study delivered that the drugs were discharged in a burst method after releasing the caffeine to an extent of 100% and the riboflavin to an extent of 40% within the time period of 60 s from the prepared PVA nanofibrous mediums. PVA and Soluplus (SP)-based electrospun nanofiber mat were employed for the delivery of Angelica gigas Nakai (AGN) extract to recover the oral cancers [74]. Then, the AGN-loaded PVA/SP electrospun nanofiber (AGN/PVA/SPNF) mat owns a usual diameter of 170 ± 35 nm and an entrapment efficiency of 84.6%. As seen from Figure 7.7a, when compared with the AGN-loaded PVA (AGN/PVA) nanofiber mat, the AGN/PVA/SP nanofiber mat showed instantaneous wetting (within 2 s) and quick breakdown (within 3 min) properties.

Since electrospun fibers not only have a high active surface area to volume ratio and a high porosity but a large quantity of entrant polymers is also accessible, the use of electrospun fiber mats in the field of transdermal drug delivery process looks possible. Transdermal delivery signifies a smart substitute to the oral drugs delivery process. As shown in Figure 7.7b, polycaprolactone nanofibers were used as a drug carrier to attain a sustained release of vitamin B12, a molecule well-suited for transdermal patch applications [75]. An electro-responsive transdermal drug delivery system was fabricated by using electrospinning PVA/poly (acrylic acid) (PAA)/multi-walled carbon nanotubes (MWCNTs) nano composites [76]. This system was installed with the uniform distribution of the oxyfluorinated MWCNTs in the nanofibers to observe the electro-responsive swelling and drug releasing behaviors.

As indicated in Figure 7.7c, fabricated electrospun PLGA fibers comprising 15 wt% dexamethasone (DEX) and 10 wt% green tea polyphenols (GTP) in the field of the active transdermal cure of keloid [58]. The fiber mesh was fixed over the keloid model in the subcutaneous tissue of athymic nude mice by suture and the fixed fiber mesh was disinfected and soaked in functional saline solution in advance in a different study. The fiber meshes were moistened by dropping saline solution into it every other day throughout the treatment process. After treating for 3 months, the results of histological analysis described that the disappearance of the hyalinized collagen fibers occurred and the loose, thin, and small collagen bundles seemed in the keloid, representing these dual drugs-loaded fiber meshes signified an operative transdermal treatment of keloid [58]. A facile method to functionalize the bio process of PLGA fibrous frameworks with different molecules such as PEG polymer during the cell revolting, the Arginine-Glycine-Aspartic (RGD) peptide for the cell linkage and bFGF

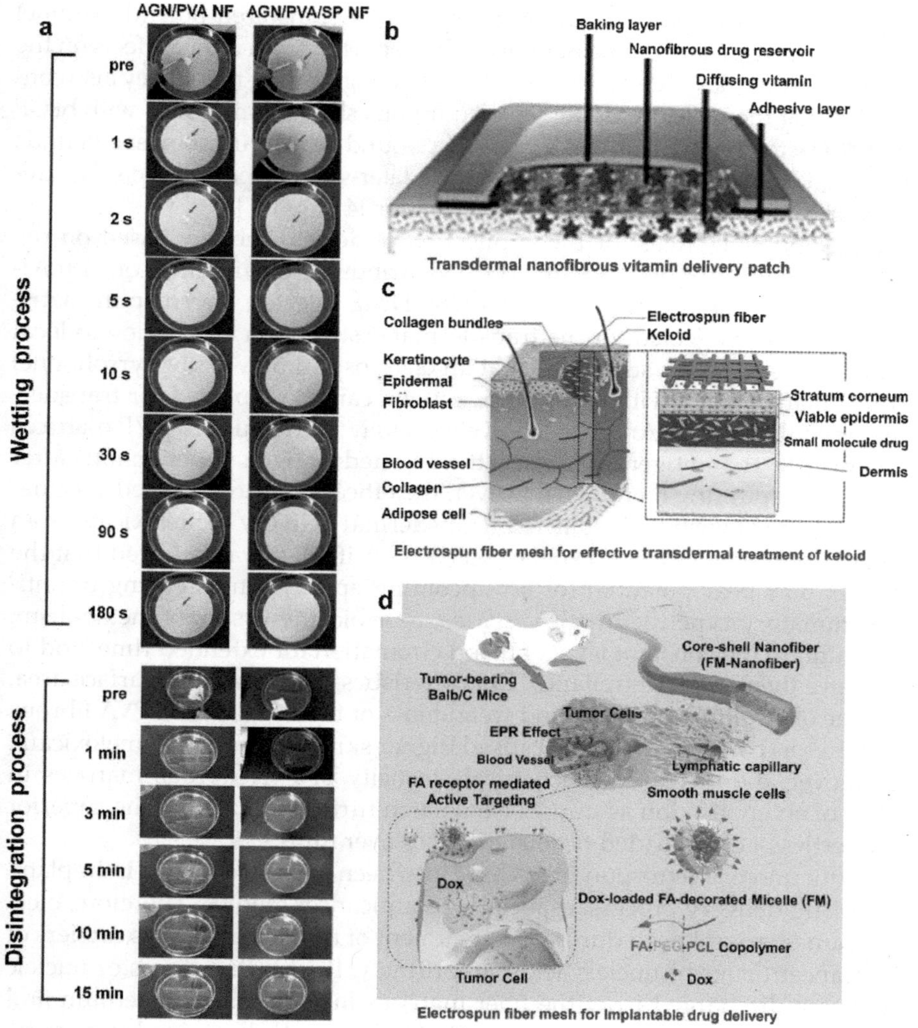

FIGURE 7.7
(a) Photographs of the wetting and disintegration process of oral drug delivery membranes. The AGN/PVA/SP NF membrane is fabricated from a PVA and soluplus (SP)-based nanofiber (NF) membrane containing the *Angelica gigas* Nakai (AGN) extract, used for oral cancer therapy. The AGN/PVA NF membrane is made from AGN-loaded PVA nanofibers. (b) Electrospun nanofibrous transdermal vitamin B_{12} delivery system. (c) Electrospun ultrafine fibrous transdermal dexamethasone and green tea polyphenols system for effective treatment of keloid. (d) An implantable active-targeting micelle-in-nanofiber device for cancer therapy. (Reprinted from *Progress in Polymer Science*, Yang, G. et al., From nano to micro to macro: Electrospun hierarchically structured polymeric fibers for biomedical applications, 81: pp. 80–113, Copyright 2018, with permission from Elsevier.)

growth factor for cell propagation through mussel inspired poly (dopamine) (PDA) coating in aqueous solution was also reported, with minor effects on the support properties such as drug activity and drug loading ratio. They also confirmed that the ginsenoside-Rg3-loaded fibrous skeleton improved with bFGF growth factor could encourage early-stage wound healing in rabbit ear wounds (bio-signal from bFGF) and co nstrain the later-stage hypertrophic scars creation (release of Rg3 drug) after imbedding for 14 days [77].

Other possible areas for the application of dermal therapy based on the drug-releasing matrices are the local distribution of vitamins or antioxidants such as skin care resolutions [78]. Drug injected electrospun matrices were also projected for the transdermal distribution in addition to local drug delivery. It was suggested that an electrospun poly (vinylpyrrolidone) (PVP) matrix laden with an anthrax antigen can be employed for transdermal inoculation [79]. An investigative cell assay with liquefied PVP matrices comprising the antigen displayed the rejected antigen functionality after the electrospinning methods; however, the efficacy of the projected vaccination path has to be yet established. Transdermal delivery of meloxicam from electrospun fibers through snake skin was verified. It was believed that the matrix may bear potential for an upcoming application as curing of anti-inflammatory type rheumatic diseases, to avoid the first-pass metabolism, to maintain the constant level of blood circulation for extended time, and to improve the patient compliance. Characteristics such as higher surface area, degree of swelling, porosity, and weight loss of the drug loaded PVA fibrous matrices as compared to films caused higher skin permeation of meloxicam. Moreover, the increased mechanical elasticity of PVA fibrous matrices in view of an application as transdermal patch further donated to the superior properties of drug loaded fibrous matrices over films [80].

Furthermore, electrospun fibers have also been extensively used in implantable drug delivery process, especially in anticancer studies. Therefore, electrospun fibers are used during the treatment of the cancer cells as carriers of anticancer drugs or nucleic acids; accordingly, the loading drugs or nucleic acids are discharged from the fiber matrices into the culture medium and impact the growth of cancer cells [81,82]. Although there are large variations between the lab-based experimental studies and the practical cancer remedies in a clinic, the drug-loaded nanofibers are helpful to screen and to preevaluate the drugs before going to lab test experiments or as a platform to explore the physiological appearances of growth cells [27]. Antitumor implants effectively confirm the drug concentration is at a relatively high level near the tumor tissues and decrease the spreading of the drug in surrounding normal tissues, thereby dropping side effects when compared with systemic administration such as intravenous injection [83]. Moreover, implants reduction of the drug management time has occurred, which reduces the discomfort felt by patients [84]. Therefore, electrospun drug-loaded fibers deliver the local drug agent due to the significant control of the release of drug and are simple to reduce in vivo.

Numerous studies regarding electrospun drug-loaded fibers as antitumor implants have been reported [85]. The antitumor efficiency of the implants was estimated mainly according to the tumor size or weight changes and the histological analysis after the therapy, while animal body weight as well as survival rate was always studied to estimate the safety of the implants. As seen in Figure 7.7d, an implantable developed active-targeting micelle-in-nanofiber device where the Dox-loaded active-targeting micelles were enclosed in the core poly vinyl alcohol (PVA) of the core-shell polymeric nanofibers, and the outer shell layer contains the cross-linked gelatin [86]. The authors implanted the system subcutaneously into murine breast cancer cell near solid tumors to verify the effect of drugs on tumor cells. From this study, the significant effects of the implantable nanofiber device groups found perfect comparable tumor growth suppression compared with the intravenously transported Dox formulations (free Dox and Dox-loaded micelles), though the intravenously carried Dox formulations were vaccinated 4 times and the nanofiber devices were entrenched only once [27].

For further assessment, the intravenously distributed Dox formulations were also injected only once to murine breast cancer cell at an equivalent Dox dose in the beginning of the process, and it is observed that after 21 days, the mean tumor volumes of all the intravenously transported Dox formulations groups were approximately 1000 mm^3, which were higher than the nanofiber groups (less than 500 mm^3), signifying that under the same dosage of Dox; implantable nanofiber groups performed much improved antitumor effect than the once-injection of free Dox or Dox-loaded micelles groups [27]. Additionally, the body weight of mice in the nanofiber devices groups was invariant and even had a slight raise compared with the control groups, such as saline grouper blank-nanofiber materials, and a relative higher survival rate over 33% after 42 days of treatment was observed for the mice in the active-targeting micelle-in-nanofiber device group, indicating this active-targeting micelle-in-nanofiber device delivers a low toxicity to the body. An orthotropic mice model of cervical/vaginal cancers was recognized to inject the drug into the vaginal submucosa of female KM mice near the cervix [87].

Nevertheless, there was a severe deterioration in the body weight of mice after injecting cisplatin drug indicating the average weight losses of 30.6% (5 mg/kg) and 12.1% (2.5 mg/kg) on day 9, thus, half of the mice in group of cisplatin/i.v group (5 mg/kg) lived past 9 days. In comparison, only a weight loss of 2.8% (5 mg/kg) was found for the mice of cisplatin/fiber group, and all animals survived. Considering these results, the cisplatin-loaded fibers were capable of favored partition of cisplatin in vaginal tract, reasonable distribution in rectum, uterus and tumor, whereas a low concentration in peripheral organs achieved a significant balance between the protection and antitumor effectiveness compared to the injection of cisplatin. Moreover, an injection of hepatoma H22 cells into Kunming mice was performed dermally to generate the solid tumors and used hydroxycamp-tothecin (HCPT)-loaded core-sheath PELA (HCPT/PELA) fibers as intra tumoral inserts against an

H22 tumor system [88]. The authors found that the tumor sizes of different groups raised after 14 days of treatment; the growth was higher for over 500% of saline group and empty PELA fibers group than about 420% of free HCPT group. The growth was minimum for the 256% HCPT/PELA fibers group, resulting in better tumor suppression for HCPT/PELA fibers, which may be owed to the sustained discharge of HCPT from fibers can continuous hinder the tumor growth.

A Lewis lung cancer model subcutaneously in the backs of C57BL/6 female mice was established successfully [89]. Therefore, when the tumors stretched up to 300 mm^3, the tumors were removed and threated the mice for stopping the local cancer reappearance in the different pathway, such as using a cisplatin-loaded super hydrophobic electrospun polycaprolactone/poly (glycerol monostearate-co-caprolactone) (PCL/PGC-C18) nanofiber mesh (13.2 mg/kg equivalent dose) implanted at their section site, and the implantation of an unloaded super hydrophobic mesh is also an important pathway, and another implantation of an unloaded super hydrophobic mesh and intraperitoneal (i.p.) cisplatin (2 doses × 6.6 mg/kg); or (4) only i.p. cisplatin 2 doses × 6.6 mg/kg process. Subsequently, the treatment is performed, the resected control animals without chemotherapy were observed to develop local reappearances within a median of 6 days, and both ways of 3 and 4 did not significantly increase median freedom from local return, whereas the median freedom from local reappearance was expressively raised for mice preserved with cisplatin-loaded meshes compared to i.p. cisplatin, with median recurrence-free existence to over 23 days.

In another study, both the multi-walled carbon nanotubes and doxorubicin-containing PLLA fiber mats were used in a recognized cervical cancer model under the mouse's armpit; then, since the carbon nanotubes were able to competently produce thermo by absorbing near-infrared radiation (NIR), they performed the NIR radiation photo thermal therapy combined with chemotherapy to successfully inhibit the growth of the tumor [90]. In addition to anticancer studies, electrospun fibers have also been used as implantable devices as a potential therapy during gout attacks, anti-viral and antibiotics, and in the field of surgical treatments [27]. Lute Olin-loaded electrospun PCL/gelatin composite fibrous mats were investigated and applied to the gout sites of the New Zealand rabbits' knees [91]. The authors found that these implantable devices improved acute gouty arthritis and may offer a possible gout therapy for overcoming repeated attacks. The application of electrospun fibers in vaginal anti-HIV drug delivery was revised [92]. Based on the authors' opinions, drug-eluting fibers can be designed to attain multiple design limitations in a single product for topical HIV inhibition, due to the significant effect to deliver miscellaneous agents and enable the controlled discharged for precoitally and sustained (coitally independent) utilization, and this matrix is technically possible for the continuous scaling-up. Challenges of using electrospun fibers are present, like issues connected to vehicle arrangement, dispersion and retaining the vaginal vault, the unknown safety and toxicity of the communication of the fibers with the mucosal environment, and problems in precisely controlled discharge of multiple drugs from a single electrospun fiber matrix [27].

7.1.2.4 Nucleic Acid Delivery

Recently, the application of electrospun scaffolds for nucleic acid delivery has received increasing attention. An initial attempt used condensed plasmidDNA that was captured with poly(lactide)-b-poly(ethylene glycol)-b-poly(lactide) and combined with electrospun PLGA fibers [93,94]. Defense and regulated discharge of DNA from the fibrous framework over one week was attained where transfection levels were low (<1%) when scaffolds were absorbed into the cell culture dish with a rising monolayer of cells, and the transfection effectiveness was improved if cells were plated on the supports. This improvement in efficacy was attributed to an increased local concentration of DNA and the direct adhesion of cells to the fibers releasing DNA.

As known from various studies on delivery systems for nucleic acids, efficient transfection of cells involves several steps, such as endosomal release, cellular uptake, and translocation into the nucleus [95]. Preferably, a delivery system should be arranged to defeat these barriers, for instance, by cell specific uptake, intrinsic endosomolytic activity, and intra cellular discharge of the cargo. It is also mentioned that chitosan is one of the common biopolymers that have been studied during the transfection of DNA or siRNA for the fulfilling of some of these requirements [96,97]. Therefore, it is not unexpected that DNA/chitosan complexes were assimilated into electrospun matrices to deliver the therapeutic genes to the targeted cells to grow on such frameworks. Various composed supports such as PLGA/HA fiber covered with naked DNA, PLGA/HA fibers treated with chitosan/DNA complexes, and chitosan/DNA complexes captured in the composite fibers were studied previously [98,99]. The duration of DNA release was faster for encapsulated chitosan/DNA complexes than the chitosan/DNA complexes on the surface and naked DNA on surface. After for up to 2 months, the scaffolds with captured chitosan/DNA complexes sustained the release of DNA. The cytotoxicity of the different used scaffolds varied significantly and the chitosan/DNA coated scaffolds produced the highest toxicity. Then, after 3 days of cultivation, the highest transfection efficiency was obtained using chitosan/DNA coated scaffolds following encapsulated chitosan/DNA scaffolds [99].

In another study, the established systems were studied in vivo to treat tibia defects in mice. Amazingly, the scaffolds coated with naked DNA performed best during the first 14 days, while scaffolds coated with chitosan/DNA complexes performed best when mice were examined after 28 days. Therefore, the conceptual advantages of encapsulated complexes couldn't be translated into practice in this model, possibly because of relatively fast self-healing of tibia defects in mice [98]. Core–shell nanofibers with DNA located in the polyethylene glycol (PEG) core and a poly(ethylene imine)-hyaluronic acid delivery vector combined with PCL forming the shell were carefully studied to evaluate treating parameters using a fractional factorial design [100]. The transfection efficiency and kinetics of drug release could be extended by varying operating parametric conditions and most of the frameworks prolonged the discharge of DNA over a period of 2 months. Moreover, DNA encapsulation can also be achieved with the help of layer-by-layer covering of PLLA fibers with PEI and DNA [101]. In this case, transgene expression could be regulated after changing the DNA concentration of the coating solution or the number of deposition layers on the fiber active surface. However, it is reported that the transgene appearance raised up to 120 h of cultivation time. Layer-by-layer coating is rather appropriate to attain a high initial release of drag reagents and may be judiciously combined with the release of DNA/polycation complexes on the fibers significantly. Moreover, the encapsulation of naked siRNA into PCL or PEG–PCL fibers was also reported [102]. Therefore, using PCL as scaffold material release from the fibers was insignificant with pure PCL and raised with rising proportion of PEG. Actually, siRNA was well secured by encapsulation and continuously released during 4 weeks, though released siRNA required to transfect cells seeded on the scaffold with high efficiency, whereas remediation occurs using transfection reagent. Furthermore, an upgraded scaffold facilitated siRNA transfection was made up using electrospinning method and blend of complexes comprising of a transfection reagent, siRNA and a copolymer of caprolactone and ethyl ethylene phosphate, and protection of siRNA and release over approximately 4 weeks was achieved [103].

7.1.3 Tissue Engineering and Regenerative Medicine

Tissue engineering is a multidisciplinary area that conglomerates both the areas of engineering and life sciences to develop the biological alternatives and also for renovation, conservation, or development of tissue function. Biomaterials research is known as an emerging engineering area that shows an essential role in tissue engineering by serving as matrices for cellular in growth, propagation, and new tissue formation in three-dimensions. When associated to other fiber forming processes, such as self-assembly and phase separation processes, electrospinning delivers a simpler and more efficient incomes to produce fibrous scaffolds with an interconnected pore structure and fiber diameters in the submicron range. Nanotechnology and nanoengineering are effective for the design, preparation, characterization, and applications of nanoscale devices, which consist of functional organizations with at least one dimension in the range from several to hundreds of nanometers. Currently, the synergy between nanoscience and tissue engineering has led to great developments in biomedical research as well as clinical practices, including the realms of bone and cartilage regeneration, blood vessel tissue engineering, wound dressing, etc. [104–106].

Probably because of the high surface/volume ratio of nanofibers that mimic an extracellular matrix structure, different types of cells, including mesenchymal stem cells, embryonic stem cells, keratinocytes, and hepatocytes, cultivated on nanofibrous meshes showed superior viability compared to other tissue engineering materials [107]. Therefore, Galactosylated moieties were immobilized onto the surfaces of highly porous nanofibers for in vitro culture systems of rat hepatocytes [108]. Acrylic acid was photo-irradiated on polymeric nanofibrous meshes and carboxylic acid groups were subsequently employed to conjugate galactopyranoside. This galactosylated scaffold composed of poly(ε-caprolactone-co-ethyl ethylene phosphate) (PCLEEP) promoted cellular functions of rat hepatocytes when the hepatocytes were cultivated into spheroid-like aggregates. Electrospun PCL nanofibers were surface-modified with calcium phosphate [109], where the nanofibers strongly enhanced osteoblastic differentiation and proliferation for a prolonged period of cell culture was observed. Therefore, the calcium phosphate layers on nanofibers displayed similar characteristics of human bones, suggesting potential application in bone tissue engineering. In another study, hematopoietic stem cells were cultivated on the nanofibers surface-modified with hydroxyl, carboxyl, and amino groups [110]. From this study, the authors have found that aminated nanofibers most effectively promoted expansion of stem cells and progenitor cells including CD34(+) and CD45(+). Moreover, electrospun polymeric nanofibrous meshes were prepared in two different ways: randomly distributed and aligned. These aligned and randomly distributed nanofibers were subsequently modified with collagen on the surface. Then, human coronary artery endothelial cells proliferated in the same direction as that of the aligned surface-modified nanofibers. Also, the endothelial cells on the nanofibers exhibited high retention of their original phenol types regardless of the degree of alignment. In a separate report, human mesenchymal stem cells were also cultivated on nanofibrous scaffolds for subcutaneous implantation of reconstructive surgery [111]. In this case, the stem cells were differentiated into chondrocytes on the nanofibrous mesh in bioreactors for efficient proliferation. Therefore, it has been confirmed that high expressions of collagen type 2 and aggrecan were attained after 42 days of in vitro cultivation. Moreover, cell adhesion proteins were also conjugated to electrospun nanofibers for cultivation of epithelial cells [112]. Fibronectin, another cell adhesive molecule, was grafted on the fibrous meshes composed of PLLC by glutaldehyde crosslinking. The morphology of the surface-modified nanofibers was the same compared to that of the unmodified ones. They also tested mechanical properties of the nanofibers, which showed a minor decrease in the strain strength after polyester aminolysis. Porcine esophageal epithelial cells cultivated on the nanofibers showed enhanced proliferation, which was confirmed by SEM, immune histology, and protein analysis. Collagen was also employed for enhancing attachments and proliferation of stem cells [113]. Chemically immobilized collagen on nanofibrous meshes strongly promoted cell growth rates as well as differentiation of neural stem cells in a dose-dependent manner. This was attributed to increase in the attachment and preserved viability of cultivated stems cells on the collagen nanofibrous meshes. Human mesenchymal stem cells were cultivated in electrospun nanofibrous mesh composed of biodegradable polymer, PLGA beads, and PLLA [114]. Presently, many approaches toward tissue engineering applications have been limited to in vitro studies, and only a few studies were performed for in vivo application because cells could not be loaded within the nanofibrous meshes in a large quantity. This limitation can be overcome by employing multilayered nanofibrous scaffolds, where layers of cells are proliferated in between the layers of mesh. Therefore, the cell layers can proliferate and differentiate according to the microenvironment of surface-functionalized nanofibrous membranes. Numerous techniques to fabricate 3D nanofibrous scaffolds have been attempted for in vivo tissue engineering applications. As seen in Figure 7.8, electrospun nanofibers for drug and gene delivery application have been used for tissue engineering to improve therapeutic efficacy. Moreover, the fibrous surface structure shows strong adhesiveness to mucous layers because their nanoporous structures instantly absorb moisture at mucous layers through nanovoid volumes [115,116]. The superior adhesiveness toward biological surfaces allows nanofibers to be an ideal candidate for topical drug delivery devices.

Today, electrospun fibers are also extensively used as scaffolds in the field of tissue engineering due to inherent structural topographies. Because of the large active surface areas, complex fibrous interfacial topological configuration, ease of functionalization, and significant and controlled mechanical properties, continuous thin fibers made by electrospinning have the capability to mimic the extracellular matrices (ECM) and may be a perfect technique

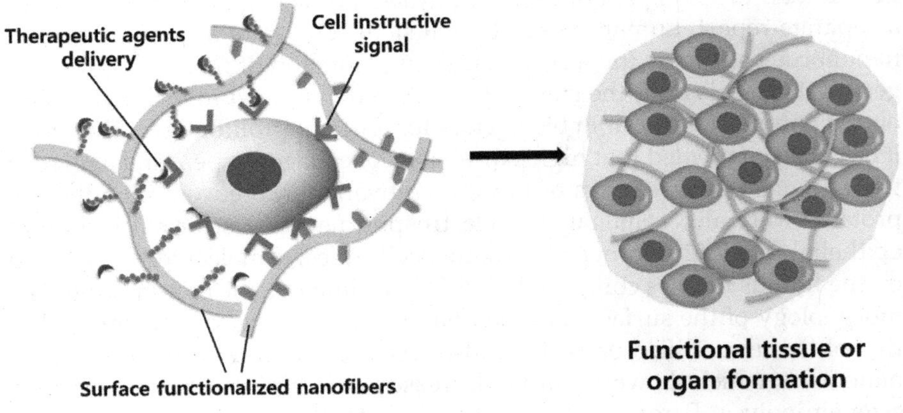

FIGURE 7.8
Tissue engineering application is often combined with drug delivery strategy. (Reprinted from *Advanced Drug Delivery Reviews*, 61(12), Yoo, H.S. et al., Surface-functionalized electrospun nanofibers for tissue engineering and drug delivery, pp. 1033–1042, Copyright 2009, with permission from Elsevier.)

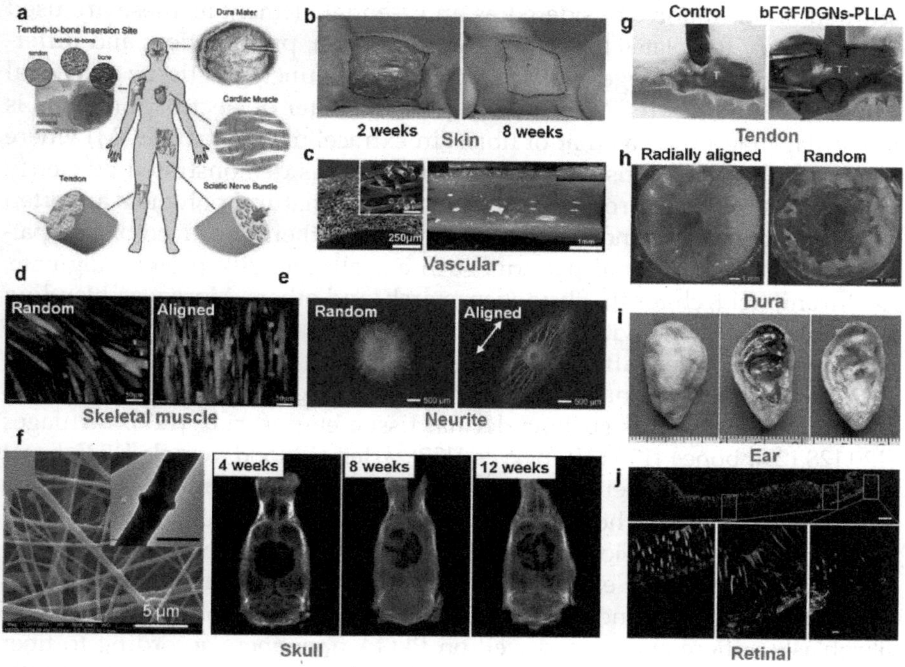

FIGURE 7.9
(a) Illustration of typical examples of tissues in the human body whose regeneration would benefit from the use the anisotropic structured electrospun nanofiber scaffolds; (b) electrospun nanofiber scaffold for skin regeneration; (c) tubular electrospun fibers for artificial blood vessels; (d) nanofibers for muscle tissue engineering; (e) nanofibers for outgrowth of neurite; (f) nanofibers for repair of rat calvarial defect; (g) nanofibers for tendon regeneration; (h) radially aligned electrospun nanofibers as dural substitutes for wound healing; (i) electrospun fibers for engineering ear-shaped cartilage; (j) nanofibers for repair of retinal. (Reprinted from *Progress in Polymer Science*, Yang, G. et al., From nano to micro to macro: Electrospun hierarchically structured polymeric fibers for biomedical applications, 81: pp. 80–113, Copyright 2018, with permission from Elsevier.)

to fabricate the tissue engineering (TE) scaffolds [117,118]. Therefore, various electrospun fibrous scaffolds that mimic native extracellular matrices should be applied in accordance with the natural organization of cells, including the use of random, aligned, and 3-D structural fibers to adapt to the demands of tissue repair [27]. In Figure 7.9a, typical examples of tissues in the human body whose regeneration would benefit from the use of anisotropic structured electrospun nanofiber scaffolds are summarized [119].

7.1.3.1 Scaffolds for Tissue Engineering

Electrospinning technique has been used increasingly to create nanofibrous scaffolds for tissue engineering as there are reports that these scaffolds positively promote cell–matrix and cell–cell interactions with the cells having a normal phenotypic shape and gene expression [120,121]. Biodegradable scaffold is commonly considered as an essential element as these are used as temporary templates for cell seeding, invasion, proliferation, and differentiation prior to the regeneration of biologically functional tissue or natural extracellular matrix (ECM). Therefore, the diameter of electrospun fibers is of similar magnitude as that of fibrils in extracellular matrix (ECM) where it mimics the natural tissue environment and has demonstrated efficiency as a substrate for cell growth [122]. Furthermore, natural polymers are often used for fabricating nanofibrous scaffolds due to their improved biocompatibility and bio-functional patterns such as collagen, silk protein, alginate, hyaluronic acid, chitosan, fibrinogen, starch, and others. Moreover, blending the natural polymers with synthetic polymers can improve the overall cytocompatibility of the scaffold [123–126].

Different polymeric nanofibers have been considered for use as scaffolds for engineering tissues such as dermal tissue engineering [127], cartilages [120,128,129], bones [130,131], nerves [132], arterial blood vessels [133], heart [134], etc. Because of their high porosity, that is, greater than 90%, their high surface area allows higher cellular attachment; and also due to the presence of multiple focal adhesion points, electrospun PLGA fiber mats are considered as ideal for tissue engineering scaffolds. These electrospun fiber mats support proliferation and growth of different cell types, for instance, mouse fibroblasts adhere and spread well on PLGA nanofibers according to fiber location [120,121]. Silk fibroin fiber scaffolds having bone morphogenetic factor 2 (BMP-2) and/or nanoparticles of hydroxypetite (nHAP) fabricated using electrospinning have been used to form in vitro bone from the mesenchymal stem cells (hMSCs) found in human bone marrow. Therefore, it was suggested that nanofibrous scaffolds of silk fibroin serve as ideal candidates for bone tissue engineering [135]. Because of their ability to differentiate into multiple cell lineages, an increase in the incorporation of marrow stromal cells (MSC) into cartilage and bone tissue engineering strategies was observed, and it was also reported that electrospun PCL scaffolds support the proliferation, attachment, and differentiation of MSCs into adipogenic, chondrogenic, or osteogenic lineages based on the culture media selected.

The cell biocompatibility of electrospun fibrous scaffolds with MSCs has been demonstrated by using PLGA electrospun nanofibrous scaffolds and B.mori silkworm silk fibroin/PEO composite electrospun scaffolds in two separate studies through the evaluation of the attachment and proliferation of MSCs in the scaffolds [120,130,136,137]. The initial anticipation about the use of nanofibrous matrix for tissue engineering applications is declining due to the fact that low pore size of the mat hinders cell infiltration inside the electrospun matrix, leading to almost two-dimensional tissues that eventually fail to simulate the physiological 3D tissue microenvironment. Therefore, a variety of attempts are being made to enhance the porosity level and improve cell infiltration. To get large pore size and high porosity it is suggested that a novel technique of electrospinning silk fibroin using different collecting parts and where the nanofibers were dropped directly on the coagulation bath containing methanol below the spinneret [138]. Wet spinning has also been suggested to improve the porosity and cellular migration [139]. Currently, novel wet electrospinning technique is used for preparation of sponge form nanofiber 3-dimensional fabric with controlled fiber density [140]. Therefore, the authors have used a combination of both wet spinning and electrospinning system for control of nanofiber fabric. Poly (glycolic acid) is used as polymer and the solvents used in wet spinning are pure water, 50% tertiary-butyl alcohol (t-BuOH), 99% t-BuOH. It is confirmed that the surface tension of the solvents influences the fiber density considerably.

7.1.3.2 Wound Dressing

For wound healing, an ideal dressing needs to have certain properties such as efficiency as bacterial barrier, haemostatic ability, absorption ability of excess exudates (wound fluid/pus), adequate gaseous exchange ability, appropriate water vapor transmission rate, ability to conform to the contour of the wound area, functional adhesion, that is, adherent to healthy tissue but non-adherent to wound tissue, painless to patient and ease of removal, and finally low cost [26]. Present efforts employing polymer nanofibrous membranes as medical dressing are still beginning; nonetheless, electrospun materials meet most of the requirements outlined for wound-healing polymer devices due to their nanofibrous and microfibrous structures providing the nonwoven textile with desirable characteristics [141–144].

The three layers, namely, the dermis, epidermis, and the hypodermis, are important layers to make human skin (the largest organ of the human body) immunologic, protective, thermoregulatory sensual, and they behave as a barrier with protective, immunologic, thermoregulatory, and sensory functions [145]. Conventional skin replacements such as auto grafts, allografts, and xenografts have been used for skin revival. However, the gold standard of skin rehabilitations is auto grafts due to the protection from the rejection, limited availability, donor site morbidity risk factor, and formation of scar function indicating the safer alternatives [146,147]. Allografts and xeno grafts are more accessible substitutes but transmit the threat of disease conduction and immunological rejection [147–149]. Recently, studies have been found on the development of the regeneration of the facilitating skin as a bioengineered substitute of promising therapeutic alternative. It can help during the acute and chronic wounds treatment or burns as wound healing.

There are various scaffold types such as films, sponges, and micro- and nanofibers fabricated by natural and synthetic polymers [150]. It is obvious that the high specific surface area of nanofibers signifies the enhancement of fluid absorption, delivery of dermal drug, and absorption of functional proteins such as albumin and fibronectin successfully, as well as surface laminin [151]. Furthermore, the high porosity of these electrospun nanofibrous scaffolds enables the exchanges of oxygen, water, and nutrient as well as the elimination of metabolic waste. In addition, the small-sized pores restrain the penetration of harmful microorganisms [152]. As seen from Figure 7.9b, freeze-dried and electrospun collagen scaffolds were compared as skin substitutes and indicated that the electrospun scaffold skin substitutes influenced optimal cellular organization, potentially decreasing wound shrinkage [153].

The coaxial electrospinning experiment was performed to make separately collagen-coated PCL nanofibers such as collagen-r-PCL in the form of a core-shell structure resulting in the linear increment of the density of human dermal fibroblasts on the core-shell nanofiber membrane up to 19.5% within 2 days, 22.9% in 4 days, and 31.8% in 6 days as compared to PCL nanofibers [154]. A system in which collagenase was stored inside PEO electrospun nanofibers was developed and released on hydration [155]. Through a series of in vitro and in vivo surveys, their results show that the partial ingestion of the wound boundary improved wound reparation by making a more obedient and porous microenvironment that accelerated cell immigration toward production at the wound margin. A three-dimensional fibrous framework was also prepared for enhanced vascularization using a photo cross-linkable natural hydrogel based on gelatin methacryloyl (GelMA) using electrospinning technology [156]. Therefore, the authors found the fibrous membranes using ultraviolet (UV) photo cross-linkable gelatin electrospun hydrogel, owning soft adaptable mechanical properties and controllable dilapidation properties. These fibrous membranes can support the endothelial cells and dermal fibroblasts bond, proliferation, and migration into the supports, which further enables for vascularization. The skin flap survival rate of a rat model was advanced above the other control group after the implementation of the fibrous scaffolds below the skin flap. More microvascular formation was detected, which was possibly cooperative for the flap tissue vascularization. It is also found that the morphology of an electrospun fiber mat impacts the development of seeded fibroblasts.

Furthermore, the arrangement of the fibers may control cell alignment and support the contact between the cell body and the fibers in a longitudinal way [157,158]. Moreover, novel sandwich-type frameworks were prepared as micro skin grafts to use in skin revival containing outward aligned nanofibers as the bottom layer, nanofiber membranes with square arrayed micro wells and nanostructured prompts as the top layer, and in the middle position, micro skin tissues were present [159]. Therefore, the micro skin tissues were inadequate in the square arrayed wells and instantaneously directed by the nano topographic cues, enabling the in vitro immigration of the cultured NIH 3T3 fibroblasts and prime rat skin cells. The distribution of the micro skin grafts was obtained in the sandwich-type transplants and the "take" rate of the micro skin tissues was improved, promoting re-epithelialization on the wound. Additionally, the exudate drainage from the wound was promoted from the void regions in the scaffolds. Assembly of electrospun three-dimensional fibrous with an irregular inner structure and engineered surfaces was reported to heal the wound, with thickness of 3.7 ± 0.1 mm and fibrous-based micro-sized conical extensions whose length and width were

1000 ± 300 m and 3800 ± 80 m, respectively, at the top side of the fibers, with an average peak density of 73 peaks per cm^2, while keeping a nonwoven mesh bottom [160]. Therefore, when the three-dimensional hypothesis was used as wound dressing, it actively supported wound healing (90 ± 0.5% of wound closure within 48 h).

Electrospun fibers can also perform as supporting material for wound dressing. For instance, a novel dissolving microneedle (DMN) delivery scheme on an electrospun pillar array (DEPA) was invented, where it was capable to rapidly graft the DMNs into the skin. Then the separation of the DMNs from the fibrous sheet was noticed to be dependent on both the pillar height and the properties of the fibrous sheet [161]. A cell-on-a-chip model with electrospun fibers was processed to simulate a cutaneous wound in vitro and screened the performance of several electrospun fibers as wound dressings. This electrospun screening process is predicted to change the path that researchers screen the candidates during the wound dressings [162].

Moreover, wound-healing properties of mats of electrospun type I collagen fibers have been investigated on wounds in mice, and the authors found that healing of the wounds was better with the nanofiber mats than with conventional wound care, especially in the early stages of the healing process [129]. Furthermore, nonwoven antibacterial poly (vinyl alcohol) (PVA) membranes are prepared by electrospinning of PVA aqueous solution with addition of Ag^+ loaded zirconium phosphate nanoparticles for potential application in wound healing materials [171]. Therefore, the antimicrobial tests showed the effectiveness of nanoparticles containing nanofibers against tested strains. Recently, fibrous poly (L-lactide) (PLLA) and bio component PLLA/poly (ethylene glycol) mats have also been prepared by electrospinning and coated with chitosan and confirmed that with the increase of chitosan content, the hemostatic activity of the mats was observed to increase [172].

A various range of natural and synthetic polymers, namely, polyurethane (PU), PLA, PCL, PLGA, polyvinyl alcohol, dextran, chitin, chitosan, cellulose acetate, gelatin, and collagen, was investigated as applicants to make dressing materials, where anti-inflammatory drugs and tissue growth agents, the bioactive agents were also merged in the polymeric nanofibers to control the drug delivery [106,173–177].

A guinea pig wound model was demonstrated using electrospun PU matrices, and it was confirmed as a suitable wound dressing material that providing raised rates of epithelialization [144]. In addition to wound coverage, the porous property of nanofibrous scaffolds support to enable wound healing by acting as a barrier for bacteria. It can absorb the wound exudates, and a moist environment can be provided, allow gas exchange that permits informal removal off the wound [178]. Moreover, loading wound dressings with drugs may further help an efficient wound healing process. Therefore, several wound dressings with drug delivery functionality are demonstrated starting from hydrocolloids and hydrogels up to electrospun nonwoven [179]. It is worth mention that electrospun scaffolds are suitable as wound dressings due to the large active surface area, nanofibrous structure, and amendable porosity [180].

7.1.3.3 Vascular Tissue Engineering

The major categories of the vascular tissue engineering preparation techniques are mainly flat fiber membranes scattered with endothelial cells or even muscle cells for a convinced time followed by the rolling technique into the vessels [181,182], then as spun micro-/nanofiber tubes utilized straight to the culture cells to make blood vessels [183–186] and as spun micro-/nanofiber tubular frameworks surrounded in vivo and performed to prompt the cells from experimental animals to migrate into the frameworks to finally attain vascular tissue [163,187]. Three-dimensional fibrous tubes composed of ultrafine electrospun fibers were prepared with an innovative static method and combined macro tubes were arranged directly [188]. A multilayer vascular graft based on collagen-mimetic proteins was also developed [186]. Therefore, the electrospun polyurethane tubular mesh was used to reinforce the tubular PEG-Scl2 hydrogel to develop the tubular hydrogel to attain suitable biomechanical properties. PEG-Scl2 hydrogel, which was skilled to bind the endothelial cells (ECs) and resist platelet linkage, was prepared with the help of conjugated collagen-mimetic protein derived from group A streptococcus, Scl2.28 (Scl2), into a PEG hydrogel [27].

Moreover, the minimal platelet interactions were observed for the prepared multilayer vascular scaffold and can support the migration of ECs. Furthermore, a bilayered tubular scaffold with an inner layer composed of small diameter fibers and an outer layer composed of large diameter fibers was fabricated, and the results confirmed that the bilayer scaffold allowed both EC adhesion on the lumen and SMC infiltration into the outer layer [183]. In addition, collective electrospinning and spin-casting approaches were employed to design the luminal surface of small-diameter polyurethane (PU) grafts with microfibers and microgrooves [185]. It is also reported that the microgrooves guided endothelial cell alignment alongside the axis of the blood vessel. Then, the tracks of the circumferentially leaning microfibers were fabricated using electrospinning polyurethane (PU) onto a mandrel rotated at high-speed velocity, while the longitudinal tracks of the microgrooves were generated using spin-casting polyurethane (PU) over a rotating poly (dimethylsiloxane) (PDMS) mold. The authors confirmed that the endothelial cells scattered onto the grafts developed merging monolayers with individual cells showing an elongated morphology parallel to the micro patterns. As-spun micro-/nanofiber tubular scaffolds were also able to induce cells to migrate into the scaffolds after in vivo implantation by replacing rat abdominal aorta. As revealed in Figure 7.9c, rat abdominal aorta was replaced with macroporous electrospun PCL vascular scaffold with thicker fibers about to 5–6 μm and larger pores approximately 30 μm [163]. After 100 days' experimentation, the authors noticed that the completed structure of the endothelium coverage was generated and the similar smooth muscle layer structured with abundant ECM is found to those in the native arteries. Analysis of the cellularization process exposed that a large number of M2 macrophages were prompted by the thicker-fiber scaffolds to penetrate into the graft wall, which further reinforced cellular penetration and vascularization.

Today, vascular tissue engineering is used to substitute the large-scale blood vessels with diameters greater than 6 mm, where the process would prompt microvasculature or neovascularization procedures inside or near the entrenched scaffolds. Therefore, to attain good blood vessel regeneration, several vascular tissue scaffolds, comprising nanoscale porous membranes and mainly electrospun polymeric nanofibrous frameworks, have been considered and prepared to make various types of blood vessels. As seen from Figure 7.10, during the fabrication of these scaffolds some considerations that should be taken include the following: they should maintain endothelial coverage to control the diverse physiological signals; they also should show suitable mechanical strength and elasticity; and the remodeling of blood vessel should be able to respond stimulatory cues [189].

In recent decades, significant efforts have been completed to make vascular frameworks with nanoscale properties, targeting to replicate the ECMs architecture. ECMs contain nanofibers with diameters in the range of 5–500 nm,

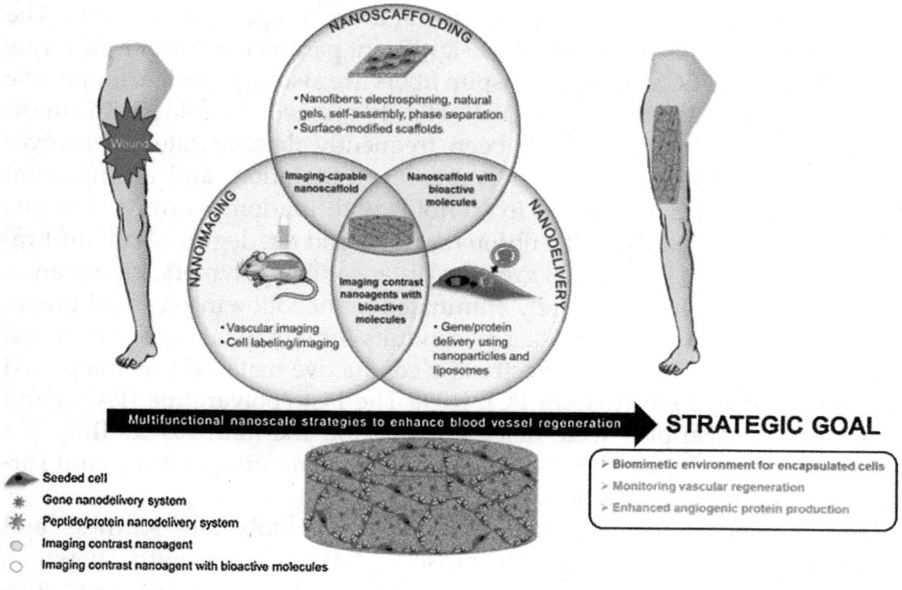

FIGURE 7.10
Multifunctional nanoscale strategies, including scaffolding, imaging, and bioactive molecule delivery systems for vascular tissue engineering. (Reprinted from *Progress in Polymer Science*, 60, Tang, Z. et al., Polymeric nanostructured materials for biomedical applications, pp. 86–128, Copyright 2016, with permission from Elsevier.)

and compositions comprising elastins, collagens, and nanoscale adhesive proteins (e.g., fibronectin and laminin) [190,191]. Therefore, the distinctive engineered frameworks are three-dimensional concepts having porous interwoven arrangements. In the literature it is stated that a nanostructured polymeric scaffold with the dimension between 200 and 400 nm was prepared to attain the excellent mechanical properties and steady tube-like networks for cell seeding [192]. Moreover, the integration of bioactive polymeric nanoparticles into the fiber frameworks is a logical approach to improve the regenerative capability of tissue engineering devices. Moreover, in blood vessel regeneration, polymeric nanoparticles can be used for this purpose and provide bioactive biomolecules such as growth factors, adhesion molecules, extracellular matrices, signaling molecules, and tight junction proteins. These molecules can penetrate into the microvasculature inside the tissue scaffolds, cells, or cell nuclei and accelerate the regeneration process [189,193].

7.1.3.4 Muscle Tissue Engineering

Human muscle tissues, comprising smooth muscle, skeletal, and cardiac, not only function as a train device but also provide the protein breakdown products and keep the normal functioning of the organs [194–196]. Though the muscle tissues have an intrinsic capability for regeneration following injury, tumor ablation or heart disease, and severe trauma (e.g., volumetric muscle loss) could overcome these natural muscle repair mechanisms. The fiber arrangement is one of the most significant parameters for muscle tissue engineering and ECM-like electrospun fibers are also appropriate for muscle cell adhesion and proliferation [197]. Highly aligned scaffolds with unidirectionally oriented fibers have been frequently demonstrated to recover myoblast adherence, myotube generation, proliferation, and arrangement of muscle cell when compared to scaffolds with randomly arranged fibers. Furthermore, the porosity of a fibrous scaffold and the degree of cell infiltration influences the muscle cell growth. Water soluble polymers, for instance, co-spun with PCL can be merely eliminated in the following scaffold preparation, ensuing in spaces that act as conduits for enhancing cell infiltration [198]. As seen in Figure 7.9d, electrically conductive material was integrated within associated electrospun PCL [199]. The PCL/polyaniline (PANi) and polyurethane scaffolds were helped to develop the matrices leading to a more operative electrical encouragement in the emerging cultures and further developing myotube maturity.

The previously demonstrated factors such as substrate stiffness and mechanical stimulation are used to impact skeletal muscle growth. Therefore, elastomeric polyurethane scaffolds were developed using the optimum stiffness process and repetitive mechanical strain during culturing the myotubes controlled to the growth of more developed, striated myotubes than with fiber alignment alone [200]. Moreover, the muscle tissue engineering, mainly in the field of cardiac tissue engineering, has continuously advanced from two-dimensional culture process to three-dimensional culture systems that are more illustrative of living systems [201]. Other methods such as microfluidics [202] and micro patterning [203] have been combined with electrospinning method to prepare different and functional three-dimensional frameworks for muscle tissue engineering. Therefore, a bottom-up methodology to assemble a modular tissue comprised of multiple tissue layers with discrete structures and functions has been reported. In this study, one layer was designed to support cardiac cells and promote their organization into a contracting tissue. Therefore, microgrooves were created, on this layer having ridge width of 120 ± 2 m, width of 115 ± 5 m, and height of 110 ± 10 m, on the thin albumin electrospun fiber layer by laser-patterning for mimicking native anisotropy of the natural ECM and increasing the surface area of the fibrous scaffold. Furthermore, micro holes with 40 ± 0.8 m in diameter were created on the edges of the fibrous layer to enhance the mass transfer among different layers. Another layer allows the arrangement of endothelial cells into blood vessels that were patterned with micro tunnels (450 m) to develop a predefined vasculature, and cage-like structures between the micro tunnels to facilitate controlled release systems to constantly supply signals for vascularization. Additionally, the third type of layers with cage-like structures comprise drug-loaded PLGA microparticles permitting the controlled release of various bio factors such as DEX (an anti-inflammatory agent) influencing the engineered tissue. Before transplantation, the tissue and micro particulate layers were combined to form thick 3D cardiac patches by ECM-based biological glue. Last, the patches were transplanted in rats and after 2 weeks, the authors found that patches without the vascular endothelial growth factor (VEGF) stayed white, whereas the VEGF-loaded patches were occupied with blood vessels. Moreover, histological study and immunostaining the extracts for smooth muscle cells further revealed the capacity of the VEGF micro particulate layer to develop blood vessel penetration into the cardiac patch [204].

7.1.3.5 Neural Tissue Engineering

The feasibility of using electrospun fibrous scaffolds as substrates for neural tissue engineering has been explored by several researchers. Therefore, comparisons among substrates prepared from polymer films vs. electrospun fibers have been examined both in vitro and in vivo to demonstrate the effectiveness of these fibers in improving nerve regeneration [205]. Therefore, in vitro, configurations of neurite cell consequence from the prime dorsal root ganglia (DRG) seeded on a scaffold of electrospun nanofibers with several order, arrangements, and surface properties were observed. As shown in Figure 7.9e, neurite out growth happened radially without specific directionality on randomly oriented nanofibers, whereas the cells differently stretched beside the longitudinal axis of the fiber on allied fibers [206,207]. In vivo, axonal addition in the electrospun fibrous nerve channels and partial functional re-joining has been demonstrated [208,209]. Furthermore, the electrospun fibers were also used to relocate and generate the human neurons in the brain. A three-dimensional (3-D) micro topographic framework was fabricated with the help of tunable electrospun microfibrous polymeric substrates for neural network development, promising in situ stem cell neuronal reprogramming, and supporting neuronal engraftment into the brain [210]. Therefore, the authors have found that the relocation of framework reinforced neuronal networks into mouse brain striatum efficiently improved survival approximately, 38-fold at the injection site, as compared to injected isolated cells, and allowed the supply of numerous neuronal subtypes.

7.1.3.6 Bone Tissue Engineering

Bone tissue engineering is an essential branch of tissue engineering where its objective is to regenerate bone tissue by using cell-based growth enhancements based on functional scaffolds. Furthermore, this technology is frequently employed to renovate the skeleton function in the process of orthopedic or oral-maxillo facial surgery [106]. Therefore, in bone regeneration, electrospun nanofiber has been receiving considerable attention, which has mainly been intended in identifying appropriate composition of materials for electrospinning process [211]. Therefore, the nanofibrous substratum could provide favorable environments for cell anchorage and development. Though the applicability of electrospun nanofibers in bone regeneration and tissue engineering is in the early stages, current research using electrospun nanofibers with new structures intended for bone growth and some processing tools for designing three-dimensional scaffolds have emphasized the possible use of electrospun materials in bone tissue engineering [27]. Therefore, to broaden the application of electrospun nanofiber in bone tissue engineering, several research groups have prepared different electrospun nanofiber scaffolds to promote either in vivo or in vitro osteogenesis, such as electrically conductive electrospun aligned PLA/MWCNTs nanofibers with a synergistic combination of topographic cues and electrical stimulation for osteoblast extension [212]; beaded nanofibers with surface nano roughness for modulating and facilitating cellular behaviors [213,214]; a 3-D nano/micro alternating multilayered scaffold loaded with rBMSCs and BMP-2 [215], and nanoparticle-embedded electrospun nanofibers for the efficient repair of critical sized rat calvarial defects, Figure 7.9f [166].

In another study, PCL scaffolds containing electrospun nanofibers in the range of 20 nm to 5 mm dimension have been used to develop the supportive in vitro diversity and mineralization of bone marrow–derived mesenchymal stem cells (BM-MSCs) from rat [130]. The authors have reported the effectiveness of the PCL scaffold with nanofiber diameter around 370 nm, where the scaffold could enable both proliferation and adhesion of MSCs. Moreover, they create higher levels of alkaline phosphatase activity, osteocalcin, mineralization, and osteopontin productions. Furthermore, the scaffold was seeded with MSCs and consequently implanted in the omenta of rat for 28 days [216], and the cells were found to effectively differentiate and infiltrate into the frameworks [217]. In addition, nanofibers comprised of PEG, silk, hydroxy-apatite nanoparticles, and bone morphogenetic protein 2(BMP-2) were prepared using electrospinning to produce composite scaffolds. Therefore, the silk/PEG nanofibers were found to be able to support the osteogenic differentiation of human MSCs (hMSCs), where the presences of BMP-2 and hydroxyapatite nanoparticles could considerably improve the bone generation in vitro [135]. Furthermore, polymeric nanoparticles have been also employed for bone tissue engineering where they are essentially to be entrapped and deliver bone morphogenetic proteins, genetic resources, and biomolecules [218,219]. Though their use in bone regeneration needs some specific considerations, polymersomes, polymeric micelles, nano gels, nano capsules, nanoparticles, and dendrimers are all possible vehicles for controlled delivery. Generally, the solid, porous, or hollow nanoparticles are appropriate for bone applications, where they can be prepared using nano manipulation, self-assembly, photochemical patterning, and bio-aggregation [220]. It is also reported that the scaffolds for bone tissue engineering comprising bioactive polymeric nanoparticles can show numerous advantages over conventional monolithic scaffolds such as enhanced control over sustained delivery of therapeutic agents; acting as compartmentalized micro reactors for dedicated biochemical processes; acting as porogen or reinforcement phase to introduce porosity and/or improve the mechanical properties of bulk scaffolds; imbedding injectable or moldable formulations to be applied in minimally invasive surgery; and acting as cell delivery vehicles [221].

There are several other possible applications of electrospun fibers in tissue engineering in addition to previously mentioned applications. For instance, aligned PCL–PEG nanofibers in porous chitosan scaffolds were developed to enhance the orientation of collagen fibers in regenerated periodontium [222]. As seen in Figure 7.9g, dextran glassy nanoparticles (DGNs) loaded with basic fibroblast growth factor (bFGF) and encapsulated these nanoparticles into poly-l-lactic acid (PLLA) copolymer fibers were prepared to protect the bioactivity of bFGF in a repeated manner, accordingly promoting tendon healing and cell proliferation both in vivo and in vitro [167]. The preparation of "aligned-to-random" electrospun nanofiber scaffolds was also demonstrated that mimic the structural organization of collagen fibers at the tendon-to-bone insertion site [223]. Therefore, tendon fibroblasts cultured on such a scaffold displayed highly organized and randomly oriented morphologies on the random and aligned portions, respectively. As seen in Figure 7.9h, the authors also fabricated scaffolds based on radially aligned electrospun nanofibers, a high potential as artificial dural substitutes were confirmed [168]. Furthermore, the fabricated scaffold was also able to direct and improve the cell migration from the periphery to the center and bring faster cellular migration and population than random nanofibers. It is also found in this study that dural fibroblast cells, which cultured on the scaffolds of radially aligned fibers, expressed extracellular matrix such as type I collagen in a high degree of organization. As seen from Figure 7.9i, a successful use of electrospun gelatin/polycaprolactone fibrous membranes was confirmed to build an ear-shaped cartilage. As seen in Figure 7.9j, electrospun fibrous scaffolds with 200 nm fiber topography enhanced retinal pigment epithelial culture were developed and sub retinal biocompatibility was shown [170].

7.1.4 Other Applications in Medicine

7.1.4.1 Humoral Diagnosis of Cancer and Other Diseases

Electrospun fibers were developed to detect the tumor markers or circulating tumor cells (CTCs) in the blood cells. It also helps to identify the cancer cells in other body fluids of a patient in early stage cancer [224]. Titanium butoxide (TBT)/PVP composite fibers were subjected to electrospun onto a silicon substrate and achieved titanium dioxide (TiO_2) nanofibers using calcination process and by eliminating organic components [225]. The nanofiber materials were modified using a biotinylated anti-epithelial cell adhesion molecule (anti-EpCAM) and lastly treated the colon cancer cells and gastric

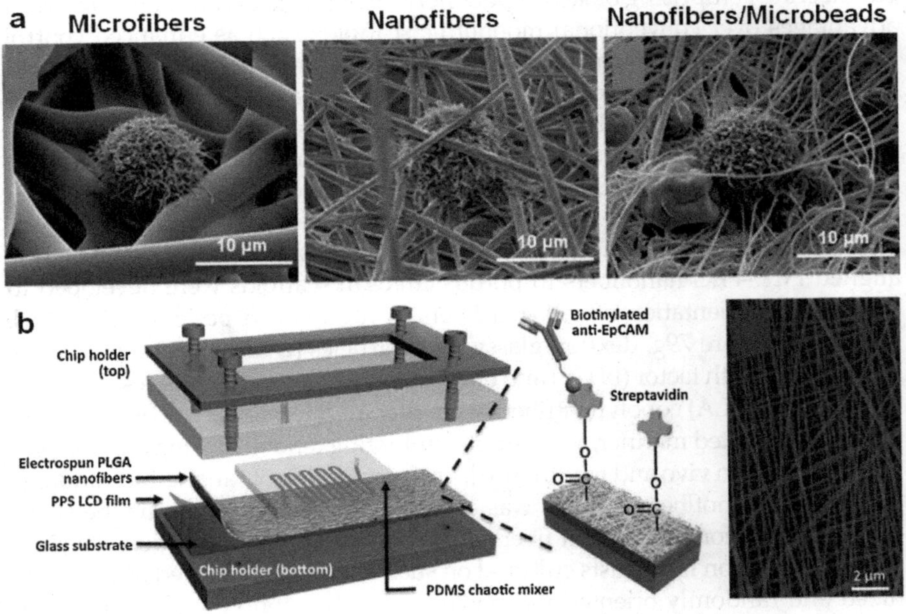

FIGURE 7.11
(a) SEM images of captured CTCs by anti-EpCAM modified PS microfibers, PS nanofibers, and nanofibers/microspheres composites; (b) combination of biotinylatedanti-EpCAM modified PLGA nanofibers and microfluidics for capturing CTCs. (a. Ma, L. et al.: Trap Effect of Three-Dimensional Fibers Network for High Efficient Cancer-Cell Capture. *Advanced Healthcare Materials.* 2015. 4(6). pp. 838–843. Copyright Wiley-VCH Verlag GmbH & Co. KGaA. Reproduced with permission; b. Zhao, L. et al.: High-Purity Prostate Circulating Tumor Cell Isolation by a Polymer Nanofiber-Embedded Microchip for Whole Exome Sequencing. *Advanced Materials.* 2013. 25(21). pp. 2897–2902. Copyright Wiley-VCH Verlag GmbH & Co. KGaA. Reproduced with permission.)

cancer cells using the fabricated device. As seen from Figure 7.11a, microfibers, nanofibers, and combined nanofibers and microbeads composite materials were developed from PS using electrospinning technique, and next anti-EpCAM was employed to update the fibers. Therefore, it was applied to capture the human breast adenocarcinoma cell (MCF7) using these fibers efficiently, where significant results were observed. However, the nanofibers/microbeads composites showed the highest capture efficiency [226].

As seen from Figure 7.11b, biotinylated anti-EpCAM modified PLGA nanofibers were attached with a microfluidic technique and efficiently detected and treated human breast cancer cells (SK-BR-3) in a PBS and in blood [227]. The authors used a confocal microscope laser to separate captured cells for additional gene sequencing detection. Furthermore, they also used a biotinylated anti-CD146 antibody-modified PLGA nanofiber mat combined with a microfluidic device to effectively detect melanoma circulating tumor cells [228]. Moreover, PEI-modified electrospun PLGA (PLAG-PEI-HA) nanofibers were developed using covalently conjugated hyaluronic acid, and it integrated the nanofibers into a microfluidic chamber [229]. Therefore, the PLGA-PEI–HA nanofibers combined microfluidic platform was used successfully to capture the HeLa, KB, A549, and MCF-7 cells that are the different categories of the $CD44^+$ carcinoma cells. Mostly, the HeLa cells were more proficiently captured by the nanofibrous membrane (PLGA-PEI-HA) under flowing circumstances than in static dish at a density of 20 cells mL^{-1}. Finally, the captured HeLa cells could grow on the nanofibrous membrane in the micro-chip for days without compromised cell viability; the electrospun fibrous membranes have also been employed, in addition to the detection of CTCs, for the detection of tumor associated factors and genes, like the p53 gene [230] or the K-RAS gene [231].

7.1.4.2 Mimicking the Tumor Microenvironment

Tumor associated macrophages (TAMs) are critical stromal components intimately involved with the progression, invasion, and metastasis of cancer cells. A malignant tumor is more than a single, mutated cell population replicating without regard to the otherwise healthy tissue within which it resides. Rather, the surrounding stroma maintains a dynamic relationship with the tumor, through which it is intimately involved in cancer initiation, growth, and progression [232]. Furthermore, cellular components of the tumor stroma include fibroblasts, myofibro blasts, endothelial cells, pericytes, macrophages, and a variety of inflammatory cells. Therefore, tumor associated macrophages (TAMs) are a macrophage subset that drew early interest due to histological observations of tumor infiltration. Macrophage content in human tumors varies from 50–80%, with one study in breast cancer quantifying infiltration as 490 and 343 macrophages/mm^2 for medullary carcinomas and ductal carcinomas, respectively [233]. Therefore, a detailed study of tumor development and its growth environment benefits enhance the management of drugs and the testing of new medicines.

Numerous materials and techniques have been employed to mimic tumor microenvironment and to develop in vitro tumor models [234]. A three-dimensional development of microenvironment for tumor cells based on electrospun PCL fibers was developed and a bioactive peptide derived from domain IV of perlecan heparan sulfate proteoglycan [235]. This microenvironment can preserve the relocation process of prostate cancer cells. Moreover, electrospun PCL fibers with various molecular weights and mouse fibrosis lung extract was used to develop a three-dimensional (3-D) microenvironment fiber matrix. With the help of the fabricated matrix, the pathogenesis of congenital pulmonary interstitial fibrosis was explored [236]. Ewing sarcoma (ES) cells onto electrospun PCL 3-D scaffolds were cultured within a flow perfusion bioreactor, and the flow-derived shear stress can provide a physiologically related mechanical stimulation [237]. The authors found that cells exposed to flow perfusion produced more insulin-like growth factor-1 (IGF1) ligand, displaying shear stress-dependent sensitivity to IGF-1 receptor targeted drugs as compared with static environments. Moreover, the flow perfusion increased nutrient supply throughout the scaffold, which enriched ES culture over static conditions. Furthermore, aligned electrospun fibers were employed to investigate breast cancer cells [238] and their invasion and metastasis ability in vitro. As seen from Figure 7.12, attractively, the apoptosis of glioblastoma tumor induced by conduit built from electrospun aligned fibers was thoroughly explored [239]. In their study, they have induced directional migration of tumor cells through the aligned fibers to a gel pool that would prompt apoptosis of cancer cells. Moreover, as seen in Figure 7.12c, the amount of tumor cells in the aligned nanofiber film conduits was the highest after one week, and the corresponding internal glioma in the animal brain was the smallest, which suggested that this conduit with aligned fibers could promote the migration of glioma cells from the glioma tumor. The inventive fibrous antitumor implants techniques are gaining attraction, and the antitumor influence is noteworthy as a new way to recover cancer cells with the help of electrospun fibers during treatment [27].

7.1.4.3 Enhancing Magnetic Resonance Imaging

Magnetic resonance imaging (MRI) is a sophisticated technology to visualize the interior structures of the body (i.e., muscle, brain, heart, and cancer regions) by using radio waves and magnetic fields. Generally, this technology provides greater contrasts between the different soft tissues of the body when compared with other imaging techniques [106]. However, the intrinsic contrast in some parts of the body is often inadequate for clear differentiation, for example, detection of small tumors. Therefore, to overcome this difficulty, it is recommended to use MRI contrast enhancing agents. Consequently, the most frequently used MRI contrast agents are low molecular weight chelates of gadolinium (Gd3+) or iron oxide particles. Frequently, biocompatible polymers such as polylysine, poly(l-glutamic acid)-cystamine, poly(ethylene

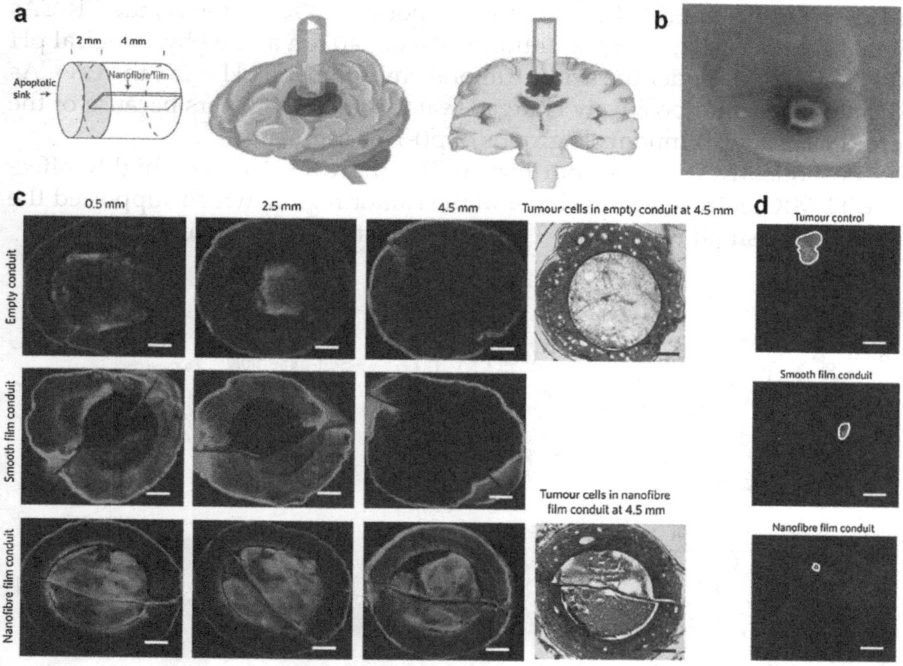

FIGURE 7.12
(a) Representation of the tumor guide containing a nanofiber film inserted into a rat brain. Three-dimensional view (left) and coronal view (right) of the brain and conduit; (b) digital image of extracted brain containing a conduit from the top view; after 7 days' implantation, (c) fluorescence images showing the presence of tumor cells throughout the cross-section of all the conduits (empty, smooth film, and aligned nanofiber film) at distances 0.5 mm, 2.5 mm, and 4.5 mm, respectively, from the interface of the tumor in the brain, scale bar = 400 m; (d) fluorescence images showing the tumor core in the brain for the tumor control, the smooth film conduit, and the nanofiber film conduit. (Reprinted by permission from Macmillan Publishers Ltd. *Nature Materials*, Jain, A. et al., *Nature Materials*, 13(3): pp. 308, 2014, copyright 2014.)

glycol), dextran, and l-cystine bisamide copolymer have been employed for the preparation of MRI contrast enhancing agents through conjugation with low-molecular weight chelate agents [240]. Moreover, the use of biocompatible polymers is reported where it is believed that it could increase the signal intensity by 121% in damaged lungs and by 118% in pulmonary arteries of healthy lungs [241]. Therefore, the biocompatible polymers in poly(l-lysine) (PLL) has been also employed to modify the pharmacokinetic characteristics of the MRI contrast agents. It was also reported that the usage of PEG-b-poly(l-lysine) could considerably prolong the circulation time of gadolinium ion in blood. Furthermore, a significant amount of biocompatible polymers, namely, PEG-P(Lys-DOTA-Gd) micelles, was detected to collect in solid tumors after 24 h intravenous injection due to the enhanced permeation retention (EPR) effect. Consequently, the MRI signal intensity of the tumor was improved 2.0-fold by the use of this polymeric micelle contrast agent [242]. A cancer-recognizable

MRI contrast agent (CR-CAs) was also reported [243]. Therefore, the CR-CAs had a spherical shape with a uniform size of ~40 nm at the physiological pH (pH = 7.4) level. Under the acidic tumoral environment, pH = 6.5, the CR-CAs disintegrated into positively charged water soluble polymers because of the protonation of the imidazole groups of p(l-His) segments.

Consequently, as seen from Figure 7.13, the CR-CAs show highly effective T1 MR contrast improvement in the tumor region, which supported the detection of small tumors of ~3 mm^3 in vivo at 1.5 T within a few minutes.

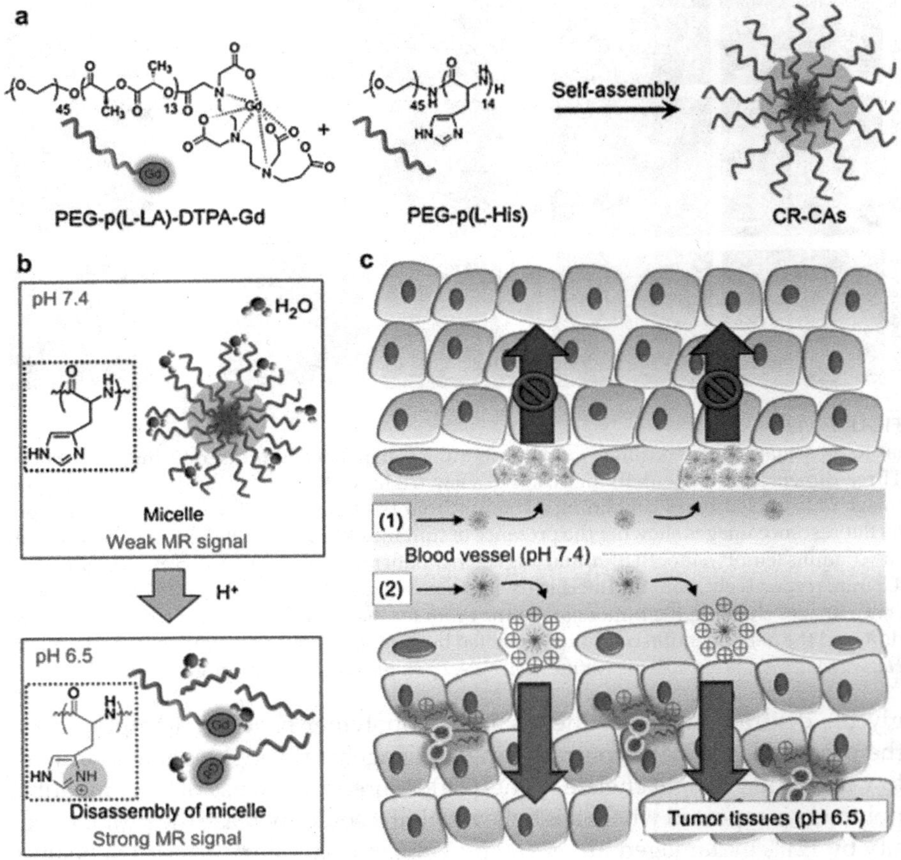

FIGURE 7.13
(a) Schematic representation of the preparation of the cancer-recognizable MRI contrast agents (CR-CAs); amphiphilic block copolymers (i.e., PEG-p(l-LA)-DTPA-Gd and PEG-p(l-His)) self-assemble into micelles in an aqueous solution at pH 7.4. (b) Schematic representation of pH-dependent structural transformation and related MR signal change in CR-CAs. Inset: Chemical structural representation of the protonation of imidazole groups in PEG-p(l-His) at acidic pH. (c) Schematic representation of the tumor-accumulation behavior of (1) conventional micelle-based CAs and (2) CR-CAs. (Reprinted from *Progress in Polymer Science*, 60, Tang, Z. et al., Polymeric nanostructured materials for biomedical applications, pp. 86–128, Copyright 2016, with permission from Elsevier.)

7.1.5 Summary

The electrospun fibers have been developed in the field of drug delivery, tissue engineering technology, disease diagnostics, and many other applications. Defined and planned drug discharge profiles can be attained as a result of the well-defined multilevel assembly of the fibers. The polymer fibers are perfect during the tissue engineering frameworks due to the fine nanofibrous structure resulting in a high active surface to volume ratio and capable of simulating the ECM of alive cells. Highly refined electrospinning methods need to be developed in accordance with the requirements of biomedical applications. For instance, for drug delivery applications such as cancer therapy, interdisciplinary investigation from material scientists and biologists are essential to additional develop multileveled fibers. Therefore, the microfluidic-combined electrospinning technology, which can regulate the materials present in the electrospinning solution more accurately, may be improved to fabricate multileveled fibers.

In tissue engineering application, a major factor is properly regulating the fiber installation, mainly on a nanometer range, to more exactly simulate an ECM. In addition, it is dire to know the cellular response to electrospun compounds with various compositions, fiber locations, and the microstructures. These aims can be informal to attain better result by integrating the use of electrospinning process, microfluidic spinning technique, and the three-dimensional (3-D) printing to build the 3-D frameworks containing wide range fiber sizes, depositions, and alignments similar to in vivo like ECM. During the diagnostics of a cell, more complex materials should be engaged during electrospinning process, or surface alteration of the fibers by changing and doping different functional groups such as polydopamine action, plasma treatment, and layer-by-layer modification technique will be needed. Furthermore, the continuous production of electrospun nanofibers is a requirement for this technology to rise profitable value, which may be positively understood through the novel needleless electrospinning. The biomedical applications of electrospun fibers such as engineered skin and tissue anti-adhesion membrane has been developed commercially, which is getting more attention in recent years. However, before the widespread commercialization of electrospun fiber materials in the field of biomedical applications, more broad studies, even clinical trials, are required to estimate the real effect of the electrospun nano and microfibers in healthcare and biomedical engineering.

7.2 Membrane Processes in Biotechnology: An Introduction

7.2.1 Introduction

Membranes have been employed conventionally for separation processes based on size with high throughput, however with comparatively low resolution requirements. These uses consist of ultrafiltration for protein concentration and buffer exchange and microfiltration for clarification and sterile filtration [244]. Morever, membrane processes are increasingly being used in reaction, clarification, and recovery schemes for the production of molecules, emulsions, and particles due to their properties such as high surface area to unit volume ratio and their potential for controlling the level of contact and/or mixing between two phases. Therefore, membranes are suitable to the processing of biological molecules because they operate at comparatively low temperatures and pressures and involve no phase changes or chemical additives, thus reducing the extent of deactivation, denaturation, and/or degradation of biological products [245].

In this section different membrane processes, such as the well established microfiltration and ultrafiltration, and emerging processes like membrane chromatography, membrane bioreactors, and membrane contactors for the preparation of emulsions and particles are explained in detail. Microfiltration and ultrafiltration are commonly used to recover macromolecules; retain suspended colloids and particles; and are being integrated into both downstream and upstream processes. Several applications of ultrafiltration and microfiltration have been reported, such as to concentrate proteins; clarify suspensions for cell harvesting; exchange buffer systems; and sterilize liquids to remove bacterias and viruses. Other membrane separation processes including membrane bioreactors, where enzymes, microorganisms, or antibodies are suspended in solution and classified by a membrane in a reaction vessel or immobilized within the membrane matrix; membrane chromatography, where it is used as an alternative to conventional resin-based chromatography columns, for a large range of chromatographic purification schemes, including ion-exchange, hydrophobic, reversed-phase, and affinity chromatography; and membrane contactor, which involves using a pressure to force the dispersed phase to permeate through a membrane into the continuous phase, for the preparation of emulsions and different types of particles, as water/oil emulsions, oil/water emulsions, and polymeric particles. The research and development efforts have been focused toward drastic developments in selectivity while maintaining the inherent high throughput characteristics of membranes. Therefore, this is important if membrane separation processes are to satisfy the new purification and process economic challenges posed in various fields such as its application in biotechnology.

7.2.2 Microfiltration and Ultrafiltration

Microfiltration and ultrafiltration are well-known membrane separation processes [245,246]. One of the common applications of ultrafiltration in downstream processing is for product concentrations like buffer and/or solvent removal. Therefore, buffer exchange and desalting can be done using diafiltration to wash the initial buffer out and to replace it with a new buffer. On the other hand, microfiltration is most commonly applicable in sterile filtration (i.e., bacterial removal) earlier to final formulation of many products, and in the initial clarification of fermentation broths to remove the suspended cell mass and other particulate debris. Moreover, sterile filtration process is done in a dead-end configuration using 0.2 μm pore size membranes that have been confirmed for the complete removal of *Brevundimonas diminuta* [247]. However, such sterilizing filters allow to pass small microorganisms under some process conditions. Consequently, some users employ 0.1 μm pore size membrane to give improved sterility assurance in pharmaceutical processes [248].

To separate viruses range from about 12 to 300 nm from proteins of 4 to 12 nm, microfiltration membrane is used for virus removal from cell cultures [244,249]. The production of recombinant proteins commonly requires mammalian cell lines that have therapeutic, prophylactic, or diagnostic applications. But cell lines are contaminated with virus or virus like particles. Viruses may also be introduced by addition of supplements and other constituents introduced into the fermentation process, or through handling and other manipulations during processing in addition to endogenous contaminants. Today, membrane manufacturers are developing membrane filters with increasing resolution for virus-protein separation purposes [244]. This has significant importance to the biotechnology industry because incidents of parvovirus contamination of cell cultures may happen. Parvoviruses are mainly complicated to remove, as these are both small (about 20 nm diameter) and highly resistant to many chemical and thermal techniques of inactivation. Therefore, to make sure that virus removal is consistent with validation studies, membrane integrity is monitor both pre and post use.

Membrane filtration is also particularly well suitable in antibiotic production where most antibiotics, such as erythromycin, benzylpenicillin (penicillin G), and medmycin, are broadly employed and also used as raw materials for semisynthetic antibiotics. Then, these are produced by fermentation and recovered from their broths by using the conventional steps, like solvent extraction (isolation and purification), filtration (removal of biomass), and subsequent crystallisation (polish). Therefore, membrane filtration is used for the primary clarification of these fermentation broths [250–252]. One of the main advantages of membrane systems for the initial recovery of antibiotics from a fermentation broth is the ability to obtain high yield using a combined filtration and diafiltration process. Complete retention of the cells and particulate matter can be achieved using membranes with pore sizes up to 0.2–0.45 μm, and essentially complete passage of the antibiotics can be achieved as long as the nominal membrane molecular cut-off is greater than about 20,000 g/mol [245]. Ultrafiltration was used to remove emulsifiers in antibiotic broths before solvent extraction to avoid emulsification and to improve extraction efficiency. Purification of benzylpenicillin filtered broths obtained from fermented broths by ultrafiltration with diafiltration was also reported by Nabais and Cardoso (1999) [251]. The authers have conducted an investigation using pilot ultrafiltration process with a tubular membrane and membranes with molecular weight cut-off of 100,000 (PVDF), 20,000 (polysulfone), and 8000 (polysulfone) were investigated. The colored substances, proteins, and other impurities were effectively removed, and high benzylpenicillin recovery in permeate was also detected. Moreover, it was demonstrated that ultrafiltration may be an alternative to the use of flocculants and anti-emulsion agents to attain good phase separation in benzylpenicillin solvent extraction.

Numerous applications of microfiltration and ultrafiltration in the biotechnology field have been reported. Some examples: concentration and purification of recombinant Brain-Derived Neurotrophic Factor (rBDNF) inclusion bodies from *E. coli* cell suspensions by cross-flow microfiltration and diafiltration [253]; recovery of heterogeneous immunoglobulins (IgG) from transgenic goat milk by microfiltration [254]; recovery and purification of yeast alcohol dehydrogenease (ADH) from bakers' yeast as typical of downstream processing for the extraction of an intracellular enzyme product [255]; and recovery of naturally glycosylated therapeutic proteins produced from animal cell cultures by microfiltration [256]. Furthermore, the optimization of monoclonal antibody recovery from transgenic goat milk by microfiltration was also reported as an interesting example by Baruah and Belfort (2004). Therefore, the optimization has involved transmembrane pressure, varying pH, membrane module type, milk feed concentration, and axial velocity. In the pressure-dependent operation regime at low uniform transmembrane pressures using permeate circulation in co-flow, at the pI of the protein is revealed to increase IgG recovery from less than 1% to over 95%. Therefore, such methodology is in general appropriate for the recovery of target proteins found in other complex suspensions of biological origin.

Currently, many efforts are being devoted to develop new membrane modules with enhanced mass transfer characteristics for microfiltration and ultrafiltration processes. Some of the newly developed modules include rotating disk filters [257,258], conically shaped rotors [256], cylindrical Taylor vortex devices [259], and helical coiled dean vortex systems [253,260]. Dean vortex devices show high mass-transfer rates, owing to the existence of centrifugal flow instabilities. Furthermore, these devices display significant rises in protein transmission and capability, though fouling remains as difficult in many applications. Alternative methods like high-frequency back-pulsing can be used to continually clean the membrane surfaces [255]. Concequently, high-frequency back-pulsing was revealed to increase flux, reduce fouling, and increase protein transmission in the purification of conjugated vaccine products [261]. Moreover, to improve ultrafiltration and microfiltration performances other techniques including pulsative flow, gas sparging, electric fields, ultrasonic fields, and combined electric/ultrasonic fields were also reported [262]. Moreover, the effects of different hydrodynamic parameters on the permeate flux provided by two different dynamic filtration systems using the same membrane materials and the same fluids were compared by Jaffrin et al. (2004) [257]. The tested systems were two homemade rotating disk modules and a VSEP pilot with a circular vibrating membrane. In this study, the flux was confirmed to be primarily governed by the maximum shear rate and not by details of internal flow and can be increased to high levels by increasing rotation speed or vibration amplitude or by equipping the disk with large vanes.

In another study, Dean vortex microfiltration with controlled centrifugal instabilities (Dean vortices produced in helical flow) was used to enhance cross-flow microfiltration and diafiltration for the concentration and purification of rBDNF inclusion bodies from *E. coli* cell suspensions [253]. For microfiltration experimentations with the feeds comprising cell and homogenate suspensions, developments in flux of about 50% and 70%, respectively, were attained with the helical module as compared with that obtained with the linear module. For diafiltration with the homogenate suspensions as feed, solute transport was from 100% to 40% higher after 40 and 100 min, respectively, with the helical module as compared with that obtained with the linear module. Continual enhancements in understanding the effects of solution environment on molecules, particles retention, and fouling have also led to further improvements in ultrafiltration and microfiltration performances [263]. Moreover, recent studies have confirmed that it is possible to control the rate of protein transport through membranes by adjusting the solution pH or the ionic strength [264–266]. The processes have to be operated at the pI of the transmitted protein and far from the pI of the retained protein. Therefore, to enhance the separation, the ionic strength has to be kept low so that the thickness of the diffuse double layer of the charged solute is noticeable, leading to high retention, while the uncharged solute readily permeates the membrane. Furthermore, using High Performance Tangential Flow Filtration (HPTFF) process these electronic interactions can also be exploited for protein separation. For instance, 99-fold purification of an antigen-binding fragment of a monoclonal antibody (Fab) from BSA were achievedby operating the membrane process near the isoelectric point of the BSA and using a positively charged membrane to attain high rejection of the positively charged Fab.

7.2.3 Membrane Bioreactors

It has been reported that membrane bioreactors are alternative methods to classical means of immobilizing biocatalysts, namely, enzymes, microorganisms, and antibodies, which are suspended in solution and partitioned by a membrane in a reaction vessel or immobilized within the membrane matrix [267–269]. As seen in Figure 7.14, first the system might consist of a conventional stirred tank reactor combined with a membrane separation unit; second, the membrane acts as a separation unit and as support for the catalyst. Furthermore, the biocatalyst can be flushed along a membrane module, segregated within a membrane module, or immobilized in/on the membrane by entrapment, gelification, physical adsorption, ionic binding,

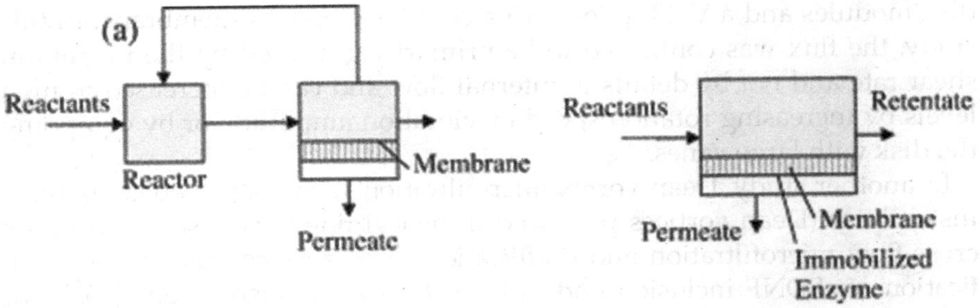

FIGURE 7.14
Membrane bioreactor configurations: (a) reactor combined with a membrane operation unit, (b) reactor with the membrane active as a catalytic and separation unit. (Reprinted from *Biotechnology Advances*, 24(5), Charcosset, C., Membrane processes in biotechnology: An overview, pp. 482–492, Copyright 2006, with permission from Elsevier.)

covalent binding, or cross-linking. Some of the advantages of immobilizing enzymes are increased reactor stability and productivity; enhanced product purity and quality; and waste reduction [269]. The effectiveness of the overall system depends on the geometric parameters (e.g., membrane configuration, morphology, and pore size distribution), biochemical (e.g., catalytic activity, reaction kinetics, concentration, viscosity of substrate and product, immobilization stability), and hydrodynamics parameters (such as transmembrane pressure and flow velocity) [270]. Furthermore, immobilized biocatalyst membrane reactors are repeatedly used in a hollow-fiber configuration due to their high packing density (large surface area per unit volume of reactor space).

Membrane bioreactors have been employed in the production of antibiotics, aminoacids, anti-inflammatories, anticancer drugs, optically pure enantiomers, vitamins, isomers, etc. Membrane bioreactors for the synthesis of lovastatin with immobilized *Candida rugosa* lipase on a nylon support [271]; the production of diltiazem chiral intermediate with a multiphase/extractive enzyme membrane reactor [272]; the synthesis of isomaltooligosaccharides and oligodextrans in a recycle membrane bioreactor by the combined use of dextransucrase and dextranase [273]; the production of a derivative of kyotorphin (analgesic) in solvent media using α-chymotrypsin as catalyst and α-alumina mesoporous tubular support [274]; and biodegradation of high-strength phenol solutions by *Pseudomonas putida* using microporous hollow fibers [275] have been reported. A forced-flow membrane enzyme reactor was developed in which the enzyme is immobilized on porous ceramic membrane [276]. The enzymes were attached to the porous membrane surface, and the mass transfer was enhanced by using convection rather than diffusion, and the convection was not limited by the significant pressure drops found in enzyme-immobilized bead-filled column reactors. In this study, a 10-fold higher efficiency was detected in their system as compared to a conventional column reactor in which the enzyme was immobilized on beads. Membrane bioreactors for immobilized whole cells have also been tested successfully, and they provide an environment for improved cell densities so as to produce higher product titre [267]. The cells are perfused using a membrane with a steady continuous flow of medium and supplied with oxygen and nutrients whereas wastes and desired products are removed. Therefore, the cells are frequently retained in the bioreactor through a membrane barrier.

Numerous membrane configurations such as flat sheet and rotating bioreactors have been evaluated, though the hollow fiber configuration is mainly interesting. In this case, cells are either developed in the extra capillary space with medium flow through the fibers, or developed within the fibers with medium flow outside or across the fibers. It was revealed that a mass transfer hinderance for oxygen and glucose was achieved and well-defined spacing among the fibers was a means of reducing this occurrence [277]. Other geometries were suggested, such as hollow fibers inserted within another to grow the cells in the annulus between the two fibers [278]. Therefore, a proper selection of outer and inner fibers diameters can limit the distance required for diffusion of medium components (oxygen, glucose, glutamine, and other nutrients) to approximately 50 μm for the furthest cells. Then, a comparison of the traditional and concentric reactors for antibodies production from hybridoma cells displayed that a much higher concentration of cells can be sustained in the concentric reactor along with higher specific productivity and cell viability. A significant aspect for membrane bioreactors with immobilized enzymes is the chemistry of enzyme immobilization. Therefore, it is frequently accomplished either directly on the membrane or via spacer arm, often through the ε-amino functionality of lysine residues on the protein [279].

Though immobilization of enzymes in general improves their stability, one major shortcoming of random immobilization of enzymes onto polymeric microfiltration type membranes is that the activity of the immobilized enzyme is often considerably reduced because the active site may be blocked from substrate accessibility, multiple point-binding may happen, or the enzyme may be denatured. Membrane bioreactors have been introduced last years, and till now their main industrial applications have been for wastewater treatment such as industrial, domestic and municipal [280]. However, in the field biotechnology, membrane bioreactors have attained only limited accomplishment. The main technological problems in using membrane bioreactors in an industrial level are associated with rate-limiting aspects and scale-up problems of this technology, together with the lifetime of the enzyme, the accessibility of pure enzyme at an acceptable cost, the necessity for biocatalysts to operate at low substrate concentrations, and microbial contamination.

7.2.4 Virus Filtration

It is reported that mammalian cell cultures are susceptible to contamination with adventitious viruses introduced during processing [244]. In addition, mammalian cells used in the manufacture of recombinant DNA products have been shown to contain endogenous virus-like particles [281]. Therefore, the safety of mammalian cell-derived products is controlled by requirements to attain less than one virus particle per million doses. Furthermore, this can be realized by designing purification processes that comprise validated viral elimination using multiple clearance and inactivation procedures. Physical inactivation (e.g., heat or UV), chemical inactivation (e.g., using chaotropes, low pH, solvents, or detergents), size separation (filters and size-exclusion chromatography), and chromatographic separation (affinity and anion exchange) offer a wide array of choices for virus clearance [244]. Therefore, filters provide the advantage of the physical removal of viruses combined with a size-based technique that complements other virus removal steps. Nevertheless, the size-based technique has the limitation of precisely separating viruses that range in size from about 12–300 nm from proteins that typically range in size from 4–12 nm [282]. Membrane manufacturers continue to successfully improve normal flow filters with increasing resolution for virus–protein separation process. This is of substantial meaning to the biotechnology industry due to incidents of parvovirus contamination of cell cultures that have happened. Parvoviruses are particularly difficult to remove, as they are both small (about 20 nm diameter) and highly resistant to many thermal and chemical methods of inactivation. Threfore, the development of higher resolution membranes with enhanced permeability is likely to continue and the application of charged membranes [283] and the optimization of solution pH and ionic strength [284] may also be used to develop resolution by increasing the effective difference in hydrodynamic volume between the protein and virus and by exploiting a charge repulsion mechanism.

7.2.5 Membrane Chromatography

Membrane chromatography was explored many years ago without any considerable commercial achievement. Current developments have generated renewed attention in this technology, and several membrane chromatography products have been brought to the market [244]. Membranes have an intrinsic advantage of not being diffusion limited in contrast to conventional bead chromatography where the binding capacity is independent of flow rate. The challenges have been to attain binding capabilities that are competitive with beads where there is an inherent trade-off between the convective flow through larger pores with limited internal surface area and high internal surface area available in small pores with diffusion limitations. Furthermore, adsorptive membranes have been investigated for decades as an alternative to conventional resin-based chromatography columns [285–288]. As seen in Figure 7.15, the advantage of adsorptive membranes is the shorter diffusion times than those gained in resin-based chromatography, as the interactions between molecules and active sites on the membrane occur in convective

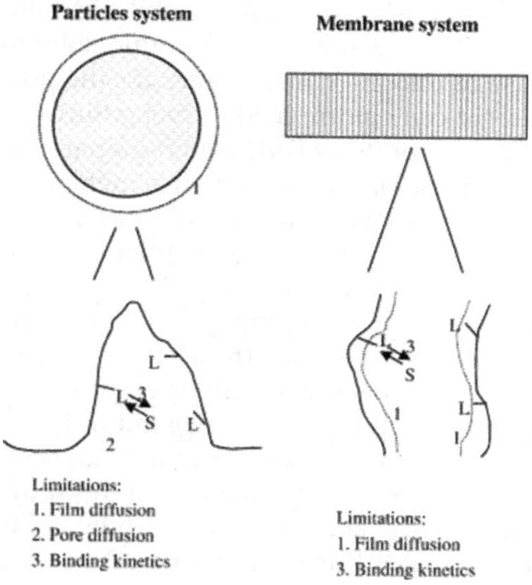

FIGURE 7.15
Comparison between (a) membrane chromatography and (b) gel bead chromatography. L: ligand, S: solute. (Reprinted from *Biotechnology Advances*, 24(5), Charcosset, C., Membrane processes in biotechnology: An overview, pp. 482–492, Copyright 2006, with permission from Elsevier.)

through-pores, rather than in stagnant fluid inside the pores of an adsorbent particle. Moreover, adsorptive membranes have the potential to keep high efficiency both at high flow-rates and for use of large biomolecules with small diffusivities. Chromatographic membranes have been employed in different configurations such as staked membranes, spiral wound membranes, hollow fibers, and various adsorptive mechanisms (i.e., ion-exchange, reversed-phase, hydrophobic, and affinity based procedures). Moreover, ion-exchange membranes have been examined with strongly basic (quaternary ammonium), strongly acidic groups (sulfonic acid), weakly acid (carboxylic acid), and weakly basic (diethylamine) types. Furthermore, affinity membranes have been investigated with a large range of ligands such as Protein A and G, immunoaffinity ligands, low-molecular-mass ligands (Cibacron Blue, histidine, tryptophan), and other ligands (peptide, Cu^{2+}) [288].

Membrane materials examined for chromatographic application includes cellulose, polysulfone, hydrazide, polyamide, and composite membranes such as blend of polyethersulphone and polyethylene oxide coated on all surfaces with a covalently bound layer of hydroxyethylcellulose [286]. Ion-exchange membranes (anion and cation exchange) are also available as well as affinity membranes (protein A and Cu^{2+} as ligands). Chromatographic membranes have been also reported for purification processes for a variety of compounds, such as DNA, proteins (monoclonal antibody, serum antibody, serum albumin, enzymes, etc.), and viruses. For example, applications in this field has been reported such as immobilized L-histidine in hollow-fiber membranes for the separation of immunoglobulin G from human serum [289], the use of thiophilic membranes for the purification of monoclonal antibodies from cell culture media [290], affinity membranes for the separation of MBP fusion proteins [291], cation-exchange membranes for the purification of alphaviruses [292], ion-exchange membranes for the isolation of antibacterial peptides from lactoferrin [285], anion-exchange membranes for the adsorption of DNA [293], and strong anion exchange membranes for reduction of endotoxin in a protein mixture [294].

In this field membrane chromatography, a hollow-fiber method was proposed for purification of fibronectin from blood plasma and purification of IgG using hollow-fiber membrane-supported protein A [295]. Therefore, the high throughput rate and the effective ligand use of this device allowed rapid bind–elute cycle times. Furthermore, the volume of a typical agarose affinity method was 100–1000 times than that of the affinity-membrane and the membrane required only about 0.1% as much ligand to handle the same throughput at the same mass transfer efficiency. An example of a successful application in the pharmaceutical manufacturing was also reported [294]. In this study, a scale-up of strong anion-exchange adsorber membranes that eliminate endotoxin from bacterial extracts while preserving enzyme activity in the protein mixture was validated. Therefore, the endotoxin removal process was directly adapted from the small-scale Q-100 MA cartridge (Sartorius Corporation, 100 cm^2 working surface area) to the large-scale Q 550 MA sheets (5500 cm^2 working surface area). Furthermore, the characteristics of endotoxin removal, protein absorption, and photolyase purification were observed to be similar in the two processes.

One of the major drawbacks with membrane chromatography is non-uniform flow distribution across the membrane, because of the large diameter-to-length ratio of the modules. However, the membrane chromatography system has not certainly gained the expected success. The reason is probably due to the reticence of potential users for this new technology. In addition, membranes for chromatography are particularly attractive for preparative chromatography, as initially developed by Sepracor Inc., to purify large amounts of molecules. Particularly, hollow fibers are mainly well suitable, more than flat sheet membrane modules [295].

7.2.6 Membrane Contactors

As per Drioli et al. (2005) recently, membrane contactors have shown growing attention [296]. They have reported that membrane emulsification for emulsion preparation [297,298] and for the preparation of precipitates [299,300], membrane contactors are used. As seen from Figure 7.16, the term "membrane contactor" can be defined as the connection of phases A and B through membrane pores. The membrane contactor involves using a pressure to force a dispersed phase A to permeate through a membrane into a continuous phase B,

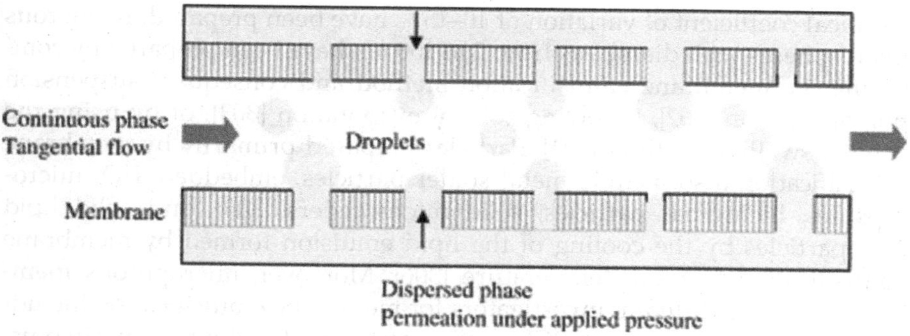

FIGURE 7.16
Schematic diagram of the membrane emulsification process. (Reprinted from *Biotechnology Advances*, 24(5), Charcosset, C., Membrane processes in biotechnology: An overview, pp. 482–492, Copyright 2006, with permission from Elsevier.)

which flows tangentially to the membrane surface. Therefore, membrane pores behave as parallel capillaries for the introduction of phase A into phase B. There may either be no reaction between the two phases (membrane emulsification) or a reaction may occur between the two phases (preparation of precipitates). In case of membrane emulsification, droplets grow at the pore outlets until, upon reaching a certain size, they detach. This is determined by the balance between the drag force on the droplet from the flowing continuous phase, the buoyancy of the droplet, the interfacial tension forces, and the driving pressure [297]. The final droplet size and size distribution are determined by the pore size and size distribution of the membrane and also by the degree of coalescence, both at the membrane surface and in the bulk solution.

A typical experimental set-up for a membrane emulsification process includes a tubular microfiltration membrane, a pump, a feed vessel, and a pressurized (N_2) oil container. The oil phase (to be dispersed) is pumped under gas pressure through the membrane pores into the aqueous continuous phase that circulates tangentially to the membrane surface. Membrane emulsification has been developed during the past 15 years [297,298]. The distinguishing feature is that the resulting droplet size is controlled primarily by the choice of the membrane and not by the generation of turbulent droplet break-up. The apparent shear stress is lower than the conventional emulsification systems, because small droplets are directly formed by permeation of the dispersed phase through the micropores, instead of disruption of large droplets in zones of high-energy density.

Besides the possibility of using shear-sensitive ingredients, emulsions with narrow droplet size distributions can be produced. Additionally, membrane emulsification processes allow production of emulsions at lower energy input (104–106 J/m^3) compared to conventional mechanical methods (106–108 J/m^3) [301]. Single (w/o or o/w) and multiple (w/o/w, o/w/o) emulsions, with various droplet sizes ranging from 0.8 to over 100 μm, and a typical coefficient of variation of 10–15%, have been prepared. Numerous microspheres with diameters from 2 to 100 μm have been prepared by combining the membrane emulsification method and consequent suspension polymerization [302], following solvent evaporation [303], or by using the droplet swelling method [304]. Particles prepared primarily by membrane emulsification also include metal solder particles, embedded TiO_2 microcapsules, bichromal particles, solid microcarriers [305], and solid lipid nanoparticles by the cooling of the lipid emulsion formed by membrane emulsification to room temperature [306]. Moreover, microporous membranes used by different investigators for membrane emulsification including Shirasu Porous Glass (SPG) membranes, coated α-alumina or zirconia, anodic porous alumina, polypropylene, polyamide, and polytetrafluoroethylene (PTFE) membranes [297]. A large range of particles, such as uniform polyurethaneurea-vinyl polymer microspheres of about 20 μm; using a Shirasu Porous Glass (SPG) membrane emulsification technique; and subsequent radical suspension polymerization were also prepared [302]. In this case, a mixture of a 40% urethane prepolymer solution of xylene and a vinyl monomer comprising an initiator was permeated through the uniform pores of the SPG membrane into a continuous phase containing a stabilizer to form uniform droplets, where the droplets were then permitted to stand for chain extension at room temperature with di- or triamines for 2 h in the absence or presence of ethyl acetate, followed by suspension polymerization at 70°C for 24 h. In another study, a reaction between phase A passing through the membrane pores with phase B flowing tangentially to the membrane surface, that is, for the preparation of nanosized $BaSO_4$ and $CaCO_3$ particles, have been reported [299,300], and polymeric nanoparticles by interfacial polymerization or nanoprecipitation reaction between an organic and an aqueous phase. A membrane contactor for the preparation of nanoparticles was also described recently [307] and the organic phase was pressed through the membrane pores by the filtrate side. Furthermore, the aqueous phase circulated inside the membrane module, and swept away the nanoparticles forming at the pore outlets. It was confirmed that nanoparticles as small as 260 nm could be attained with a 1000-Da nanofiltration membrane, a transmembrane pressure of 3 bar, and a cross-flow rate of 1.7 m s^{-1}. Moreover, high fluxes were attained with the 0.1 μm pore size microfiltration membrane (1.6 m^3/h m^2), leading to the preparation of 1.8×10^{-3} m^3 nanoparticles, with an average diameter of 360 nm, in 4 min. Therefore, the advantage of this membrane contactor compared to other processes for nanoparticle preparation validated its scale-up capacity and the possibility to control nanoparticle size by a suitable choice of membrane. However, the possible disadvantage of direct membrane emulsification is the relatively low maximum disperse phase flux through the membrane (typically 0.01 to 0.1 $m^3/(m^2$ h)) needed to avoid the transition from a "size-stable" to "continuous outflow" zone and to avoid steric hindrance among droplets that may be formed simultaneously at the adjacent pores. Therefore, numerous operating techniques were introduced such as rotating membranes [308] and repeated membrane extrusion of coarsely preemulsified feeds [309]. For instance, multiple w/o/w emulsions by multistage (repeated) premix emulsification using Shirasu-porous glass (SPG) membranes with a mean pore size of 10.7 μm have prepared [309]. A coarse emulsion containing droplets with a mean particle size of about 100 μm was homogenized 5–6 times through the same membrane at a constant pressure difference of 20–300 kPa to achieve additional droplet homogenization and size reduction. However, membrane contactors will have to face competition from other emulsification and particle preparation processes, to find their place in industrial practice [310].

7.2.7 Summary

Membrane processes are currently used throughout downstream purification processes. Recent advances in ultrafiltration membranes and process designs can provide the high concentration factors and greater overall yield required for high-dose products. Developments in normal flow filtration provide significant capabilities for virus removal. Therefore, membranes have long been an integral part of biotechnology processes. The most well-known examples are ultrafiltration and microfiltration, which have become routine methods. The sterile filtration of fermentation media, purification of buffers, and proteins are now standard practices. Other applications of membrane processes have been introduced more recently, as membrane bioreactors, membrane chromatography, and membrane contactors.

Although ultrafiltration and microfiltration are well-established processes, membrane bioreactors and membrane chromatography have certainly not obtained the expected levels of success. The reasons are related to a reticence of users for applying new technologies, and also to their potential advantages compared to other processes, such as classical reactors for biological reactions and conventional bead chromatography. In this sense, the examples of ultrafiltration and microfiltration may be advantageous, for which a continuous improvement in understanding the physicochemical phenomena governing the processes has been shown to enhance considerably the separation of proteins using their inherent electrostatic properties. The membranes biosensors development and molecularly imprinted polymeric membranes for separation of molecules are the emerging membrane process applications. The permeation, reaction and mixing mechanisms are the most striking features of membrane processes. For this reason, there is no doubt that new successful applications will continue to appear in the near future. Their industrial success will again depend on their advantages over other competing processes and on their acceptance. Theoretical understanding of the physical and chemical phenomena governing these processes will also help to improve their performance and to facilitate their introduction in the biotechnology-based industries. Continued efforts to develop improved membrane materials, modules, and process designs should enable membrane systems to play an important role in the next generation of biotechnology processes.

References

1. Zhao, N., Wang, Z., Cai, C., Shen, H., Liang, F., Wang, D., Wang, C., Zhu, T., Guo, J., Wang, Y., and Liu, X., Bioinspired materials: From low to high dimensional structure. *Advanced Materials*, 2014. 26(41): pp. 6994–7017.
2. Zan, G., and Wu, Q., Biomimetic and bioinspired synthesis of nanomaterials/nanostructures. *Advanced Materials*, 2016. 28(11): pp. 2099–2147.
3. Liu, Y., Luo, D., and Wang, T., Hierarchical structures of bone and bioinspired bone tissue engineering. *Small*, 2016. 12(34): pp. 4611–4632.
4. Mitragotri, S., and Lahann, J., Physical approaches to biomaterial design. *Nature Materials*, 2009. 8(1): pp. 15.
5. Hu, X., Hu, J., Tian, J., Ge, Z., Zhang, G., Luo, K., and Liu, S., Polyprodrug amphiphiles: Hierarchical assemblies for shape-regulated cellular internalization, trafficking, and drug delivery. *Journal of the American Chemical Society*, 2013. 135(46): pp. 17617–17629.
6. Zeng, W., Shu, L., Li, Q., Chen, S., Wang, F., and Tao, X.M., Fiber-based wearable electronics: A review of materials, fabrication, devices, and applications. *Advanced Materials*, 2014. 26(31): pp. 5310–5336.
7. Wang, X.D., and Wolfbeis, O.S., Fiber-optic chemical sensors and biosensors (2008–2012). *Analytical Chemistry*, 2012. 85(2): pp. 487–508.
8. Senthamizhan, A., Balusamy, B., and Uyar, T., Glucose sensors based on electrospun nanofibers: A review. *Analytical and Bioanalytical Chemistry*, 2016. 408(5): pp. 1285–1306.
9. Anis, S.F., Khalil, A., Singaravel, G., and Hashaikeh, R., A review on the fabrication of zeolite and mesoporous inorganic nanofibers formation for catalytic applications. *Microporous and Mesoporous Materials*, 2016. 236: pp. 176–192.
10. Li, X., Chen, Y., Huang, H., Mai, Y.W., and Zhou, L., Electrospun carbon-based nanostructured electrodes for advanced energy storage–A review. *Energy Storage Materials*, 2016. 5: pp. 58–92.
11. Wang, X., Yu, J., Sun, G., and Ding, B., Electrospun nanofibrous materials: A versatile medium for effective oil/water separation. *Materials Today*, 2016. 19(7): pp. 403–414.
12. Zhu, M., Han, J., Wang, F., Shao, W., Xiong, R., Zhang, Q., Pan, H., Yang, Y., Samal, S.K., Zhang, F., and Huang, C., Electrospun nanofibers membranes for effective air filtration. *Macromolecular Materials and Engineering*, 2017. 302(1).
13. Xue, J., Xie, J., Liu, W., and Xia, Y., Electrospun nanofibers: New concepts, materials, and applications. *Accounts of Chemical Research*, 2017. 50(8): pp. 1976–1987.
14. Wu, J., Wang, N., Zhao, Y., and Jiang, L., Electrospinning of multilevel structured functional micro-/nanofibers and their applications. *Journal of Materials Chemistry A*, 2013. 1(25): pp. 7290–7305.

15. Peng, S., Jin, G., Li, L., Li, K., Srinivasan, M., Ramakrishna, S., and Chen, J., Multi-functional electrospun nanofibres for advances in tissue regeneration, energy conversion & storage, and water treatment. *Chemical Society Reviews*, 2016. 45(5): pp. 1225–1241.
16. Li, D., and Xia, Y., Electrospinning of nanofibers: Reinventing the wheel?. *Advanced Materials*, 2004. 16(14): pp. 1151–1170.
17. Doshi, J., and Reneker, D.H., Electrospinning process and applications of electrospun fibers. *Journal of Electrostatics*, 1995. 35(2–3): pp. 151–160.
18. Kim, J.S., and Reneker, D.H., Mechanical properties of composites using ultrafine electrospun fibers. *Polymer Composites*, 1999. 20(1): pp. 124–131.
19. How, T.V., University of Liverpool, and Ethicon Inc., Synthetic vascular grafts, and methods of manufacturing such grafts. U.S. Patent 4552707, 1985.
20. Reneker, D.H., Yarin, A.L., Fong, H., and Koombhongse, S., Bending instability of electrically charged liquid jets of polymer solutions in electrospinning. *Journal of Applied Physics*, 2000. 87(9): pp. 4531–4547.
21. Stitzel, J.D., Pawlowski, K.J., Wnek, G.E., Simpson, D.G., and Bowlin, G.L., Arterial smooth muscle cell proliferation on a novel biomimicking, biodegradable vascular graft scaffold. *Journal of Biomaterials Applications*, 2001. 16(1): pp. 22–33.
22. Smith, L.A., and Ma, P.X., Nano-fibrous scaffolds for tissue engineering. *Colloids and Surfaces B: Biointerfaces*, 2004. 39(3): pp. 125–131.
23. Ramakrishna, S., Fujihara, K., Teo, W.E., Yong, T., Ma, Z., and Ramaseshan, R., Electrospun nanofibers: Solving global issues. *Materials Today*, 2006. 9(3): pp. 40–50.
24. Pinchuk, L., Martin Jr, J.B., and Maurin, A.A., Corvita Corp., Breast prostheses. U.S. Patent 5376117, 1994.
25. Burger, C., Hsiao, B.S., and Chu, B., Nanofibrous materials and their applications. *Annual Review of Materials Research*, 2006. 36: pp. 333–368.
26. Bhardwaj, N., and Kundu, S.C., Electrospinning: A fascinating fiber fabrication technique. *Biotechnology Advances*, 2010. 28(3): pp. 325–347.
27. Yang, G., Li, X., He, Y., Ma, J., Ni, G., Zhou, S., From nano to micro to macro: Electrospun hierarchically structured polymeric fibers for biomedical applications. *Progress in Polymer Science*, 2018. 81: pp. 80–113.
28. Kost, J., and Langer, R., Responsive polymeric delivery systems. *Advanced Drug Delivery Reviews*, 2012. 64: pp. 327–341.
29. Kenawy, E.R., Bowlin, G.L., Mansfield, K., Layman, J., Simpson, D.G., Sanders, E.H., and Wnek, G.E., Release of tetracycline hydrochloride from electrospun poly (ethylene-co-vinylacetate), poly (lactic acid), and a blend. *Journal of Controlled Release*, 2002. 81(1–2): pp. 57–64.
30. Verreck, G., Chun, I., Peeters, J., Rosenblatt, J., and Brewster, M.E., Preparation and characterization of nanofibers containing amorphous drug dispersions generated by electrostatic spinning. *Pharmaceutical Research*, 2003. 20(5): pp. 810–817.
31. Zong, X., Kim, K., Fang, D., Ran, S., Hsiao, B.S., and Chu, B., Structure and process relationship of electrospun bioabsorbable nanofiber membranes. *Polymer*, 2002. 43(16): pp. 4403–4412.
32. Kim, K., Luu, Y.K., Chang, C., Fang, D., Hsiao, B.S., Chu, B., and Hadjiargyrou, M., Incorporation and controlled release of a hydrophilic antibiotic using poly (lactide-co-glycolide)-based electrospun nanofibrous scaffolds. *Journal of Controlled Release*, 2004. 98(1): pp. 47–56.

33. Jiang, H., Fang, D., Hsiao, B., Chu, B., and Chen, W., Preparation and characterization of ibuprofen-loaded poly (lactide-co-glycolide)/poly (ethylene glycol)-g-chitosan electrospun membranes. *Journal of Biomaterials Science, Polymer Edition*, 2004. 15(3): pp. 279–296.
34. Sanders, E.H., Kloefkorn, R., Bowlin, G.L., Simpson, D.G., and Wnek, G.E., Two-phase electrospinning from a single electrified jet: Microencapsulation of aqueous reservoirs in poly (ethylene-co-vinyl acetate) fibers. *Macromolecules*, 2003. 36(11): pp. 3803–3805.
35. Chew, S.Y., Wen, J., Yim, E.K., and Leong, K.W., Sustained release of proteins from electrospun biodegradable fibers. *Biomacromolecules*, 2005. 6(4): pp. 2017–2024.
36. Zong, X.H., Fang, D.F., Kim, K.S., Ran, S.F., Hsiao, B.S., Chu, B., Brathwaite, C., Li, S., and Chen, E., Nonwoven nanofiber membranes of poly (lactide) and poly (glycolide-co-lactide) via electrospinning and applications for anti-adhesions. In ABSTRACTS OF PAPERS OF THE AMERICAN CHEMICAL SOCIETY (Vol. 224, pp. U466–U466). 1155 16TH ST, NW, WASHINGTON, DC 20036 USA: AMER CHEMICAL SOC, 2002, August.
37. Zhang, Y.Z., Venugopal, J., Huang, Z.M., Lim, C.T., and Ramakrishna, S., Crosslinking of the electrospun gelatin nanofibers. *Polymer*, 2006. 47(8): pp. 2911–2917.
38. Zeng, J., Xu, X., Chen, X., Liang, Q., Bian, X., Yang, L., and Jing, X., Biodegradable electrospun fibers for drug delivery. *Journal of Controlled Release*, 2003. 92(3): pp. 227–231.
39. Jiang, H., Hu, Y., Li, Y., Zhao, P., Zhu, K., and Chen, W., A facile technique to prepare biodegradable coaxial electrospun nanofibers for controlled release of bioactive agents. *Journal of Controlled Release*, 2005. 108(2–3): pp. 237–243.
40. Xie, J., and Wang, C.H., Electrospun micro- and nanofibers for sustained delivery of paclitaxel to treat C6 glioma in vitro. *Pharmaceutical Research*, 2006. 23(8): pp. 1817.
41. Luong-Van, E., Grøndahl, L., Chua, K.N., Leong, K.W., Nurcombe, V., and Cool, S.M., Controlled release of heparin from poly (ε-caprolactone) electrospun fibers. *Biomaterials*, 2006. 27(9): pp. 2042–2050.
42. Zeng, J., Aigner, A., Czubayko, F., Kissel, T., Wendorff, J.H., and Greiner, A., Poly (vinyl alcohol) nanofibers by electrospinning as a protein delivery system and the retardation of enzyme release by additional polymer coatings. *Biomacromolecules*, 2005. 6(3): pp. 1484–1488.
43. Casper, C.L., Yamaguchi, N., Kiick, K.L., and Rabolt, J.F., Functionalizing electrospun fibers with biologically relevant macromolecules. *Biomacromolecules*, 2005. 6(4): pp. 1998–2007.
44. Chou, S.F., Carson, D., and Woodrow, K.A., Current strategies for sustaining drug release from electrospun nanofibers. *Journal of Controlled Release*, 2015. 220: pp. 584–591.
45. Yohe, S.T., Colson, Y.L., and Grinstaff, M.W., Superhydrophobic materials for tunable drug release: Using displacement of air to control delivery rates. *Journal of the American Chemical Society*, 2012. 134(4): pp. 2016–2019.
46. Heunis, T.D.J., and Dicks, L.M.T., *Nanofibers offer alternative ways to the treatment of skin infections*. BioMedical Research International, 2010. 2010.
47. Carson, D., Jiang, Y., and Woodrow, K.A., Tunable release of multiclass anti-HIV drugs that are water-soluble and loaded at high drug content in polyester blended electrospun fibers. *Pharmaceutical Research*, 2016. 33(1): pp. 125–136.

48. Huang, Z.M., He, C.L., Yang, A., Zhang, Y., Han, X.J., Yin, J., and Wu, Q., Encapsulating drugs in biodegradable ultrafine fibers through co-axial electrospinning. *Journal of Biomedical Materials Research Part A*, 2006. 77(1): pp. 169–179.
49. Zhang, Y.Z., Wang, X., Feng, Y., Li, J., Lim, C.T., and Ramakrishna, S., Coaxial electrospinning of (fluorescein isothiocyanate-conjugated bovine serum albumin)-encapsulated poly (ε-caprolactone) nanofibers for sustained release. *Biomacromolecules*, 2006. 7(4): pp. 1049–1057.
50. Qi, H., Hu, P., Xu, J., and Wang, A., Encapsulation of drug reservoirs in fibers by emulsion electrospinning: Morphology characterization and preliminary release assessment. *Biomacromolecules*, 2006. 7(8): pp. 2327–2330.
51. Falde, E.J., Freedman, J.D., Herrera, V.L., Yohe, S.T., Colson, Y.L., and Grinstaff, M.W., Layered superhydrophobic meshes for controlled drug release. *Journal of Controlled Release*, 2015. 214: pp. 23–29.
52. Hu, C., and Cui, W., Hierarchical structure of electrospun composite fibers for long-term controlled drug release carriers. *Advanced Healthcare Materials*, 2012. 1(6): pp. 809–814.
53. Mickova, A., Buzgo, M., Benada, O., Rampichova, M., Fisar, Z., Filova, E., Tesarova, M., Lukas, D., and Amler, E., Core/shell nanofibers with embedded liposomes as a drug delivery system. *Biomacromolecules*, 2012. 13(4): pp. 952–962.
54. Ionescu, L.C., Lee, G.C., Sennett, B.J., Burdick, J.A., and Mauck, R.L., An anisotropic nanofiber/microsphere composite with controlled release of biomolecules for fibrous tissue engineering. *Biomaterials*, 2010. 31(14): pp. 4113–4120.
55. DeVolder, R.J., Bae, H., Lee, J., and Kong, H., Directed blood vessel growth using an angiogenic microfiber/microparticle composite patch. *Advanced Materials*, 2011. 23(28): pp. 3139–3143.
56. Tan, S.T., Wendorff, J.H., Pietzonka, C., Jia, Z.H., and Wang, G.Q., Biocompatible and biodegradable polymer nanofibers displaying superparamagnetic properties. *Chem Phys Chem*, 2005. 6(8): pp. 1461–1465.
57. Shao, S., Li, L., Yang, G., Li, J., Luo, C., Gong, T., and Zhou, S., Controlled green tea polyphenols release from electrospun PCL/MWCNTs composite nanofibers. *International Journal of Pharmaceutics*, 2011. 421(2): pp. 310–320.
58. Li, J., Fu, R., Li, L., Yang, G., Ding, S., Zhong, Z., and Zhou, S., Co-delivery of dexamethasone and green tea polyphenols using electrospun ultrafine fibers for effective treatment of keloid. *Pharmaceutical Research*, 2014. 31(7): pp. 1632–1643.
59. Dong, B., Smith, M.E., and Wnek, G.E., Encapsulation of multiple biological compounds within a single electrospun fiber. *Small*, 2009. 5(13): pp. 1508–1512.
60. Jo, E., Lee, S., Kim, K.T., Won, Y.S., Kim, H.S., Cho, E.C., and Jeong, U., Core-Sheath Nanofibers Containing Colloidal Arrays in the Core for Programmable Multi-Agent Delivery. *Advanced Materials*, 2009. 21(9): pp. 968–972.
61. Friedemann, K., Turshatov, A., Landfester, K., and Crespy, D., Characterization via two-color STED microscopy of nanostructured materials synthesized by colloid electrospinning. *Langmuir*, 2011. 27(11): pp. 7132–7139.
62. Hou, Z., Li, X., Li, C., Dai, Y., Ma, P.A., Zhang, X., Kang, X., Cheng, Z., and Lin, J., 2013. Electrospun upconversion composite fibers as dual drugs delivery system with individual release properties. *Langmuir*, 29(30). 9473–9482.
63. Yang, G., Wang, J., Li, L., Ding, S., and Zhou, S., Electrospun micelles/drug-loaded nanofibers for time-programmed multi-agent release. *Macromolecular Bioscience*, 2014. 14(7): pp. 965–976.

64. Su, Y., Su, Q., Liu, W., Jin, G., Mo, X., and Ramakrishn, S., Dual-drug encapsulation and release from core–shell nanofibers. *Journal of Biomaterials Science, Polymer Edition,* 2012. 23(7): pp. 861–871.
65. Okuda, T., Tominaga, K., and Kidoaki, S., Time-programmed dual release formulation by multilayered drug-loaded nanofiber meshes. *Journal of Controlled Release,* 2010. 143(2): pp. 258–264.
66. Chen, J., Zhang, S., Sun, F., Li, N., Cui, K., He, J., Niu, D., and Li, Y., Multi-stimuli responsive supramolecular polymers and their electrospun nanofibers. *Polymer Chemistry,* 2016. 7(17): pp. 2947–2954.
67. Wang, Y., Kotsuchibashi, Y., Uto, K., Ebara, M., Aoyagi, T., Liu, Y., and Narain, R., pH and glucose responsive nanofibers for the reversible capture and release of lectins. *Biomaterials Science,* 2015. 3(1): pp. 152–162.
68. Yuan, H., Li, B., Liang, K., Lou, X., and Zhang, Y., Regulating drug release from pH-and temperature-responsive electrospun CTS-g-PNIPAAm/poly (ethylene oxide) hydrogel nanofibers. *Biomedical Materials,* 2014. 9(5): pp. 055001.
69. Kim, Y.J., Ebara, M., and Aoyagi, T., A smart hyperthermia nanofiber with switchable drug release for inducing cancer apoptosis. *Advanced Functional Materials,* 2013. 23(46): pp. 5753–5761.
70. Li, L., Yang, G., Zhou, G., Wang, Y., Zheng, X., and Zhou, S., Thermally switched release from a nanogel-in-microfiber device. *Advanced Healthcare Materials,* 2015. 4(11): pp. 1658–1663.
71. Ignatious, F., Sun, L., Lee, C.P., and Baldoni, J., Electrospun nanofibers in oral drug delivery. *Pharmaceutical Research,* 2010. 27(4): pp. 576–588.
72. Yu, D.G., Shen, X.X., Branford-White, C., White, K., Zhu, L.M., and Bligh, S.A., Oral fast-dissolving drug delivery membranes prepared from electrospun polyvinylpyrrolidone ultrafine fibers. *Nanotechnology,* 2009. 20(5): pp. 055104.
73. Li, X., Kanjwal, M.A., Lin, L., and Chronakis, I.S., Electrospun polyvinyl-alcohol nanofibers as oral fast-dissolving delivery system of caffeine and riboflavin. *Colloids and Surfaces B: Biointerfaces,* 2013. 103: pp. 182–188.
74. Nam, S., Lee, J.J., Lee, S.Y., Jeong, J.Y., Kang, W.S., and Cho, H.J., Angelica gigas Nakai extract-loaded fast-dissolving nanofiber based on poly (vinyl alcohol) and Soluplus for oral cancer therapy. *International Journal of Pharmaceutics,* 2017. 526(1–2): pp. 225–234.
75. Madhaiyan, K., Sridhar, R., Sundarrajan, S., Venugopal, J.R., and Ramakrishna, S., Vitamin B_{12} loaded polycaprolactone nanofibers: A novel transdermal route for the water soluble energy supplement delivery. *International Journal of Pharmaceutics,* 2013. 444(1–2): pp. 70–76.
76. Yun, J., Im, J.S., Lee, Y.S., and Kim, H.I., Electro-responsive transdermal drug delivery behavior of PVA/PAA/MWCNT nanofibers. *European Polymer Journal,* 2011. 47(10): pp. 1893–1902.
77. Cheng, L., Sun, X., Zhao, X., Wang, L., Yu, J., Pan, G., Li, B., Yang, H., Zhang, Y., and Cui, W., Surface biofunctional drug-loaded electrospun fibrous scaffolds for comprehensive repairing hypertrophic scars. *Biomaterials,* 2016. 83: pp. 169–181.
78. Taepaiboon, P., Rungsardthong, U., and Supaphol, P., Vitamin-loaded electrospun cellulose acetate nanofiber mats as transdermal and dermal therapeutic agents of vitamin A acid and vitamin E. *European Journal of Pharmaceutics and Biopharmaceutics,* 2007. 67(2): pp. 387–397.

79. Knockenhauer, K.E., Sawicka, K.M., Roemer, E.J., and Simon, S.R., Protective antigen composite nanofibers as a transdermal anthrax vaccine. In *Engineering in Medicine and Biology Society*, 2008. EMBS 2008. 30th Annual International Conference of the IEEE (1040–1043), 2008, August.
80. Ngawhirunpat, T., Opanasopit, P., Rojanarata, T., Akkaramongkolporn, P., Ruktanonchai, U., and Supaphol, P., Development of meloxicam-loaded electrospun polyvinyl alcohol mats as a transdermal therapeutic agent. *Pharmaceutical Development and Technology*, 2009. 14(1): pp. 73–82.
81. Park, Y., Kang, E., Kwon, O.J., Hwang, T., Park, H., Lee, J.M., Kim, J.H., and Yun, C.O., Ionically crosslinked Ad/chitosan nanocomplexes processed by electrospinning for targeted cancer gene therapy. *Journal of Controlled Release*, 2010. 148(1): pp. 75–82.
82. Yan, E., Fan, Y., Sun, Z., Gao, J., Hao, X., Pei, S., Wang, C., Sun, L., and Zhang, D., Biocompatible core–shell electrospun nanofibers as potential application for chemotherapy against ovary cancer. *Materials Science and Engineering: C*, 2014. 41: pp. 217–223.
83. Ho, E.A., Soo, P.L., Allen, C., and Piquette-Miller, M., Impact of intraperitoneal, sustained delivery of paclitaxel on the expression of P-glycoprotein in ovarian tumors. *Journal of Controlled Release*, 2007. 117(1): pp. 20–27.
84. De Souza, R., Zahedi, P., Allen, C.J., and Piquette-Miller, M., Polymeric drug delivery systems for localized cancer chemotherapy. *Drug Delivery*, 2010. 17(6): pp. 365–375.
85. Ranganath, S.H., and Wang, C.H., Biodegradable microfiber implants delivering paclitaxel for post-surgical chemotherapy against malignant glioma. *Biomaterials*, 2008. 29(20): pp. 2996–3003.
86. Yang, G., Wang, J., Wang, Y., Li, L., Guo, X., and Zhou, S., An implantable active-targeting micelle-in-nanofiber device for efficient and safe cancer therapy. *ACS Nano*, 2015. 9(2): pp. 1161–1174.
87. Zong, S., Wang, X., Yang, Y., Wu, W., Li, H., Ma, Y., Lin, W., Sun, T., Huang, Y., Xie, Z., and Yue, Y., The use of cisplatin-loaded mucoadhesive nanofibers for local chemotherapy of cervical cancers in mice. *European Journal of Pharmaceutics and Biopharmaceutics*, 2015. 93: pp. 127–135.
88. Luo, X., Xie, C., Wang, H., Liu, C., Yan, S., and Li, X., Antitumor activities of emulsion electrospun fibers with core loading of hydroxycamptothecin via intratumoral implantation. *International Journal of Pharmaceutics*, 2012. 425(1–2): pp. 19–28.
89. Kaplan, J.A., Liu, R., Freedman, J.D., Padera, R., Schwartz, J., Colson, Y.L., and Grinstaff, M.W., Prevention of lung cancer recurrence using cisplatin-loaded superhydrophobic nanofiber meshes. *Biomaterials*, 2016. 76: pp. 273–281.
90. Zhang, Z., Liu, S., Xiong, H., Jing, X., Xie, Z., Chen, X., and Huang, Y., Electrospun PLA/MWCNTs composite nanofibers for combined chemo-and photothermal therapy. *Acta Biomaterialia*, 2015. 26: pp. 115–123.
91. Wang, Y., Luo, C., Yang, G., Wei, X., Liu, D., and Zhou, S., A Luteolin-Loaded Electrospun Fibrous Implantable Device for Potential Therapy of Gout Attacks. *Macromolecular Bioscience*, 2016. 16(11): pp. 1598–1609.
92. Blakney, A.K., Ball, C., Krogstad, E.A., and Woodrow, K.A., Electrospun fibers for vaginal anti-HIV drug delivery. *Antiviral Research*, 2013. 100: pp. S9–S16.
93. Luu, Y.K., Kim, K., Hsiao, B.S., Chu, B., and Hadjiargyrou, M., Development of a nanostructured DNA delivery scaffold via electrospinning of PLGA and PLA–PEG block copolymers. *Journal of Controlled Release*, 2003. 89(2): pp. 341–353.

94. Liang, D., Luu, Y.K., Kim, K., Hsiao, B.S., Hadjiargyrou, M., and Chu, B., In vitro non-viral gene delivery with nanofibrous scaffolds. *Nucleic Acids Research*, 2005. 33(19): pp. e170–e170.
95. Nishikawa, M., and Huang, L., Nonviral vectors in the new millennium: Delivery barriers in gene transfer. *Human Gene Therapy*, 2001. 12(8): pp. 861–870.
96. Howard, K.A., Rahbek, U.L., Liu, X., Damgaard, C.K., Glud, S.Z., Andersen, M.Ø., Hovgaard, M.B., Schmitz, A., Nyengaard, J.R., Besenbacher, F., and Kjems, J., RNA interference in vitro and in vivo using a novel chitosan/siRNA nanoparticle system. *Molecular Therapy*, 2006. 14(4): pp. 476–484.
97. Mao, H.Q., Roy, K., Troung-Le, V.L., Janes, K.A., Lin, K.Y., Wang, Y., August, J.T., and Leong, K.W., Chitosan-DNA nanoparticles as gene carriers: Synthesis, characterization and transfection efficiency. *Journal of Controlled Release*, 2001. 70(3): pp. 399–421.
98. Nie, H., Ho, M.L., Wang, C.K., Wang, C.H., and Fu, Y.C., BMP-2 plasmid loaded PLGA/HAp composite scaffolds for treatment of bone defects in nude mice. *Biomaterials*, 2009. 30(5): pp. 892–901.
99. Nie, H., and Wang, C.H., Fabrication and characterization of PLGA/HAp composite scaffolds for delivery of BMP-2 plasmid DNA. *Journal of Controlled Release*, 2007. 120(1–2): pp. 111–121.
100. Saraf, A., Baggett, L.S., Raphael, R.M., Kasper, F.K., and Mikos, A.G., Regulated non-viral gene delivery from coaxial electrospun fiber mesh scaffolds. *Journal of Controlled Release*, 2010. 143(1): pp. 95–103.
101. Sakai, S., Yamada, Y., Yamaguchi, T., Ciach, T., and Kawakami, K., Surface immobilization of poly (ethyleneimine) and plasmid DNA on electrospun poly (L-lactic acid) fibrous mats using a layer-by-layer approach for gene delivery. *Journal of Biomedical Materials Research Part A*, 2009. 88(2): pp. 281–287.
102. Cao, H., Jiang, X., Chai, C., and Chew, S.Y., RNA interference by nanofiber-based siRNA delivery system. *Journal of Controlled Release*, 2010. 144(2): pp. 203–212.
103. Rujitanaroj, P.O., Wang, Y.C., Wang, J., and Chew, S.Y., Nanofiber-mediated controlled release of siRNA complexes for long term gene-silencing applications. *Biomaterials*, 2011. 32(25): pp. 5915–5923.
104. Nerem, R.M., and Sambanis, A., Tissue engineering: From biology to biological substitutes. *Tissue Engineering*, 1995. 1(1): pp. 3–13.
105. Hubbell, J.A., Biomaterials in tissue engineering. *Nature Biotechnology*, 1995. 13(6): pp. 565.
106. Tang, Z., He, C., Tian, H., Ding, J., Hsiao, B.S., Chu, B., and Chen, X., Polymeric nanostructured materials for biomedical applications. *Progress in Polymer Science*, 2016. 60: pp. 86–128.
107. Agarwal, S., Wendorff, J.H., and Greiner, A., Use of electrospinning technique for biomedical applications. *Polymer*, 2008. 49(26): pp. 5603–5621.
108. Chua, K.N., Lim, W.S., Zhang, P., Lu, H., Wen, J., Ramakrishna, S., Leong, K.W., and Mao, H.Q., Stable immobilization of rat hepatocyte spheroids on galactosylated nanofiber scaffold. *Biomaterials*, 2005. 26(15): pp. 2537–2547.
109. Araujo, J.V., Martins, A., Leonor, I.B., Pinho, E.D., Reis, R.L., and Neves, N.M., Surface controlled biomimetic coating of polycaprolactone nanofiber meshes to be used as bone extracellular matrix analogues. *Journal of Biomaterials Science, Polymer Edition*, 2008. 19(10): pp. 1261–1278.
110. Chua, K.N., Chai, C., Lee, P.C., Tang, Y.N., Ramakrishna, S., Leong, K.W., and Mao, H.Q., Surface-aminated electrospun nanofibers enhance adhesion and

expansion of human umbilical cord blood hematopoietic stem/progenitor cells. *Biomaterials*, 2006. 27(36): pp. 6043–6051.

111. Janjanin, S., Li, W.J., Morgan, M.T., Shanti, R.M., and Tuan, R.S., Mold-shaped, nanofiber scaffold-based cartilage engineering using human mesenchymal stem cells and bioreactor. *Journal of Surgical Research*, 2008. 149(1): pp. 47–56.
112. Zhu, Y., Leong, M.F., Ong, W.F., Chan-Park, M.B., and Chian, K.S., Esophageal epithelium regeneration on fibronectin grafted poly (L-lactide-co-caprolactone) (PLLC) nanofiber scaffold. *Biomaterials*, 2007. 28(5): pp. 861–868.
113. Li, W., Guo, Y., Wang, H., Shi, D., Liang, C., Ye, Z., Qing, F., and Gong, J., Electrospun nanofibers immobilized with collagen for neural stem cells culture. *Journal of Materials Science: Materials in Medicine*, 2008. 19(2): pp. 847–854.
114. Xin, X., Hussain, M., and Mao, J.J., Continuing differentiation of human mesenchymal stem cells and induced chondrogenic and osteogenic lineages in electrospun PLGA nanofiber scaffold. *Biomaterials*, 2007. 28(2): pp. 316–325.
115. Spolenak, R., Gorb, S., and Arzt, E., Adhesion design maps for bio-inspired attachment systems. *Acta Biomaterialia*, 2005. 1(1): pp. 5–13.
116. Yoo, H.S., Kim, T.G., and Park, T.G., Surface-functionalized electrospun nanofibers for tissue engineering and drug delivery. *Advanced Drug Delivery Reviews*, 2009. 61(12): pp. 1033–1042.
117. Jiang, T., Carbone, E.J., Lo, K.W.H., and Laurencin, C.T., Electrospinning of polymer nanofibers for tissue regeneration. *Progress in Polymer Science*, 2015. 46: pp. 1–24.
118. Agarwal, S., Wendorff, J.H., and Greiner, A., Progress in the field of electrospinning for tissue engineering applications. *Advanced Materials*, 2009. 21(32–33): pp. 3343–3351.
119. Liu, W., Thomopoulos, S., and Xia, Y., Electrospun nanofibers for regenerative medicine. *Advanced Healthcare Materials*, 2012. 1(1): pp. 10–25.
120. Li, W.J., Laurencin, C.T., Caterson, E.J., Tuan, R.S., and Ko, F.K., Electrospun nanofibrous structure: A novel scaffold for tissue engineering. *Journal of Biomedical Materials Research Part A*, 2002. 60(4): pp. 613–621.
121. Gong, Y., He, L., Li, J., Zhou, Q., Ma, Z., Gao, C., and Shen, J., Hydrogel-filled polylactide porous scaffolds for cartilage tissue engineering. *Journal of Biomedical Materials Research Part B: Applied Biomaterials*, 2007. 82(1): pp. 192–204.
122. Friess, W., Collagen–biomaterial for drug delivery1. *European Journal of Pharmaceutics and Biopharmaceutics*, 1998. 45(2): pp. 113–136.
123. Pavlov, M.P., Mano, J.F., Neves, N.M., and Reis, R.L., Fibers and 3D mesh scaffolds from biodegradable starch-based blends: Production and characterization. *Macromolecular Bioscience*, 2004. 4(8): pp. 776–784.
124. Almany, L. and Seliktar, D., Biosynthetic hydrogel scaffolds made from fibrinogen and polyethylene glycol for 3D cell cultures. *Biomaterials*, 2005. 26(15): pp. 2467–2477.
125. Wayne, J.S., McDowell, C.L., Shields, K.J., and Tuan, R.S., In vivo response of polylactic acid–alginate scaffolds and bone marrow-derived cells for cartilage tissue engineering. *Tissue Engineering*, 2005. 11(5–6): pp. 953–963.
126. Yoo, H.S., Lee, E.A., Yoon, J.J., and Park, T.G., Hyaluronic acid modified biodegradable scaffolds for cartilage tissue engineering. *Biomaterials*, 2005. 26(14): pp. 1925–1933.
127. Venugopal, J. and Ramakrishna, S., Biocompatible nanofiber matrices for the engineering of a dermal substitute for skin regeneration. *Tissue Engineering*, 2005. 11(5–6): pp. 847–854.

128. Fertala, A., Han, W.B., and Ko, F.K., Mapping critical sites in collagen II for rational design of gene-engineered proteins for cell-supporting materials. *Journal of Biomedical Materials Research Part A*, 2001. 57(1): pp. 48–58.
129. Rho, K.S., Jeong, L., Lee, G., Seo, B.M., Park, Y.J., Hong, S.D., Roh, S., Cho, J.J., Park, W.H., and Min, B.M., Electrospinning of collagen nanofibers: Effects on the behavior of normal human keratinocytes and early-stage wound healing. *Biomaterials*, 2006. 27(8): pp. 1452–1461.
130. Yoshimoto, H., Shin, Y.M., Terai, H., and Vacanti, J.P., A biodegradable nanofiber scaffold by electrospinning and its potential for bone tissue engineering. *Biomaterials*, 2003. 24(12). pp. 2077–2082.
131. Rezwan, K., Chen, Q.Z., Blaker, J.J., and Boccaccini, A.R., Biodegradable and bioactive porous polymer/inorganic composite scaffolds for bone tissue engineering. *Biomaterials*, 2006. 27(18): pp. 3413–3431.
132. Silva, E.A., Mooney, D.J., and Gerald, P.S., 8 Synthetic Extracellular Matrices for Tissue Engineering and Regeneration. *Current Topics in Developmental Biology*, 2004. 64: pp. 182–207.
133. Mo, X.M., Xu, C.Y., Kotaki, M.E.A., and Ramakrishna, S., Electrospun P (LLA-CL) nanofiber: A biomimetic extracellular matrix for smooth muscle cell and endothelial cell proliferation. *Biomaterials*, 2004. 25(10): pp. 1883–1890.
134. Zong, X., Bien, H., Chung, C.Y., Yin, L., Fang, D., Hsiao, B.S., Chu, B., and Entcheva, E., Electrospun fine-textured scaffolds for heart tissue constructs. *Biomaterials*, 2005. 26(26): pp. 5330–5338.
135. Li, C., Vepari, C., Jin, H.J., Kim, H.J., and Kaplan, D.L., Electrospun silk-BMP-2 scaffolds for bone tissue engineering. *Biomaterials*, 2006. 27(16): pp. 3115–3124.
136. Tuan, R.S., Boland, G., and Tuli, R., Adult mesenchymal stem cells and cell-based tissue engineering. *Arthritis Research & Therapy*, 2002. 5(1): pp. 32.
137. Jin, H.J., Chen, J., Karageorgiou, V., Altman, G.H., and Kaplan, D.L., Human bone marrow stromal cell responses on electrospun silk fibroin mats. *Biomaterials*, 2004. 25(6): pp. 1039–1047.
138. Ki, C.S., Kim, J.W., Hyun, J.H., Lee, K.H., Hattori, M., Rah, D.K., and Park, Y.H., Electrospun three-dimensional silk fibroin nanofibrous scaffold. *Journal of Applied Polymer Science*, 2007. 106(6): pp. 3922–3928.
139. Kobayashi H., Yokoyama Y., Takato T., Koyama H., and Ichioka S., Spongiform structured materials and it's manufacturing methods. Japanese patent Application number 2007–103201.
140. Yokoyama, Y., Hattori, S., Yoshikawa, C., Yasuda, Y., Koyama, H., Takato, T., and Kobayashi, H., Novel wet electrospinning system for fabrication of spongiform nanofiber 3-dimensional fabric. *Materials Letters*, 2009. 63(9–10): pp. 754–756.
141. Smith, D.J., Reneker, D.H., McManus, A.T., Schreuder-Gibson, H.L., Mello, C., and Sennett, M.S., University of Akron, Electrospun fibers and an apparatus therefor. U.S. Patent 6,753,454, 2004.
142. Wnek, G.E., Carr, M.E., Simpson, D.G., and Bowlin, G.L., Electrospinning of nanofiber fibrinogen structures. *Nano Letters*, 2003. 3(2): pp. 213–216.
143. Huang, Z.M., Zhang, Y.Z., Kotaki, M., and Ramakrishna, S., A review on polymer nanofibers by electrospinning and their applications in nanocomposites. *Composites Science and Technology*, 2003. 63(15): pp. 2223–2253.
144. Khil, M.S., Cha, D.I., Kim, H.Y., Kim, I.S., and Bhattarai, N., Electrospun nanofibrous polyurethane membrane as wound dressing. *Journal of Biomedical Materials Research Part B: Applied Biomaterials*, 2003. 67(2): pp. 675–679.

145. Norouzi, M., Boroujeni, S.M., Omidvarkordshouli, N., and Soleimani, M., Advances in skin regeneration: Application of electrospun scaffolds. *Advanced Healthcare Materials*, 2015. 4(8): pp. 1114–1133.
146. Kuppan, P., Vasanthan, K.S., Sundaramurthi, D., Krishnan, U.M., and Sethuraman, S., Development of poly (3-hydroxybutyrate-co-3-hydroxyvalerate) fibers for skin tissue engineering: Effects of topography, mechanical, and chemical stimuli. *Biomacromolecules*, 2011. 12(9): pp. 3156–3165.
147. Lin, H.Y., Chen, H.H., Chang, S.H., and Ni, T.S., Pectin-chitosan-PVA nanofibrous scaffold made by electrospinning and its potential use as a skin tissue scaffold. *Journal of Biomaterials Science, Polymer Edition*, 2013. 24(4): pp. 470–484.
148. Kumbar, S.G., Nukavarapu, S.P., James, R., Nair, L.S., and Laurencin, C.T., Electrospun poly (lactic acid-co-glycolic acid) scaffolds for skin tissue engineering. *Biomaterials*, 2008. 29(30): pp. 4100–4107.
149. Krishnan, R., Rajeswari, R., Venugopal, J., Sundarrajan, S., Sridhar, R., Shayanti, M., and Ramakrishna, S., Polysaccharide nanofibrous scaffolds as a model for in vitro skin tissue regeneration. *Journal of Materials Science: Materials in Medicine*, 2012. 23(6): pp. 1511–1519.
150. Sundaramurthi, D., Krishnan, U.M., and Sethuraman, S., Electrospun nanofibers as scaffolds for skin tissue engineering. *Polymer Reviews*, 2014. 54(2): pp. 348–376.
151. Babaeijandaghi, F., Shabani, I., Seyedjafari, E., Naraghi, Z.S., Vasei, M., Haddadi-Asl, V., Hesari, K.K., and Soleimani, M., Accelerated epidermal regeneration and improved dermal reconstruction achieved by polyethersulfone nanofibers. *Tissue Engineering Part A*, 2010. 16(11): pp. 3527–3536.
152. Ramakrishna, S., An introduction to electrospinning and nanofibers. *World Scientific*, 2005. 396.
153. Powell, H.M., Supp, D.M., and Boyce, S.T., Influence of electrospun collagen on wound contraction of engineered skin substitutes. *Biomaterials*, 2008. 29(7): pp. 834–843.
154. Zhang, Y.Z., Venugopal, J., Huang, Z.M., Lim, C.T., and Ramakrishna, S., Characterization of the surface biocompatibility of the electrospun PCL-collagen nanofibers using fibroblasts. *Biomacromolecules*, 2005. 6(5): pp. 2583–2589.
155. Qu, F., Pintauro, M.P., Haughan, J.E., Henning, E.A., Esterhai, J.L., Schaer, T.P., Mauck, R.L., and Fisher, M.B., Repair of dense connective tissues via biomaterial-mediated matrix reprogramming of the wound interface. *Biomaterials*, 2015. 39: pp. 85–94.
156. Sun, X., Lang, Q., Zhang, H., Cheng, L., Zhang, Y., Pan, G., Zhao, X., Yang, H., Zhang, Y., Santos, H.A., and Cui, W., Electrospun photocrosslinkable hydrogel fibrous scaffolds for rapid in vivo vascularized skin flap regeneration. *Advanced Functional Materials*, 2017. 27(2).
157. Zhong, S., Teo, W.E., Zhu, X., Beuerman, R.W., Ramakrishna, S., and Yung, L.Y.L., An aligned nanofibrous collagen scaffold by electrospinning and its effects on in vitro fibroblast culture. *Journal of Biomedical Materials Research Part A*, 2006. 79(3): pp. 456–463.
158. Kurpinski, K.T., Stephenson, J.T., Janairo, R.R.R., Lee, H., and Li, S., The effect of fiber alignment and heparin coating on cell infiltration into nanofibrous PLLA scaffolds. *Biomaterials*, 2010. 31(13): pp. 3536–3542.
159. Ma, B., Xie, J., Jiang, J., and Wu, J., Sandwich-type fiber scaffolds with square arrayed microwells and nanostructured cues as microskin grafts for skin regeneration. *Biomaterials*, 2014. 35(2): pp. 630–641.

160. Reis, T.C., Castleberry, S., Rego, A.M., Aguiar-Ricardo, A., and Hammond, P.T., Three-dimensional multilayered fibrous constructs for wound healing applications. *Biomaterials Science*, 2016. 4(2): pp. 319–330.
161. Yang, H., Kim, S., Huh, I., Kim, S., Lahiji, S.F., Kim, M., and Jung, H., Rapid implantation of dissolving microneedles on an electrospun pillar array. *Biomaterials*, 2015. 64: pp. 70–77.
162. Zhao, Q., Wang, S., Xie, Y., Zheng, W., Wang, Z., Xiao, L., Zhang, W., and Jiang, X., A Rapid screening method for wound dressing by cell-on-a-chip device. *Advanced Healthcare Materials*, 2012. 1(5): pp. 560–566.
163. Wang, Z., Cui, Y., Wang, J., Yang, X., Wu, Y., Wang, K., Gao, X., Li, D., Li, Y., Zheng, X.L., and Zhu, Y., The effect of thick fibers and large pores of electrospun poly (ε-caprolactone) vascular grafts on macrophage polarization and arterial regeneration. *Biomaterials*, 2014. 35(22): pp. 5700–5710.
164. Chen, M.C., Sun, Y.C., and Chen, Y.H., Electrically conductive nanofibers with highly oriented structures and their potential application in skeletal muscle tissue engineering. *Acta Biomaterialia*, 2013. 9(3): pp. 5562–5572.
165. Xie, J., MacEwan, M.R., Li, X., Sakiyama-Elbert, S.E., and Xia, Y., Neurite outgrowth on nanofiber scaffolds with different orders, structures, and surface properties. *ACS Nano*, 2009. 3(5): pp. 1151–1159.
166. Li, L., Zhou, G., Wang, Y., Yang, G., Ding, S., and Zhou, S., Controlled dual delivery of BMP-2 and dexamethasone by nanoparticle-embedded electrospun nanofibers for the efficient repair of critical-sized rat calvarial defect. *Biomaterials*, 2015. 37: pp. 218–229.
167. Liu, S., Qin, M., Hu, C., Wu, F., Cui, W., Jin, T., and Fan, C., Tendon healing and anti-adhesion properties of electrospun fibrous membranes containing bFGF loaded nanoparticles. *Biomaterials*, 2013. 34(19): pp. 4690–4701.
168. Xie, J., MacEwan, M.R., Ray, W.Z., Liu, W., Siewe, D.Y., and Xia, Y., Radially aligned, electrospun nanofibers as dural substitutes for wound closure and tissue regeneration applications. *ACS Nano*, 2010. 4(9): pp. 5027–5036.
169. Xue, J., Feng, B., Zheng, R., Lu, Y., Zhou, G., Liu, W., Cao, Y., Zhang, Y., and Zhang, W.J., Engineering ear-shaped cartilage using electrospun fibrous membranes of gelatin/polycaprolactone. *Biomaterials*, 2013. 34(11): pp. 2624–2631.
170. Liu, Z., Yu, N., Holz, F.G., Yang, F., and Stanzel, B.V., Enhancement of retinal pigment epithelial culture characteristics and subretinal space tolerance of scaffolds with 200 nm fiber topography. *Biomaterials*, 2014. 35(9): pp. 2837–2850.
171. Jia, J., Duan, Y.Y., Wang, S.H., Zhang, S.F., and Wang, Z.Y., Preparation and characterization of antibacterial silver-containing nanofibers for wound dressing applications. *Journal of US-China Medical Science*, 2007. 4(2): pp. 52–54.
172. Spasova, M., Paneva, D., Manolova, N., Radenkov, P., and Rashkov, I., Electrospun chitosan-coated fibers of poly (L-lactide) and poly (L-lactide)/poly (ethylene glycol): Preparation and characterization. *Macromolecular Bioscience*, 2008. 8(2): pp. 153–162.
173. Goh, Y.F., Shakir, I., and Hussain, R., Electrospun fibers for tissue engineering, drug delivery, and wound dressing. *Journal of Materials Science*, 2013. 48(8): pp. 3027–3054.
174. Singh, A.V., AS, A., N Gade, W., Vats, T., Lenardi, C., and Milani, P., Nanomaterials: New generation therapeutics in wound healing and tissue repair. *Current Nanoscience*, 2010. 6(6): pp. 577–586.

175. Abdelgawad, A.M., Hudson, S.M., and Rojas, O.J., Antimicrobial wound dressing nanofiber mats from multicomponent (chitosan/silver-NPs/polyvinyl alcohol) systems. *Carbohydrate Polymers*, 2014. 100: pp. 166–178.
176. Huang, Y., Zhong, Z., Duan, B., Zhang, L., Yang, Z., Wang, Y., and Ye, Q., Novel fibers fabricated directly from chitin solution and their application as wound dressing. *Journal of Materials Chemistry B*, 2014. 2(22): pp. 3427–3432.
177. Shahverdi, S., Hajimiri, M., Esfandiari, M.A., Larijani, B., Atyabi, F., Rajabiani, A., Dehpour, A.R., Gharehaghaji, A.A., and Dinarvand, R., Fabrication and structure analysis of poly (lactide-co-glycolic acid)/silk fibroin hybrid scaffold for wound dressing applications. *International Journal of Pharmaceutics*, 2014. 473(1–2): pp. 345–355.
178. Zhang, Y., Lim, C.T., Ramakrishna, S., and Huang, Z.M., Recent development of polymer nanofibers for biomedical and biotechnological applications. *Journal of Materials Science: Materials in Medicine*, 2005. 16(10): pp. 933–946.
179. Matthews, K.H., Drug delivery dressings D. Farrar (Ed.), *Advanced Wound Repair Therapies*, Woodhead Publishing Ltd., Cambridge, UK. 2011. pp. 361–394.
180. Zahedi, P., Rezaeian, I., Ranaei-Siadat, S.O., Jafari, S.H., and Supaphol, P., A review on wound dressings with an emphasis on electrospun nanofibrous polymeric bandages. *Polymers for Advanced Technologies*, 2010. 21(2): pp. 77–95.
181. Rayatpisheh, S., Heath, D.E., Shakouri, A., Rujitanaroj, P.O., Chew, S.Y., and Chan-Park, M.B., Combining cell sheet technology and electrospun scaffolding for engineered tubular, aligned, and contractile blood vessels. *Biomaterials*, 2014. 35(9): pp. 2713–2719.
182. Lee, Y.B., Jun, I., Bak, S., Shin, Y.M., Lim, Y.M., Park, H., and Shin, H., Reconstruction of vascular structure with multicellular components using cell transfer printing methods. *Advanced Healthcare Materials*, 2014. 3(9): pp. 1465–1474.
183. Ju, Y.M., San Choi, J., Atala, A., Yoo, J.J., and Lee, S.J., Bilayered scaffold for engineering cellularized blood vessels. *Biomaterials*, 2010. 31(15): pp. 4313–4321.
184. Ahn, H., Ju, Y.M., Takahashi, H., Williams, D.F., Yoo, J.J., Lee, S.J., Okano, T., and Atala, A., Engineered small diameter vascular grafts by combining cell sheet engineering and electrospinning technology. *Acta Biomaterialia*, 2015. 16: pp. 14–22.
185. Uttayarat, P., Perets, A., Li, M., Pimton, P., Stachelek, S.J., Alferiev, I., Composto, R.J., Levy, R.J., and Lelkes, P.I., Micropatterning of three-dimensional electrospun polyurethane vascular grafts. *Acta Biomaterialia*, 2010. 6(11): pp. 4229–4237.
186. Browning, M.B., Dempsey, D., Guiza, V., Becerra, S., Rivera, J., Russell, B., Höök, M., Clubb, F., Miller, M., Fossum, T., and Dong, J.F., Multilayer vascular grafts based on collagen-mimetic proteins. *Acta Biomaterialia*, 2012. 8(3): pp. 1010–1021.
187. Wang, Z., Zheng, W., Wu, Y., Wang, J., Zhang, X., Wang, K., Zhao, Q., Kong, D., Ke, T., and Li, C., Differences in the performance of PCL-based vascular grafts as abdominal aorta substitutes in healthy and diabetic rats. *Biomaterials Science*, 2016. 4(10): pp. 1485–1492.
188. Zhang, D., and Chang, J., Electrospinning of three-dimensional nanofibrous tubes with controllable architectures. *Nano Letters*, 2008. 8(10): pp. 3283–3287.
189. Chung, E., Ricles, L.M., Stowers, R.S., Nam, S.Y., Emelianov, S.Y., and Suggs, L.J., Multifunctional nanoscale strategies for enhancing and monitoring blood vessel regeneration. *NanoToday*, 2012. 7(6): pp. 514–531.

190. Dvir, T., Timko, B.P., Kohane, D.S., and Langer, R., Nanotechnological strategies for engineering complex tissues. *Nature Nanotechnology,* 2011. 6(1): p. 13.
191. Tsang, K.Y., Cheung, M.C., Chan, D., and Cheah, K.S., The developmental roles of the extracellular matrix: Beyond structure to regulation. *Cell and Tissue Research,* 2010. 339(1): p. 93.
192. Zhang, G., Drinnan, C.T., Geuss, L.R., and Suggs, L.J., Vascular differentiation of bone marrow stem cells is directed by a tunable three-dimensional matrix. *Acta Biomaterialia,* 2010. 6(9): pp. 3395–3403.
193. Hung, H.S., Chen, H.C., Tsai, C.H., and Lin, S.Z., Novel approach by nano-biomaterials in vascular tissue engineering. *Cell Transplantation,* 2011. 20(1): pp. 63–70.
194. Brozovich, F.V., Nicholson, C.J., Degen, C.V., Gao, Y.Z., Aggarwal, M., and Morgan, K.G., Mechanisms of vascular smooth muscle contraction and the basis for pharmacologic treatment of smooth muscle disorders. *Pharmacological Reviews,* 2016. 68(2): pp. 476–532.
195. Shimizu, N., Maruyama, T., Yoshikawa, N., Matsumiya, R., Ma, Y., Ito, N., Tasaka, Y., Kuribara-Souta, A., Miyata, K., Oike, Y., and Berger, S., A muscle-liver-fat signalling axis is essential for central control of adaptive adipose remodelling. *Nature Communications,* 2015. 6: p. 6693.
196. Senyo, S.E., Lee, R.T., and Kühn, B., Cardiac regeneration based on mechanisms of cardiomyocyte proliferation and differentiation. *Stem Cell Research,* 2014. 13(3): pp. 532–541.
197. Wolf, M.T., Dearth, C.L., Sonnenberg, S.B., Loboa, E.G., and Badylak, S.F., Naturally derived and synthetic scaffolds for skeletal muscle reconstruction. *Advanced Drug Delivery Reviews,* 2015. 84: pp. 208–221.
198. Baker, B.M., Gee, A.O., Metter, R.B., Nathan, A.S., Marklein, R.A., Burdick, J.A., and Mauck, R.L., The potential to improve cell infiltration in composite fiber-aligned electrospun scaffolds by the selective removal of sacrificial fibers. *Biomaterials,* 2008. 29(15): pp. 2348–2358.
199. McKeon-Fischer, K.D., Flagg, D.H., and Freeman, J.W., Coaxial electrospun poly (ε-caprolactone), multiwalled carbon nanotubes, and polyacrylic acid/polyvinyl alcohol scaffold for skeletal muscle tissue engineering. *Journal of Biomedical Materials Research Part A,* 2011. 99(3): pp. 493–499.
200. Liao, I.C., Liu, J.B., Bursac, N., and Leong, K.W., Effect of electromechanical stimulation on the maturation of myotubes on aligned electrospun fibers. *Cellular and Molecular Bioengineering,* 2008. 1(2–3): pp. 133–145.
201. Emmert, M.Y., Hitchcock, R.W., and Hoerstrup, S.P., Cell therapy, 3D culture systems and tissue engineering for cardiac regeneration. *Advanced Drug Delivery Reviews,* 2014. 69: pp. 254–269.
202. Visone, R., Gilardi, M., Marsano, A., Rasponi, M., Bersini, S., and Moretti, M., Cardiac meets skeletal: What's new in microfluidic models for muscle tissue engineering. *Molecules,* 2016. 21(9): p. 1128.
203. Jana, S., Levengood, S.K.L., and Zhang, M., Anisotropic materials for skeletal-muscle-tissue engineering. *Advanced Materials,* 2016. 28(48): pp. 10588–10612.
204. Fleischer, S., Shapira, A., Feiner, R., and Dvir, T., Modular assembly of thick multifunctional cardiac patches. *Proceedings of the National Academy of Sciences,* 2017. 114: pp. 1898–1903.
205. Cao, H., Liu, T., and Chew, S.Y., The application of nanofibrous scaffolds in neural tissue engineering. *Advanced Drug Delivery Reviews,* 2009. 61(12): pp. 1055–1064.

206. Li, X., Li, M., Sun, J., Zhuang, Y., Shi, J., Guan, D., Chen, Y., and Dai, J., Radially aligned electrospun fibers with continuous gradient of SDF1α for the guidance of neural stem cells. *Small*, 2016. 12(36): pp. 5009–5018.
207. Xie, J., MacEwan, M.R., Liu, W., Jesuraj, N., Li, X., Hunter, D., and Xia, Y., Nerve guidance conduits based on double-layered scaffolds of electrospun nanofibers for repairing the peripheral nervous system. *ACS Applied Materials & Interfaces*, 2014. 6(12): pp. 9472–9480.
208. Panseri, S., Cunha, C., Lowery, J., Del Carro, U., Taraballi, F., Amadio, S., Vescovi, A., and Gelain, F., Electrospun micro-and nanofiber tubes for functional nervous regeneration in sciatic nerve transections. *BMC Biotechnology*, 2008. 8(1): p. 39.
209. Meiners, S., Ahmed, I., Ponery, A.S., Amor, N., Harris, S.L., Ayres, V., Fan, Y., Chen, Q., Delgado-Rivera, R., and Babu, A.N., Engineering electrospun nanofibrillar surfaces for spinal cord repair: A discussion. *Polymer International*, 2007. 56(11): pp. 1340–1348.
210. Carlson, A.L., Bennett, N.K., Francis, N.L., Halikere, A., Clarke, S., Moore, J.C., Hart, R.P., Paradiso, K., Wernig, M., Kohn, J., and Pang, Z.P., Generation and transplantation of reprogrammed human neurons in the brain using 3D microtopographic scaffolds. *Nature Communications*, 2016. 7: p. 10862.
211. Jang, J.H., Castano, O., and Kim, H.W., Electrospun materials as potential platforms for bone tissue engineering. *Advanced Drug Delivery Reviews*, 2009. 61(12): pp. 1065–1083.
212. Shao, S., Zhou, S., Li, L., Li, J., Luo, C., Wang, J., Li, X., and Weng, J., Osteoblast function on electrically conductive electrospun PLA/MWCNTs nanofibers. *Biomaterials*, 2011. 32(11): pp. 2821–2833.
213. Luo, C., Li, L., Li, J., Yang, G., Ding, S., Zhi, W., Weng, J., and Zhou, S., Modulating cellular behaviors through surface nanoroughness. *Journal of Materials Chemistry*, 2012. 22(31): pp. 15654–15664.
214. Ding, S., Li, J., Luo, C., Li, L., Yang, G., and Zhou, S., Synergistic effect of released dexamethasone and surface nanoroughness on mesenchymal stem cell differentiation. *Biomaterials Science*, 2013. 1(10): pp. 1091–1100.
215. Ding, S., Li, L., Liu, X., Yang, G., Zhou, G., and Zhou, S., A nano-micro alternating multilayer scaffold loading with rBMSCs and BMP-2 for bone tissue engineering. *Colloids and Surfaces B: Biointerfaces*, 2015. 133: pp. 286–295.
216. Ruckh, T.T., Kumar, K., Kipper, M.J., and Popat, K.C., Osteogenic differentiation of bone marrow stromal cells on poly (ε-caprolactone) nanofiber scaffolds. *Acta Biomaterialia*, 2010. 6(8): pp. 2949–2959.
217. Shin, M., Yoshimoto, H., and Vacanti, J.P., In vivo bone tissue engineering using mesenchymal stem cells on a novel electrospun nanofibrous scaffold. *Tissue Engineering*, 2004. 10(1–2): pp. 33–41.
218. Tautzenberger, A., Kovtun, A., and Ignatius, A., Nanoparticles and their potential for application in bone. *International Journal of Nanomedicine*, 2012. 7: p. 4545.
219. Arora, P., Sindhu, A., Dilbaghi, N., Chaudhury, A., Rajakumar, G., and Rahuman, A.A., Nano-regenerative medicine towards clinical outcome of stem cell and tissue engineering in humans. *Journal of Cellular and Molecular Medicine*, 2012. 16(9): pp. 1991–2000.
220. Cade, D., Ramus, E., Rinaudo, M., Auzély-Velty, R., Delair, T., and Hamaide, T., Tailoring of bioresorbable polymers for elaboration of sugar-functionalized nanoparticles. *Biomacromolecules*, 2004. 5(3): pp. 922–927.

221. Wang, H., Leeuwenburgh, S.C., Li, Y., and Jansen, J.A., The use of micro-and nanospheres as functional components for bone tissue regeneration. *Tissue Engineering Part B: Reviews*, 2011. 18(1): pp. 24–39.
222. Jiang, W., Li, L., Zhang, D., Huang, S., Jing, Z., Wu, Y., Zhao, Z., Zhao, L., and Zhou, S., Incorporation of aligned PCL–PEG nanofibers into porous chitosan scaffolds improved the orientation of collagen fibers in regenerated periodontium. *Acta Biomaterialia*, 2015. 25: pp. 240–252.
223. Xie, J., Li, X., Lipner, J., Manning, C.N., Schwartz, A.G., Thomopoulos, S., and Xia, Y., "Aligned-to-random" nanofiber scaffolds for mimicking the structure of the tendon-to-bone insertion site. *Nanoscale*, 2010. 2(6): pp. 923–926.
224. Chen, Z., Chen, Z., Zhang, A., Hu, J., Wang, X., and Yang, Z., Electrospun nanofibers for cancer diagnosis and therapy. *Biomaterials Science*, 2016. 4(6): pp. 922–932.
225. Zhang, N., Deng, Y., Tai, Q., Cheng, B., Zhao, L., Shen, Q., He, R., Hong, L., Liu, W., Guo, S., and Liu, K., Electrospun TiO_2 nanofiber-based cell capture assay for detecting circulating tumor cells from colorectal and gastric cancer patients. *Advanced Materials*, 2012. 24(20): pp. 2756–2760.
226. Ma, L., Yang, G., Wang, N., Zhang, P., Guo, F., Meng, J., Zhang, F., Hu, Z., Wang, S., and Zhao, Y., Trap effect of three-dimensional fibers network for high efficient cancer-cell capture. *Advanced Healthcare Materials*, 2015. 4(6): pp. 838–843.
227. Zhao, L., Lu, Y.T., Li, F., Wu, K., Hou, S., Yu, J., Shen, Q., Wu, D., Song, M., OuYang, W.H., and Luo, Z., High-purity prostate circulating tumor cell isolation by a polymer nanofiber-embedded microchip for whole exome Sequencing. *Advanced Materials*, 2013. 25(21): pp. 2897–2902.
228. Hou, S., Zhao, L., Shen, Q., Yu, J., Ng, C., Kong, X., Wu, D., Song, M., Shi, X., Xu, X., and OuYang, W.H., Polymer nanofiber-embedded microchips for detection, isolation, and molecular analysis of single circulating melanoma cells. *Angewandte Chemie International Edition*, 2013. 52(12): pp. 3379–3383.
229. Xu, G., Tan, Y., Xu, T., Yin, D., Wang, M., Shen, M., Chen, X., Shi, X., and Zhu, X., Hyaluronic acid-functionalized electrospun PLGA nanofibers embedded in a microfluidic chip for cancer cell capture and culture. *Biomaterials Science*, 2017. 5(4): pp. 752–761.
230. Wang, X., Wang, X., Wang, X., Chen, F., Zhu, K., Xu, Q., and Tang, M., Novel electrochemical biosensor based on functional composite nanofibers for sensitive detection of p53 tumor suppressor gene. *Analytica Chimica Acta*, 2013. 765: pp. 63–69.
231. Wang, X., Shu, G., Gao, C., Yang, Y., Xu, Q., and Tang, M., Electrochemical biosensor based on functional composite nanofibers for detection of K-ras gene via multiple signal amplification strategy. *Analytical Biochemistry*, 2014. 466: pp. 51–58.
232. Tevis, K.M., Colson, Y.L., and Grinstaff, M.W., Embedded spheroids as models of the cancer microenvironment. *Advanced Biosystems*, 2017. 1(10).
233. Sica, A., Larghi, P., Mancino, A., Rubino, L., Porta, C., Totaro, M.G., Rimoldi, M., Biswas, S.K., Allavena, P., and Mantovani, A., October. Macrophage polarization in tumour progression. *Seminars in Cancer Biology*, 2008, 18: pp. 349–355.
234. Thoma, C.R., Zimmermann, M., Agarkova, I., Kelm, J.M., and Krek, W., 3D cell culture systems modeling tumor growth determinants in cancer target discovery. *Advanced Drug Delivery Reviews*, 2014. 69: pp. 29–41.

235. Hartman, O., Zhang, C., Adams, E.L., Farach-Carson, M.C., Petrelli, N.J., Chase, B.D., and Rabolt, J.F., Biofunctionalization of electrospun PCL-based scaffolds with perlecan domain IV peptide to create a 3-D pharmacokinetic cancer model. *Biomaterials*, 2010. 31(21): pp. 5700–5718.
236. Fischer, S.N., Johnson, J.K., Baran, C.P., Newland, C.A., Marsh, C.B., and Lannutti, J.J., Organ-derived coatings on electrospun nanofibers as ex vivo microenvironments. *Biomaterials*, 2011. 32(2): pp. 538–546.
237. Santoro, M., Lamhamedi-Cherradi, S.E., Menegaz, B.A., Ludwig, J.A., and Mikos, A.G., Flow perfusion effects on three-dimensional culture and drug sensitivity of Ewing sarcoma. *Proceedings of the National Academy of Sciences*, 2015. 112(33): pp. 10304–10309.
238. Nelson, M.T., Short, A., Cole, S.L., Gross, A.C., Winter, J., Eubank, T.D., and Lannutti, J.J., Preferential, enhanced breast cancer cell migration on biomimetic electrospun nanofiber 'cell highways'. *BMC Cancer*, 2014. 14(1): pp. 825.
239. Jain, A., Betancur, M., Patel, G.D., Valmikinathan, C.M., Mukhatyar, V.J., Vakharia, A., Pai, S.B., Brahma, B., MacDonald, T.J., and Bellamkonda, R.V., Guiding intracortical brain tumour cells to an extracortical cytotoxic hydrogel using aligned polymeric nanofibres. *Nature Materials*, 2014. 13(3): pp. 308.
240. Kim, J.H., Park, K., Nam, H.Y., Lee, S., Kim, K., and Kwon, I.C., Polymers for bioimaging. *Progress in Polymer Science*, 2007. 32(8–9): pp. 1031–1053.
241. Böck, J.C., Kaufmann, F., and Felix, R., Comparison of gadolinium-DTPA and macromolecular gadolinium-DTPA-polylysine for contrast-enhanced pulmonary time-of-flight magnetic resonance angiography. *Investigative Radiology*, 1996. 31(10): pp. 652–657.
242. Shiraishi, K., Kawano, K., Minowa, T., Maitani, Y., and Yokoyama, M., Preparation and in vivo imaging of PEG-poly (L-lysine)-based polymeric micelle MRI contrast agents. *Journal of Controlled Release*, 2009. 136(1): pp. 14–20.
243. Kim, K.S., Park, W., Hu, J., Bae, Y.H., and Na, K., A cancer-recognizable MRI contrast agents using pH-responsive polymeric micelle. *Biomaterials*, 2014. 35(1): pp. 337–343.
244. Van Reis, R., and Zydney, A., Membrane separations in biotechnology. *Current Opinion in Biotechnology*, 2001. 12(2): pp. 208–211.
245. Steiner, R., Microfiltration and ultrafiltration-principles and applications. *Chemie Ingenieur Technik*, 1997. 69(10): pp. 1479–1479.
246. McGregor, W.C. *Membrane separation in biotechnology*. New York: Marcel Dekker, 1986.
247. Kuriyel, R., and Zydney, A.L., Sterile filtration and virus filtration. In *Downstream Processing of Proteins*. Humana Press, 2000. pp. 185–194.
248. Sundaram, S., Auriemma, M., Howard, G., Brandwein, H., and Leo, F., Application of membrane filtration for removal of diminutive bioburden organisms in pharmaceutical products and processes. *PDA Journal of Pharmaceutical Science and Technology*, 1999. 53(4): pp. 186–201.
249. Liu, S., Carroll, M., Iverson, R., Valera, C., Vennari, J., Turco, K., Piper, R., Kiss, R., and Lutz, H., Development and qualification of a novel virus removal filter for cell culture applications. *Biotechnology Progress*, 2000. 16(3): pp. 425–434.
250. Alves, A.B., Morao, A., and Cardoso, J.P., Isolation of antibiotics from industrial fermentation broths using membrane technology. *Desalination*, 2002. 148(1–3): pp. 181–186.

251. Nabais, A.M.A., and Cardoso, J.P., Purification of benzylpenicillin filtered broths by ultrafiltration and effect on solvent extraction. *Bioprocess Engineering*, 1999. 21(2): pp. 157–163.
252. Morao, A., Alves, A.B., and Cardoso, J.P., Ultrafiltration of demethylchlortetracycline industrial fermentation broths. *Separation and Purification Technology*, 2001. 22: pp. 459–466.
253. Schutyser, M., Rupp, R., Wideman, J., and Belfort, G., Dean vortex membrane microfiltration and diafiltration of rbdnf e. coliinclusion bodies. *Biotechnology Progress*, 2002. 18(2): pp. 322–329.
254. Baruah, G.L., and Belfort, G., Optimized recovery of monoclonal antibodies from transgenic goat milk by microfiltration. *Biotechnology and Bioengineering*, 2004. 87(3): pp. 274–285.
255. Levesley, J.A., and Hoare, M., The effect of high frequency backflushing on the microfiltration of yeast homogenate suspensions for the recovery of soluble proteins. *Journal of Membrane Science*, 1999. 158(1–2): pp. 29–39.
256. Vogel, J.H., and Kroner, K.H., Controlled shear filtration: A novel technique for animal cell separation. *Biotechnology and Bioengineering*, 1999. 63(6): pp. 663–674.
257. Jaffrin, M.Y., Ding, L.H., Akoum, O., and Brou, A., A hydrodynamic comparison between rotating disk and vibratory dynamic filtration systems. *Journal of Membrane Science*, 2004. 242(1–2): pp. 155–167.
258. Lee, S.S., Burt, A., Russotti, G., and Buckland, B., Microfiltration of recombinant yeast cells using a rotating disk dynamic filtration system. *Biotechnology and Bioengineering*, 1995. 48(4): pp. 386–400.
259. Parnham, C.S., and Davis, R.H., Protein recovery from cell debris using rotary and tangential crossflow microfiltration. *Biotechnology and Bioengineering*, 1995. 47(2): pp. 155–164.
260. Luque, S., Mallubhotla, H., Gehlert, G., Kuriyel, R., Dzengeleski, S., Pearl, S., and Belfort, G., A new coiled hollow-fiber module design for enhanced microfiltration performance in biotechnology. *Biotechnology and Bioengineering*, 1999. 65(3): pp. 247–257.
261. Meacle, F., Aunins, A., Thornton, R., and Lee, A., Optimization of the membrane purification of a polysaccharide–protein conjugate vaccine using backpulsing. *Journal of Membrane Science*, 1999. 161(1–2): pp. 171–184.
262. Wakeman, R.J., and Williams, C.J., Additional techniques to improve microfiltration. *Separation and Purification Technology*, 2002: pp. 26(1). 3–18.
263. Aimar, P., Meireles, M., Bacchin, P., and Sanchez, V., Fouling and concentration polarisation in ultrafiltration and microfiltration. In *Membrane processes in separation and purification*. Springer, Dordrecht, 1994. pp. 27–57.
264. van Reis, R., Gadam, S., Frautschy, L.N., Orlando, S., Goodrich, E.M., Saksena, S., Kuriyel, R., Simpson, C.M., Pearl, S., and Zydney, A.L., High performance tangential flow filtration. *Biotechnology and Bioengineering*, 1997. 56(1): pp. 71–82.
265. Nyström, M., Aimar, P., Luque, S., Kulovaara, M., and Metsämuuronen, S., Fractionation of model proteins using their physiochemical properties. *Colloids and Surfaces A: Physicochemical and Engineering Aspects*, 1998. 138(2–3): pp. 185–205.
266. Baruah, G.L., Venkiteshwaran, A., and Belfort, G., Global model for optimizing crossflow microfiltration and ultrafiltration processes: A new predictive and design tool. *Biotechnology Progress*, 2005. 21(4): pp. 1013–1025.

267. Belfort, G., and Heath, C.A. New developments in membrane bioreactors. In: Crespo JG, Böddeker KW, editors. *Membrane processes in separation and purification.* NATO ASI Series E: Applied Science Dordrecht: Kluwer Academic Publishers, 1993. pp. 127–149.
268. Cheryan, M., and Mehaia, M.A. Membrane bioreactors. In: Mc Gregor, editor. *Membrane separation in biotechnology.* New York: Marcel Dekker, 1986. pp. 255–301.
269. Giorno, L., and Drioli, E. Biocatalytic membrane reactors: Applications and perspectives. *TIBTECH.* 2000. 18: pp. 339–349.
270. Giorno, L., De Bartolo, L., and Drioli, E., Membrane bioreactors for biotechnology and medical applications. *Membrane Science and Technology,* 2003. 8: pp. 187–217.
271. Yang, F., Weber, T.W., Gainer, J.L., and Carta, G., Synthesis of lovastatin with immobilized Candida rugosa lipase in organic solvents: Effects of reaction conditions on initial rates. *Biotechnology and Bioengineering,* 1997. 56(6): pp. 671–680.
272. Lopez, J.L., and Matson, S.L., A multiphase/extractive enzyme membrane reactor for production of diltiazem chiral intermediate. *Journal of Membrane Science,* 1997. 125(1): pp. 189–211.
273. Goulas, A.K., Cooper, J.M., Grandison, A.S., and Rastall, R.A., Synthesis of isomaltooligosaccharides and oligodextrans in a recycle membrane bioreactor by the combined use of dextransucrase and dextranase. *Biotechnology and Bioengineering,* 2004. 88(6): pp. 778–787.
274. Belleville, M.P., Lozano, P., Iborra, J.L., and Rios, G.M., Preparation of hybrid membranes for enzymatic reaction. *Separation and Purification Technology,* 2001. 25(1–3): pp. 229–233.
275. Chung, T.P., Wu, P.C., and Juang, R.S., Use of microporous hollow fibers for improved biodegradation of high-strength phenol solutions. *Journal of Membrane Science,* 2005. 258(1–2): pp. 55–63.
276. Nakajima, M., Watanabe, A., Jimbo, N., Nishizawa, K., and Nakao, S.I., Forced-flow bioreactor for sucrose inversion using ceramic membrane activated by silanization. *Biotechnology and Bioengineering,* 1989. 33(7): pp. 856–861.
277. Heath, C. A., and Belfort, G., Immobilization of suspended mammalian cells: Analysis of hollow fiber and microcapsule bioreactors. *Advances in Biochemical Engineering/Biotechnology,* 1988. 34: pp. 1–31.
278. Custer, L., Physiological studies of hybridoma cultivation in hollow fiber bioreactors, PhD Thesis, University of California at Berkley, CA, 1988.
279. Butterfield, D.A., Bhattacharyya, D., Daunert, S., and Bachas, L., Catalytic biofunctional membranes containing site-specifically immobilized enzyme arrays: A review. *Journal of Membrane Science,* 2001. 181(1): pp. 29–37.
280. Yang, W., Cicek, N., and Ilg, J., State-of-the-art of membrane bioreactors: Worldwide research and commercial applications in North America. *Journal of Membrane Science,* 2006. 270(1–2): pp. 201–211.
281. Adamson, S.R., Experiences of virus, retrovirus and retrovirus-like particles in Chinese hamster ovary (CHO) and hybridoma cells used for production of protein therapeutics. *Developments in Biological Standardization,* 1998. 93: pp. 89–96.
282. DiLeo, A.J., Allegrezza Jr, A.E., and Builder, S.E., High resolution removal of virus from protein solutions using a membrane of unique structure. *Nature Biotechnology,* 1992. 10(2): p. 182.

283. Van Reis, R., Brake, J.M., Charkoudian, J., Burns, D.B., and Zydney, A.L., High-performance tangential flow filtration using charged membranes. *Journal of Membrane Science,* 1999. 159(1–2): pp. 133–142.
284. Burns, D.B., and Zydney, A.L., Effect of solution pH on protein transport through ultrafiltration membranes. *Biotechnology and Bioengineering,* 1999. 64(1): pp. 27–37.
285. Recio, I., and Visser, S., Two ion-exchange chromatographic methods for the isolation of antibacterial peptides from lactoferrin: In situ enzymatic hydrolysis on an ion-exchange membrane. *Journal of Chromatography A,* 1999. 831(2): pp. 191–201.
286. Charcosset, C., Purification of proteins by membrane chromatography. *Journal of Chemical Technology and Biotechnology,* 1998. 71(2): pp. 95–110.
287. Klein, E., Affinity membranes: A 10-year review. *Journal of Membrane Science,* 2000. 179(1–2): pp. 1–27.
288. Ghosh, R., Protein separation using membrane chromatography: Opportunities and challenges. *Journal of Chromatography A,* 2002. 952(1–2): pp. 13–27.
289. Bueno, S.M., Haupt, K., and Vijayalakshmi, M.A., Separation of immunoglobulin G from human serum by pseudobioaffinity chromatography using immobilized L-histidine in hollow fibre membranes. *Journal of Chromatography B: Biomedical Sciences and Applications,* 1995. 667(1): pp. 57–67.
290. Finger, U.B., Thömmes, J., Kinzelt, D., and Kula, M.R., Application of thiophilic membranes for the purification of monoclonal antibodies from cell culture media. *Journal of Chromatography B: Biomedical Sciences and Applications,* 1995. 664(1): pp. 69–78.
291. Cattoli, F., and Sarti, G.C., Separation of MBP fusion proteins through affinity membranes. *Biotechnology Progress,* 2002. 18(1): pp. 94–100.
292. Karger, A., Bettin, B., Granzow, H., and Mettenleiter, T.C., Simple and rapid purification of alphaherpes viruses by chromatography on a cation exchange membrane. *Journal of Virological Methods,* 1998. 70(2): pp. 219–224.
293. Charlton, H.R., Relton, J.M., and Slater, N.K., Characterisation of a generic monoclonal antibody harvesting system for adsorption of DNA by depth filters and various membranes. *Bioseparation,* 1999. 8(6): pp. 281–291.
294. Belanich, M., Cummings, B., Grob, D., Klein, J., O'Connor, A., Yarosh, D., Reduction of endotoxin in a protein mixture using strong anion-exchange membrane absorption. *Pharmaceutical Technology,* 1996. pp. 142–150.
295. Brandt, S., Goffe, R.A., Kessler, S.B., O'Connor, J.L., and Zale, S.E., Membrane-based affinity technology for commercial scale purifications. *Nature Biotechnology,* 1988. 6(7): pp. 779.
296. Drioli, E., Curcio, E., and Di Profio, G., State of the art and recent progresses in membrane contactors. *Chemical Engineering Research and Design,* 2005. 83(3): pp. 223–233.
297. Joscelyne, S.M., and Trägårdh, G., Membrane emulsification—A literature review. *Journal of Membrane Science,* 2000. 169(1): pp. 107–117.
298. Charcosset, C., Limayem, I., and Fessi, H., The membrane emulsification process—A review. *Journal of Chemical Technology and Biotechnology,* 2004. 79(3): pp. 209–218.
299. Jia, Z., Liu, Z., and He, F., Synthesis of nanosized $BaSO_4$ and $CaCO_3$ particles with a membrane reactor: Effects of additives on particles. *Journal of Colloid and Interface Science,* 2003. 266(2): pp. 322–327.

300. Wang, Y., Zhang, C., Bi, S., and Luo, G., Preparation of ZnO nanoparticles using the direct precipitation method in a membrane dispersion micro-structured reactor. *Powder Technology*, 2010. 202(1–3): pp. 130–136.
301. Altenbach-Rehm, J., Suzuki, K., Schubert, H., Production of O/W-emulsions with narrow droplet size distribution by repeated premix membrane emulsification. 3ième Congrès Mondial de l'Emulsion, 24–27 September 2002, Lyon, France.
302. Ma, G.H., An, C.J., Yuyama, H., Su, Z.G., and Omi, S., Synthesis and characterization of polyurethaneurea–vinyl polymer (PUU–VP) uniform hybrid microspheres by SPG emulsification technique and subsequent suspension polymerization. *Journal of Applied Polymer Science*, 2003. 89(1): pp. 163–178.
303. Liu, R., Ma, G., Meng, F.T., and Su, Z.G., Preparation of uniform-sized PLA microcapsules by combining Shirasu Porous Glass membrane emulsification technique and multiple emulsion-solvent evaporation method. *Journal of Controlled Release*, 2005. 103(1): pp. 31–43.
304. Omi, S., Taguchi, T., Nagai, M., and Ma, G.H., Synthesis of 100 μm uniform porous spheres by SPG emulsification with subsequent swelling of the droplets. *Journal of Applied Polymer Science*, 1997. 63(7): pp. 931–942.
305. Vladisavljević, G.T., and Williams, R.A., Recent developments in manufacturing emulsions and particulate products using membranes. *Advances in Colloid and Interface Science*, 2005. 113(1): pp. 1–20.
306. Charcosset, C., and Fessi, H., Preparation of solid lipid nanoparticles with a membrane contactor. *Journal of Control Release*, 2005a. 108: pp. 112–120.
307. Charcosset, C., and Fessi, H., Preparation of nanoparticles with a membrane contactor. *Journal of Membrane Science*, 2005b. 266(1–2): pp. 115–120.
308. Zhu, J., and Barrow, D., Analysis of droplet size during crossflow membrane emulsification using stationary and vibrating micromachined silicon nitride membranes. *Journal of Membrane Science*, 2005. 261(1–2): pp. 136–144.
309. Vladisavljević, G.T., Shimizu, M., and Nakashima, T., Preparation of monodisperse multiple emulsions at high production rates by multi-stage premix membrane emulsification. *Journal of Membrane Science*, 2004. 244(1–2): pp. 97–106.
310. Charcosset, C., Membrane processes in biotechnology: An overview. *Biotechnology Advances*, 2006. 24(5): pp. 482–492.

Appendix

Specific Surface Area Analysis

The nitrogen adsorption and desorption isotherm for electrospun PVA nanofiber membranes are shown in Figure A.1. Adsorption isotherms show the quantity of molecules adsorbed on the surface of a solid as a function of equilibrium pressure at a given (constant) temperature.

It is observed from Figure A.1 that initially the adsorption volume quickly raises at lower relative pressure, once the monolayer development of the adsorbed molecules is achieved and multilayer development begins to occur in matching to the sharp-knee of the isotherm that indicates that the isotherm graphs are of Type IV, which is characteristics of the mesoporous structure.

The shapes of the hysteresis loop are also associated with the different pore shapes, for the case of nonporous material, those desorption isotherm curves repeat the adsorption curves. Nevertheless, for the macroporous and mesoporous, desorption isotherm curve does not repeat the adsorption curve causing in a wide loop as seen from the isotherm graphs (Figure A.1a and A.1b). The pore volume decreased as the electrospun deposition time increases from 25 to 60 min. In addition, the BET study indicated that the specific surface area of the e-PVA membranes was 21.8, 27.3, 34.2, and 53.5 m^2/g for electrospinning durations of 25, 35, 45, and 60, respectively. The specific surface areas were increased considerably from 21.8 to 53.5m^2/g as the electrospinning duration was increased from 25 to 60 min. This improvement in surface area was explained due to increasing in the entanglements of the fibers and due to an increasing of the depth/flowing channel within the membrane matrix as the thickness of the membrane increased. In other words, increasing the electrospinning duration allows the entanglement of more fibers onto the collector, which in turn permits increase in the thickness as well as the depth of the membrane pores. So, the key factor to increase the specific surface in this study was explained as the increase in membrane channel due to the deposition of more entangled fibers onto the collector plate. The total pore volume of the membranes has found to be in the range from 0.06 to 0.28 cm^3/g.

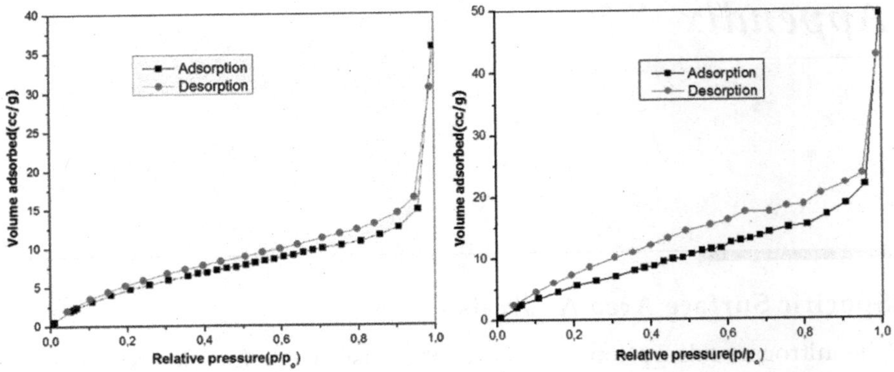

FIGURE A.1
Adsorption desorption isotherm graph of ePVA membranes at (a) 45 min and (b) 60 min process times.

X-Ray Powder Diffractometer (XRD)

XRD patterns of the ePVA nanofiber membranes deposited at various times are shown in Figure A.2. No sharp peak is detected for ePVA 25 and ePVA 35 on the XRD pattern. This indicates that these membranes are amorphous. As shown in Figure A.2, two weak peaks appeared for ePVA 45 and ePVA 60 at

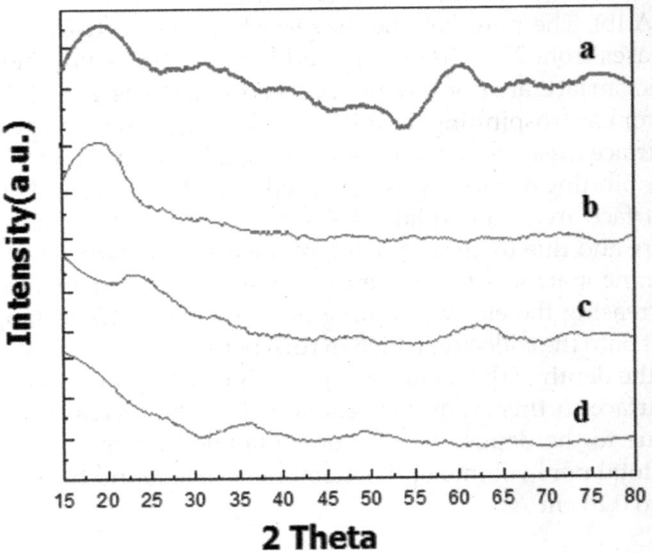

FIGURE A.2
XRD patterns of ePVA fibers using 10% w/v at different deposition times, (a) 60 min, (b) 45 min, (c) 35 min, and (d) 25 min.

around 2θ = 20°, analogous with (101) plane of semi-crystalline of polyvinyl alcohol membranes [1], and this may be due to the entanglement of more fibers as the time of deposition increases.

Cross-Linking e-PVA Nanofiber Membranes

The second objective of this work was to evaluate the effect of cross-linking on the properties the membranes. When the electrospun PVA membrane is immersed in water or used for water treatment applications, it can be dissolved slowly. In other words, the specific nanofibrous structures of e-PVA membranes are not stable in aqueous condition. It is already well-known that polyvinyl alcohol can be cross-linked chemically with a range of aldehydes, such as glutaraldehyde and glyoxal [2]. The interaction is because of the development of acetal-bridges among the –OH within PVA and the aldehyde molecules [3,4]. Crosslinking of electrospun membrane was done at room temperature. Acetone, which is a water-miscible and a non-solvent for PVA, was mixed with hydrochloric acid (35 w%) and gultaraldehyde aqueous solutions (25 wt.%) to prepare the crosslinking solution. The procedure of crosslinking solution preparation is clearly presented in Figure A.3. No evidence of shrinkages was shown after the e-PVA membrane was dipped into crosslinking solutions.

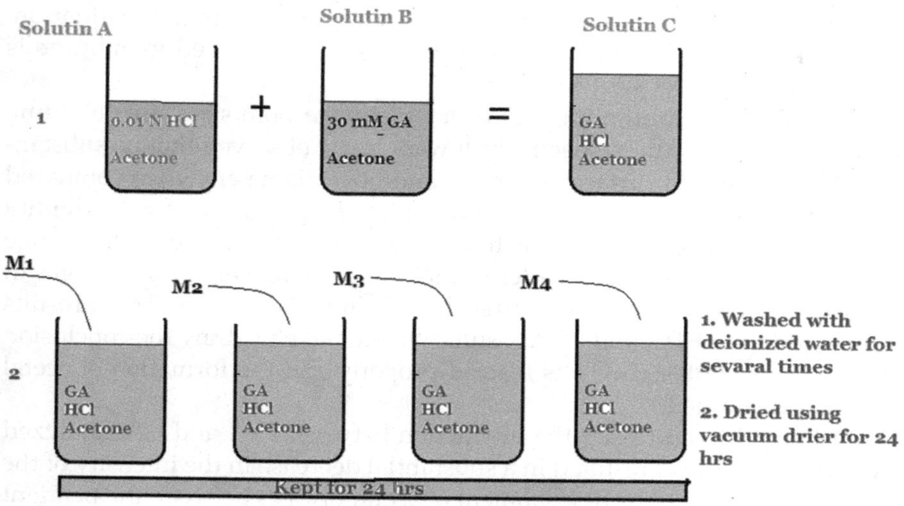

FIGURE A.3
Preparation of cross-linking solution (1) and immersing the membranes into the solution (2).

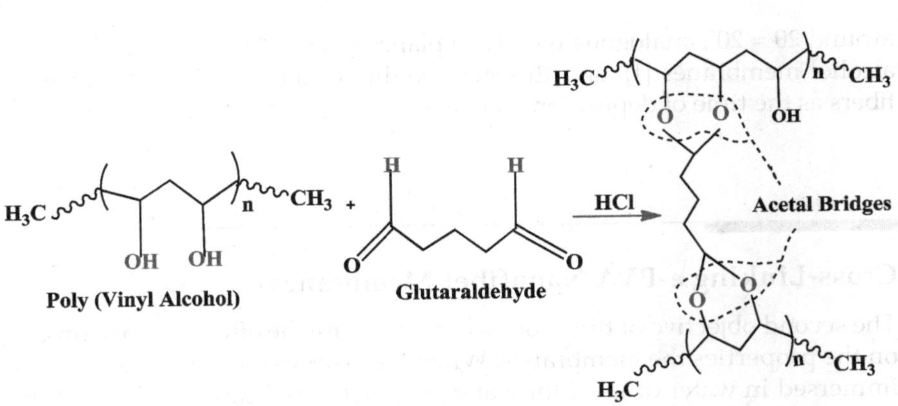

FIGURE A.4
Schematic of cross-linked PVA formed by chemical reaction of PVA and glutaraldehyde catalysed by hydrochloric acid.

The tested ePVA membranes were cross-linked using a mixture of acetone and 30 mM glutaraldehyde at 25 °C for 24 hrs [5]. Figure A.4 shows the predicted chemical cross-linking reaction among the polyvinyl alcohol chains and glutaraldehyde catalysed using HCl [6].

The crosslinked ePVA membranes were rinsed numerous times and soaked in water for 48 hours and then dried. As observed from the images Figure A.5a–d, the surface pore size and the diameters of each fiber were measured by using Image J from FESM images. The results indicated that the diameters of fibers were ranging from 69 to 200 nm with the average fiber diameter being 98 nm. The membrane fabricated at 45 min electrospinning duration was selected for Image J evaluation. The surface pore diameter distribution of the cross-linked membrane is presented in Figure A.6.

The maximum, minimum, and average surface pore sizes are 261.7 nm, 55.5 nm, and 125.9 nm, respectively. It was clearly observed that no substantial changes both in surface pore sizes and fiber diameters when compared with the non-cross-linked fibers, but uniform arrangement and rigidity of the fibers was observed, which may be due to the strong acetal bridge between the PVA monomers after crosslinking. Moreover, no shrinkages in the membranes were observed after the crosslinking process. These results agreed with the FTIR and TGA results, which make us draw the conclusion that the cross linker (GA) has reacted properly and the formation of acetal bridge is confirmed.

As shown in Figure A.7, the interaction between PVA and GA catalyzed by hydrochloric acid resulted in a substantial decrease in the intensity of the O–H peak, showing the development of acetal-bridges between the pendant hydroxyl groups of PVA chains [6]. Table A.1 shows the typical band assignment of ePVA cross-linked with GA. As shown in Figure A.7, the broad

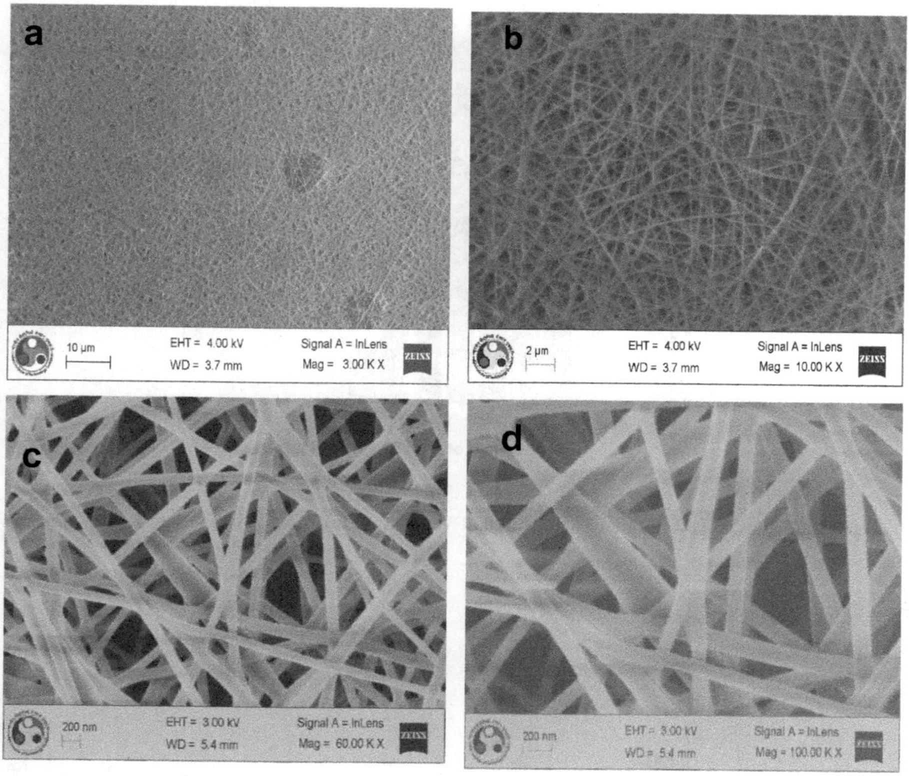

FIGURE A.5
FESEM images of crosslinked ePVA membranes with glutaraldehyde (ePVA/GA); (a) 3 KX magnified, (b) 10 KX magnified, (c) 50 KX magnified, and (d) 100 KX magnified.

bands observed at 3363 cm^{-1} are assigned to –OH stretching because of the presence of the strong bond (hydrogen bonding) of intramolecular and intermolecular type. The distinguishing bands at 1095, 1430, and 2947 cm^{-1} were attributed to the C–O stretching, C–H bending, and C–H stretching of PVA, respectively. The band observed at 1714 cm^{-1} may be due to the C = O stretching bands of remaining acetate group, residual after the synthesis of PVA during polyvinyl acetate hydrolysis process. The spectra show that there is no change in the molecular species and their interconnectivity when the process time is varied from 25 to 60 min.

The presence of aldehyde peaks (ν = 1643 cm^{-1}) could be because of the partial reaction of glutaraldehyde with the hydroxyl groups within the polyvinyl alcohol throughout the cross-linked network development. One aldehyde group could react with –OH groups within the polymeric chain by means of developing hemi-acetal structures due to its bifunctional cross-linker property, whereas the other one does not interact could be related with some kinetics drawbacks.

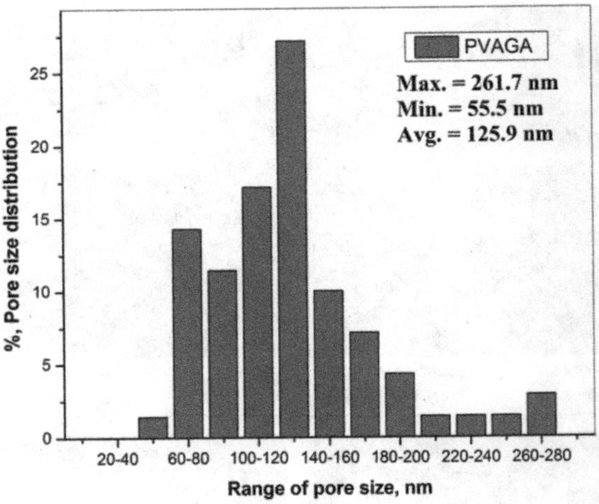

FIGURE A.6
% Surface pore size diameter distribution of ePVA/GA membranes obtained at 45 min electrospinning duration.

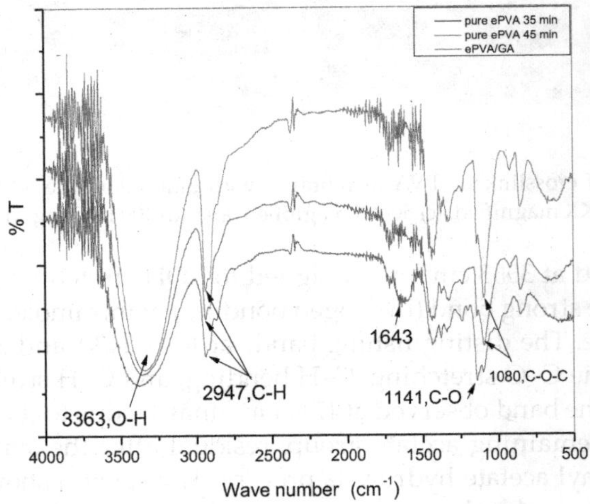

FIGURE A.7
FTIR results of electrospun PVA cross-linked with glutaraldehyde (ePVA/GA).

TABLE A.1

Characteristic Bands of ePVA Crosslinked with GA and Their Assignments

Range of Wavelengths (cm^{-1})	Assignment	Wave Number (cm^{-1})	Reference
3000–2850	C–H stretch	2947	–
3500–3200	O–H stretch, H–bonded	3350	–
1670–1640	Carboxylic groups	1643	–
1150–1085	C–O–C	1080	[3]
1320–1000	C–O (Crystallinity)	1141	[6,7]

The Thickness Measurement of the Nanofiber Membranes

The images of the ePVA fibers are captured at 25 min, 35 min, 45 min and 60 min. These are shown in Figure A.8.

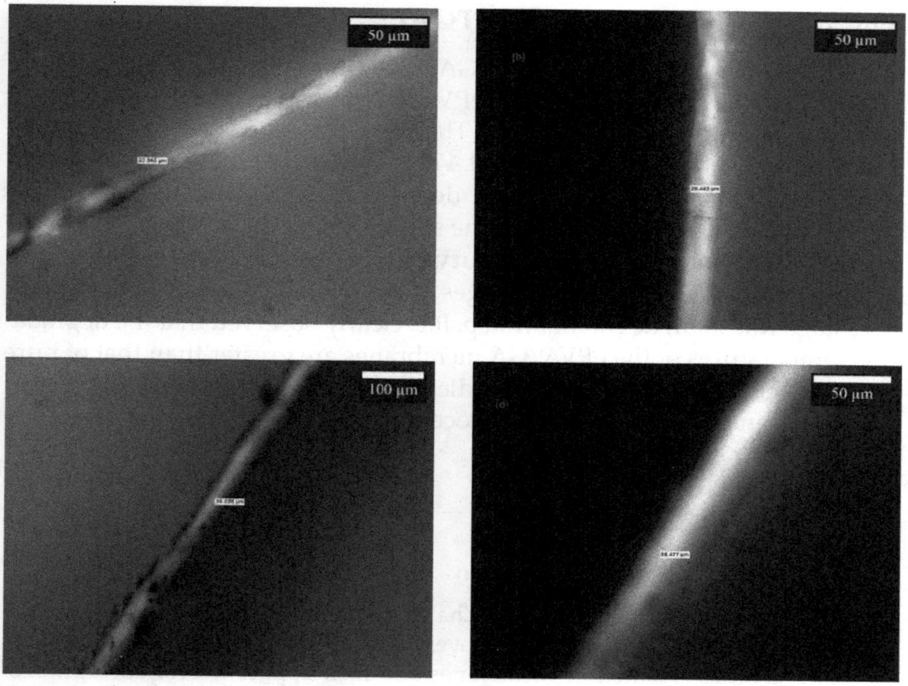

FIGURE A.8
Microscopic images of the ePVA fibers at different process times, (a) 25 min, (b) 35 min, (c) 45 min, and (d) 60 min.

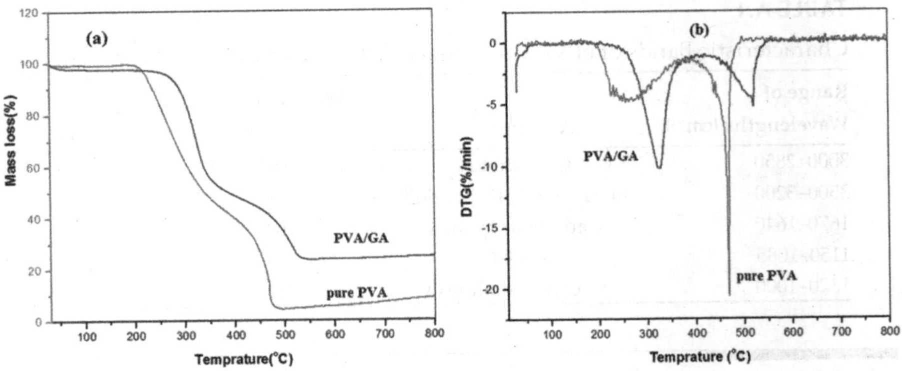

FIGURE A.9
(a) TGA and (b) DTG curves of pure ePVA and ePVA/GA membranes.

Thermogravimetric Analysis (TGA)

The Thermogravimetric Analysis (TGA) and differential Thermogravimetry (DTG) curves for pure ePVA and ePVA/GA membranes are presented in Figure A.9a and A.9b, respectively. The Pure PVA showed two main degradation steps at around 220 °C and 440 °C where the first stage was main degradation step and indicates the decomposition of side PVA chain [8]. The second smaller step related to the splintering of the central chain of the pure ePVA membrane. The TGA curve for the ePVA/GA membrane also indicated two main degradation stages at around 291 °C and 480 °C, and the first stage was main degradation step. It is clearly observed that the degradation temperatures of the ePVA/GA membranes are greater than that of pure ePVA membranes, which further indicates a rise in the thermal stability of the membrane after cross-linking process.

References

1. S. Wu, F. Li, H. Wang, L. Fu, B. Zhang, G. Li, Effects of poly (vinyl alcohol) (PVA) content on preparation of novel thiol-functionalized mesoporous PVA/SiO_2 composite nanofiber membranes and their application for adsorption of heavy metal ions from aqueous solution, Polymer, 2010. 51(26): pp. 6203–6211.
2. E. Yang, X. Qin, S. Wang, Electrospun crosslinked polyvinyl alcohol membrane, Material Letters, 2008. 62: pp. 3555–3557.
3. C.-H. Zhang, F.-l. Yang, W.-J. Wang, B. Chen, Preparation and characterization of hydrophilic modification of polypropylene non-woven fabric by dip-coating PVA (polyvinyl alcohol), Separation and Purification Technology, 2008. 61: pp. 276–286.

4. R.W. Yang Liu, Hongyang Ma, Benjamin S. Hsiao, Benjamin Chu, High-flux microfiltration filters based on electrospun polyvinyl alcohol nanofibrous membranes, Polymer, 2013. 54: pp. 548–556.
5. X. Wang, X. Chen, K. Yoon, D. Fang, B.S. Hsiao, B. Chu, High flux filtration medium based on nanofibrous substrate with hydrophilic nanocomposite coating. Environmental Science and Technology, 39 (2005) pp. 7684–7691.
6. H.S. Mansur, C.M. Sadahira, A.N. Souza, A.A.P. Mansur, FTIR spectroscopy characterization of poly (vinyl alcohol) hydrogel with different hydrolysis degree and chemically crosslinked with glutaraldehyde, Material Science and Engineering, 2008. C 28: pp. 539–548.
7. A.K. Deepak A. Musale, Effects of surface crosslinking on sieving characteristics of chitosan: Poly(acrylonitrile) composite nanofiltration membranes, Separation and Purification Technology, 2000. 21: pp. 27–38.
8. C. Santos, C.J. Silva, Z. Büttel, R. Guimarães, S.B. Pereira, P. Tamagnini, A. Zille, Preparation and characterization of polysaccharides/PVA blend nanofibrous membranes by electrospinning method, Carbohydydrate Polymer, 99 (2014) pp. 584–592.

List of Abbreviations and Symbols

Nomenclature

A_m	Effective area of the membrane (m2)
b	Langmuir constant (L mg^{-1})
C_e	Equilibrium concentration of metal ions (mg/L)
C_f	Concentration in the feed (mg/L)
C_o	Initial concentrations of heavy metal ions (mg/L)
C_p	Concentration in permeate (mg/L)
F_{ir}	Irreversible fouling
F_r	Reversible fouling
F_t	Total fouling
J_{B0}	BSA flux in the first run (L/m^2 h)
J_{B1}	BSA flux in the second run (L/m^2 h)
J_{B2}	BSA flux in the third run (L/m^2 h)
J_{BS}	Steady-state BSA flux (L/m^2 h)
J_{w0}	Pure water flux (L/m^2 h)
J_{w1}	Initial water flux (L/m^2 h)
J_{w2}	Water flux in second run (L/m^2 h)
J_{w3}	Water flux in third run (L/m^2 h)
k_1	Pseudo first-order model rate constant (min^{-1})
k_2	Pseudo second-order model rate constant (g/mg min^{-1})
K_{DR}	Activity coefficient (mol^2 J^{-2})
k_f	Freundlich capacity factor (mg/g) (mg/L)n
M	Mass of adsorbent membrane (g)
M_W	Molecular weight (Da)
n	Intensity parameter
NFR	Normalized flux ratio (%)
P_m	Hydraulic permeability (L/m^2 h kPa)
q_{DR}	Highest adsorption capability (mmol g^{-1})
q_e	Adsorption capacities at equilibrium time (mg/g)
q_{max}	Highest quantity of the metal ions per unit mass of adsorbents (mg/g)
q_t	Adsorption capacities at time *t* (mg/g)
Q_W	Water flow (m^3 s^{-1})
R	Regression value
R	Solute rejection (%)

R_f	Resistance due to fouled membrane
r_m	Average pore radius (nm)
R_m	Membrane resistance (m^{-1})
R_m	Resistance due to membrane
R_p	Resistance due to concentration polarization
t	Adsorption time (h)
t	Experimental time interval (h)
V	Total volume permeated during an experimental time interval (L)
V	Volume of the liquid in the solution (mL)
W_D	Weight of dry membranes (g)
W_W	Weight of wet membranes (g)
Δ_P	Operating pressure (kPa)

Greek letters

ε	Membrane porosity (%)
ε	Polanyi potential
ζ	Membrane thickness (µm)
μ	Dynamic water viscosity (Pa s)
ρ_P	Density of the polymer (g cm^{-3})
ρ_W	Density of pure water at operating conditions (g cm^{-3})

Abbreviations

AAS	atomic absorption spectrophotometer
ABR	aerobic biofilm reactor
AC	acetone
AFM	atomic force microscopy
BET	Brunauer–Emmet–Teller
BIS	Bureau of Indian Standards
BOD	biological oxygen demand
BPO	benzoyl peroxide
BSA	bovine serum albumin
BTEAC	benzyl triethylammonium chloride
CA	cellulose acetate
CAN	ceric ammonium nitrate
CAS	conventional activated sludge
CB	Cibacron blue
CBB	Coomassie Brilliant Blue

CBT	coagulation bath temperature
CCD	central composite design
CE	cellulose esters
CF	compaction factor
CMC	carboxy methyl cellulose
CMC	critical micelle concentration
CN	nitrocellulose
COD	chemical oxygen demand
Cu	copper
CS	ceric sulfate
CTCs	circulating tumor cells
DEX	dexamethasone
DGNs	dextran glassy nanoparticles
DI	deionized
DMAc	N,N-dimethylacetamide
DMF	dimethyl formamide
DMN	dissolving microneedle
DOE	design of experiment
D-R	Dubinin–Radushkevich
DRG	dorsal root ganglia
EC	extracellular
ECM	extracellular matrix
EDA	ethylenediamine
EDL	electrical double layer
EDS	energy dispersive x-ray spectroscopy
EGSB	expanded granular sludge bed
EPR	enhanced permeation retention
EPS	polymeric substances
EWC	equilibrium water content
FAO	Food and Agriculture Organization
FD	fiber diameter
FESEM	field emission scanning electron microscopy
FS	flat sheet
FTIR	Fourier transform infrared
GA	glutaraldehyde
GMA	glycidyl methacrylate
GTL	gas to liquid
GTP	green tea polyphenols
HF	hollow fiber
HMTA	hexamethylenetetramine
HRT	hydraulic retention time
ICMR	Indian Council of Medical Research
IMBr	immersed membrane bioreactor
LNG	liquefied natural gas
MA	maleic anhydride

MAA	methacrylic acid
MABR	membrane aerated biofilm reactor
MB	methylene blue
MBR	membrane bioreactor
MCL	maximum contaminant level
MEUF	micellar enhanced ultrafiltration
MF	microfiltration
MLSS	mix liquor suspended solid
MMA	methyl methacrylate
MMS	modified mesoporous silica
MRI	magnetic resonance imaging
MWCO	molecular weight cut off
NF	nanofiltration
NIR	near-infrared radiation
NMP	N-methyl-2-pyrrolidone
NMR	nuclear magnetic resonance
PAA	polyacrylic acid
PAN	polyacrilonitrile
Pb	lead
PC	polycarbonate
PCL	polycaprolactone
PE	polyethylene
PEG	polyethylene glycol
PEGDMA	polyethylene glycol dimethacrylate
PEI	polyethylenimine
PEO	poly(ethylene oxide)
PES	polyethersulfone
PET	polyethylene terephthalate
PEUF	polyelectrolyte enhanced ultrafiltration
PP	polypropylene
PS	polysulfone
PTFE	polytetrafluoroethylene
PU	polyurethane
PVA	polyvinyl alcohol
PVC	polyvinylchloride
PVDF	polyvinylidene fluoride
PVP	polyvinylpyrrolidone
PWF	pure water flux
RO	reverse osmosis
RSM	response surface methodology
SAN	styrene-acrylonitrile
SBR	sequencing batch reactor
SEM	scanning electron microscope
SMBr	submerged membrane bioreactor

SMP	soluble microbial products
SPD	surface pore diameter
SPG	Shirasu porous glass
SRT	solid retention time
TAMs	tumor associated macrophages
TDS	total dissolved solids
TEM	transmission electron microscopy
TEPA	tetraethylenepentamine
TFC	thin film composite
TFNC	thin film nanofibrous composite
TGA	thermogravimetric analysis
THF	tetrahydrofuran
TiO_2	titanium dioxide
TKN	total kjeldahl nitrogen
TMP	trans membrane pressure
TOC	total organic carbon
TP	total phosphorous
TPEE	thermal plastic elastomer ester
TS	total solids
TSS	total suspended solids
UF	ultrafiltration
USEPA	United States Environmental Protection Agency
UV	ultraviolet
VCF	volume concentration factors
WC	water content
WHO	World Health Organization
XRD	X-ray powder diffractometer
ZLD	zero liquid discharge

Index

E

F

G

H

N

Q

R

S

V

W

X

Z

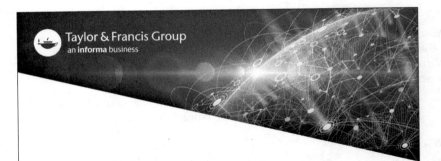